Graphene-based Carbocatalysts: Synthesis, Properties and Applications

(Volume 1)

Edited by

Pinki Bala Punjabi
Department of Chemistry, University College of Science, Mohanlal Sukhadia University, Udaipur-313001, Rajasthan, India

Rakshit Ameta
Department of Chemistry, J.R.N. Rajasthan Vidyapeeth University, Udaipur-313001, Rajasthan, India

&

Sharoni Gupta
Department of Chemistry, University College of Science, Mohanlal Sukhadia University, Udaipur-313001, Rajasthan, India

Graphene-based Carbocatalysts: Synthesis, Properties and Applications (Volume 1)

Editors: Pinki Bala Punjabi, Rakshit Ameta & Sharoni Gupta

ISBN (Online): 978-981-5050-89-9

ISBN (Print): 978-981-5050-90-5

ISBN (Paperback): 978-981-5050-91-2

First published in 2023.

Bentham Science Publishers Pte. Ltd.
80 Robinson Road #02-00
Singapore 068898
Singapore
Email: subscriptions@benthamscience.net

BENTHAM SCIENCE

CONTENTS

FOREWORD

Graphene is an allotrope of carbon, which is in the form of a thin layer with a two-dimensional honeycomb-like structure. It exhibits unique properties such as lightweight, excellent thermal and electrical conductivities, large specific surface area, easy preparation and functionalization, high intrinsic mobility, chemical stability, simple recovery, recyclability, *etc*. Therefore, it has emerged as the most successful entity with a wide range of applications in various medical, chemical and industrial processes, such as flexible electric/photonics circuits, solar cells, drug delivery, tissue engineering, bioimaging, optoelectronics, photodetectors, generation and storage of energy, biosensors, removal of contaminants, catalyst, water and sound proofing, and many more. The editors have made a very judicious choice in selecting graphene and its applications as a topic covering major fields of interest. I appreciate their efforts in compiling the different areas related to graphene and putting them all in a single arena. I believe that this book will get an overwhelming response from the readers.

Suresh C. Ameta
Past President & Life Time Advisor
Indian Chemical Society, Kolkata
Professor of Eminence (Distinguished Professor)
Faculty of Science
PAHER University
Udaipur-313003 (Raj.) India

PREFACE

Catalysis is a fundamental and multidisciplinary phenomenon that has been ruling the energy sector and chemical industry for centuries impacting the world's economy. Not only production processes but even biological and natural reactions are catalytically controlled by enzymes and other substances to maintain life on Earth. Various manufacturing units including petrochemicals, pharmaceuticals, food, polymers, materials, and fine chemicals-based industries along with pollution-abating firms are highly dependent on catalysts. This is because catalysts offer green means for accelerating chemical transformations *via* energy-saving and atom economic pathways.Untill today, multitudes of homogeneous and heterogeneous catalysts have been explored for carrying out several conversions and enhancing the feasibility of reactions. However, the issues of catalyst recoverability and efficiency have been a cause of concern across the globe. More recently, the necessity of environmental conservation has further accentuated the search for sustainable catalysts. In such a scenario, graphene-based catalysts or carbocatalysts have emerged as a boon to meet the growing demand for efficacious, benign and inexpensive heterogeneous catalysts.

Graphene with its distinguished opto-electronic, thermo-mechanical, surface and chemical characteristics is renowned as the most invincible nanomaterial. Ever since the path-breaking discovery of graphene in 2004, the two-dimensional, honeycomb lattice-based material has enthralled the scientific community throughout the world. The exceptional conductivity, tensile strength, stability, large surface area, recoverability, recyclability and ease of functionalization of graphene materials have especially captivated the researchers working in the field of catalysis. Owing to the surge in demand for graphene-based catalysts, graphene research is being carried out at a very rapid pace. Every year new additions to the knowledge and scope of graphene carbocatalysts appear at a considerably large scale. Consequently, this book is an attempt to acquaint readers with the recent advances in the field of graphene carbocatalysis.

The book encapsulates the recent developments involving the syntheses, properties, characterizations, functionalization, and catalytic applications of graphene, its derivatives and composites. The book is in two volumes. The first volume is divided into 10 chapters. In Chapter 1, a brief introduction to carbocatalysis has been laid out. The properties, syntheses and scope of carbocatalysts have been discussed to highlight their significance. Chapter 2 discusses the fundamental structure and properties of graphene and chemically modified graphene contributing to their applications in diverse fields. Chapter 3 describes the diverse synthetic strategies for the preparation of graphene and its derivatives. The advantages of present methods and future challenges related to industrial-scale synthesis have also been outlined in this chapter. Chapter 4 focuses on the latest and most commonly employed characterization techniques used for investigating the morphological, structural, and thermal properties of graphene materials. In Chapter 5, recent trends in functionalization and its role in the catalytic activity of graphene have been put forward. Chapter 6 summarizes the recent progress in the synthesis of graphene-based composites along with their properties and applications in catalytic reactions. The future prospects and challenges towards the design and development of graphene-based nanocomposites for catalytic reactions have also been addressed in the chapter. Chapter 7 reviews the recent advances in graphene-supported palladium catalysts for coupling reactions. It also underscores the synthesis of these catalysts and their mechanistic aspects spanning across a variety of cross-coupling reactions. A comparison of graphene-supported catalysts with traditional catalysts has also been included in this chapter. Chapter 8 provides an in-depth review of recent applications of graphene-based catalysts in multicomponent and domino reactions. In Chapter 9, current progress made in the field of oxidation and reduction reactions of organic molecules catalyzed by graphene materials has been explored. Chapter 10 accounts for the contemporary trends in the area of graphene-based biocatalysts.

The second volume includes six chapters. Chapter 1 of the second volume incorporates the most recent advances in photocatalytic applications of graphene-based materials such as graphene-based semiconductor photocatalysts for the degradation of various contaminants (treatment of waste water), production of hydrogen, and photocatalytic reduction of carbon dioxide to energy-rich synthetic fuels (combating against global warming and energy crisis), *etc.* Chapter 2 discusses the latest advances in electrocatalysis by graphene materials with a special focus on the electrocatalytic activities of non-metal doped graphene, graphene-2D materials heterostructures, and graphene-plasmonic nanostructures. Chapter 3 provides an overview of the recent advancement made by graphene-based materials including graphene oxide, reduced graphene oxide and graphene oxide quantum dots for hydrogen evolution from light-driven water splitting and future prospects. Chapter 4 highlights the modern trends in the fabrication of graphene-based smart energy materials for applications in various energy storage systems. The future trends and challenges have also been underlined. Chapter 5 underscores the potential utility of graphene materials in electrochemical sensing devices. Chapter 6 concludes the book and reports state-of-the-art graphene carbocatalysis with the future challenges accompanying graphene-based catalysts.

The book covers multidimensional applications of graphene-based materials cutting across various fields ranging from energy generation, chemical synthesis, and electrochemical sensing to photocatalysis. Hopefully, this book will serve as a reference work for all those researchers, students, industry workers and engineers who are interested in graphene research as well as its emerging applications in catalysis and beyond.

At last, we would like to thank all the authors of this book for their invaluable contribution to enriching the content of this book. We are also extremely indebted to the managers, editors, and reviewers of Bentham Science Publications for their magnanimous help throughout the creation and publication of this book. Finally, we are highly grateful to our families for their constant support and inspiration.

Pinki B Punjabi
Department of Chemistry,
University College of Science,
Mohanlal Sukhadia University,
Udaipur-313001, Rajasthan,
India

Rakshit Ameta
Department of Chemistry,
J.R.N. Rajasthan Vidyapeeth University,
Udaipur-313001, Rajasthan, India

&

Sharoni Gupta
Department of Chemistry,
University College of Science,
Mohanlal Sukhadia University,
Udaipur-313001, Rajasthan, India

DEDICATION

In fond memory of my beloved uncle Mr. Anil Kothari who taught me how to smile through difficult times.

Dr. Sharoni Gupta

List of Contributors

Abhilasha Jain	Department of Chemistry, St. Xavier's College (Autonomus), Mumbai, Maharashtra-400001, India
Akella Sivaramakrishna	Department of Chemistry, School of Advanced Sciences, Vellore Institute of Technology (VIT), Vellore 632014, Tamil Nadu, India
Anamika Srivastava	Department of Chemistry, Banasthali Vidyapith, Banasthali, Rajasthan, India
Anjali Banger	Department of Chemistry, Banasthali Vidyapith, Banasthali, Rajasthan, India
Chetna Ameta	Photochemistry Laboratory, Department of Chemistry, University College of Science, Mohanlal Sukhadia University, Udaipur (Raj.), India
Dharmendra Dharmendra	Photochemistry Laboratory, Department of Chemistry, University College of Science, Mohanlal Sukhadia University, Udaipur (Raj.), India
Dinesh Kumar Yadav	Department of Chemistry, Mohanlal Sukhadia University, Udaipur, 313001, India
G.S. Sahoo Navneet	Department of Chemistry, School of Advanced Sciences, Vellore Institute of Technology (VIT), Vellore 632014, Tamil Nadu, India
H.S. Sushma	Department of Chemistry, School of Advanced Sciences, Vellore Institute of Technology (VIT), Vellore 632014, Tamil Nadu, India
Ivaturi Sai Vighnesh	Department of Chemistry, School of Advanced Sciences, Vellore Institute of Technology (VIT), Vellore 632014, Tamil Nadu, India
K. M. Sreeranjan	Department of Chemistry, School of Advanced Sciences, Vellore Institute of Technology (VIT), Vellore 632014, Tamil Nadu, India
Manish Srivastava	Department of Chemistry, Banasthali Vidyapith, Banasthali, Rajasthan, India
Monika Jangid	Department of Chemistry, PAHER University, 313001-Udaipur, Rajasthan, India
Munsaf Ali	College of Nanoscale Science and Engineering, State University of New York, Polytechnic Institute - CNSE, 257 Fuller Rd., Albany, NY 12203, New York
Nirmala Kumari Jangid	Department of Chemistry, Banasthali Vidyapith, Banasthali, Rajasthan, India
Nitinkumar Satyadev Upadhyay	College of Nanoscale Science and Engineering, State University of New York, Polytechnic Institute - CNSE, 257 Fuller Rd., Albany, NY 12203, New York
Priyanka Chundawat	Photochemistry Laboratory, Department of Chemistry, University College of Science, Mohanlal Sukhadia University, Udaipur (Raj.), India
Pinki Bala Punjabi	School of Basic and Applied Sciences, Sangam University, Bhilwara-311001, Rajasthan, India
Pradeep Kumar Jaiswal	Department of Biochemistry and Biophysics, Texas A&M University, College Station, Texas 77843, USA

Priy Brat Dwivedi	National University of Science and Technology, AlHail South, Seen 111, Muscat, Oman
Reema Agarwal	Photochemistry Laboratory, Department of Chemistry, University College of Science, Mohanlal Sukhadia University, Udaipur (Raj.), India
Rajagopal Desikan	Department of Chemistry, School of Advanced Sciences, Vellore Institute of Technology (VIT), Vellore 632014, Tamil Nadu, India
Rakesh R Panicker	Department of Chemistry, School of Advanced Sciences, Vellore Institute of Technology (VIT), Vellore 632014, Tamil Nadu, India
Rakshit Ameta	Department of Chemistry, J.R.N. Rajasthan Vidyapeeth University, Udaipur-313001, Rajasthan, India
Ramesh L. Gardas	Department of Chemistry, Indian Institute of Technology Madras, Chennai - 600 036, India
S. Ahallya	Department of Chemistry, School of Advanced Sciences, Vellore Institute of Technology (VIT), Vellore 632014, Tamil Nadu, India
S. Haripriya	Department of Chemistry, School of Advanced Sciences, Vellore Institute of Technology (VIT), Vellore 632014, Tamil Nadu, India
S. Sathish	Department of Physics, MVJ College of Engineering, Bangalore – 560067, India
Sagar Ulkesh Patil	Department of Chemistry, School of Advanced Sciences, Vellore Institute of Technology (VIT), Vellore 632014, Tamil Nadu, India
Siddharth Sharma	Department of Chemistry, Mohanlal Sukhadia University, Udaipur, 313001, India
Sharoni Gupta	Department of Chemistry, University College of Science, Mohanlal Sukhadia University, Udaipur-313001, Rajasthan, India
Sunita Panchawat	Department of Phatmaceutical Sciences, Mohanlal Sukhadia University, Udaipur, Rajasthan-313001, India
Vepa Jagannadha Rao	Department of Chemistry, Anil Neerukonda Institute of Technology and Sciences, Affiliated to Andhra University, Sanghivalasa, Visakhapatnam 531162, Andhra Pradesh, India
Vickramjeet Singh	Department of Chemistry, Dr B R Ambedkar National Institute of Technology, Jalandhar-144011, Punjab, India
Yogeshwari Vyas	Photochemistry Laboratory, Department of Chemistry, University College of Science, Mohanlal Sukhadia University, Udaipur (Raj.), India

CHAPTER 1

Introduction to Carbocatalysis

Pinki Bala Punjabi[1,*] and **Sharoni Gupta**[1]

[1] *Department of Chemistry, University College of Science, Mohanlal Sukhadia University, Udaipur-313001, Rajasthan, India*

Abstract: Carbocatalysis has emerged as a promising field of catalysis. The exceptional surface morphology, pore distribution, thermal conductivity, chemical inertness, electrical property and renewability of carbon materials have rendered them suitable for various catalytic processes namely, photocatalysis, electrocatalysis, biocatalysis and chemical catalysis. Therefore, the introductory chapter on carbocatalysis describes the useful properties of carbonaceous materials which govern their catalytic behaviour. Moreover, synthetic approaches for the fabrication of diverse carbon polymorphs such as active carbon, graphite, fullerene, glassy carbon, carbon black, carbon nanotubes, carbon nanofibres, nanodiamonds, carbon nano-onions, and graphene have also been briefly discussed in this chapter. The scope of carbocatalysts over broad areas has also been elucidated by quoting instances.

Keywords: Carbocatalysis, Electrical Properties, Surface Properties, Sustainability, Synthesis, Thermal Conductivity.

INTRODUCTION

Catalysis is the backbone of chemical transformations which facilitates a sustainable and efficacious means to convert starting raw materials to useful chemical compounds. One of the twelve principles of green chemistry states that the "use of catalysts should be encouraged as these enhance selectivity, reduce the formation of by-products, reaction time and energy requirements" [1]. Therefore, catalytic transformations have become an integral part of environmentally benign chemical processes. Homogenous and heterogeneous catalysts are the two main classes of catalysts that over the past several decades have served as a driving force in achieving high performance conversions by abating the activation energy. Both homogeneous and heterogeneous catalysts have their high points, for instance, homogeneous catalysts have excellent catalytic efficiency owing to

* **Corresponding author Pinki Bala Punjabi:** Department of Chemistry, University College of Science, Mohanlal Sukhadia University, Udaipur-313001, Rajasthan, India; E-mail: pb_punjabi@yahoo.com

Pinki Bala Punjabi, Rakshit Ameta & Sharoni Gupta (Eds.)

readily accessible active sites [2] while heterogeneous catalysts are quite stable and easy to separate and recover [3]. However, with growing awareness of environmentally friendly protocols, the use of conventional catalysts has become a debatable topic. This is primarily because the traditional catalysis employs acids, bases, transition and non-transition metals which are toxic, expensive, get easily deactivated by impurities and non-reusable.

Sustainability being an essential element of modern chemistry has coerced researchers and chemists globally to look for renewable and cost-effective catalytic systems. The recent emergence of graphene based catalysts has paved way for a green alternative to conventional catalysts. Carbocatalysis *viz.* catalysis driven by carbonaceous materials is not a novel avenue of study. Several carbon based catalyst have been used in the past. Perhaps the beginning of carbocatalysis can be traced back to the year 1854 when Stenhouse observed that activated carbon was capable of oxidizing a mixture of organic gases coming from putrefied biomass [4]. Since then there have been several reports where carbon allotropes were used as catalytic materials. In 1925, Rideal and Wright employed charcoal as a catalyst for the oxidation of oxalic acid [5]. In the 1930s, Kutzelnigg and Kolthoff separately confirmed that activated charcoals were highly suitable for catalyzing the aerobic oxidation of ferrocyanide to ferricyanide [6, 7]. Thereafter in 1979, charcoal was used to catalyze the oxidative dehydrogenation (ODH) reaction of ethylbenzene to styrene [8]. Also, graphite catalyst was employed for the first time by Lucking and co-workers for the oxidative cleavage of 4-chlorophenol, which yielded CO_2, H_2O, and HCl [9]. In another instance, Fortuny *et al.* have also reported oxidation of phenol using active carbon as a catalyst [10]. Despite the good catalytic performance of carbonaceous materials attributable to their versatile porosity and surface properties, carbocatalysis did not achieve the desired level of attention until the discovery of fullerene, carbon nanotubes, and graphene. Further, mostly the carbon polymorphs such as charcoal [11], graphite [12], carbon black [13], and diamond [14] have been used as support in catalytic systems because of their advantages including ease of metal immobilization on carbon, resistance to attack by acids and bases, simple as well as cost-effective preparation, high thermal stability and straightforward recovery of the active phase. Regardless of the beneficial features, some shortcomings such as lack of active sites, microporous structure, hydrophobic character and poor resistance to oxidation stalled the growth of carbon based catalysts in earlier days [15]. However, the recent developments in the synthetic chemistry of carbon have allowed easy modification of the pore size distribution, surface chemistry, and hydrophilicity to afford carbocatalyst with desirable electronic and physicochemical properties, large surface to weight ratio, and better biocompatibility.

The discovery of fullerenes in 1985 was a landmark in the field of carbon science [16]. The unique three dimensional structure of fullerene comprises intricately arranged five and six membered rings of covalently bonded carbon atoms arranged to form spherical shape. The high conductivity (narrow band gap) of fullerene arises from the pi electron delocalization within carbon layers which favours its catalytic behaviour. Functionalization of fullerenes also aids in tuning their electrical properties and serves as a precursor for the preparation of fullerene based nanocatalyst. Ever since their discovery, fullerene based catalysts have been used to catalyze several reactions such as reduction [17], oxidation [18], photodegradation [19], dehydration [20], proton transfer [21], deallylation [22], *etc*.

The birth of nanotechnology proved to be a boon for carbocatalysis. The designing of nanodimensional carbon tubes with readily accessible sites on large surface area revolutionized the field of catalysis. The breakthrough invention of carbon nanotubes in 1991 by Iijima [23] followed by its use as catalyst support in 1994 [24] led to a new generation of catalysts dominated by carbon nanomaterials. Carbon nanotubes belong to the fullerene family and are basically graphite sheets rolled in the form of cylindrical shapes. The excellent electrical conductivity, striking mechanical and thermal stabilities, tolerance against poisoning effect, easily modifiable surface area, and ease of functionalization of carbon nanotubes make them promising catalytic systems [25]. Carbon nanotubes have been widely applied and employed in heterogeneous catalysis [26], photocatalysis [27], and electrocatalysis [28].

One of the lightest and strongest known materials, graphene is a single layered allotrope of carbon that has honeycomb structure and is the most famous carbocatalyst of recent times. The discovery of graphene dates back to 1859 when Brodie while examining the structure of graphite, observed graphene and named it "graphone" [29]. In 2004, the isolation of graphene sheets by Novoselov *et al.* brought about a storm in the field of carbonaceous materials [30]. The extraordinary physicochemical and thermoelectrical properties of graphene such as large specific surface area (theoretically 2630 m^2/g for single-layer graphene) [31], amazing electronic properties and electron transport potential [32, 33], unparalleled flexibility and impermeability [34, 35], sturdy mechanical property [36] and excellent thermal and electrical conductivities [37, 38] attracted enormous interest and resulted in an explosion of research studies in the domain of graphene based materials. In 2009, Lu *et al.* highlighted the use of gold embedded graphene based catalyst for oxidation of CO [39]. Jafri *et al.* also reported nitrogen doped graphene nanoplatelets as a potent catalyst for oxygen reduction reaction [40]. In 2010, Dreyer *et al.* demonstrated graphene oxide as a compatible carbocatalyst for oxidation and hydration reactions [41]. In the same year, Li and his research group reported electro-oxidation of methanol promoted by Pt nanoparticles on reduced graphene oxide [42]. In another instance, Gao and co-workers used reduced graphene oxide as carbocatalyst for the reduction of nitrobenzene [43]. Since then, a plethora of literature has been available which showcases the abilities of graphene and its derivatives as potent carbocatalysts.

Thus, a bountiful of reusable and economically viable carbonaceous materials are available today which serve as heterogeneous and semi-homogeneous catalysts as well catalytic supports for a range of reactions that find significance in different areas namely, chemical manufacturing, environmental remediation, biomass production, water splitting, energy conversion and storage devices. In the present chapter, an effort has been made to acquaint the readers with the properties, syntheses and scope of carbocatalysts (Fig. **1**).

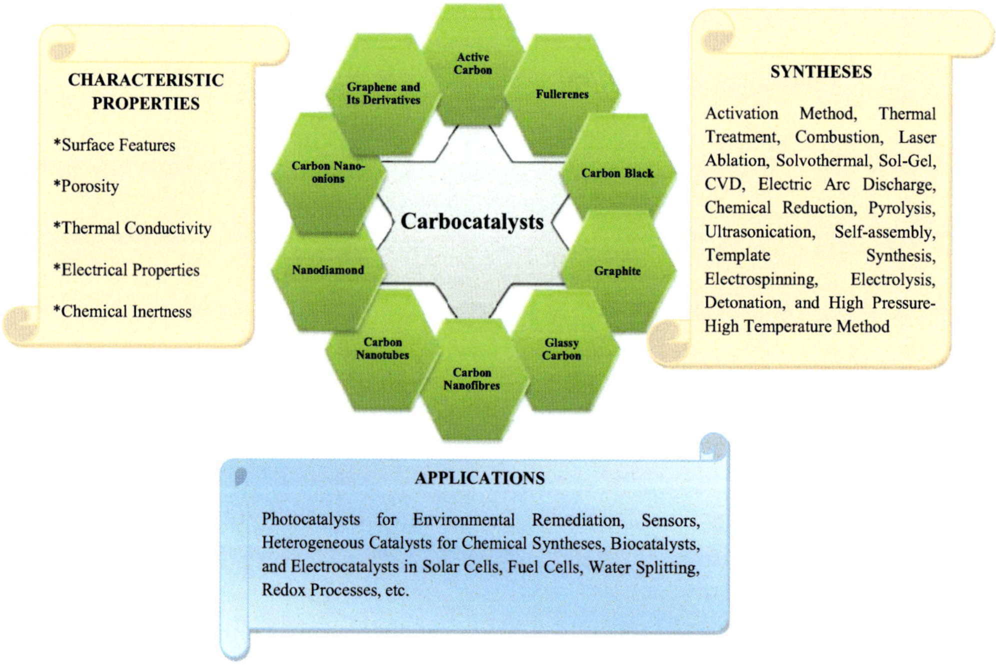

Fig. (1). Properties, syntheses and scope of carbocatalysts.

PROPERTIES OF CARBOCATALYSTS

Now let us delve into the properties of carbon materials that render them suitable as catalytic systems.

Surface Properties

It is well known that for any catalyst, surface features are crucial in governing its catalytic activity. The surface properties of carbon materials depend on the spatial arrangement of carbon atoms which in turn are based on the method of preparation or treatment used to synthesize these materials. The presence of a high density of pores and large surface area on carbonaceous materials play a key role in the adsorption of reactant molecules [44]. The introduction of defects in carbon atom layers present in carbon based materials and eases of their functionalization make their surface apt for catalysis. These structural defects particularly on the edges serve as active sites for binding of reaction molecules being easily accessible.

The defects in carbon materials can be introduced by thermal treatment or doping. For instance, Qiu *et al.* have reported that the amount of surface defects is directly related to the catalytic efficiency of activated carbons [45]. They used a simple thermal treatment method aided by a nitrate solid-phase oxidative method to increase the amount of surface defects on the activated carbons and observed their catalytic effect on the hydrochlorination of acetylene. It was reported that this treatment removed oxygen moieties and introduced defects on the edges of active carbons. Armchair and zigzag edges were observed in active carbon due to the stacking arrangement of nanosized graphitic layers, thereby populating the edges with active sites. In another study, it was reported that selective doping of carbon materials with heteroatoms induces catalytically beneficial defects. Chen *et al.* doped carbon nanosheets with B-N pairs and employed them as a catalyst to promote ammonia synthesis [46]. It was highlighted that the B-N doped carbon nanosheets showed superior activity compared to metallic catalysts for the nitrogen reduction reaction. The better electrocatalytic behavior of B-N doped carbon nanosheets was attributed to the fact that doping resulted in generating sufficient active sites at the edge carbon atoms.

The functionalization of carbon surfaces causes changes in the surface energy as well as chemical reactivity of the materials. Surface functional groups may increase free valences on the edges of the external surface, thereby increasing the catalytic activity [47]. The functional groups also assist in tuning the hydrophilicity of carbon materials and are capable of altering the acidic and basic character of these materials [48, 49]. Oxygen, nitrogen, hydrogen, sulphur, phosphorous and halogens functionalities are commonly used for this purpose. Chen *et al.* functionalized carbon nanotubes with oxygen functional groups [50]. Several oxidants like H_2O_2, $KMnO_4$, HNO_3/H_2SO_4 were used to functionalize the nanotubes. After functionalization, the pH_{pzc} (pH at which net surface charge on carbon is zero) was measured by the chemical titration method [51]. It was reported that after functionalization the carbon surface became acidic due to the presence of functional groups like carboxyls which decreased the pH_{pzc} value making the surface more hydrophilic [52 - 54]. It was also reported that the presence of carboxylic functional groups also enhanced the dispersion of Pd and Pt metals on CNTs *via* the formation of chemical bonds, which hindered the agglomeration metal particles on CNTs and minimized the sintering propensity of the formed Pd–Pt particles across the CNT surface in the calcination and reduction process [55, 56]. This increased uniform dispersion of Pd-Pt on functionalized CNTs resulted in improved catalytic efficiency for hydrogenation of naphthalene to tetralin. In another study, it was reported that the functionalization of sulphonic acid group on graphene oxide improved its hydrophilicity as well as permeability [57].

It has been also reported that surface functionalization also improves the adsorption property of carbon materials. Sun *et al.* functionalized carbon beads *via* surface amination treatment and used it to study CO_2 capture [58]. It was demonstrated that the surface modifications enhanced the adsorption of CO_2 without compromising the desirable spherical morphology and mechanical strength of the activated carbon beads. This was attributed to surface modification of carbon, greater number of N-H functionalities resulted in stronger binding between CO_2 and C-N-H groups than that of C=N-C.

A study conducted by Arrigo *et al.* showed that the functionalization of nanocarbons aids in the regulation of their acidic/basic character [49]. They treated oxidized carbon nanofibres with ammonia and evaluated the surface acid-base properties. It was reported that the samples contained both O and N species on the surface which served as acidic and basic sites. It was further illustrated that the population of O and N functionalities depended on temperature and could be controlled to modify the acid-base properties of the carbon surface. At 473 K, nitrogen moieties were incorporated on the surface as a consequence of the reaction of carboxylic acid sites with ammonia, thereby generating a bifunctional hydrophilic material with both acidic and basic sites on the carbon surface. Further raising the temperature to 873 K resulted in strong basic sites leading to a hydrophobic carbon surface. Thus, it was proved that altering the surface functionalization once can modify the acid-base nature of the carbon surface.

The specific surface area of materials is one the most imperative factors that directly affect catalysis. Greater surface area means better ease of access to active sites. The size reduction of particles causes the enhancement in the surface area of a substance. The surfacing of nanotechnology has resulted in the development of methods through which one can get nanodimensional materials with extremely large surface area. The nanocarbon materials with higher surface to volume ratio have therefore emerged as very effective carbocatalysts. For example, Pan *et al.* prepared porous N-doped carbon nanotubes (NCNTs) by pyrolysis of polypyrrole nanotubes and KOH activation at different temperatures 800 °C, 900 °C and 1000 °C [59]. The prepared nanotubes were used as electrocatalysts for the oxygen reduction reaction. A BET analysis was conducted to deduce the surface area of the nanotubes. It was reported that the specific surface area of polypyrrole nanotubes, NCNT-800, NCNT-900, and NCNT-1000 was 26, 1057, 1402, and 1137 m^2/g, respectively. The assessment of ORR in a standard three electrodes cell showed that the NCNT-900 and NCNT-1000 were the most efficient catalysts in promoting the reactions. Furthermore, the nanotubes exhibited superior catalytic activities as compared to the commercially available Pt/C catalyst. It was thus highlighted that the high surface areas in porous nanotubes afforded active sites for diffusing O_2 and KOH electrolyte thereby enhancing the rate of ORR.

Pore Size Distribution

The porosity of carbon materials is another important factor that influences their catalytic performance. Porous carbon materials can be classified as microporous (<2 nm), mesoporous (2-50 nm), and macroporous (>50 nm) on the basis of their pore diameters [60]. The pore volume and pore-size distribution play a crucial role in shape-selective catalysis [61]. Porosity in a material enhances catalytic activity not only by increasing accessibility to active sites albeit also by facilitating the mass transfer. Thus, by tuning the porosity one can induce desired catalytic properties in a material. Ferrero *et al.* investigated nitrogen-doped carbon microspheres with tunable porosity as electrocatalysts for the oxygen reduction reaction (ORR) [62]. It was reported that the catalyst was prepared by nanocasting using pyrrole as N-containing carbon precursor and porous silica microspheres as a template. Using two different silica samples as templates, the porosity of the microspheres was tuned from a micro- to a mesoporous network. The prepared catalytic system was applied to ORR and it was found that mesoporous electrocatalyst was more efficient than microporous one. This result was attributed to better accessibility of the active sites in the case of the mesoporous microspheres compared to the microporous ones. This is because in microporous materials the diffusion of reactant molecules becomes restricted due to small pore size and volume. It was further deciphered that the improved accessibility to the active sites on the surface of the mesoporous microspheres led to increase in mass-transfer reactions during the electrochemical process, thereby increasing the rate of catalytic reduction. Consequently, it can be said that pore size and volume are significant contributing factors in carbon based catalysis.

Thermal Conductivity

Thermal conductivity is another important criterion in carbocatalysis. Thermal conductivity is the ability of an object to conduct heat. Heat transfer is an important phenomenon in catalysis, particularly for strongly endo- and exothermic reactions in which fast heat management is advantageous for active sites [63]. Heat conduction in materials usually occurs due to the transmission of lattice vibrations known as phonons [64]. Carbonaceous materials like graphite, diamond, graphene, carbon fibres, and carbon nanotubes are well known for their thermal properties. The arrangement and coordination of carbon atoms along with defects in these materials are the main reasons for high thermal conductivities [65, 66]. A thermally conductive cobalt loaded graphite based catalyst for low temperature based highly exothermic Fischer-Tropsch synthesis (FTS) was reported [67]. It was reported that the use of this catalyst resulted in high productivity of hydrocarbons. This is because the incorporation of exfoliated graphite in the catalyst provides an amalgamation of extensive surface with high thermal conductivity and durability for the gasification process. Thermal analyses of graphite and aluminium based catalysts showed that the thermal conductivity of graphite is twice more than aluminum based catalyst and 30 times greater than a cobalt catalyst without heat conducting agents. It was further revealed that due to the high thermal stability of graphite, the catalyst survives overheating while the aluminum based counterpart is limited to 450 °C or even lower temperature because of irreversible effects such as metal pre-melting and oxidation. Hence, it is clear that thermally conductive as well as stable catalysts are perquisite for promoting reactions that require conduction of heat energy or heat removal.

Electrical Properties

To understand the electrical properties of carbonaceous materials, first, it is imperative to clarify the concepts of electrical resistivity and conductivity. Electric resistivity is a fundamental property of a material due to which it opposes the flow of electric current [68]. Mathematically, electric resistivity (ρ, Ωm) is defined as the ratio of electrical field (E, Vm^{-1}) inside the material to the electric current density (J, Am^{-1}), eq. (1). In other words, the resistivity is the resistance (R) offered by a material with unit cross sectional area (A) and unit length (l), eq. (2),

$$\rho = \frac{E}{J} \quad \textbf{(1)}$$

$$\rho = R.\frac{A}{l} \quad \textbf{(2)}$$

Electric conductivity (σ, $\Omega^{-1}m^{-1}$ or Sm^{-1}) is the property of a material by the virtue of which it conducts electricity. It is the reciprocal of electric resistivity. Consequently, both the terms resistivity and conductivity are interrelated and form the basis of the classification of materials [68]. On the basis of electric conductivity or resistivity, materials can be classified as conductors (materials that conduct electric current), insulators (materials that resist electric current) and semi-conductors (materials that intermediate between conductors and insulators).

The electrical properties of carbon materials are very much dependent on the structural attributes and type of treatment given to them. Different allotropes of carbon show different degrees of conductivities due to structural differences [69]. For instance, diamond exhibits low conductivity because each carbon atom is sp^3 hybridized and bonded to four carbon atoms by strong covalent bonds. Therefore, no electrons are available for conduction. It has a wide band gap of 5.5 eV [70]. While in graphite the sp^2 hybridized carbon atoms are arranged as two-dimensional hexagonal lattice structure with carbon sheets stacked one above another and each carbon has one free electron. The sheets are held by weak van der Waals forces, electrons freely move throughout the layered structure. Graphite, thus, has a small band gap (40 meV) between valence and conduction bands and behaves as a conductor similar to metal [71].

Fullerene, a carbon polymorph, is a large caged polyhedral molecule with either fcc or bcc packing of atoms. In fullerene sp^2 carbons are directly bonded to three neighbors in an arrangement of five- and six-membered rings. Fullerenes are basically semiconductors, however, by doping heteroatoms, superconductivity can be induced [72].

Carbon nanotubes (CNTs) are rolled-up graphene sheets with cylindrical structure. These are of two kinds, single-walled carbon nanotubes (SWCNT) with a diameter in the nanometer range and multi-walled carbon nanotubes (MWCNT) are nested single-walled carbon nanotubes held together by van der Waals forces with a diameter of less than 100 nm. The interlayer distance in multi-walled carbon nanotubes is 0.34-0.39 nm [73]. SWCNTs can have an armchair, chiral, and zigzag geometries depending on the wrapping of cylinders while MWCNTs are of two types Russian Doll model and the Parchment model [74]. When the diameter of the outer nanotube is greater than the inner nanotube then, it is called the Russian Doll model. Whereas, when a single graphene sheet wraps around itself multiple times just like a scroll of paper, it is called the Parchment model. On the basis of structural symmetry, carbon nanotubes can exhibit conducting or semiconducting nature [75].

Graphene is a two dimensional single sheet of carbon atoms arranged like a honeycomb lattice and displays unusual electrical properties. In graphene, each C atom is linked to three other carbon atoms, leaving an electron freely available for electronic conduction. These mobile electrons are called pi (π) electrons and are located above and below the graphene sheet. Graphene is a zero-gap semiconductor because its conduction and valence bands meet at the Dirac points, where the effective mass of electrons (or holes) is zero. The free electrons in graphene interact with the periodic potential of the honeycomb lattice, generating new quasiparticles that have lost their mass or rest mass and hence, called massless Dirac fermions. Thus, graphene conducts electrons with high velocity and the lattice symmetry inhibits direct backscattering of electrons, enhancing the electrical mobility and conductivity [76].

The remarkable electric conductivities of graphite, fullerenes, nanotubes, and graphene make them highly suitable as a catalyst for various redox reactions. The electrical properties of carbon materials arising from their unique structures can further be tuned by functionalization, heteroatom doping and controlling the temperature of graphitization or carbonization [77].

Functionalization of carbon surface alters the electrical behaviour of carbon material by modifying their structure. In one such instance, Suslova *et al.* have reported that the electrical conductivity of nitrogen doped carbon nanotubes increased considerably with an increase in oxygen functionalities on their surfaces owing to elevation in the density of defects in their structural arrangement [78]. In another example, it was shown that graphitized carbon nanofibre bearing oxygen functional groups demonstrated high electrocatalytic activity as bifunctional electrode for vanadium redox flow battery [79]. The high catalytic performance of carbon nanofibres was attributed to an increase in ion diffusion and electron transfer processes owing to improved electrical conductivity and surface area due to presence of surface functionalities.

Doping is a method by which introduction of heteroatom results in generation of holes or addition of an extra electron in a substance. This process leads to decrease in band gap of valence and conduction bands, thereby, improving the conductivity of semiconductors. Heteroatom doping is therefore a strong tool for tuning the electrical conductivities of carbon materials. Han *et al.* studied the effect of N-doping in carbon nano-onions [80]. The N-doped carbon nano-onions (N-CNOs) were prepared by a one-step, *in situ* flame synthesis technique using clarified butter and acetonitrile as precursors. The prepared N-CNO was than employed as catalyst for ORR process in a microbial fuel cell. It was reported that the catalytic activity of N-CNO was 5.4 times higher than CNO. This high performance was ascribed to the presence of nitrogen atoms in carbon framework.

Through electrochemical impedance spectroscopy, it was elucidated that the conductivity of N-CNO is higher than CNO due to N-doping. It has been explained that the sp^2–C atoms adjacent to N atoms and N atoms stimulate the electrostatic adsorption of oxygen molecules leading to sharp reduction in the kinetic barrier for the ORR process [81]. Simultaneously, N-doping causes electron excess in the delocalized π-system which results in enhancement of electrical conductivity. Finally, the high electrical conductivity induces electron transfer during catalysis, assisting the contribution of C-N active sites and improving the ORR activity on the whole [82]. As a result, the high ORR catalytic performance of N-CNO can be attributed to the high electrical conductivity arising from N-doping.

It has also been reported that thermal [83] and graphitization treatments [84] generally result in metallic conductivity in carbon materials while oxidation induces semi-conductivity due to π-electron delocalization [85]. Therefore, by applying different treatments one can tune the electrical properties of carbocatalysts.

SYNTHETIC STRATEGIES FOR CARBOCATALYSTS

As is well known that synthetic methods greatly influence the properties of a material, therefore, in this section some strategies employed for synthesis of carbocatalyst have been discussed. So many methods for synthesis of carbon materials are reported that a whole chapter can be devoted to the synthetic aspects of carbon allotropes. However, since the book deals with graphene here we have highlighted some important methods only.

Active Carbon

Activated carbon is highly porous form of carbon. Coal, wood, agricultural wastes or biomass, ionic liquids, salt solutions, *etc.* are used as precursors to prepared activated carbon or charcoal [86]. Physical and chemical activation methods are commonly used to prepare active carbon materials. Before activation, carbonization and de-ashing is performed as pre-treatment. In carbonization the raw materials are subjected to heat treatment or pyrolysis at high temperatures around 573.15 K to 1173.15 K in order to eliminate non-carbon materials and volatile impurities. This result in formation of char after rearrangement of carbon atoms into graphitic sheets stacked above one another to form rigid microcrystals with voids which constitute the pore structure [44]. Ahead of activation, ash and mineral content of the carbon materials is reduced by de-ashing or demineralization technique *via* chemical leaching. In this process, acid and alkali solutions are used to demineralize the carbon raw materials [87]. For instance,

Ahmad *et al.* performed acid treatment of cocoa shell pellets with 1M HCl which reduced the ash content to less than 10% [88]. In another study, it was reported that high ash content reduced the adsorption property of activated charcoal prepared from stone coal ore hence, to improve the adsorption. The stone coal ore was pre-treated with HF and H_2SO_4. The acid treatment reduced the contents of mineral atoms like Fe, Mg, Ca, Si, *etc.* resulting in stone coal ore with 74.76% carbon content [89]. Similarly, KOH promoted de-ashing of rice husk was reported by Shen and Fu which resulted in highly porous biochar [90].

Next step involved in formation of activated carbon is activation. Activation can be carried out in two methods: physical and chemical. Physical activation includes use of carbon dioxide, air, steam or other gaseous agents at 1073.15 to 1273.15 K to remove oxides of carbon from char and improving the porosity of carbonaceous materials. Zhou and co-workers have reported preparation of highly porous activated carbon from waste tea *via* physical activation with steam [91]. It was further highlighted that typically micropores were developed when the activation temperature was below 800 °C while activation above 800 °C enriched the carbon with both micropores and mesopores.

Chemical activation involves single-step carbonization and activation of carbon materials by impregnating them with chemical agents such as KOH, $ZnCl_2$, H_3PO_4, $FeCl_3$, $CaCl_2$, H_2SO_4, NaOH, $MgCl_2$, K_2CO_3, HNO_3, *etc.*, followed by thermal treatment [86]. The chemical agents due to dehydrating property result in generation of a rigid framework which is less prone to reduction in volume during carbonization, thereby generating excellent yields of activated carbon [92, 93]. Moreover, since chemical activation is performed at lower temperature than physical activation, it offers better control of porous structure [94]. However, from environment point of view physical activation is better as it avoids use of chemical agents while chemically activated materials contain some amount of inorganic matter [95]. Recently, Chen *et al.* have demonstrated synthesis of activated carbon with large surface area from tobacco stem [96]. Chemical activation was performed using KOH, K_2CO_3 and $ZnCl_2$ as activating agents. It was reported that $ZnCl_2$ caused higher activation of tobacco stem compared to KOH and K_2CO_3. Further, increasing the concentration of $ZnCl_2$ enhanced the development of new pores and the widening of existing pores. Furthermore, activation of carbon with K_2CO_3 resulted in high oxygen functionalities on the surface. Thus, it is evident that nature of chemical agent and concentration affects the activation process as well as surface properties.

Graphite

Synthetically, graphite is prepared *via* thermal treatment of petroleum coke or coal-tar. Hydrothermal methods involving treatment of polymers and metal carbides can also be used to produce graphitic materials. Of late, graphite encapsulated nanoparticles have also emerged as potent carbocatalysts. These can be fabricated by several methods such as pyrolysis, template synthesis, chemical vapor deposition, arc discharge, laser ablation technique, *etc.* Li *et al.* have demonstrated hydrothermal synthesis of graphite encapsulated molybdenum carbides [97]. The graphite encased molybdenum carbide nanoparticles were prepared by the hydrothermal carbonization of a mixture of glucose solution and ammonium molybdate followed by temperature programmed reduction under mild conditions. The prepared graphite encapsulated nanoparticles showed high catalytic efficiency for conversion of guaiacol to phenolic compounds in methanolic medium. In another study, graphite encapsulated gold nanoparticles displaying excellent electrocatalytic activities towards oxidation reduction reaction were prepared by laser ablation method. The process of laser ablation involved installation of Au target in a vessel containing graphite precursor-acetone (16 mL). Thereafter, the revolving Au target was ablated by a focused beam of laser with wavelength of 1064 nm, 10 Hz pulse repetition rate and 100 mJ pulse energy for 30 min. Finally, the graphite encased Au nanoparticles were obtained by centrifugation followed by freeze drying. Thus, laser ablation technique afforded a facile and efficient route to synthesis of graphite encased nanoparticles.

Recently, expanded graphite has been employed as catalyst for oxidation reaction [98]. Expanded graphite is a type of graphite with interlayer spaces. The expanded graphite is prepared by oxidation of natural graphite flakes using acids like nitric acid and sulfuric acid and oxidants like potassium permanganate, chromic acid, hydrogen peroxide, *etc.* Then, the mixture is subjected to high temperature which causes expansion and swelling of graphite resulting in separation of layers leading to formation of its expanded form.

In recent past, hydroxylated high surface area graphite has also been used as carbocatalyst for ozonation. Usually, oxygen functionalized graphite are prepared *via* oxidation using strong acids and mechanical or thermal exfoliation methods. In a method reported by Bernat-Quesada *et al.*, the water soluble graphite was synthesized by oxidation of graphite by nitric acid [99]. Dispersion of graphite in nitric acid followed by heating at 83 °C for 20 h followed by filtration and drying resulted in formation of oxygen functionalized graphite on edges as well as basal planes [48]. In another method, the workers used ball milling technique to fabricate hydroxylated graphite. In this method, graphite, KOH and water were loaded in a grinding jar of planetary ball mill. The jar was agitated at 300 rpm for 10 hours at ambient temperature. Later, the mixture was filtered and dried to obtain graphite functionalized with oxygen moieties on the edges.

Xu *et al.* have also reported the synthesis of functionalized graphite felts as catalyst for oxidation of hydrogen sulphide [100]. The commercial graphite felts bearing oxygen functionalities were prepared by treatment with nitric acid. According to the method, graphite felts were first heated at 500 °C for 1 hour in air. Then, the felts were loaded in a tubular reactor and heated at 250 °C using an external electric furnace [101]. The reactor loaded with the felts was connected to a round bottom flask (RBF) containing 65% HNO_3 solution (150 mL). The temperature of the RBF was maintained at 125 °C and the solution was continuously stirred magnetically. The gaseous acid passed through the felts for several hours causing their oxidation. The sample was then washed using deionized water and dried at 130 °C overnight. The acid treated functionalized graphite felts exhibited large surface area and it was concluded that the structural defects in graphite felts as well as oxygen groups on them served as active sites for oxidation of hydrogen sulphide.

Fullerenes

Fullerenes and their composites are widely used as catalysts. Fullerenes are giant caged carbon allotropes that are generally prepared by arc discharge and laser ablation of graphite, and combustion of hydrocarbons. In a typical arc discharge method, electric current is struck between graphite or carbon anode and cathode electrode in an inert atmosphere. The electric discharge produces soot which is collected and fullerene is then extracted from the soot by solvent extraction method, sublimation or supercritical fluid extraction technique. Generally for fullerene synthesis, an alternating current between the graphite electrodes in He atmosphere of about 200 Torr is used to produce soot containing C_{60} and C_{70} molecules [102].

The laser ablation technique makes use of a pulsed laser beam which is focused on the surface of carbon sample at high temperature (usually 1000 °C). Thus, the ablation of carbon results in fragments which form fullerenes [103]. Also, hydrocarbon combustion in presence of electric field acting on the flame is widely used to achieve soot-containing fullerene [104].

Additionally, several fullerene based complexes are also well known as catalysts and can be prepared by different methods. For instance, Sabounchei and co-workers have prepared mono- and bidentate palladium(0)- [60]fullerene based complexes as catalyst for Mizoroki-Heck coupling reaction [105]. The synthetic strategy involved a Schlenk tube containing C_{60} (0.05 mmol, 0.036 g) and

$Pd(dba)_2$ (0.05 mmol, 0.029 g) in 15 mL of toluene. The reaction mixture was stirred for 15 min at room temperature during which the color changed from purple to black. Thereafter, a phosphorus ylide solution (0.05 mmol) in toluene (5 mL) was injected into the reaction solution *via* a syringe pump. Further, a 2 hour stirring was continued at ambient temperature which resulted in deep green solution. Dark green crystals were obtained by treating the solution with n-hexane overnight and later washing with diethyl ether.

Further, several fullerene based composites fabricated by sol-gel method, chemical reduction, ultrasonication, hydro- or solvothermal technique, *etc.* serve as useful catalysts [106]. For example, Liu *et al.* have reported synthesis of silver (I) doped fullerene by liquid- liquid interfacial precipitation method [107]. The method involved mixing of saturated solution of fullerene (approx. 2.8 mg/mL) in toluene with isopropyl alcohol saturated with silver nitrate at ambient temperature. The bilayer so formed was kept still under room conditions for 3 hours. Thereafter, centrifugation of the mixture at 7000 rpm for 5 minutes resulted in precipitation of crystalline solid which was dried under nitrogen stream. Then, chemical reduction the $AgNO_3$ complex encapsulated fullerene crystals with hydrazine hydrate solution (1.3% wt.) resulted in formation of Ag(I)-fullerene composite. Finally, the composite was washed with deionized water and ethanol to remove excess silver ions. The prepared Ag(I)-fullerene composite was used as a catalytic system for reduction of 4-nitrophenol and photo-degradation of orange G dye.

A self assembled method based on hydrothermal mode for synthesis of amine functionalized fullerene decorated with palladium nanoflowers was reported by Li and Han [108]. Initially, amine functionalized fullerene was prepared by dissolving ethylenediamine in ethanol-water system (5:1). Then this solution was added drop by drop to toluene saturated with C_{60} fullerene and sodium hydroxide with continuous stirring. The stirring was continued for 5 days in argon atmosphere at room temperature, later the solution was filtered, centrifuged, and washed with ethanol and distilled water for a number of times. Eventually, dark brown solid products were obtained after drying in vacuum for 24 hours at 60 °C. For preparation of composite of fullerene, palladium chloride was dissolved in HCl and DMF and mixed with functionalized fullerene. The reaction mixture was then placed in autoclave for 6 hours at 140 °C in nitrogen atmosphere. The obtained product was washed with ethanol and vacuum dried overnight to yield amine functionalized fullerene decorated with palladium nanoflowers which displayed excellent electrocatalytic effect for reduction of *p*-nitrophenol.

Glassy Carbon

It is a type of carbon material which combines characteristics of both glass and ceramic materials with that of graphite. It is a commonly used electrocatalyst or electrode material. The strategy for fabrication of glassy carbon is based on pyrolysis of polymeric materials at elevated temperatures of about 3000 °C [44, 109]. The mechanism basically involves breakdown of carbon-heteroatom bonds in polymers and generation of C-C bonds [110 - 112]. Initially, hydrocarbon radicals are generated with their highest concentration at around 600 °C [113] and after 800 °C or higher temperature, a network of graphene fragments develop [114]. Further, annealing process results in formation of cross-linked carbon network and removal of volatile by-products like oxides of carbon, water, methane and small hydrocarbons [115]. For instance, Jurkiewicz *et al.* prepared glassy carbon by carrying out pyrolysis of polymeric furfuryl alcohol under Ar gas flow at varied temperatures ranging from 600 to 2500 °C at different heating rates [116]. The carbonized samples were then cooled in argon atmosphere resulting in glassy carbon with fullerene-like or nanotube-like elements in its structure. It was also reported that pyrolysis up to 1000 °C resulted in increase in fullerene-like non-planar sp^2 carbon bonds. But on further increase in pyrolysis temperature the amount of non-planar sp^2 bond content decreased.

Recently, a nitrogen enriched glassy carbon was prepared as an electrocatalyst for hydrogen evolution reaction by Thirukumaran and his research group [117]. Initially polybenzoxazine was synthesized by step-wise polymerization of benzoxazine. The carbonization of polybenzoxazine at 600 °C for 5 hours under nitrogen atmosphere was performed. The resultant carbonaceous material was then activated by adding KOH solution followed by water elimination through evaporation at 120 °C. Subsequently, the carbon material was heated in a tubular furnace for an hour at 600 °C to obtain nitrogen enriched glassy carbon. This method thus offered an efficient method for synthesis of glassy carbon electrocatalyst.

Carbon Black

Carbon black is composed of colloidal carbon particles and is usually synthesized *via* partial combustion or thermal decomposition of petroleum products. Carbon blacks are extensively used as electrocatalysts for redox processes. Diverse thermal methods for carbon black synthesis are summarized below [118].

Channel Black Process

In this process, combustion of natural gases using a continuous small wide luminous fan-shaped flame occurs. This process yields carbon black with a particle size of approximately 10-30 nm.

Gas Black Process

It involves thermal oxidative decomposition, where the formation of carbon black takes place in diffusion flame.

Thermal Black Process

In the thermal black process, thermal combustion of natural gas into elemental carbon and hydrogen takes place at high temperatures to produce carbon black and tail gas. The sizes of obtained carbon black particles are about 180 nm.

Acetylene Black Process

The method involves constant thermal decomposition of acetylene gas in the absence of oxygen gas to generate carbon black as aerosols.

Lamp Black Process

It includes partial combustion of carbon materials such as resins, fatty oils and acids, coal tar, *etc*., resulting in the deposition of carbon in the form of soot. Carbon black particles so obtained range in the size from 110 to 120 nm.

Furnace Black Process

In this proccss, aromatic oils are pyrolized by exposing them to a hot gas stream to produce carbon blacks with particle sizes of about 20-100 nm.

Besides these methods, carbon black is also obtained by plasma synthesis. An arc discharge plasma method coupled with thermal pyrolysis for the synthesis of carbon black was demonstrated by Sun *et al.* [119]. An apparatus composed of a plasma generator, thermal pyrolysis device, gas injector, power supply and collecting device was used for this purpose. Thermal plasma was generated by electrodes supplied with 10 kV power. Argon was injected into the device for 5 min. to firstly drain out oxygen and then mixed with propane at definite rate of volume. As the propane gas flowed into the plasma region, carbon black was produced instantly. Then the thermal pyrolysis furnace pyrolyzed the un-reacted propane. Finally, carbon black was collected from the bottom of sand core funnel connected with a vacuum pump.

Functionalized carbon black has also gained attention as catalyst. Recently, oxidized carbon black has been employed as a catalyst for the synthesis of benzodiazepines [119]. The oxygen functionalized carbon black was prepared by Hummer's method. In this method, sulphuric acid (120 mL) and sodium nitrate (2.5 g) were mixed with 5g of carbon black in a round bottom flask kept on an ice bath. The suspension was magnetically stirred and then 15 g potassium permanganate was added gradually. The mixture was subjected to heating at 35 °C and stirred for about 24 hours. Then, 700 mL deionized water was added in small portions resulting in black slurry for carbon black under stirring. Finally, 5 mL of hydrogen peroxide solution (30 wt%) was added to the slurry. The obtained sample was poured into deionized water and then centrifuged for 15 min at 10000 rpm. The sample was then washed with deionized water and HCl (5 wt %) and later dried to obtain oxidized carbon black. The optimal conditions for highly graphitized carbon black production were a discharge current of 0.8 A, 120 L/h of argon flow rate, 80 ml/min of propane flow rate and pyrolysis temperature at 700 °C.

Also, carbon black based composites to have found applications as electrocatalysts. In a work conducted by Lim *et al.*, TiO_2/carbon black composite was prepared and used as a catalytic electrode in dye-sensitized solar cells [120]. The composite formation involved pulverization of carbon black followed by its calcination in a muffle furnace at 500°C for 2 hours by placing the carbon black in 80-unit mesh. The calcinated carbon black was again ground and thermally treated at 300 °C for 2 hours to obtain carbon black powder with a particle size of about 80 nm. Then, nano-TiO_2 was prepared by the modified Burnside method. Finally, TiO_2/carbon black composite was fabricated by mixing carbon black powder with TiO_2 NPs followed by ultrasonication for 10 min. at 750 W. Then, the addition of 100 μl of Triton X-100 to the mixture followed ultrasonic treatment for 10 min. yielded the TiO_2/carbon black composite. The prepared composite was subsequently used as the counter electrode material in solar cells.

Carbon Nanomaterials

Carbon Nanofibres

Carbon fibres are carbon containing fibres of diameter of the order of few micrometers that find applications in industries due to their enormous tensile strength and thermo-chemical conductivity and stability. In last few years nanodimensional carbon fibres have attracted huge attention in the field of catalysis due to their advanced chemical and electrical activities along with large surface area. Carbon nanofibres are usually prepared by chemical vapour deposition method (CVD), catalytically promoted CVD (CCVD), electrospinning and template synthesis techniques. CVD is a well acknowledged method for synthesis of 2-dimensional nanomaterials. The method is based on decomposition of vapour or gaseous precursors on a substrate for preparation of materials at elevated temperature in presence of vacuum. The CVD method was explored by Hulicova-Jurcakova *et al.* for synthesis of carbon nanofibre [121]. Carbon precursors like methane and acetylene were used for this process and nickel supported on γ-alumina was used as substrate. The reaction was carried out in tubular furnace with a heating rate of 10 °C/min at 500°C and 700°C, respectively. The obtained carbon samples were purified in reflux system using 5 M nitric acid solution for 6 h, the obtained solution was filtered and dried to obtain carbon nanofibres. Highly graphitic carbon nanofibres were obtained by this CVD method.

Another important method that has emerged for nanofibre synthesis is CCVD, in which catalyst based chemical vapour decomposition, is carried out. Miniach and co-workers have reported a CCVD method for synthesis of carbon nanofibres [122]. They used hydroxyapatite-supported nickel catalyst (Ni/HAp) on which methane decomposition occurred resulting in the formation of herringbone bamboo-like carbon nanofibres. In this method, first Ni/Hap was loaded in a quartz tube placed inside a horizontal furnace. First the catalyst was reduced under H_2 flow (150 ml min^{-1}) at 450-650°C for 10-120 min. Afterwards, the methane and hydrogen gases were made to flow through the tubes at 650°C for 3-180 min. Later the nanofibres were cooled down to room temperature in N_2 atmosphere and then treated with 5% HCl in a sonication bath for 2 hours to remove the excess catalyst. Finally, the carbon nanofibres were washed with deionized water and dried.

Electrospinning is a commonly used method for the fabrication of nanofibres from polymeric materials using high power electric force. Recently, Alarifi *et al.* prepared carbon nanofibres from polyacrylonitrile (PAN) by electrospinning technique [123]. It was reported that a homogeneous solution of PAN was prepared by dissolving PAN in DMF and stirring it at 500 rpm for an hour. The solution was then transferred into a 10 mL syringe and using electrospinning technique with the electric field of 25kVat injection rate of 1 ml/h and a spinneret-to-collector distance of 25 cm, the nanofibres were formed. Thereafter, thermal treatment of nanofibres in oxygen atmosphere at 270°C for an hour followed by carbonization in argon atmosphere at different temperatures 750, 850, and 950°C was carried out resulting in high carbon content nanofibres.

Template synthesis is also one widely used method for synthesis of nanofibres. The strategy involves use of a nanoporous membrane as a template for synthesis of nanofibers from diverse materials such as carbon, polymers, semiconductors, and metals [124 - 126] (Huczko, 2000; Hulteen and Martin, 1997; Toro and Buriak, 2014). A template based synthesis for carbon nanofibres were reported by Li *et al.* [127]. The method involved addition of anhydrous zinc acetate to ethylene glycol followed by stirring of the mixture at 150 ° C for 90 min. Then, the mixture was subjected to centrifugation and the obtained slurry was calcinated at 600 ° C for 2 hours in argon atmosphere. Later, the composite was immersed in a 2.0 M HCl solution with overnight stirring to obtain the nanofibres. The carbon nanofibres were then purified by washing with water and ethanol and then dried. Mesoporous carbon nanofibers were obtained after being washed with water and ethanol, and then dried at room temperature. The carbon nanofibres produced by this method were highly mesoporous and were used as electrodes.

Carbon Nanotubes

There are several methods for fabrication of carbon nanotubes (CNTs) namely, CVD, arc discharge, laser ablation, electrolysis, ultrasonication techniques, *etc.* Verma *et al.* have demonstrated CNT fabrication using CVD approach [128]. An automated CVD reactor was used and iron catalyst supported over magnesium oxide was loaded on quartz boats of the reactor. Argon and hydrogen gases were flown into the reactor at a rate of 100 ml/min. at 25 °C for 5 min. Then, temperature was raised to 750 °C and after 10 minutes acetylene along with hydrogen was flown in the reactor for 15 min. at the rate of 300 ml/min. and 100 ml/min., respectively, over the catalyst to develop CNTs.

Arc discharge method for CNTs is carried out at elevated temperatures usually above 1000 °C. Cu and Ni containing carbon nanotubes were prepared by Saravanan *et al.* using DC arc discharge method [129]. Firstly, the graphite anode was filled with mixture of graphite and metal powder in the ratio 1:1. A graphite rotating disc with a speed of 8-10 rpm was used as cathode on which soot deposition occurred. The electric arc was generated *via* a DC source of 20 V and 150 A. Owing to high temperature and energy, vaporization of graphite and metals in air took place. The graphene sheets rolled into CNTs due to elevated temperature and enormous energy while the metals upon condensation deposited in the inner walls of CNTs. The CNTs so formed were washed with distilled water and dried. Finally, treatment with toluene for 5 hours and annealation for an hour at 650°C led to metal encapsulated CNTs.

Laser ablation technique is primarily used to prepare single-walled CNTs (SWCNTs). But nowadays multi-walled CNTs (MWCNTs) are also prepared by this technique. Recently, Ismail *et al.* have demonstrated fabrication of MWCNTs using pulsed laser ablation method in aqueous medium [130]. In this method high grade graphite pellet was placed in the glass cell filled with double distilled water.

The laser beam from Q-switched Nd:YAG of two wavelengths: 532 and 1064 nm with pulse width of 9 ns were focused on the graphite target by using positive lens of focal length of 5 cm. The ablation time and laser repletion rate selected for the process were 30 min and 2 Hz, respectively. The ablation cell rotated during the ablation process by using DC motor with 25 rpm and was used to prevent the vapour from depositing on the lens. Finally, colloidal CNTs were collected in the glass cell. Thus, MWCNTs with diamond like structure were formed.

Electrolysis of carbon materials in molten salts are widely used to prepare CNTs. Thin walled CNTs from CO_2 using molten carbonate was prepared *via* electrolysis by Wang *et al.* [131]. Electrodes (Muntz brass sheet-cathode and Nichrome-anode) were immersed in a mixture of electrolytes (lithium carbonate, lithium oxide, calcium oxide, and boric acid) and a constant current of 0.1 A/cm^2 was applied for 4 hours. The electrolysis temperature was 770 °C and impure sample was collected at cathode which was later washed. The washed product was recovered by vacuum filtration and dried overnight at 60 °C to obtain pure CNTs. The method resulted in ultra-thin walled CNTs (22–42 nm diameter) consisting of about 25 concentric, cylindrical graphene walls.

Sonochemical treatment of carbonaceous materials also results in CNTs. For instance, Wang and his research group have developed CNTs from graphene oxide *via* ultrasonication method [132]. The method involved dispersion of graphene oxide into HNO_3 solution (70%) which was subjected to sonication in ultrasonic bath at ambient temperature for 4 hours. Thereafter, water was poured into the suspension and centrifugation was carried out at 15,000 rpm to obtain CNTs. The CNTs were further purified their dispersion in water followed by centrifugation at 12,000 rpm for an hour to obtain pure CNTs.

Nanodiamonds

Nanodiamonds are nanosized carbon particles with dimensions less than 100 nm and were first discovered in 1963 [133]. The nanodiamonds due to their remarkable features like high solidity, large surface area, chemical inertness, enormous strength, optical clarity, biocompatibility, and thermal conductivity are quite useful carbon materials and can be fabricated by diverse techniques [134]. Detonation, CVD, laser irradiation and high pressure-high temperature methods are commonly used strategies for nanodiamond synthesis.

A bulk production of nanodiamonds from tetryl, a low cost explosive, was demonstrated by Panich and co-workers [135]. The detonation method proceeded with pressing of tetryl or its mixture with TNT and RDX at 1000 to 2000 kg/cm^3 into a cylindrical disc of 60 mm diameter and 110 mm height to form an explosive charge. The explosive charge was kept in an explosive chamber in a plastic bag and was detonated. Further, it was reported that after every fifth detonation the valve at the bottom opened and the diamond soot was washed off with water. The diamond soot was then filtered, magnetically separated and centrifuged to remove impurities. Then, the soot was dried and crushed to obtain nanodiamonds. Thus, it was elucidated that tetryl serves as an excellent explosive material for fabrication of nanodiamonds.

Recently, Tallaire and his research group have synthesized loose nanodiamonds by plasma-assisted CVD method [136]. The synthesis was carried out in a quartz bell jar microwave reactor. The plasma was generated by microwaves at 2.45 GHz in resonant cavity. The reactor operated using a mixture of gases ($H_2/CH_4/O_2$) along with high microwave energy to produce nanodiamonds which agglomerated on water cooled molybdenum holder exposed to plasma at 450-500 °C. The agglomerated nanodiamonds were then washed away by ethanol and collected in cavity of bell jar reactor. The nanodiamonds thus obtained by the CVD technique showed superior crystallinity and lower impurities, which make this method advantageous.

In another method, pulsed laser induced ablation was used for synthesis of nanodiamonds from graphite. Basso *et al.* reported that laser ablation of graphite with KrF excimer laser was carried out [137]. The graphite target was kept in vial made of glass filled with liquid nitrogen. The laser beam of 500 mJ energy was focused on the target. After completion of ablation, liquid nitrogen was evaporated and the nanodiamond powders were formed. Consequently, laser ablation process offered an inexpensive and controlled approach for nanodiamond synthesis.

High pressure and high temperature technique for fabrication of nanodiamonds from hydrocarbons has attained great significance nowadays. This is because this method affords nanodimensional diamonds with highest possible structural perfection and crystallinity. Ekimov and co-workers have recently synthesized nanodiamonds from adamantane under high temperature/pressure conditions [138]. The process was performed at 7.7 and 9.4 GPa pressure and temperatures at 300 to 1600°C. The adamantane sample pressed in the form of a disc of 5 mm diameter and 3 mm thickness was placed in a titanium capsule in Torrid chamber. The temperature was controlled using Chromel-Alumel thermocouple at <1350°C and tungsten-rhenium thermocouple at temperatures up to 1600°C. The thermocouple junction was held on the outer wall at the mid-height of the capsule. The high pressure/temperature treatment of adamantane caused carbonization of thus leading to formation of nanodiamonds. It was reported that at 7.7 GPa only diamond microcrystals were formed but on increasing the pressure to 9.4 GPa, nanodiamonds were formed. Thus, high pressure conditions are very important for synthesis of nanodimensional diamonds.

Das and Saikia have highlighted a novel method of synthesis of nanodiamonds from low grade coal under ultrasound exposure [139]. In this method, coal (2 g) was added to 100 mL hydrogen peroxide solution in a Teflon beaker and exposed to ultrasound-waves for 3 hours. A senor probe was used to measure the reaction temperature. After 3 hours of exposure, the reaction mixture was filtered to remove sulphur and other mineral contaminants. The filtrate was then poured into a beaker full of crushed ice and neutralized. Then the neutral suspension was filtered *via* a polytetrafluoroethylene membrane (0.22 μm) and dialyzed for 5 days in a 1 kDa dialysis bag. Thereafter, the solution was concentrated in a rotator evaporator and collected. The nanodiamonds synthesized by this method were obtained as monocrystalline and polycrystalline carbon nanocrystals with 4-15 nm dimensions.

Carbon Nano-onions

Carbon nano-onions are concentric shell like nanostructures consisting of nested fullerenes. The carbon onions were discovered for the first time by Ugarte in 1992 [140] and ever since their emergence several methods have been developed for their synthesis. Some of these include arc discharge method, chemical vapour deposition, pyrolysis, laser technique and thermal treatment of nanodiamonds.

Alessandro and co-workers have demonstrated synthesis of carbon nano-onions by water based arc discharge method [141]. It was reported that in an electric discharge apparatus, two graphite electrodes (anode of 5 nm and cathode of 10 nm diameters) were immersed in water and a potential of 12 V and current 30 A were applied between them. This resulted in formation of a powder that was dispersed in water and floated or settled at the bottom in the beaker. The floating powder like sample was collected on a microscope slide, whereas the bottom sample was recovered by water evaporation at 80 °C. The samples were then subjected to ultrasonication in ethanol for half an hour at 100 W. Finally, annealation of the samples at 400°C in air resulted in carbon nano-onions free from amorphous carbon residues. Highly stable, polyhedral carbon nano-onions were obtained by this method.

CVD method was employed by Zhang and his research group to prepare carbon nano-onions [142]. The synthesis was carried out by decomposition of methane gas over Ni-Fe alloy based catalyst at 850 °C. The process yielded highly graphitic carbon nano-onions.

In another method, carbon nano-onions were prepared from waste frying oil by pyrolysis technique. Jung *et al.* pyrolyzed the oil using a cotton wick under stable air conditions [143]. Then, to collect the sooty material, a beaker was inverted above the flame and the powder was gathered by scrapping off the deposited material. Next, the soot was calcined at 900 °C for an hour in a muffle furnace which led to carbonization and graphitization of carbon nano-onions. The method thus offered an economical approach for preparation of carbon nano-onions for application in energy storage devices.

A laser ablation induced carbon nano-onions synthesis was demonstrated by Dorobantu and co-workers [144]. A pulsed laser from Compex Pro 205 (KrF) excimer laser with a wavelength of 248 nm, pulse length of 25 ns and repetition rate of 10 Hz was used. The laser beam was focused on a pure graphite target under an ambient pressure of 7 Torr. During the process argon gas flowed at about 300 L/h through the reactor and temperature was maintained at 900°C. The ablation products then condensed on the water cooled cold finger to finally obtain the carbon nano-onions. The carbon nano-onions formed by this technique were about 10-25 nm in diameter.

Recently, Bartolome and Fragoso have reported synthesis of carbon nano-onions *via* thermal annealing of nanodiamonds [145]. The method involved annealing of 300 mg of nanodiamond at 1200°C at a heating rate of 50°C min^{-1} under argon atmosphere for 6 hours. Thereafter, the annealed sample cooled down to room temperature and a portion of this sample was then treated with 30% H_2O_2 for 3 hours, washed and dried to obtain pure carbon nano-onions. The protocol afforded carbon nano-onions with a diameter of 3-4 nm and comprising 5-6 graphitic shells.

Graphene and Its Derivatives

A multitude of top-down and bottom-up strategies for fabrication of graphene and its derivatives are reported in literature. These methods have been described in detail in Chapter 3.

SCOPE OF CARBOCATALYSTS

As mentioned earlier the carbon materials due to their several interesting properties find applications as catalysts in various fields such as environmental remediation, solar cells, fuel cells, energy devices, water splitting, redox reactions, sensors, biocatalysis, synthesis of fine chemicals, *etc*. For instance, Christoforidis and co-workers have lately reported the use of carbon nanotubes supported on graphitic nitride as photocatalyst for hydrogen production [146]. It was elucidated that the carbon nitrides functionalized with SWCNTs were found to be about 2-5 times more efficient in catalyzing hydrogen evolution as compared to DWCNTs (double-walled carbon nanotubes) and MWCNTs under both the illumination of solar as well as visible light. This was attributed to the fact that CNTs worked as acceptors of electrons and further it was reported that the charge transfer varied inversely with the number of walls in CNTs. Thus, the transfer of photogenerated electrons occurred more readily when SWCNTs were used, thereby promoting H_2 production by providing more electrons.

Removal of pollutants from environment is a big concern and nowadays carbon catalysts are being widely used to remediate environment. Pausova *et al.* have recently employed active carbon (AC)-TiO_2 based composites as catalyst for photodegradation of benzoic acid in water and toluene in air [147]. It was reported that the incorporation of active carbon in composite enhanced photocatalytic activity due to greater adsorption ability of AC which facilitates adhesion of the pollutants. Further, it was also illustrated that the enhanced catalytic efficiency of AC was due to the milling process which increased its surface area thereby causing better adsorption of the substrates. Thus, active carbon based composites have merged as efficient catalytic systems for elimination of water and air pollutants.

The high conductivity, surface property and high melting point of carbon materials favour their application as electrodes in solar cells and fuel cells. Recently, Don and his research group have demonstrated use of acetylene carbon black-graphite composite as electrode for dye sensitized solar cells (DSSCs) [148]. It was highlighted that the composite was prepared in different ratio of carbon black and graphite *viz.* 0:1, 1:0, 1:1, 1:3, and 3:1 and the highest conversion efficacy was found to 5.06% for DSSCs containing acetylene carbon black-graphite in the ratio 3:1 compared to conventional expensive Pt counter electrode. This was because the carbon black particles offered greater surface area for ekectrochemical reaction in DSSC. Further, it was elucidated that the incorporation of acetylene carbon black particles on graphite enhanced the electrocatalytic efficiency of triiodide ion reduction by enhancing the electric conductivity *via* linking of the graphite flakes. Thus, acetylene carbon black-graphite composite can be used as cost effective alternative to Pt counter electrode in DSSCs.

Similarly, Bhavani *et al.* have reported use of fullerene based electrocatalyst in direct alcohol fuel cells (DAFCs) [149]. They prepared bimetallic Pd-W nanoparticles anchored on fullerene-C_{60} catalyst (Pd-W@Fullerene-C_{60}) for electrooxidation of methanol and ethanol for DAFCs applications. It was reported that owing to large electrochemical surface area, excellent conductivity, high reproducibility and long standing durability, the electrocatalysts showed enormous catalytic activity for oxidation reaction compared to monometallic catalyst and commonly used Pt/C catalyst. It was revealed that the remarkable activity of the catalyst for oxidation of alcohols in fuel cells was attributed to the synergistic relation between Pd-W nanoparticles and fullerene. The uniform distribution of nanoparticles on fullerene prevented their aggregation resulting in enhanced surface area and stability. Moreover, the outstanding electron conductivity of fullerenes contributed to the overall catalytic activity of the Pd-W@Fullerene-C_{60} electrocatalyst, making them highly suitable electrode material for fuel cells.

In recent past, Chen and co-workers have reported the use of PT-Ni nanoparticles decorated on carbon nanofibres (PtNi/CNFs) as potent electrocatalyst for water splitting [150]. The catalyst was fabricated by electrospinning technique and carbonization. The catalyst was used as electrode for water splitting. It was reported that for hydrogen evolution in acidic medium (0.5 M H_2SO_4), the electrocatalyst achieved a 10 mA cm^{-2} current density at low overpotentials of 34 mV and a Tafel slope of 31 mV dec^{-1} similar to that of commercially available Pt/C (20 wt%) electrocatalysts. In alkaline medium (1 M KOH), the PtNi/CNFs catalyst showed a low overpotential of 82 mV to attain a current density of 10 mA cm^{-2} a small Tafel slope corresponding to 34 mV dec^{-1}. In addition, the PtNi/CNFs also show good activity for oxygen evolution in alkaline medium displaying a small overpotential of 151 mV to reach a current density of 10 mA cm^{-2} and a Tafel slope at 159 mV dec^{-1}. In acidic medium, the PtNi/CNFs exhibited an overpotential of 451 mV and Tafel slope at 198 mV dec^{-1}. Thus, on the whole the PtNi/CNFs electrocatalysts showed tolerable electrocatlaytic activity for water splitting. The catalytic efficiency of the PtNi/CNFs was ascribed to the synergism which arises due to homogeneous dispersion of Pt and Ni nanoparticles on carbon nanofibres, elevated conductivity of the nanofibres and three-dimensional interlinked frameworks of the carbon nanofiber mats. Consequently, carbon nanofibres have surfaced as significant electrocatlytic material for water splitting.

Owing to the extraordinary electrochemical properties of carbon nanomaterials, they have become a commonly employed as sensors. Mohapatra and his research group have designed carbon nano-onions (CNOs) as electrochemical sensor for detection of glucose [151]. The carbon onions were obtained by combustion of natural oils and a hybrid nanocomposite bearing CNOs, Nafion and glucose oxidase enzyme was fabricated to be used as sensor. The enzymatic CNO composite displayed good sensitivity of 26.5 μA mM^{-1} cm^{-2} with 0.21 mM detection limit. Non-enzymatic CNO nanocomposite bearing Pt nanoparticles were also fabricated which showed superior sensitivity of about 21.6 μA mM^{-1} cm^{-2} and a lower detection limit of 0.09 mM than enzymatic CNOs for detection of glucose. The good sensitivities of CNO based nanocomposites were attributed to due to the high surface area, excellent conductivity and catalytic behaviour of CNOs. Overall, the research findings highlighted that CNOs serve as facile, economical and effective carbon materials for development of glucose detecting sensors.

The immobilization of enzymes on carbon materials has resulted in a new class of biocatalysts which is used for various purposes such as biosensing [152], removal of pollutants [153], dye degradation [154], synthesis of organic compounds [155], *etc*. In a study carried out by Wang and co-workers a lipase-immobilized carbon nanotubes were reported as biocatalyst for production of biodiesel [156]. The multiwalled carbon nanotubes bearing *Candida antarctica* lipase B were designed by pickering emulsification method. The prepared biocatalyst was then used to manufacture biodiesel from *J. curcas* L. seed oil. The esterification and trans-esterification of *J. curcas* L. seed oil with methanol in presence of biocatalyst resulted in 95.2% yield of biodiesel in 11 hours and 6 seconds at low temperature of about 35 °C. Thus, the liapse immobilized MWCNTs showed excellent biocatalytic activity. Moreover, the study revealed that the use of MWCNTs provided stability to the enzyme as well as rendered the catalyst recyclable up to 10 runs without significant loss in catalytic performance. Hence, it can be concluded that use of carbon nanostructures in biocatalytic systems can add features such as enlarged surface area, thermo-chemical stability, conductivity and mechanical properties which are beneficial for catalytic materials.

Over the past few decades, the carbocatalysts have been extensively used for chemical transformations. Recently, Dam *et al.* have employed graphite oxide as a heterogeneous catalyst for synthesis of pyrimidines [157]. The pyrimidine derivatives were afforded from one pot reaction between 1 mmol each of substituted benzaldehyde/heteroaldehyde, dimedone and 2-amino-benzothiazole or 2-amino-imidazole under solvent-less condition in presence of graphite oxide (20 mg) at 60 °C. It was also revealed that the catalyst generated good yields in short time and was reusable for 9 cycles. The use of graphite oxide carbocatalyst offered several advantages such as efficient reaction, easy recovery, high recyclability of catalyst, facile work-up and inexpensive synthetic protocol with avoidance of toxic solvents and metal based catalyst.

Considering the above examples it is clear that carbon materials exhibit wide range of catalytic applications. The broad scope of carbocatalysts cover diverse spheres including photocatalysis, electrocatalysis, biocatalysis and heterogeneous chemical catalysis owing to their merits such as interesting surface features, porosity, thermal conductivity, stability and electrical properties. Moreover, the sustainability factor due to their renewability also renders carbon materials as suitable catalytic materials.

CONCLUDING REMARKS

Carbon and its allotropes are incredible materials that have been an integral part of chemical and physical sciences for ages. The extraordinary surface properties, conductivity, chemical inertness and renewability of carbon materials have turned them into sustainable catalytic systems. Ever since the first use of activated carbon as a catalyst for oxidation reaction, there has been a continuous development in the field of carbocatalysis. The diversely synthesized carbon materials with exorbitantly useful properties exhibit broad applications as photocatalysts for the removal of pollutants in the environment, sensors, heterogeneous catalysts for chemical syntheses, biocatalysts, and electrocatalysts in solar cells, fuel cells, water splitting, redox processes, *etc*. Consequently, the properties, synthetic methods and the scope of carbocatalysts have been highlighted in this chapter. The carbon polymorphs due to their unique surface characteristics and porosity have received immense attention as catalyst or catalyst support. The presence of large surface area and pores on carbonaceous materials facilitate the binding of substrates. Additionally, by suitable functionalization, doping or thermal treatment, the surface features and porosity of carbon facets can further be tuned to introduce beneficial defects which serve as active sites for reactant molecules. The emergence of nanodimensional carbon materials has also revolutionized the field of carbocatalysis because a reduction in size enhances the surface area which in turn improves their catalytic activity. Furthermore, the exceptional thermal conductivities and electrical properties of active carbon, graphite, fullerene, glassy carbon, carbon black, carbon nanotubes, carbon nanofibres, nanodiaomonds, carbon nano-onions, graphene and its derivatives have contributed enormously to their catalytic applications. The thermal conductivity and electrical property of a carbon material depend on the coordination and arrangement of C atoms as well as defects in the structural framework. By controlling the degree of carbonization or structural defects one can achieve desired properties in a carbocatalyst.

To realize the desirable characteristics of a carbon material, one needs a proper synthetic strategy. Diverse synthetic approaches for the synthesis of carbocatalysts have been described in this chapter. Active carbons are mainly procured by physical or chemical activation of raw materials like wood, coal, biomass, *etc*. The activation process governs the surface features and pores of active carbon. Graphite is a natural allotrope of carbon but synthetically it is fabricated by thermal treatment of coke or coal tar and hydrothermal treatment of polymers or metal carbides. The thermal treatment controls the graphitization process. Nowadays, graphite encapsulated nanoparticles are also prepared by pyrolysis, template synthesis, chemical vapor deposition, arc discharge, laser ablation technique, *etc*. Fullerenes are macro-caged structures that are prepared by arc discharge, laser ablation and combustion techniques. Further, functionalization of fullerenes can also be carried out by sol-gel, solvo-thermal, chemical reduction, self-assembly, and ultrsonication techniques, to develop composites with catalytic properties. Glassy carbon electrocatalysts are fabricated by pyrolysis of polymers.

Besides, there are several thermal methods such as furnace black method, lampblack method, *etc*. for the preparation of carbon black particles of nanosize. Plasma synthesis has also been employed for the synthesis of glassy carbon composites. Chemical vapor deposition (CVD), electrospinning and template synthesis are commonly used strategies for the synthesis of carbon nanofibre based catalysts. Carbon nanotubes with relevant features can be synthesized *via* CVD, arc discharge, laser ablation, ultrasonication, electrolysis, *etc*. Detonation, high pressure-high temperature, laser irradiation and CVD are typical methods for the synthesis of nanodiamonds with exceptional surface, mechanical, electrical and thermal properties. Several bottom-up strategies including pyrolysis, arc discharge, CVD, laser ablation and thermal treatment of nanodiamonds are available for designing catalytic carbon nano-onions.

Considering the ever growing interest in carbocatalysis and the emergence of numerous fabrication strategies, it is evident that the field of carbocatalysts will continue to bloom in the coming days. However, there are certain concerns that need to be settled. The industrial scale syntheses of carbon based catalysts and their application need to be more cost-effective and feasible. The incorporation of desired properties using more benign methods also needs to be looked into. Further, identification of specific active sites on the carbon surface and detailed study of structural intricacies of carbon materials, thermal and electrical properties that essentially drive the catalytic behaviour of carbon catalysts also need to be dealt with at atomic as well as molecular levels. Overcoming these challenges will surely help in the expansion of the application spectrum and scalability of carbocatalysts.

LIST OF ABBREVIATIONS

AC	Active carbon
bcc	Body centered cubic
BET	Brunauer-Emmett-Teller
CVD	Chemical vapour deposition
CCVD	Catalytically-promoted chemical vapour deposition

CNF	Carbon nanofibres
CNO	Carbon nano-onions
CNT	Carbon nanotube
DAFCs	Direct alcohol fuel cells
DC	Direct current
DMF	Dimethyl formamide
DWCNTS	Double-walled carbon nanotubes
fcc	Face centered cubic
FTS	Fischer-Tropsch synthesis
MWCNT	Multi-walled carbon nanotube
N-CNOs	Nitrogen doped carbon nano-onions
NCNTs	Nitrogen doped carbon nanotubes
ODH	Oxidative dehydrogenation
ORR	Oxygen reduction reaction
PAN	Polyacrylonitrile
RBF	Round bottom flask
RDX	Research Department Explosive
RPM	Rotations per minute
SWCNT	Single-walled carbon nanotubes
TNT- 2	4,6-Trinitrotoluene

CONSENT FOR PUBLICATION

Not applicable.

CONFLICT OF INTEREST

The author declares no conflict of interest, financial or otherwise.

ACKNOWLEDGEMENTS

Declared none.

REFERENCES

[1] Anastas, P.T.; Warner, J.C. *Green Chemistry: Theory and Practice*; Oxford University Press: New York, **1998**, p. 30.

[2] Cole-Hamilton, D.J. Homogeneous catalysis--new approaches to catalyst separation, recovery, and recycling. *Science,* **2003**, *299*(5613), 1702-1706. [http://dx.doi.org/10.1126/science.1081881] [PMID: 12637737]

[3] Sheldon, R.A.; Downing, R.S. Heterogeneous catalytic transformations for environmentally friendly

production. *Appl. Catal. A Gen.*, **1999**, *189*(2), 163-183.
[http://dx.doi.org/10.1016/S0926-860X(99)00274-4]

[4] Stenhouse, J. Charcoal: On the deodorizing and disinfecting properties of charcoal, with the description of a charcoal respirator for purifying the air by filtration. *Chem. Gaz,* **1854**.

[5] Rideal, E.K.; Wright, W.M. CCCCXXX.—Low temperature oxidation at charcoal surfaces. Part III. The behaviour of blood charcoal and the influence of temperature on the reaction rate. *J. Chem. Soc.*, **1926**, *129*(0), 3182-3190.
[http://dx.doi.org/10.1039/JR9262903182]

[6] Kutzelnigg, A. Zur Kenntnis der sauerstoff-übertragenden Wirkung von Stoffen mit großer spezifischer Oberfläche. *Ber. Dtsch. Chem. Ges. B,* **1930**, *63*(7), 1753-1758.
[http://dx.doi.org/10.1002/cber.19300630721]

[7] Kolthoff, I.M.; Tomsicek, W.J. The oxidation potential of the system potassium ferrocyanide–potassium ferricyanide at various ionic strengths. *J. Phys. Chem.*, **1935**, *39*(7), 945-954.
[http://dx.doi.org/10.1021/j150367a004]

[8] Alkhazov, T.G.; Lisovskii, A.E.; Gulakhmedova, T.K. Oxidative dehydrogenation of ethylbenzene over a charcoal catalyst. *React. Kinet. Catal. Lett.*, **1979**, *12*(2), 189-193.
[http://dx.doi.org/10.1007/BF02071909]

[9] Lücking, F.; Köser, H.; Jank, M.; Ritter, A. Iron powder, graphite and activated carbon as catalysts for the oxidation of 4-chlorophenol with hydrogen peroxide in aqueous solution. *Water Res.*, **1998**, *32*(9), 2607-2614.
[http://dx.doi.org/10.1016/S0043-1354(98)00016-5]

[10] Fortuny, A.; Font, J.; Fabregat, A. Wet air oxidation of phenol using active carbon as catalyst. *Appl. Catal. B,* **1998**, *19*(3-4), 165-173.
[http://dx.doi.org/10.1016/S0926-3373(98)00072-1]

[11] Saha, J.; Kumar, A.; Pm, A.; Jakhad, V. Oxidised charcoal: an efficient support for NiFe layered double hydroxide to improve electrochemical oxygen evolution. *Chem. Commun. (Camb.),* **2020**, *56*(62), 8770-8773.
[http://dx.doi.org/10.1039/D0CC02880K] [PMID: 32618289]

[12] Xu, J.; Li, R.; Qian, X.; Ba, J.; Wu, Q.; Luo, W.; Meng, D. Nanoarray-structured nitrogen-doped graphite foil as the support of NiFe layered double hydroxides for enhancing oxygen evolution reaction. *J. Power Sources,* **2020**, *469*, 228419.
[http://dx.doi.org/10.1016/j.jpowsour.2020.228419]

[13] Celorrio, V.; Montes de Oca, M.G.; Plana, D.; Moliner, R.; Fermín, D.J.; Lázaro, M.J. Electrochemical performance of Pd and Au–Pd core–shell nanoparticles on surface tailored carbon black as catalyst support. *Int. J. Hydrogen Energy,* **2012**, *37*(8), 7152-7160.
[http://dx.doi.org/10.1016/j.ijhydene.2011.12.014]

[14] La-Torre-Riveros, L.; Guzman-Blas, R.; Méndez-Torres, A.E.; Prelas, M.; Tryk, D.A.; Cabrera, C.R. Diamond nanoparticles as a support for Pt and PtRu catalysts for direct methanol fuel cells. *ACS Appl. Mater. Interfaces,* **2012**, *4*(2), 1134-1147.
[http://dx.doi.org/10.1021/am2018628] [PMID: 22270177]

[15] Zhang, L.H.; Shi, Y.; Wang, Y.; Shiju, N.R. Nanocarbon catalysts: recent understanding regarding the active sites. *Adv. Sci. (Weinh.),* **2020**, *7*(5), 1902126.
[http://dx.doi.org/10.1002/advs.201902126] [PMID: 32154069]

[16] Kroto, H.W.; Heath, J.R.; O'Brien, S.C.; Curl, R.F.; Smalley, R.E. C60: Buckminsterfullerene. *Nature,* **1985**, *318*(6042), 162-163.
[http://dx.doi.org/10.1038/318162a0]

[17] María Girón, R.; Marco-Martínez, J.; Bellani, S.; Insuasty, A.;. Comas Rojas, H.; Tullii, G.; Antognazza, M.R.; Filippone, S.; Martín, N. Synthesis of modified fullerenes for oxygen reduction

reactions. *J. Mater. Chem. A Mater. Energy Sustain.*, **2016**, *4*(37), 14284-14290. [http://dx.doi.org/10.1039/C6TA06573B]

[18] Fukuzumi, S.; Imahori, H.; Okamoto, K.; Yamada, H.; Fujitsuka, M.; Ito, O.; Guldi, D.M. Uphill photooxidation of NADH analogues by hexyl viologen catalyzed by zinc porphyrin linked fullerenes. *J. Phys. Chem. A,* **2002**, *106*(10), 1903-1908. [http://dx.doi.org/10.1021/jp011613g]

[19] Rogozea, E.A.; Meghea, A.; Olteanu, N.L.; Bors, A.; Mihaly, M. Fullerene-modified silica materials designed for highly efficient dyes photodegradation. *Mater. Lett.*, **2015**, *151*, 119-121. [http://dx.doi.org/10.1016/j.matlet.2015.03.069]

[20] Ghosh, A.; Bhaduri, K.; Shah, S.; Auroux, A.; Pandey, J.K.; Chowdhury, B. Dehydration of isopropanol to propylene over fullerene[C_{60}] containing niobium phosphate catalyst: Study on catalyst recyclability. *Molecular Catalysis,* **2019**, *475*, 110470. [http://dx.doi.org/10.1016/j.mcat.2019.110470]

[21] Varadwaj, P.R.; Varadwaj, A.; Marques, H.M. $_{70}$ Fullerene cage as a novel catalyst for efficient proton transfer reactions between small molecules: A theoretical study. *Sci. Rep.*, **2019**, *9*(1), 10650. [http://dx.doi.org/10.1038/s41598-019-46725-4] [PMID: 31337790]

[22] Wang, S.; Dai, T.; Lu, Y.; Chen, Q.; Feng, L.; Sui, Z. Fullerene-bearing porous polymer *via* ball-milling approach and its palladium composite for catalytic deallylation. *Microporous Mesoporous Mater.*, **2020**, *302*, 110187. [http://dx.doi.org/10.1016/j.micromeso.2020.110187]

[23] Iijima, S. Synthesis of Carbon Nanotubes. *Nature,* **1991**, *354*, 56-58. [http://dx.doi.org/10.1038/354056a0]

[24] Planeix, J.M.; Coustel, N.; Coq, B.; Brotons, V.; Kumbhar, P.S.; Dutartre, R.; Geneste, P.; Bernier, P.; Ajayan, P.M. Application of carbon nanotubes as supports in heterogeneous catalysis. *J. Am. Chem. Soc.*, **1994**, *116*(17), 7935-7936. [http://dx.doi.org/10.1021/ja00096a076]

[25] Yan, Y.; Miao, J.; Yang, Z.; Xiao, F.X.; Yang, H.B.; Liu, B.; Yang, Y. Carbon nanotube catalysts: recent advances in synthesis, characterization and applications. *Chem. Soc. Rev.*, **2015**, *44*(10), 3295-3346. [http://dx.doi.org/10.1039/C4CS00492B] [PMID: 25855947]

[26] Abdolmohammadi, S.; Mirza, B.; Vessally, E. Immobilized TiO_2 nanoparticles on carbon nanotubes: an efficient heterogeneous catalyst for the synthesis of chromeno[*b*]pyridine derivatives under ultrasonic irradiation. *RSC Advances,* **2019**, *9*(71), 41868-41876. [http://dx.doi.org/10.1039/C9RA09031B] [PMID: 35557875]

[27] Gopannagari, M.; Kumar, D.P.; Park, H.; Kim, E.H.; Bhavani, P.; Reddy, D.A.; Kim, T.K. Influence of surface-functionalized multi-walled carbon nanotubes on CdS nanohybrids for effective photocatalytic hydrogen production. *Appl. Catal. B,* **2018**, *236*, 294-303. [http://dx.doi.org/10.1016/j.apcatb.2018.05.009]

[28] Xu, Y.; Deng, P.; Chen, G.; Chen, J.; Yan, Y.; Qi, K.; Liu, H.; Xia, B.Y. 2D Nitrogen-doped carbon nanotubes/graphene hybrid as bifunctional oxygen electrocatalyst for long-life rechargeable Zn-air batteries. *Adv. Funct. Mater.*, **2020**, *30*(6), 1906081. [http://dx.doi.org/10.1002/adfm.201906081]

[29] Brodie, B.C. XIII. On the atomic weight of graphite. *Philos. Trans. R. Soc. Lond.,* **1859**, *149*, 249-259. [http://dx.doi.org/10.1098/rstl.1859.0013]

[30] Novoselov, K.S.; Geim, A.K.; Morozov, S.V.; Jiang, D.; Zhang, Y.; Dubonos, S.V.; Grigorieva, I.V.; Firsov, A.A. Electric field effect in atomically thin carbon films. *Science,* **2004**, *306*(5696), 666-669. [http://dx.doi.org/10.1126/science.1102896] [PMID: 15499015]

[31] Zhu, Y.; Murali, S.; Cai, W.; Li, X.; Suk, J.W.; Potts, J.R.; Ruoff, R.S. Graphene and graphene oxide:

synthesis, properties, and applications. *Adv. Mater.,* **2010**, *22*(35), 3906-3924.
[http://dx.doi.org/10.1002/adma.201001068] [PMID: 20706983]

[32] Castro Neto, A.H.; Guinea, F.; Peres, N.M.R.; Novoselov, K.S.; Geim, A.K. The electronic properties of graphene. *Rev. Mod. Phys.,* **2009**, *81*(1), 109-162.
[http://dx.doi.org/10.1103/RevModPhys.81.109]

[33] Akturk, A.; Goldsman, N. Electron transport and full-band electron-phonon interactions in graphene. *J. Appl. Phys.,* **2008**, *103*(5), 053702.
[http://dx.doi.org/10.1063/1.2890147]

[34] Berry, V. Impermeability of graphene and its applications. *Carbon,* **2013**, *62*, 1-10.
[http://dx.doi.org/10.1016/j.carbon.2013.05.052]

[35] Vekeman, J.; Sánchez-Marín, J.; Sánchez de Merás, A.; Garcia Cuesta, I.; Faginas-Lago, N. Flexibility in the graphene sheet: The influence on gas adsorption from molecular dynamics studies. *J. Phys. Chem. C,* **2019**, *123*(46), 28035-28047.
[http://dx.doi.org/10.1021/acs.jpcc.9b07503]

[36] Papageorgiou, D.G.; Kinloch, I.A.; Young, R.J. Mechanical properties of graphene and graphene-based nanocomposites. *Prog. Mater. Sci.,* **2017**, *90*, 75-127.
[http://dx.doi.org/10.1016/j.pmatsci.2017.07.004]

[37] Fugallo, G.; Cepellotti, A.; Paulatto, L.; Lazzeri, M.; Marzari, N.; Mauri, F. Thermal conductivity of graphene and graphite: collective excitations and mean free paths. *Nano Lett.,* **2014**, *14*(11), 6109-6114.
[http://dx.doi.org/10.1021/nl502059f] [PMID: 25343716]

[38] Murata, H.; Nakajima, Y.; Saitoh, N.; Yoshizawa, N.; Suemasu, T.; Toko, K. oko, K. High-electrica--conductivity multilayer graphene formed by layer exchange with controlled thickness and interlayer. *Sci. Rep.,* **2019**, *9*(1), 4068.
[http://dx.doi.org/10.1038/s41598-019-40547-0] [PMID: 30858422]

[39] Lu, Y.H.; Zhou, M.; Zhang, C.; Feng, Y.P. Metal-embedded graphene: a possible catalyst with high activity. *J. Phys. Chem. C,* **2009**, *113*(47), 20156-20160.
[http://dx.doi.org/10.1021/jp908829m]

[40] Imran Jafri, R.; Rajalakshmi, N.; Ramaprabhu, S. Nitrogen doped graphene nanoplatelets as catalyst support for oxygen reduction reaction in proton exchange membrane fuel cell. *J. Mater. Chem.,* **2010**, *20*(34), 7114-7117.
[http://dx.doi.org/10.1039/c0jm00467g]

[41] Dreyer, D.R.; Jia, H.P.; Bielawski, C.W. Graphene oxide: a convenient carbocatalyst for facilitating oxidation and hydration reactions. *Angew. Chem. Int. Ed.,* **2010**, *49*(38), 6686.
[http://dx.doi.org/10.1002/anie.201003238] [PMID: 20602388]

[42] Li, Y.; Gao, W.; Ci, L.; Wang, C.; Ajayan, P.M. Catalytic performance of Pt nanoparticles on reduced graphene oxide for methanol electro-oxidation. *Carbon,* **2010**, *48*(4), 1124-1130.
[http://dx.doi.org/10.1016/j.carbon.2009.11.034]

[43] Gao, Y.; Ma, D.; Wang, C.; Guan, J.; Bao, X. Reduced graphene oxide as a catalyst for hydrogenation of nitrobenzene at room temperature. *Chem. Commun. (Camb.),* **2011**, *47*(8), 2432-2434.
[http://dx.doi.org/10.1039/C0CC04420B] [PMID: 21170437]

[44] Serp, P.; Machado, B. Carbon (Nano)materials for Catalysis. In: *Nanostructured Carbon Materials for Catalysis*; Royal Society of Chemistry: London, **2015**; pp. 1-45.

[45] Qiu, Y.; Ali, S.; Lan, G.; Tong, H.; Fan, J.; Liu, H.; Li, B.; Han, W.; Tang, H.; Liu, H.; Li, Y. Defect-rich activated carbons as active and stable metal-free catalyst for acetylene hydrochlorination. *Carbon,* **2019**, *146*, 406-412.
[http://dx.doi.org/10.1016/j.carbon.2019.01.102]

[46] Chen, C.; Yan, D.; Wang, Y.; Zhou, Y.; Zou, Y.; Li, Y.; Wang, S. B-N Pairs enriched defective carbon

nanosheets for ammonia synthesis with high efficiency. *Small,* **2019**, *15*(7), 1805029. [http://dx.doi.org/10.1002/smll.201805029]

[47] Vidic, R.D.; Tessmer, C.H.; Uranowski, L.J. Impact of surface properties of activated carbons on oxidative coupling of phenolic compounds. *Carbon,* **1997**, *35*(9), 1349-1359. [http://dx.doi.org/10.1016/S0008-6223(97)00071-7]

[48] Speranza, G. The role of functionalization in the applications of carbon materials: an overview. *C- J Carbon Res,* **2019**, *5*(4), 84. [http://dx.doi.org/10.3390/c5040084]

[49] Arrigo, R.; Hävecker, M.; Wrabetz, S.; Blume, R.; Lerch, M.; McGregor, J.; Parrott, E.P.J.; Zeitler, J.A.; Gladden, L.F.; Knop-Gericke, A.; Schlögl, R.; Su, D.S. Tuning the acid/base properties of nanocarbons by functionalization *via* amination. *J. Am. Chem. Soc.,* **2010**, *132*(28), 9616-9630. [http://dx.doi.org/10.1021/ja910169v] [PMID: 20583786]

[50] Chen, J.; Chen, Q.; Ma, Q. Influence of surface functionalization *via* chemical oxidation on the properties of carbon nanotubes. *J. Colloid Interface Sci.,* **2012**, *370*(1), 32-38. [http://dx.doi.org/10.1016/j.jcis.2011.12.073] [PMID: 22280791]

[51] Boehm, H.P. Surface oxides on carbon and their analysis: a critical assessment. *Carbon,* **2002**, *40*(2), 145-149. [http://dx.doi.org/10.1016/S0008-6223(01)00165-8]

[52] Ros, T.G.; van Dillen, A.J.; Geus, J.W.; Koningsberger, D.C. Surface oxidation of carbon nanofibres. *Chemistry,* **2002**, *8*(5), 1151-1162. [http://dx.doi.org/10.1002/1521-3765(20020301)8:5<1151::AID-CHEM1151>3.0.CO;2-#] [PMID: 11891903]

[53] Moreno-Castilla, C.; Ferro-Garcia, M.A.; Joly, J.P.; Bautista-Toledo, I.; Carrasco-Marin, F.; Rivera-Utrilla, J. Ferro-Garcia, M.A. Joly, J.P. Bautista-Toledo, I. Carrasco- Marin, F. Rivera-Utrilla, J. Activated carbon surface modifications by nitric acid, hydrogen peroxide, and ammonium peroxydisulfate treatments. *Langmuir,* **1995**, *11*(11), 4386-4392. [http://dx.doi.org/10.1021/la00011a035]

[54] Boehm, H.P. Some aspects of the surface chemistry of carbon blacks and other carbons. *Carbon,* **1994**, *32*(5), 759-769. [http://dx.doi.org/10.1016/0008-6223(94)90031-0]

[55] Serp, P.; Corrias, M.; Kalck, P. Carbon nanotubes and nanofibers in catalysis. *Appl. Catal. A Gen.,* **2003**, *253*(2), 337-358. [http://dx.doi.org/10.1016/S0926-860X(03)00549-0]

[56] Rodríguez-reinoso, F. The role of carbon materials in heterogeneous catalysis. *Carbon,* **1998**, *36*(3), 159-175. [http://dx.doi.org/10.1016/S0008-6223(97)00173-5]

[57] Ayyaru, S.; Ahn, Y.H. Application of sulfonic acid group functionalized graphene oxide to improve hydrophilicity, permeability, and antifouling of PVDF nanocomposite ultrafiltration membranes. *J. Membr. Sci.,* **2017**, *525*, 210-219. [http://dx.doi.org/10.1016/j.memsci.2016.10.048]

[58] Sun, N.; Sun, C.; Liu, H.; Liu, J.; Stevens, L.; Drage, T.; Snape, C.E.; Li, K.; Wei, W.; Sun, Y. Synthesis, characterization and evaluation of activated spherical carbon materials for CO_2 capture. *Fuel,* **2013**, *113*, 854-862. [http://dx.doi.org/10.1016/j.fuel.2013.03.047]

[59] Pan, T.; Liu, H.; Ren, G.; Li, Y.; Lu, X.; Zhu, Y. Metal-free porous nitrogen-doped carbon nanotubes for enhanced oxygen reduction and evolution reactions. *Sci. Bull. (Beijing),* **2016**, *61*(11), 889-896. [http://dx.doi.org/10.1007/s11434-016-1073-3]

[60] Li, O.L.; Ishizaki, T. Development, challenges, and prospects of carbon-based electrode for lithium-air

batteries. In: *Emerging Materials for Energy Conversion and Storage*; Cheong, K.Y.; Impellizzeri, G.; Fraga, M.A., Eds.; Elsevier: Amsterdam, **2018**; pp. 115-152. [http://dx.doi.org/10.1016/B978-0-12-813794-9.00004-1]

[61] Čejka, J.; Mintova, S. Perspectives of micro/mesoporous composites in catalysis. *Catal. Rev., Sci. Eng.*, **2007**, *49*(4), 457-509. [http://dx.doi.org/10.1080/01614940701583240]

[62] Ferrero, G.A.; Preuss, K.; Fuertes, A.B.; Sevilla, M.; Titirici, M.M. The influence of pore size distribution on the oxygen reduction reaction performance in nitrogen doped carbon microspheres. *J. Mater. Chem. A Mater. Energy Sustain.*, **2016**, *4*(7), 2581-2589. [http://dx.doi.org/10.1039/C5TA10063A]

[63] Vaccaro, S.; Malangone, L. Catalytic combustion for supplying energy for endothermic reaction. *J. Adv. Chem. Eng.*, **2014**, *4*, 1000107.

[64] Koh, Y.K. Thermal conductivity and phonon transport. In: *Encyclopedia of Nanotechnology*; Bhushan, B., Ed.; Springer: Berlin, **2012**.

[65] Liu, Y.; Hu, C.; Huang, J.; Sumpter, B.G.; Qiao, R. Tuning interfacial thermal conductance of graphene embedded in soft materials by vacancy defects. *J. Chem. Phys.*, **2015**, *142*(24), 244703. [http://dx.doi.org/10.1063/1.4922775] [PMID: 26133445]

[66] Hung, N.T.; Nugraha, A.R.T.; Saito, R. N.; R. T. Nugraha, A.; Saito, R. Thermoelectric properties of carbon nanotubes. *Energies,* **2019**, *12*(23), 4561. [http://dx.doi.org/10.3390/en12234561]

[67] Asalieva, E.; Sineva, L.; Sinichkina, S.; Solomonik, I.; Gryaznov, K.; Pushina, E.; Kulchakovskaya, E.; Gorshkov, A.; Kulnitskiy, B.; Ovsyannikov, D.; Zholudev, S.; Mordkovich, V. Exfoliated graphite as a heat-conductive frame for a new pelletized Fischer–Tropsch synthesis catalyst. *Appl. Catal. A Gen.*, **2020**, *601*, 117639. [http://dx.doi.org/10.1016/j.apcata.2020.117639]

[68] Bogeat, A.B. Understanding and tuning the electrical conductivity of activated carbon: A state-of-the-art review. *Crit. Rev. Solid State,* **2019**, *46*(1), 1-37. [http://dx.doi.org/10.1080/10408436.2019.1671800]

[69] Nasir, S.; Hussein, M.; Zainal, Z.; Yusof, N. Carbon-based nanomaterials/allotropes: a glimpse of their synthesis, properties and some applications. *Materials (Basel),* **2018**, *11*(2), 295. [http://dx.doi.org/10.3390/ma11020295] [PMID: 29438327]

[70] Hausmann, B.J.M.; Bulu, I.; Venkataraman, V.; Deotare, P.; Lončar, M. Diamond nonlinear photonics. *Nat. Photonics,* **2014**, *8*(5), 369-374. [http://dx.doi.org/10.1038/nphoton.2014.72]

[71] García, N.; Esquinazi, P.; Barzola-Quiquia, J.; Dusari, S. Evidence for semiconducting behavior with a narrow band gap of Bernal graphite. *New J. Phys.*, **2012**, *14*(5), 053015. [http://dx.doi.org/10.1088/1367-2630/14/5/053015]

[72] Maruszewski, B.T. Fullerenes: Thermomechanics, Doping, Electrical Conductivity. In: *Encyclopedia of Thermal Stresses*; Hetnarski, R.B., Ed.; Springer: Dordrecht, **2014**. [http://dx.doi.org/10.1007/978-94-007-2739-7_1030]

[73] Eatemadi, A.; Daraee, H.; Karimkhanloo, H.; Kouhi, M.; Zarghami, N.; Akbarzadeh, A.; Abasi, M.; Hanifehpour, Y.; Joo, S.W. Carbon nanotubes: properties, synthesis, purification, and medical applications. *Nanoscale Res. Lett.*, **2014**, *9*(1), 393. [http://dx.doi.org/10.1186/1556-276X-9-393] [PMID: 25170330]

[74] Gupta, N.; Gupta, S.M.; Sharma, S.K. Carbon nanotubes: synthesis, properties and engineering applications. *Carbon Letters*, **2019**, *29*(5), 419-447. [http://dx.doi.org/10.1007/s42823-019-00068-2]

[75] Bandaru, P.R. Electrical properties and applications of carbon nanotube structures. *J. Nanosci.*

Nanotechnol., **2007**, *7*(4), 1239-1267.
[http://dx.doi.org/10.1166/jnn.2007.307] [PMID: 17450889]

[76] Sreeprasad, T.S.; Berry, V. How do the electrical properties of graphene change with its functionalization? *Small,* **2013**, *9*(3), 341-350.
[http://dx.doi.org/10.1002/smll.201202196] [PMID: 23169614]

[77] Borchardt, L.; Zhu, Q.L.; Casco, M.E.; Berger, R.; Zhuang, X.; Kaskel, S.; Feng, X.; Xu, Q. Toward a molecular design of porous carbon materials. *Mater. Today,* **2017**, *20*(10), 592-610.
[http://dx.doi.org/10.1016/j.mattod.2017.06.002]

[78] Suslova, E.V.; Arkhipova, E.A.; Kalashnik, A.V.; Ivanov, A.S.; Savilov, S.V.; Xia, H.; Lunin, V.V. Effect of the functionalization of nitrogen-doped carbon nanotubes on electrical conductivity. *Russ. J. Phys. Chem. A. Focus Chem.*, **2019**, *93*(10), 1952-1956.
[http://dx.doi.org/10.1134/S0036024419100303]

[79] Jiang, Y.; Cheng, G.; Li, Y.; He, Z.; Zhu, J.; Meng, W.; Zhou, H.; Dai, L.; Wang, L. Superior electrocatalytic performance of porous, graphitic, and oxygen-functionalized carbon nanofiber as bifunctional electrode for vanadium redox flow battery. *Appl. Surf. Sci.*, **2020**, *525*, 146453.
[http://dx.doi.org/10.1016/j.apsusc.2020.146453]

[80] Han, F.D.; Yao, B.; Bai, Y.J. Preparation of carbon nano-onions and their application as anode materials for rechargeable lithium-ion batteries. *J. Phys. Chem. C,* **2011**, *115*(18), 8923-8927.
[http://dx.doi.org/10.1021/jp2007599]

[81] Li, X.; Cai, W.; An, J.; Kim, S.; Nah, J.; Yang, D.; Piner, R.; Velamakanni, A.; Jung, I.; Tutuc, E.; Banerjee, S.K.; Colombo, L.; Ruoff, R.S. Large-area synthesis of high-quality and uniform graphene films on copper foils. *Science,* **2009**, *324*(5932), 1312-1314.
[http://dx.doi.org/10.1126/science.1171245] [PMID: 19423775]

[82] Sawant, S.Y.; Somani, R.S.; Newalkar, B.L.; Choudary, N.V.; Bajaj, H.C. Synthesis of submicron size hollow carbon spheres by a chemical reduction — solvothermal method using carbon tetrachloride as carbon source. *Mater. Lett.*, **2009**, *63*(27), 2339-2342.
[http://dx.doi.org/10.1016/j.matlet.2009.07.066]

[83] Wang, Y.; Wang, Y.; Fei, H.; Cai, X. The effect of heat treatment on the electrical conductivity of highly conducting graphene films. *Carbon,* **2013**, *51*, 436.
[http://dx.doi.org/10.1016/j.carbon.2012.08.034]

[84] Mattia, D.; Rossi, M.P.; Kim, B.M.; Korneva, G.; Bau, H.H.; Gogotsi, Y. Effect of graphitization on the wettability and electrical conductivity of CVD-carbon nanotubes and films. *J. Phys. Chem. B,* **2006**, *110*(20), 9850-9855.
[http://dx.doi.org/10.1021/jp061138s] [PMID: 16706438]

[85] Špitalský, Z.; Krontiras, C.A.; Georga, S.N.; Galiotis, C. Effect of oxidation treatment of multiwalled carbon nanotubes on the mechanical and electrical properties of their epoxy composites. *Compos., Part A Appl. Sci. Manuf.*, **2009**, *40*(6-7), 778-783.
[http://dx.doi.org/10.1016/j.compositesa.2009.03.008]

[86] Iwanow, M.; Gärtner, T.; Sieber, V.; König, B. Activated carbon as catalyst support: precursors, preparation, modification and characterization. *Beilstein J. Org. Chem.*, **2020**, *16*, 1188-1202.
[http://dx.doi.org/10.3762/bjoc.16.104] [PMID: 32550932]

[87] Behera, S.K.; Chakraborty, S.; Meikap, B.C. Demineralization mechanism and influence of parameters on high ash Indian coal by chemical leaching of acid and alkali solution. *Int. J. Coal Sci. Technol.*, **2018**, *5*(2), 142-155.
[http://dx.doi.org/10.1007/s40789-018-0208-3]

[88] Ahmad, F.; Daud, W.M.A.W.; Ahmad, M.A.; Radzi, R. Cocoa (*Theobroma cacao*) shell-based activated carbon by CO_2 activation in removing of Cationic dye from aqueous solution: Kinetics and equilibrium studies. *Chem. Eng. Res. Des.*, **2012**, *90*(10), 1480-1490.
[http://dx.doi.org/10.1016/j.cherd.2012.01.017]

[89] Xie, T.; Liao, T.; Liu, C.; Xu, L.; Yang, J.; Zhu, Q.; Wang, J.; Zhang, X. Synthesis of multifunctional photocatalyst vanadium oxide/activated carbon *via in situ* utilization of stone coal ore. *Ceram. Int.*, **2019**, *45*(4), 4934-4944.
[http://dx.doi.org/10.1016/j.ceramint.2018.11.193]

[90] Shen, Y.; Fu, Y. KOH-activated rice husk char *via* CO_2 pyrolysis for phenol adsorption. *Mater. Today Energy,* **2018**, *9*, 397-405.
[http://dx.doi.org/10.1016/j.mtener.2018.07.005]

[91] Zhou, J.; Luo, A.; Zhao, Y. Preparation and characterisation of activated carbon from waste tea by physical activation using steam. *J. Air Waste Manag. Assoc.*, **2018**, *68*(12), 1269-1277.
[http://dx.doi.org/10.1080/10962247.2018.1460282] [PMID: 29667515]

[92] Hsu, L.Y.; Teng, H. Influence of different chemical reagents on the preparation of activated carbons from bituminous coal. *Fuel Process. Technol.*, **2000**, *64*(1-3), 155-166.
[http://dx.doi.org/10.1016/S0378-3820(00)00071-0]

[93] Ahmadpour, A.; Do, D.D. The preparation of active carbons from coal by chemical and physical activation. *Carbon,* **1996**, *34*(4), 471-479.
[http://dx.doi.org/10.1016/0008-6223(95)00204-9]

[94] Saleem, J.; Shahid, U.B.; Hijab, M.; Mackey, H.; McKay, G. Production and applications of activated carbons as adsorbents from olive stones. *Biomass Convers. Biorefin.*, **2019**, *9*(4), 775-802.
[http://dx.doi.org/10.1007/s13399-019-00473-7]

[95] Maciá-Agulló, J.A.; Moore, B.C.; Cazorla-Amorós, D.; Linares-Solano, A. Activation of coal tar pitch carbon fibres: Physical activation *vs.* chemical activation. *Carbon,* **2004**, *42*(7), 1367-1370.
[http://dx.doi.org/10.1016/j.carbon.2004.01.013]

[96] Chen, R.; Li, L.; Liu, Z.; Lu, M.; Wang, C.; Li, H.; Ma, W.; Wang, S. Preparation and characterization of activated carbons from tobacco stem by chemical activation. *J. Air Waste Manag. Assoc.*, **2017**, *67*(6), 713-724.
[http://dx.doi.org/10.1080/10962247.2017.1280560] [PMID: 28121516]

[97] Li, R.; Shahbazi, A.; Wang, L.; Zhang, B.; Hung, A.M.; Dayton, D.C. Graphite encapsulated molybdenum carbide core/shell nanocomposite for highly selective conversion of guaiacol to phenolic compounds in methanol. *Appl. Catal. A Gen.*, **2016**, *528*, 123-130.
[http://dx.doi.org/10.1016/j.apcata.2016.09.016]

[98] Lan, R.; Su, W.; Li, J. Preparation and catalytic performance of expanded graphite for oxidation of organic pollutant. *Catalysts,* **2019**, *9*(3), 280.
[http://dx.doi.org/10.3390/catal9030280]

[99] Bernat-Quesada, F.; Espinosa, J.C.; Barbera, V.; Álvaro, M.; Galimberti, M.; Navalón, S.; García, H. Catalytic ozonation using edge-hydroxylated graphite-based materials. *ACS Sustain. Chem.& Eng.*, **2019**, *7*(20), 17443-17452.
[http://dx.doi.org/10.1021/acssuschemeng.9b04646]

[100] Xu, Z.; Duong-Viet, C.; Ba, H.; Li, B.; Truong-Huu, T.; Nguyen-Dinh, L.; Pham-Huu, C. Gaseous nitric acid activated graphite felts as hierarchical metal-free catalyst for selective oxidation of H_2S. *Catalysts,* **2018**, *8*(4), 145.
[http://dx.doi.org/10.3390/catal8040145]

[101] Duong-Viet, C.; Liu, Y.; Ba, H.; Truong-Phuoc, L.; Baaziz, W.; Nguyen-Dinh, L.; Nhut, J.M.; Pham-Huu, C. Carbon nanotubes containing oxygenated decorating defects as metal-free catalyst for selective oxidation of H_2S. *Appl. Catal. B,* **2016**, *191*, 29-41.
[http://dx.doi.org/10.1016/j.apcatb.2016.03.018]

[102] Hariharan, Y.; Bharathi, A.; Sundar, C.S.; Sastry, V.S.; Yousuf, M.; Radhakrishnan, T.S.; Rao, G.V.N. GeethaKumary, T.; Subramanian, N.; Sahu, P.Ch.; Raghunathan V.S.; Valsakumar M.C. Synthesis and characterization of fullerenes C_{60} and C_{70} and superconductivity in K and Rb-doped C_{60}. *Curr. Sci.,*

1992, *63*, 25-33.

[103] Itagaki, H.; Fujiwara, Y.; Minowa, Y.; Ikehara, Y.; Kaneko, T.; Okazaki, T.; Iizumi, Y.; Kim, J.; Sakakita, H. Synthesis of endohedral-fullerenes using laser ablation plasma from solid material and vaporized fullerenes. *AIP Adv.*, **2019**, *9*(7), 075324. [http://dx.doi.org/10.1063/1.5100980]

[104] Goel, A.; Hebgen, P.; Vander Sande, J.B.; Howard, J.B. Combustion synthesis of fullerenes and fullerenic nanostructures. *Carbon,* **2002**, *40*(2), 177-182. [http://dx.doi.org/10.1016/S0008-6223(01)00170-1]

[105] Sabounchei, S.J.; Hosseinzadeh, M.; Zarepour-jevinani, M.; Ghanbari, B. Monodentate palladium(0)–[60]fullerene complexes of diphosphine ligands as efficient and sustainable nanocatalysts for the Mizoroki–Heck coupling reaction of aryl chlorides. *New J. Chem.*, **2017**, *41*(18), 9701-9709. [http://dx.doi.org/10.1039/C7NJ00741H]

[106] Yao, S.; Yuan, X.; Jiang, L.; Xiong, T.; Zhang, J. Recent progress on fullerene-based materials: synthesis, properties, modifications, and photocatalytic applications. *Materials (Basel),* **2020**, *13*(13), 2924. [http://dx.doi.org/10.3390/ma13132924] [PMID: 32629789]

[107] Liu, R.; Hou, Y.; Jiang, S.; Nie, B. Ag(I)-hived fullerene microcube as an enhanced catalytic substrate for the reduction of 4-nitrophenol and the photodegradation of Orange G dye. *Langmuir,* **2020**, *36*(19), 5236-5242. [http://dx.doi.org/10.1021/acs.langmuir.0c00580] [PMID: 32306732]

[108] Li, Z.; Han, C. Palladium nanoflowers supported on amino-fullerene as novel catalyst for reduction of 4-nitrophenol. *Chin. Chem. Lett.*, **2020**, *31*(3), 818-820. [http://dx.doi.org/10.1016/j.cclet.2019.06.047]

[109] Sharma, S. Glassy carbon: A promising material for micro- and nanomanufacturing. *Materials (Basel),* **2018**, *11*(10), 1857. [http://dx.doi.org/10.3390/ma11101857] [PMID: 30274225]

[110] Jenkins, G.M.; Kawamura, K. *Polymeric carbons-carbon fibre, glass and char*; Cambridge University Press: Cambridge, UK: New York, **1976**.

[111] Dunn, J.G.; Chamberlain, A.C. The recovery of gold from refractory arsenopyrite concentrates by pyrolysis-oxidation. *Miner. Eng.*, **1997**, *10*(9), 919-928. [http://dx.doi.org/10.1016/S0892-6875(97)00074-5]

[112] Hoffman, W.P.; Vastola, F.J.; Walker, P.L., Jr Pyrolysis of propylene over carbon active sites II. Pyrolysis products. *Carbon,* **1988**, *26*(4), 485-499. [http://dx.doi.org/10.1016/0008-6223(88)90147-9]

[113] Sharma, S.; Rostas, A.M.; Bordonali, L.; MacKinnon, N.; Weber, S.; Korvink, J.G. Micro and nano patternable magnetic carbon. *J. Appl. Phys.*, **2016**, *120*(23), 235107. [http://dx.doi.org/10.1063/1.4972476]

[114] Sharma, S.; Kumar, C.N.S.; Korvink, J.G.; Kübel, C. Evolution of glassy carbon microstructure: *In situ* transmission electron microscopy of the pyrolysis process. *Scientific Reports,* 2018, 8(1).

[115] Sharma, S.; Kamath, R.; Madou, M. Porous glassy carbon formed by rapid pyrolysis of phenol-formaldehyde resins and its performance as electrode material for electrochemical double layer capacitors. *J. Anal. Appl. Pyrolysis,* **2014**, *108*, 12-18. [http://dx.doi.org/10.1016/j.jaap.2014.05.025]

[116] Jurkiewicz, K.; Pawlyta, M.; Zygadło, D.; Chrobak, D.; Duber, S.; Wrzalik, R.; Ratuszna, A.; Burian, A. Evolution of glassy carbon under heat treatment: correlation structure–mechanical properties. *J. Mater. Sci.*, **2018**, *53*(5), 3509-3523. [http://dx.doi.org/10.1007/s10853-017-1753-7]

[117] Thirukumaran, P.; Balasubramanian, R.; Atchudan, R.; Shakila Parveen, A.; Lee, Y.R.; Kim, S.C. Metal-free nitrogen-rich glassy carbon as an electrocatalyst for hydrogen evolution reaction. *Mater. Res. Bull.*, **2020**, *124*, 110734. [http://dx.doi.org/10.1016/j.materresbull.2019.110734]

[118] Bhagabati, P.; Rahaman, M.; Bhandari, S.; Roy, I.; Dey, A.; Gupta, P.; Ansari, M.A.; Dutta, A.; Chattopadhyay, D. Synthesis/Preparation of Carbon Materials. In: *Carbon-Containing Polymer Composites*; Rahaman, M.; Ali, D.K.; Aldalbahi, K., Eds.; Springer: Singapore, **2019**. [http://dx.doi.org/10.1007/978-981-13-2688-2_1]

[119] Villano, R.; Acocella, M.R.; Guerra, G. Oxidized carbon black as catalyst for the enamine formation in solvent-free conditions: A green strategy to build the benzodiazepine scaffold. *ChemistrySelect*, **2017**, *2*(32), 10559-10564. [http://dx.doi.org/10.1002/slct.201701711]

[120] Lim, J.; Ryu, S.Y.; Kim, J.; Jun, Y. A study of TiO_2/carbon black composition as counter electrode materials for dye-sensitized solar cells. *Nanoscale Res. Lett.*, **2013**, *8*(1), 227. [http://dx.doi.org/10.1186/1556-276X-8-227] [PMID: 23672498]

[121] Hulicova-Jurcakova, D.; Li, X.; Zhu, Z.; de Marco, R.; Lu, G.Q. Graphitic carbon nanofibers synthesized by the chemical vapor deposition (CVD) method and their electrochemical performances in supercapacitors. *Energy Fuels*, **2008**, *22*(6), 4139-4145. [http://dx.doi.org/10.1021/ef8004306]

[122] Miniach, E.; Śliwak, A.; Moyseowicz, A.; Gryglewicz, G. Growth of carbon nanofibers from methane on a hydroxyapatite-supported nickel catalyst. *J. Mater. Sci.*, **2016**, *51*(11), 5367-5376. [http://dx.doi.org/10.1007/s10853-016-9839-1]

[123] Alarifi, I.M.; Khan, W.S.; Asmatulu, R. Synthesis of electrospun polyacrylonitrile- derived carbon fibers and comparison of properties with bulk form. *PLoS One*, **2018**, *13*(8), e0201345. [http://dx.doi.org/10.1371/journal.pone.0201345] [PMID: 30091992]

[124] Huczko, A. Template-based synthesis of nanomaterials. *Appl. Phys., A Mater. Sci. Process.*, **2000**, *70*(4), 365-376. [http://dx.doi.org/10.1007/s003390051050]

[125] Hulteen, J.C.; Chen, H.X.; Chambliss, C.K.; Martin, C.R. Template synthesis of carbon nanotubule and nanofiber arrays. *Nanostruct. Mater.*, **1997**, *9*(1-8), 133-136. [http://dx.doi.org/10.1016/S0965-9773(97)00036-6]

[126] Toro, C.; Buriak, J.M. Template synthesis approach to nanomaterials. *Charles Martin. Chem. Mater.*, **2014**, *26*, 4889-4890.

[127] Li, W.; Zhang, F.; Dou, Y.; Wu, Z.; Liu, H.; Qian, X.; Gu, D.; Xia, Y.; Tu, B.; Zhao, D. A self-template strategy for the synthesis of mesoporous carbon nanofibers as advanced supercapacitor electrodes. *Adv. Energy Mater.*, **2011**, *1*(3), 382-386. [http://dx.doi.org/10.1002/aenm.201000096]

[128] Verma, B.; Sewani, H.; Balomajumder, C. Synthesis of carbon nanotubes *via* chemical vapor deposition: an advanced application in the Management of Electroplating Effluent. *Environ. Sci. Pollut. Res. Int.*, **2020**, *27*(12), 14007-14018. [http://dx.doi.org/10.1007/s11356-020-08002-0] [PMID: 32036530]

[129] Saravanan, M.S.S.; Nagaraj, T.; Mohan, S.; Singh, S.R.K.; Sivaprasad, K.; Babu, S.P.K.; Sajin, J.B. Synthesis of metal catalyst carbon nanotubes by arc-discharge method used for energy efficient applications. *AIP Conf. Proc.*, **2018**, *2039*, 020072. [http://dx.doi.org/10.1063/1.5079031]

[130] Ismail, R.A.; Kadhim, R.G.; Abdulridha, W.M. Effect of multiwalled carbon nanotubes incorporation on the performance of porous silicon photodetector. *Optik (Stuttg.)*, **2016**, *127*(19), 8144-8152. [http://dx.doi.org/10.1016/j.ijleo.2016.06.018]

[131] Wang, X.; Liu, X.; Licht, G.; Licht, S. Calcium metaborate induced thin walled carbon nanotube syntheses from CO_2 by molten carbonate electrolysis. *Sci. Rep.*, **2020**, *10*(1), 15146. [http://dx.doi.org/10.1038/s41598-020-71644-0] [PMID: 32934276]

[132] Wang, S.; Tang, L.A.; Bao, Q.; Lin, M.; Deng, S.; Goh, B.M.; Loh, K.P. Room-temperature synthesis of soluble carbon nanotubes by the sonication of graphene oxide nanosheets. *J. Am. Chem. Soc.*, **2009**, *131*(46), 16832-16837. [http://dx.doi.org/10.1021/ja905968v] [PMID: 19873970]

[133] Chauhan, S.; Jain, N.; Nagaich, U. Nanodiamonds with powerful ability for drug delivery and biomedical applications: Recent updates on *in vivo* study and patents. *J. Pharm. Anal.*, **2020**, *10*(1), 1-12. [http://dx.doi.org/10.1016/j.jpha.2019.09.003] [PMID: 32123595]

[134] Basso, L.; Cazzanelli, M.; Orlandi, M.; Miotello, A. Nanodiamonds: Synthesis and application in sensing, catalysis, and the possible connection with some processes occurring in space. *Appl. Sci. (Basel)*, **2020**, *10*(12), 4094. [http://dx.doi.org/10.3390/app10124094]

[135] Panich, A.M.; Shames, A.I.; Mogilyansky, D.; Goren, S.D.; Dolmatov, V.Y. Detonation nanodiamonds fabricated from tetryl: Synthesis, NMR, EPR and XRD study. *Diamond Related Materials*, **2020**, *108*, 107918. [http://dx.doi.org/10.1016/j.diamond.2020.107918]

[136] Tallaire, A.; Brinza, O.; De Feudis, M.; Ferrier, A.; Touati, N.; Binet, L.; Nicolas, L.; Delord, T.; Hétet, G.; Herzig, T.; Pezzagna, S.; Goldner, P.; Achard, J. Synthesis of loose nanodiamonds containing nitrogen-vacancy centers for magnetic and thermal sensing. *ACS Appl. Nano Mater.*, **2019**, *2*(9), 5952-5962. [http://dx.doi.org/10.1021/acsanm.9b01395]

[137] Basso, L.; Bazzanella, N.; Cazzanelli, M.; Miotello, A. On the route towards a facile fluorescent nanodiamonds laser-synthesis. *Carbon*, **2019**, *153*, 148-155. [http://dx.doi.org/10.1016/j.carbon.2019.07.025]

[138] Ekimov, E.A.; Kondrina, K.M.; Mordvinova, N.E.; Lebedev, O.I.; Pasternak, D.G.; Vlasov, I.I. High-pressure, high-temperature synthesis of nanodiamond from adamantane. *Inorg. Mater.*, **2019**, *55*(5), 437-442. [http://dx.doi.org/10.1134/S0020168519050042]

[139] Das, T.; Saikia, B.K. Nanodiamonds produced from low-grade indian coals. *ACS Sustain. Chem.& Eng.*, **2017**, *5*(11), 9619-9624. [http://dx.doi.org/10.1021/acssuschemeng.7b02500]

[140] Ugarte, D. Curling and closure of graphitic networks under electron-beam irradiation. *Nature*, **1992**, *359*(6397), 707-709. [http://dx.doi.org/10.1038/359707a0] [PMID: 11536508]

[141] Alessandro, F.; Scarcello, A.; Basantes Valverde, M.D.; Coello Fiallos, D.C.; Osman, S.M.; Cupolillo, A.; Arias, M.; Arias de Fuentes, O.; De Luca, G.; Aloise, A.; Curcio, E.; Nicotra, G.; Spinella, C.; Caputi, L.S. Selective synthesis of turbostratic polyhedral carbon nano-onions by arc discharge in water. *Nanotechnology*, **2018**, *29*(32), 325601. [http://dx.doi.org/10.1088/1361-6528/aac4ca] [PMID: 29761791]

[142] Zhang, C.; Li, J.; Shi, C.; Liu, E.; Du, X.; Feng, W.; Zhao, N. The efficient synthesis of carbon nano-onions using chemical vapor deposition on an unsupported Ni–Fe alloy catalyst. *Carbon*, **2011**, *49*(4), 1151-1158. [http://dx.doi.org/10.1016/j.carbon.2010.11.030]

[143] Jung, S.; Myung, Y.; Das, G.S.; Bhatnagar, A.; Park, J.W.; Tripathi, K.M.; Kim, T. Carbon nano-onions from waste oil for application in energy storage devices. *New J. Chem.*, **2020**, *44*(18), 7369-7375.

[http://dx.doi.org/10.1039/D0NJ00699H]

[144] Dorobantu, D.; Bota, P.M.; Boerasu, I.; Bojin, D.; Enachescu, M. Pulse laser ablation system for carbon nano-onions fabrication. *Surg. Eng. Appl. Electrochem.*, **2014**, *50*(5), 390-394. [http://dx.doi.org/10.3103/S1068375514050044]

[145] Bartolome, J.P.; Fragoso, A. Preparation and characterization of carbon nano-onions by nanodiamond annealing and functionalization by radio-frequency Ar/O $_2$ plasma. *Fuller. Nanotub. Carbon Nanostruct.*, **2017**, *25*(5), 327-334. [http://dx.doi.org/10.1080/1536383X.2017.1303604]

[146] Christoforidis, K.C.; Syrgiannis, Z.; La Parola, V.; Montini, T.; Petit, C.; Stathatos, E.; Godin, R.; Durrant, J.R.; Prato, M.; Fornasiero, P. Metal-free dual-phase full organic carbon nanotubes/g-C_3N_4 heteroarchitectures for photocatalytic hydrogen production. *Nano Energy,* **2018**, *50*, 468-478. [http://dx.doi.org/10.1016/j.nanoen.2018.05.070]

[147] Paušová, Š.; Pacileo, L.; Baudys, M.; Hrubantová, A.; Neumann-Spallart, M.; Dvoranová, D.; Brezová, V.; Krýsa, J. Active carbon/TiO_2 composites for photocatalytic decomposition of benzoic acid in water and toluene in air. *Catal. Today,* 2022 (9), 417-423. [http://dx.doi.org/10.1016/j.cattod.2020.06.048]

[148] Don, M.F.; Ekanayake, P.; Nakajima, H.; Mahadi, A.H.; Lim, C.M.; Atod, A. Acetylene carbon black-graphite composite as low-cost and efficient counter electrode for dye-sensitized solar cells (DSSCs). *Ionics,* **2019**, *25*(11), 5585-5593. [http://dx.doi.org/10.1007/s11581-019-03071-9]

[149] Bhavani, K.S.; Anusha, T.; Shanmukha Kumar, J.V.; Brahman, P.K. Enhanced electrocatalytic activity of methanol and ethanol oxidation in alkaline medium at bimetallic nanoparticles electrochemically decorated fullerene-C_{60} nanocomposite electrocatalyst: An efficient anode material for alcohol fuel cell applications. *Electroanalysis,* **2020**, *33*(1), 388-389. [http://dx.doi.org/10.1002/elan.202060154]

[150] Chen, J.; Wang, J.; Chen, J.; Wang, L. A bifunctional electrocatalyst of PtNi nanoparticles immobilized on three-dimensional carbon nanofiber mats for efficient and stable water splitting in both acid and basic media. *J. Mater. Sci.,* **2017**, *52*(22), 13064-13077. [http://dx.doi.org/10.1007/s10853-017-1410-1]

[151] Mohapatra, J.; Ananthoju, B.; Nair, V.; Mitra, A.; Bahadur, D.; Medhekar, N.V.; Aslam, M. Enzymatic and non-enzymatic electrochemical glucose sensor based on carbon nano-onions. *Appl. Surf. Sci.,* **2018**, *442*, 332-341. [http://dx.doi.org/10.1016/j.apsusc.2018.02.124]

[152] Lawrence, N.S.; Deo, R.P.; Wang, J. Biocatalytic carbon paste sensors based on a mediator pasting liquid. *Anal. Chem.,* **2004**, *76*(13), 3735-3739. [http://dx.doi.org/10.1021/ac049943v] [PMID: 15228348]

[153] Ji, C.; Hou, J.; Chen, V. Cross-linked carbon nanotubes-based biocatalytic membranes for micro-pollutants degradation: Performance, stability, and regeneration. *J. Membr. Sci.,* **2016**, *520*, 869-880. [http://dx.doi.org/10.1016/j.memsci.2016.08.056]

[154] Singh, R.S.; Chauhan, K.; Kennedy, J.F. Fructose production from inulin using fungal inulinase immobilized on 3-aminopropyl-triethoxysilane functionalized multiwalled carbon nanotubes. *Int. J. Biol. Macromol.,* **2019**, *125*, 41-52. [http://dx.doi.org/10.1016/j.ijbiomac.2018.11.281] [PMID: 30529206]

[155] Oliveira, S.F.; da Luz, J.M.R.; Kasuya, M.C.M.; Ladeira, L.O.; Correa Junior, A. Enzymatic extract containing lignin peroxidase immobilized on carbon nanotubes: Potential biocatalyst in dye decolourization. *Saudi J. Biol. Sci.,* **2018**, *25*(4), 651-659. [http://dx.doi.org/10.1016/j.sjbs.2016.02.018] [PMID: 29740229]

[156] Wang, L.; Liu, X.; Jiang, Y.; Zhou, L.; Ma, L.; He, Y.; Gao, J. Biocatalytic pickering emulsions stabilized by lipase-immobilized carbon nanotubes for biodiesel production. *Catalysts,* **2018**, *8*(12),

587.
[http://dx.doi.org/10.3390/catal8120587]

[157] Dam, B.; Jamatia, R.; Gupta, A.; Pal, A.K. Metal-free greener syntheses of pyrimidine derivatives using a highly efficient and reusable graphite oxide carbocatalyst under solvent-free reaction conditions. *ACS Sustain. Chem.& Eng.,* **2017**, *5*(12), 11459-11469.
[http://dx.doi.org/10.1021/acssuschemeng.7b02626]

CHAPTER 2

Structure and Properties of Graphene and Chemically Modified Graphene Materials

Manish Srivastava[1], **Anjali Banger**[2], **Anamika Srivastava**[2], **Nirmala Kumari Jangid**[2,*] and **Priy Brat Dwivedi**[3]

[1] *Department of Chemistry, Central University of Allahabad, Allahabad – 211002, Uttar Pradesh, India*

[2] *Department of Chemistry, Banasthali Vidyapith, Banasthali – 304022, Rajasthan, India*

[3] *National University of Science and Technology, AlHail South, Seen 111, Muscat, Oman*

Abstract: Graphene is an allotrope of carbon that is made up of very strongly bonded carbon atoms. The structure of graphene is a hexagonal lattice. Graphene shows sp^2 hybridization and an extremely thin atomic thickness of approximately 0.345Nm. This chapter deals with graphene structure, including hybridization, critical parameters of the unit cell, the formation of σ and π bonds, electronic band structure, edge orientations, and the number and stacking order of graphene layers. The remarkable characteristics of graphene occur because of the extended chain of π conjugation that results in high charge mobility, high conductivity & high Young's modulus value. Due to these attractive properties, graphene has gained much attention. Graphene, with the unique combination of bonded carbon atom structures with its myriad and complex physical properties is balanced to have a big impact on the future of material sciences, electronics, and nanotechnology. Graphene is converted to Graphene nanoparticles, Graphene oxide nanoparticles; Polymer-based graphene composite materials and Graphene nanoribbons, *etc* by chemical methods. Some of the application areas are batteries and ultracapacitors for energy storage and fuel cell and solar cell for energy generation and some of the possible future directions of research have been discussed.

Keywords: Actuators, Application of Graphene, Chemical Properties, Electrical Properties, Electrochemical Capacitors, Fabrication of Graphene Oxide, Functionality and Solubility, Graphene Aerogels, Graphene Composites, Graphene Derived Nanofiller, Graphene Oxide, Hybrid Graphene, Mechanical Properties, Optical Properties, Properties of Graphene, Quantum Dots of Graphene, Storage Material, Surface-Enhanced Raman Spectroscopy, Thermal Properties.

* **Corresponding author Nirmala Kumari Jangid:** Department of Chemistry, Banasthali Vidyapith, Banasthali – 304022, Rajasthan, India; E-mail:nirmalajangid.111@gmail.com

Pinki Bala Punjabi, Rakshit Ameta & Sharoni Gupta (Eds.)

1. INTRODUCTION

The 6th element of the periodic table *i.e.*, Carbon has always attracted researchers and technologists and the most studied allotrope of the Carbon family is Graphene. It is a single atomic planar layer of graphite [1]. By chemical vapor deposition, chemical or mechanical methods, can be obtained from graphite [2, 3].

Theoretically, for the last sixty years, graphene or two-dimensional graphite had been studied [4]. Earlier, it was assumed that graphene does not exist in the free state. Peierls and Landau contended that since two-dimensional crystals were chemically not stable so they do not exist [5, 6]. Hence, it was proclaimed that graphene has large three-dimensional structures [7]. Without a three-dimensional base, two-dimensional structures do not exist, later in 2004, Novoselov *et al.* accidentally discovered graphene. Using a micromechanical cleavage technique single-layer graphene was established [8].

After several broadcasts on nanotubes of carbon (1991) [9], fullerenes (1985) [10], and intercalated graphite [11], graphene gained immense interest [12, 13]. It is a flat sp^2 hybridized planar sheet of C atoms densely packed in a 2D hexagonal pattern. It can be coated into zero-dimensional fullerenes, twirled into a one-dimensional nanotube, or can be bundled into three-dimensional graphite.

The uncommon characteristics of graphene involve that its charge carriers behave as Dirac fermions, and under encompassing situations they scatter. This distinct attitude resulted in exceptional phenomenal changes [12]. Firstly, it is a two-dimensional semiconductor with minimal overlapping between the conduction band and valence band and has a bandgap of zero. Secondly, for measuring deviation in room temperature (up to 10000 cm^{-2} s^{-1}) and charge carrier concentration (up to 1013 cm^{-2}) it shows a strong bipolar electric field effect. Thirdly, by modifying chemical potential with the help of the electric field effect, exceptional half-integer QHE for hole and electrons carrier can be seen [13, 14]. At room temperature, a single layer of a sheet, shows great thermal conductivity (5000 WmK^{-1}). Additionally, it's transparent with 2.3% absorption towards visible light [15, 16]. Transistors can be produced with narrower (1-2 nm thick) ribbons of graphene having variable band gaps [17 - 19].

In past years, graphene is being used instead of carbon-based nanofiller in preparing nanocomposites of polymer and showed great thermal, electrical, and mechanical properties [20 - 27]. Chemically unstable indium tin oxide in displays and screens can be replaced by graphene [28, 29]. Hence, it is assumed that a single layer is responsible for the uniqueness of graphene although its preparation is severe in encompassing conditions. Due to inappropriate separation of graphene sheets, it aggregates and restacks leading to the formation of graphite *via* Vander Waals interaction and p–p stacking [30, 31]. Graphene has vast applications in the fields of microwave absorption shields, nanocomposites of polymer, super-capacitor devices, drug delivery, biosensing, *etc*.

According to IUPAC, graphene can be explained as a single sheet of C with a graphitic structure and similar to a polycyclic aromatic hydrocarbon. During a discussion of various properties of monatomic sheets, graphene is prescribed [32, 33]. In 1987, for the first time, the word "graphene" was used by Mouras *et al.* [34], although in 1840 Schafhaeutl (German scientist) narrated the exfoliation and intercalation of graphite by making use of HNO_3 and H_2SO_4. From that time researchers predict the presence of graphene. But it was not until 2004 that two physicists at the University of Manchester, Andre Geim and Konstantin Novoselov, obtained graphene in the laboratory and could prove it. These researchers developed an easy procedure for obtaining graphene from graphite. Using tape, a thin sample of graphite was extracted from a greater one.

This procedure was repeated several times to finally obtain a monoatomic layer, which was then electrically isolated with a thin layer of silicon dioxide on a silicon wafer. The importance of this event lies in having isolated graphene at room temperature since in the past this material was considered thermodynamically unstable. For their innovative experiments, in 2010 they received the Nobel Prize in Physics [35].

The important discoveries of a few scientists have been summarized below (Table **1**).

Table 1. Scientists with their discoveries.

S.No.	SCIENTISTS	DISCOVERIES
1.	Schafhaeutl (1840)	Narrated exfoliation and intercalation of graphite by making use of HNO_3 and H_2SO_4.
2.	B.C. Brodie (1859)	First discovered graphite oxide properties [97].
3.	Brodie, Staudenmaier, Hummers and Offeman (1958)	First synthesized graphene oxide through graphitic oxidation [109 - 111].
4.	Mouras *et al.* (1987)	For the first time, used the word "graphene" [34].
5.	Andre Geim and Konstantin Novoselov (2004)	Fabricated graphene from graphite in the laboratory and received Nobel Prize in Physics (2010) for this.
6.	Novoselov *et al.* (2004)	Discovered graphene using a micromechanical cleavage technique [8].
7.	Wu *et al.* (2011)	Established remote control graphene-based robot in the presence and absence of IR radiations [89].

2. FABRICATION OF GRAPHENE

2.1. Bottom-up Method

Various synthetic criteria can be used to prepare graphene and its derivatives but usually bottom-up and top-down methods are used (Kim, 2010a). From the past ten years, productive work has been done for the advancement of bottom-up criteria to atomically brief graphene nanoribbons [36, 37]. Usually, chemical and epitaxial methods are used to generate large-sized, defect-free sheets of graphene in minimal quantities. With the help of the CVD method's low resistance, monolayered graphene films (30-inches) had been produced having higher potential and great optical transmission properties replacing commercially available transparent electrodes (Bae, 2010). They have useful applications in electronic and fundamental fields but are inapplicable in the case of nanocomposites of polymer since they require altered surface graphene sheets. By selecting the starting material of nanotubes size can be controlled by unzipped graphene ribbons but preparing composites of polymer is still an expensive procedure.

2.2. Top-down Method

In this method, graphene is prepared by the exfoliation of graphite and its derivatives. Since globally graphite is produced in bulk (1-2 million tons per annum) so starting from it (like from graphite oxide) is favorable and economically more advantageous than a bottom-up approach.

3. STRUCTURE

Carbon has six electrons in total, out of which four are in the outer orbit and two in the inner orbit. Outer four electrons are capable of forming chemical bonds but in graphene, one C atom is attached to three C atoms in a 2-D plane leaving one electron free for electronic conduction. Such movable electrons are known as pi (π) electrons present above and below the plane and their overlapping boost C-C bonding (Fig. **1**). It also governs the electronic properties of graphene and in general, these properties depend on the bonding and anti-bonding of this π orbital.

Graphene has a 2D structure (of monatomic thickness) and is formed by a regular honeycomb network of joined C atoms. The bond length of the C-C bond is about 0.142 nm. The interplanar distance of graphene sheets is 0.335 nm which further stacks and leads to the formation of graphite. It is light with 0.77 mg /m^2 surface density.

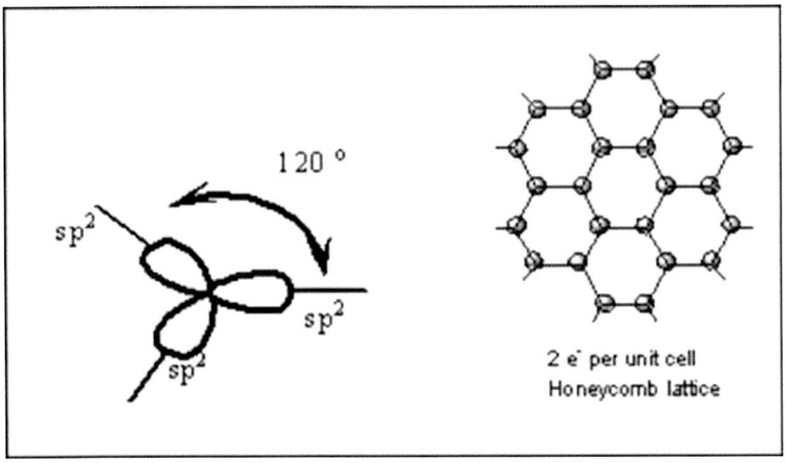

Fig. (1). Structure of graphene.

It is sp^2 hybridized and 0.345 nm thick. Since it shows normal electronic spectra, they are regarded as zero-band gap semiconductors. They can be rolled into a one-dimensional nanotube or in three-dimensional graphite. Its preparation is severe in encompassing conditions due to inappropriate separation of graphene sheets; it aggregates and restacks leading to the formation of graphite *via* Vander Waals interaction and p–p stacking [38, 39]. The aggregation *via* large-size or polar-polar interaction can be prevented if hydrophilic or hydrophobic groups are present [40].

Graphene can spontaneously repair voids present in its film when exposed to carbon-containing molecules [41]. Graphene sheets can be used to form other more complex carbon structures [42], like buckminsterfullerene, *etc.*

4. PROPERTIES

Due to its monatomic layer as well as the nature of the covalent bonds that can be extended making it no longer remain on the same plane, graphene is very flexible. Furthermore, it can be stretched up to 20% size [43].

The specific heat of graphene cannot be measured directly and is identical to that of graphite from 100 K, while for low temperatures graphene exceeds the contribution of phonons (vibrational modes of the crystal structure) of low frequency. At high temperatures, the specific heat is almost constant, approaching 25 $Jmol^{-1}K^{-1}$ (Dulong-Petit limit) [44].

Another feature to note is its impermeability to aqueous solutions, vapors, and gaseous medium, along with He (the smallest radius atom, 32 pm). However, it has been observed that water passes through the membranes of graphene [45].

Since graphene is flexible so it can be used in roller ball computers, heat-sensitive clothing, and flexible phones. It has a tendency of conducting electricity that's too better than Cu which is a boon to electronic applications for the next generation. It is transparent and thin (one-atom-thick *i.e.*, one million times thinner compared to a human hair). Graphene is associated with various properties shown in Fig. (**2**).

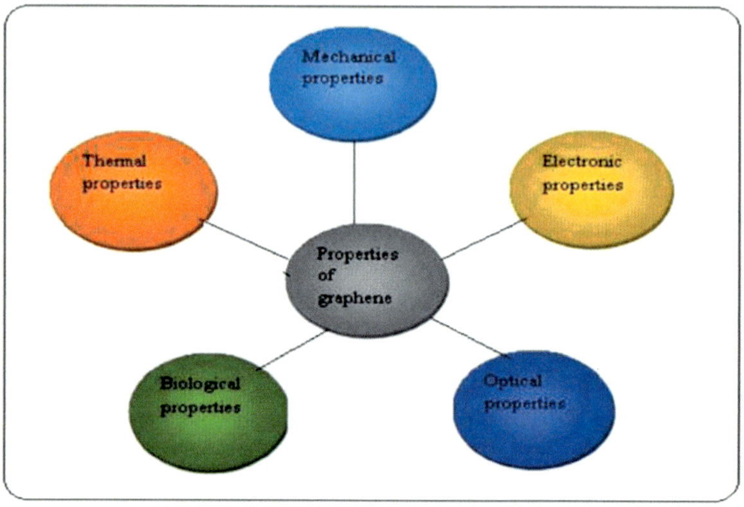

Fig. (2). Properties of graphene.

4.1. Electrical Properties

The symmetry & 2-D structure of graphene makes it interesting for research. It is illustrated as a zero-band gap semiconductor because, at the Brillouin zone's area, the meeting of its conduction & valance bond takes place. The charge carrier in graphene acts as relativistic or mass-less particles [46]. An ambipolar effect of the electric field is exhibited along with high concentrations [1013 cm^{-2}] in between electrons and the holes and have current mobility up to ~15000 $cm^2/(Vs)$ with interaction and electrical conduction between the substrate and the layers of graphene.

4.2. Mechanical Properties

Frank and his co-workers firstly reported the mechanical property of graphene. They used an atomic force microscope and calculated the spring constant. The young modulus was found to be 0.5TPa and the spring constant between 1-5 N/m. Graphene is a highly used compound in the world. Graphene is a hundred times extra influential than steel and is harder than diamonds. Few mechanical properties include Poisson's ratio, tensile strength, and young's modulus. The mechanical conception should be placed in the continuous standard structure. Using a hypothesis, one can calculate its thickness. Graphene consists of a single layer of C atoms that makes its mechanical properties more significant. By using a traditional microscope or technique, it is hard to get efficient mechanical property while doing experiments because of its two-dimensional nature. So, for this, a new arrangement called atomic force nanoindentation has been achieved [47]. Generally, graphene is used to improve the mechanical properties of composites of polymer.

4.3. Thermal Properties

It generally causes vibrations that are created by the layered structure of graphene. Based on the dispersion curve of acoustic phonon & optical phonon, generally, six categories of polar phonons are recognized. These are the following [48]:

1. ZA mode phonons [Out of plane acoustic phonons] & ZO [optical phonons].

2. TA mode phonons [In plane transverse acoustic phonons] & TO [transverse optical phonons].

3. LA mode phonons [In plane longitudinal acoustic phonons] & LO [longitudinal optical phonons].

Currently, the most commonly used thermal conductive materials are copper, aluminum foil, carbon nano tubes, *etc*. Copper shows thermal conductivity of 380 W/mK, conductivity shown by Al foil is 160$Wm^{-1}K$. Thermal conductivities of single-walled CNT & multiwalled carbon nano-tubes are 3500$Wm^{-1}K$ and 3000$Wm^{-1}K$ respectively. Diamond has the thermal conductivity ranging from 1000-2200$Wm^{-1}K$ while that of single-layer graphene is 5000 $Wm^{-1}K$ [49].

Interplanar heat transfer processes within the two-dimensional plane of graphene are among the highest for known materials. It has been reported a value of (4.84 ± 0.44)103 and (5.30 ± 0.48)103 $Wm^{-1}K^{-1}$ at ambient temperature, significantly higher than other carbon structures [50]. Processes for heat transfer perpendicular are to the plane; weak Vander Waals interactions limit the thermal conductivity. An approximate value of 6 $Wm^{-1}K^{-1}$ has been measured [51]. But in any case, both the presence of a substrate and defects in the structure of the graphene, condition their thermal properties significantly.

4.4. Optical Properties

The optical and outstanding charge carrier characteristics make it interesting. Over a vivid range of wavelengths, monolayer graphene can pass on 97% of the total incident light [52]. As the layers of graphene increase, the optical imaging & absorption of light is also increased [53]. Modulators, charge injectors, IR detectors, and emitters can also be formed from graphene [54]. When the light gets absorbed on the surface of graphene, some electron-hole pairs are generated that are based on the density of holes & electrons. These pairs can combine again and again very quickly, but the internal or external field that is generated close to the graphene-electron boundary can separate them [55]. To develop bioimaging techniques, this property of recombining & separating is very useful. Graphene can be cut into nano-ribbons which can make it luminous [56 - 58]. The recombining characteristic of the electron-hole pair can also make graphene a photo-luminescent material. Graphene has several applications in biomedical imaging & magnetic resonance imaging due to its high charge mobility, photo-luminescence, and high light transmittance [59, 60].

4.5. Chemical Properties

Graphene, the only allotrope of C whose every atom present on the two sides is responsible for chemical reactions because it is two-dimensional. As compared to other materials present in the environment graphene consists of maximum no. of C atoms present on the edge and they show specific reactivity towards chemical reactions and defects within them increase their reactive power.

4.6. Biological Properties

Sanchez and co-workers described the biological properties and the toxicity of nanocomposites of graphene [61]. Bianco also described this property recently [62]. A brief discussion of the interaction of cells and biomolecules with graphene is focused here. The nanomaterials based on graphene show a different kind of interaction with RNA & DNA. So, it is very interesting to use it in RNA or DNA delivery & sensing. Graphene oxide has better adsorption for single-strand DNA than double-strand DNA. It protects adsorbed nuclide against nuclease enzyme [63 - 66]. In high ion concentrated solution at low ph, the adsorption for small oligomers is increased because of negative charge interaction between graphene oxide and DNA [67]. Titov and his workers used coarse-grained molecules simulation of lipids and estimated that functional & stable hybrid structures can be formed of graphene [68]. Although these simulations have to be supported by some data from experiments. For understanding the interaction between graphene & lipid bi-layer in the membrane of a cell, these studies are significant as shown in Fig. (**3**).

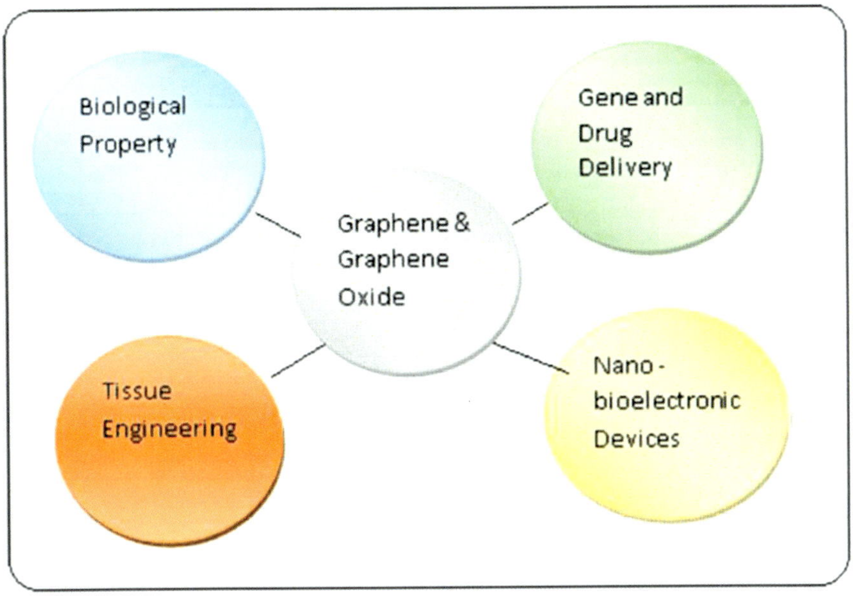

Fig. (3). Biological application of graphene.

Additional Properties

4.7. Functionality and Solubility

Graphene is insoluble in aqueous mediums like water, polymer resins. So, they require the attachment of some specific functional groups without affecting their physical and chemical properties. The functionalization of graphene has been carried out by different groups using various novel strategies [69]. Niyogy *et al.* *via* covalent bonding functionalized graphene [70]. To dissolve graphene in a non-polar solvent like CCl_4 and THF amidation was performed. When pristine graphene is treated with nitrating agents (sulfuric acid and nitric acid) it leads to the formation of aqueous graphene which remains stable for months. Si and Samulski performed a scalable and facile method to fabricate isolated, sparingly sulfonated liquid graphene [71].

Non-covalent bonding can also be performed for functionalizing graphene. This can be done by wrapping it with surfactants or *via* π-π bonding with aromatic molecules like PYBS [72]. This does not affect the electronic structure of graphene. When graphene is treated with surfactants like SDS and CTAB, it leads to the formation of aqueous graphene or it can also be obtained by the interaction of acidified graphene with excessive PEG and concentrated hydrochloric acid under solvothermal conditions [73].

Liu and co-workers performed a single-step ionic liquid-aided electrochemical procedure to obtain ionic liquid functionalized graphene directly from graphite [74]. Here graphite works as cathode dipped in phase-separated water and ionic liquid (imidazolium-based) mixture with 10–20 V potential across electrodes. After half an hour graphitic anode produced ionic liquid functionalized graphene sheets with an average length of 700 nm and breadth of 500 nm. TEM images and AFM analysis confirmed the results.

4.8. Surface Area

Since graphene has a higher specific surface area and is highly porous, it makes it highly adsorptive for various gases like H_2, CH_4, and CO_2. Theoretically, the surface area of one graphene sheet is 2600 mg [75] but experimental results came around 270–1550 mg. Calculations performed using First-principle showed that carbon monoxide is present in a parallel configuration in hexagonal rings [75]. Calculations performed using First-principle showed that the hydrogen is present in parallel and perpendicular dimensions on cyclic rings of graphene and a single sheet of graphene use 7.7 weight% of H.

4.9. Fluorescence Quenching

Graphene is capable of light emission. It helps in the selective detection of biomolecules. This is directly linked with photo enhanced transfer of electrons. Quenching of porphyrin by graphene has been submitted [76].

Ramakrishna *et al.* worked on the light-emitting property of graphene [77]. During quenching of dye molecule when graphene was present, long-range energy transfer is considered.

4.10. Surface-Enhanced Raman Spectroscopy

SERS and simple Raman spectroscopy give almost similar information except that it provides enhanced signals. Fleischmann and co-workers (in 1974) enhanced a weak signal by introducing SERS [78]. After that, it came under focus and facilitates its identity in science. This enhancement effect follows two basic mechanisms *i.e.*, electromagnetic enhancement mechanism (EM) and chemical enhancement mechanism (CM). The first one *i.e.*, EM works based on exciting surface plasmons while CM works based on the formation of a complex of the adsorbed molecule and its charge transfer. Experimentally separation of SERS and Raman spectroscopy is quite challenging. Preparation of SERS is done only in SERS research labs. By chemical reduction and electrochemical oxidation-reduction methods, active substrates of SERS can be obtained. Fabrication of nanostructures and nanowires too can be used as SERS substrates. After working on it, it was found that only "free-electron-like" metals *e.g.*, silver, copper, and gold can be used as substrates for colloidal particles and rough surfaces. Limited substrates declined its applications. The SERS substrates are less biocompatible so, there is an urgent need to develop new ones. For an ideal SERS substrate, it should be cost-effective, easily accessible, biologically compatible, and chemically inert. Professor Ling and his co-workers reported the use of graphene as an efficient SERS substrate since it is chemically inert and biologically compatible. On depositing protoporphyrin IX, phthalocyanine (Pc), and crystal violate (CV) on the graphene surface, they found it useful as a substrate [79]. Further, the Raman Effect was examined on various layers and single-layered graphene was found to be most appropriate. With the increase in layers of graphene, the Raman signals decrease. The distance between the substrate and probe molecule should be lower than 0.2 nm in the case of the chemical enhancement mechanism. For transferring charge from molecules to metals and vice-versa the Fermi level of the substrate matches should be symmetrical to HOMO and LUMO of the probe. In the visible region, the optical transmission *via* the surface of graphene is greater than 95% and terahertz plasmon excitation. Because of these characteristics, it does not support the EM process. Now only CM is left for enhancing Raman signals. This is based on the fact that researchers worked on the possibility of charge transfer between molecule and graphene [80]. If graphene works as a SERS substrate, separation of CM and EM will be easily possible compared to other SERS substrates. Graphene can be easily synthesized, biocompatible, and cheap.

5. APPLICATIONS

Different types of applications in numerous fields associated with graphene as electronics, biomedicine, photonics, sensors and metrology, composites and coatings, generation & storage of energy, *etc.* shown in Fig. (**4**).

It possesses remarkable electronic properties and could replace silicon shortly by making use of itself in computer chips, ultrafast or quantum computers working at terahertz speed. Like photons, the electrons of graphene behave as massless particles. Unluckily, the complete study of electronic property is tricky.

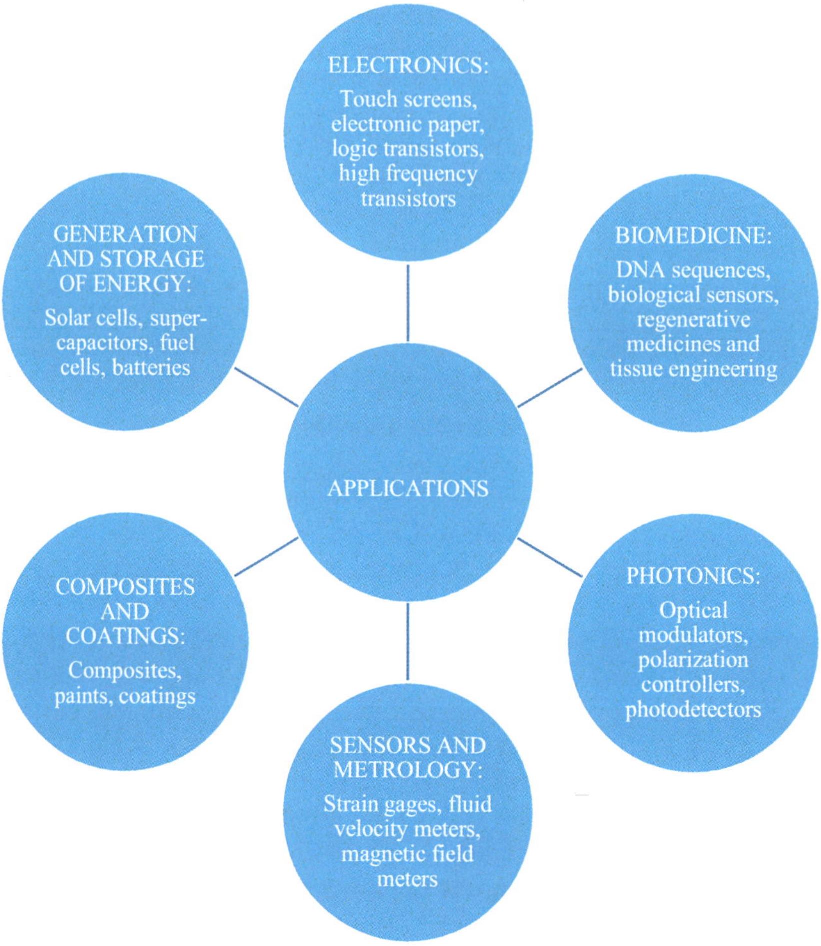

Fig. (4). Application of graphene.

Under the FET configuration, the transportation of electrons of graphene can be understood. FET is done by placing a sample of graphene on an oxidized silicon substrate and connecting it to the drain electrode and a source. The capacitance of the oxide layer is less than that of FET devices complicating the measurement of quantum capacitance.

The capability of a substance to store energy in the form of charge is known as capacitance. In physics, capacitance is resisted by electrically charged electrons. Although this capacitance is dominated by that of 2-D graphene. Pauli's Exclusion Principle explains quantum capacitance as a group of particles (which includes neutrons, electrons, and protons) that are unable to occupy the same energy stage at the same time. Only when an energy level is filled particles can excite to higher energy levels. Xia *et al.* reported direct measurement of quantum capacitance of graphene in electrolytic ionic liquid media by making use of three electrodes of electrochemical configuration. It was published in Nature Nanotechnology in 2010 [81]. He attached two electrodes to graphene, and applied voltage across it with the help of the third electrode. The use of electrochemical gate and ionic liquid electrolyte reduced the Debye screening length to zero dominating quantum capacitance. The experimental data of Tao's experiment showed small graphene's capacitance. The plot of gate potential against quantum capacitance showed an asymmetric V-shape with a small Dirac point. This data grants theoretical prediction that signifies the quantum origin of interfacial capacitance in C-based electrodes.

The data provided by Tao declared the charged impurities are responsible for the declination of capacitance. This produces a low signal-to-noise ratio and found its application in chemical sensors and biosensors. The geometry of a single layer of graphene and low capacitance are boon to biosensors. By altering the electronic properties of the graphene, it can show when external biological substances interacted with the graphene layer. The study of antibody-antigen interaction by functionalizing the graphene's surface with antibodies can act as a useful application in biosensing. M. Shekhirev and co-workers reported that graphene nanoribbons having uniform electrical conductance can be produced on arbitrary high-temperature-resistant substrates, like metals, silicon dioxide, silicon, or its glasses, *via* the CVD method on basis of thermal decomposition of commercially accessible PTCDA molecules [82]. Polymer composites having a high thermal conductivity coefficient (λ) and excellent electrical and magnetic interference shielding efficiency have theoretical as well as potential practical applications [83, 84].

5.1. Storage Material

The energy density by mass of His high and that's why it is regarded as a basic carrier for energy. H storage is cost-effective and needs fast reaction kinetics, low adsorption temperature, high gravimetric and volumetric densities, and good reversibility. Since it is explosive so it requires stable storage conditions. It can be stored in solid materials (as adsorbed gas), pressurized cylinders (in gaseous form), and cryogenic liquid. The dissolution of hydrogen at high temperatures in metal hydrides is problematic so the chemical process is the safest method for storing H. Development in the field of nanomaterials can be useful because of the large surface-to-volume ratio and decreased diffusion path lengths of H.

Because of properties like high surface area, chemically stable and light graphene can be used. By chemical processes like chemisorption and physisorption, it can be stored in graphene. Although practically it encompassed and normal temperature its storage is difficult. Therefore, to intensify the uptake of hydrogen C atom needs chemical modification. Palladium can strongly adsorb hydrogen and its role can be intensified three times under variable temperature and pressure. Graphene has a higher potential for storing H and can be used in fuel cell technology. Maidhily and co-workers made use of graphene for the proton exchange [85]. Subrahmanyam *et al.* reported the storage of H in graphene chemically [86].

5.2. Actuators

Researchers used the unique properties (optical, mechanical, and electrical) of graphene to study actuators. Park and co-workers prepared a bi-layer paper (graphene-based actuators) made of carbon nanotubes (multi-walled) and graphene oxide [87]. The curls of paper depend on temperature. Xie and co-workers by modifying two edges of a monolithic film of graphene with O_2 plasma and a six-membered ring prepared the surface asymmetrically graphene film actuator [88]. Based on graphene on organic hybrid film, a bimorph micro-actuator was established. It develops bending and capping motion that varies with frequency and potential. Wu and co-workers established remote control graphene-based robot. In the presence and absence of IR radiations, it can curl or uncurl [89]. It can pick and move any object in any direction. This is an exquisite manifestation of photochemical and thermal energy conversion by graphene-based actuators.

5.3. Electrochemical Capacitors

Nowadays, because of higher security, power output, fast charging and discharging efficiency, and ultra-long service life the electrochemical supercapacitors have been greatly advanced and contemplated as present generation devices [90 - 92].

They are also known as ultracapacitors/supercapacitors. They use fast surface redox reactions to store energy. In need of high-power delivery, they can replace batteries. A supercapacitor is high power and low-energy device and is useful in high-power pulse needs. They have a long-life cycle and higher power density and fill the space between dielectric capacitors and batteries. There are three types of supercapacitors divided based on conducting polymers, transition metal oxides, and C-C based. For higher capacitance blocking and electronically conducting electrodes are being used. Graphite-based carbon is electrochemically stable, highly conductive, and porous. The most used electrode material is activated carbon since it is cheaper, capable, and has a high surface area. Graphene-based materials like fullerenes (0D), carbon nanotubes (1D), graphene (2D), and graphite (3D) are in focus because of their unique properties and structure. Since carbon nanotubes possess high electrical conductivity and surface area so they are being synthesized for supercapacitors. Although, CNT-based supercapacitors do not meet expectations but even after those experiments are being performed to improve their performance by growing them directly on bulk metals to remove resistance [93]. For supercapacitor applications, the graphene-based electrode material is being used [94] and on studying graphene supercapacitor, data revealed specific capacitance of 135 Fg^{-1} in aqueous electrolyte [95] and 117 Fg^{-1} in aqueous H_2SO_4 [94] on basis of multi-layered graphene. The effective surface area of graphene depends on the layer. So, graphene possesses high supercapacitor performance and surface area. Wang and co-workers [96] synthesized graphene-based supercapacitors and examined them. At room temperature, by the usage of gas-based hydrazine reduction of a single layer of graphene was performed. It showed low aggregation as compared to chemically modified liquid graphene. The capacitance of 205 F g^{-1} at 1.0 V in an aqueous electrolyte with an energy density of 28.5W h kg^{-1} has been observed and is the best data till now. Further, its power density was found to be 10 kW kg^{-1} greater than carbon nanotubes-based supercapacitors.

Poly pyrrole, polyaniline, polythiophene, and their derivatives are examined for supercapacitor electrodes. Wang and co-workers employed *in-situ* anodic electro-polymerization to synthesize flexible and free-standing graphene composite paper [96]. This paper electrode uses graphene sheets as building blocks having a tensile strength of 12.6 MPa and 233 F g^{-1} gravimetric capacitance and 135 F cm^{-1} volumetric capacitance respectively. This data is excellent as compared to other carbon-based flexible electrodes.

6. GRAPHENE OXIDE

6.1. Fabrication of Graphene Oxide

B.C. Brodie first discovered graphite oxide properties [97]. Although graphene oxide is synthesized by the treatment of graphite with strong oxidizing agents still it requires a reduction step which can be thermal or chemical to attain the properties of graphene.

The surface of GO was investigated under STM to obtain information regarding its structure [98, 99]. Due to bulky oxygen-containing functional groups, the width of the mono-layered graphene sheet is more than ~ 0.34 nm of ideal mono-layered graphene. Mono-layered graphene oxide has a width of ~ 1 nm; double-layered has ~ 2 nm. Although their width varies with the degree of oxidation [100 - 102]. Xue and co-workers synthesized 3D framework structures with reduced GO on melamine foam (MF) frameworks [103].

Graphene oxide can be used in the synthesis of the ER dispersed phase since it has reduced electrical conductance and great polarizable property because of functional groups present on edges and planes. Although it has poor stability of dispersion because of the restacking of graphene sheets and in turn constrains utilization in engineering fields.

Hong and Jang reported a solvent exchange method for dispersing graphene oxide sheets in Si to intensify of stable dispersion of graphene sheets in non-polar solvents [104]. Through the drying process, the sheets agglomerated and restacked. Hummers' method was used to synthesize graphene sheets primarily. While for type II, the graphene oxide mixture was first dispersed in alcohol along with Si oil, and then it was centrifugated. They get stable in Si oil resulting in immensely stable graphene oxide suspension. Even after three months span, no specific sedimentation was seen [105].

Fabrication of graphene oxide involves two methods *i.e.*, bottom-up approach where molecules of carbon form pristine graphene, and the other one is a top-down approach where layers of derivatives of graphene are synthesized from C-source, generally from graphite [106, 107]. The bottom-up approach (like epitaxial growth on silicon carbide wafers, chemical vapor deposition, *etc.*) is found to be time-consuming and unstably scalable [108]. Therefore, top-down approaches are much preferable which initially supports the generation of graphene oxide and reduced GO. Brodie [109], Staudenmaier [110], Hummers and Offeman [111] first synthesized graphene oxide through graphitic oxidation *via* different procedures. Hummers and Offeman further worked on it and improved it by using potassium permanganate as an oxidizing agent instead of potassium hypochlorite and adding $NaNO_3$. Usually, Hummers' method is being used because it is scalable and safe for the generation of graphene oxide. Any fabrication process which alters or shows improvement is termed as the "modified Hummers' method" [112]. Usually, a C-source is kept into a protonated solvent (like H_2SO_4, PO_4^{2-}, *etc.*) in the presence of a strong oxidizer (like potassium permanganate). For removing metal ions from it, dilution is performed resulting in a yellow-brown solution followed by separation and addition of HCl. Lastly washing and centrifugation are performed to neutralize its pH.

It is altered according to the researcher's needs. The morphology of GO depends on the dimensions of the carbon source obtained through modified Hummer's method [112]. In other words, the dimensions of synthesized graphene oxide sheets depend on the diameter of the graphite powder. Huang and co-workers fabricated nanoribbons by unfastening carbon nanotubes (multi-walled) [113]. To lower the C/O ratio of the product often graphite is oxidized before the fabrication of graphene oxide. Few researchers follow Kovtyukhova and his co-workers [114], who treat graphite powder in potassium persulphate and phosphorous pentaoxide before using Hummers' method [115 - 117]. Others alter graphite into expanded graphite by thermal treatment or exposure to strong oxidizing agents [118, 119]. The famous alteration of this method is the "improved Hummers' method". It removes $NaNO_2$ *via* the addition of PO_4^{2-} and $KMnO_4$. It provides GO powders with a higher degree of oxidation at ambient temperature conditions and without the evolution of any toxic gas [120]. Novel work has been reported regarding the fabrication of graphene oxide. Zhang and his co-workers fabricated large graphene oxide sheets with an average size of 108 μm *via* stationary oxidation method [121]. The GO prepared by Hummer's method is relatively smaller than its graphite source. For synthesizing graphene oxide flakes of relatable size and morphology to graphite, sulfuric acid and potassium permanganate are added "without external mechanical agitations" hence forming a symmetrical three-dimensional structure later converting it into enlarged graphene oxide through "mild agitation". Enlarged graphene sheets are beneficial to composite materials. Dong and his co-workers reported an explanation for dispensable graphene oxide by forming GO slurries which could be scaled off and stored into flakes [122]. Single graphene oxide sheets bound *via* π-π stacking; although ion adsorption controls the aggregation of graphene oxide. By oxidizing their graphite pre-cursor and releasing it in a highly alkaline medium (pH 14), Vander Waals interaction is dominated by electrostatic repulsion hence, exfoliation leads to a slurry of low-viscosity instead of dispersion. It can be re-dispersed in N-methyl-2-pyrrolidone or alkaline medium, depicting a reliable method for storing and transporting graphene oxide flakes. Properties of graphene oxide and reduced form of graphene oxide given in Table **2**.

Table 2. Properties of graphene oxide and reduced form of graphene oxide.

Properties	GO	GO (in reduced form)
Fracture strength	120 MPa	-
Thermal conductivity	2000Wm^{-1}K for pure 600W m^{-1}K on Si/SiO_2	0.14-0.87W m^{-1}K
Electrical conductivity	10^{-1}S/cm	200-35000S/cm
Young's modulus	220 GPa	-
Optical transmittance	-	depends 60-90% on the synthesizing method and reducing agent

7. QUANTUM DOTS OF GRAPHENE

7.1. Fabrication of Quantum Dots of Graphene

Considering edge effects and quantum confinement of nano-graphene, its new derivatives have been evolved. Nanosheets of graphene having a lateral dimension below 100 nm and multi-layered stacks of graphene together are termed, graphene quantum dots [123]. Because of their flexible bandgap and fluorescence stability, they are of keen interest [123, 124]. Fabrication of these quantum dots requires alteration of graphene oxide and reduced graphene oxide *via* bottom-up and top-down approaches which include microwave-assisted thermal treatment, high-power ultrasonication, hydrothermal and liquid-phase exfoliation [125, 126]. Lu and co-workers reported the eco-friendly one-pot hydrothermal process for the fabrication of quantum dots (90 minutes) using black C [127]. By making use of H_2O_2 they can be made efficiently photo-stable and biologically compatible. Wang and co-workers demonstrated a single-step, one-pot hydrothermal procedure to fabricate quantum dots by making use of rice husk biomass. With the help of light-emitting quenchers, these quantum dots sensed iron's 3þ orbital [128]. A derivative of graphene is too diverse and involves varied fabrication processes and for fabricating graphene oxide, reduced graphene oxide, or quantum dots of graphene there is no "standard" procedure. It opens gates for many exclusive prospects.

8. GRAPHENE AND DERIVED NANOFILLER

8.1. Fabrication of Graphene and Derived Nanofiller

Because of the simple method of preparation, tunable property and wide application conductive polymer composites (CPCs) are of great interest (Fig. **5**). For synthesizing conductive polymer composites from non-conducting polymers, conducting fillers are inserted into them. They increase the conductivity of the material. This is known as an electrical percolation threshold (Pc). Both theoretically and experimentally it was found that with an increase in filler aspect ratio electrical percolation threshold decreases. So, by employing large aspect ratio conducting fillers conducting networks in a matrix can be synthesized [129]. Graphene composites possess great mechanical properties, extraordinary thermal conductivity, and higher electron portability at room temperature. For corresponding to electrical conductivity polymer composites use a 3D network. Wu and co-workers used a hot press method for the preparation of polystyrene composites having a three-dimensional network [130]. Chen and co-workers fabricated an excellent conducting composite film with the help of three-dimensional graphene foam *via* template-assisted chemical vapor deposition method [131].

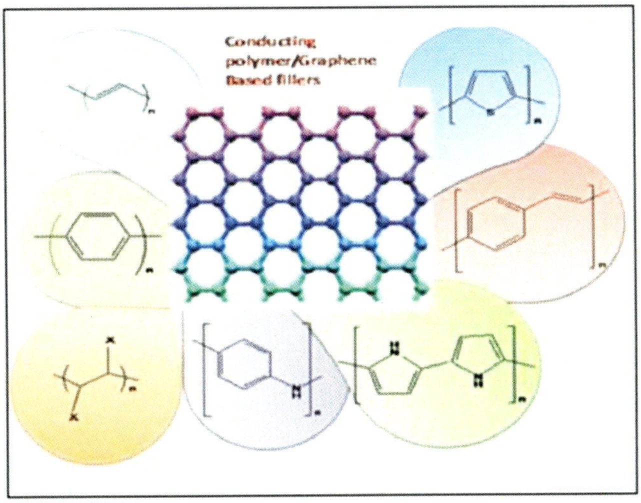

Fig. (5). Conducting graphene composite.

8.2. Significance of Graphene and Derived Nanofiller

As a nanofiller, graphene is preferred over other common nanofillers like exfoliated graphite, sodium montmorillonite, carbon nanotube, carbon nanofiber, and because of its higher aspect ratio, great surface area, low coefficient of thermal expansion tensile strength, electrical and thermal conductivity, transparency, flexibility, and shielding ability [132].

8.3. Application of Graphene and Derived Nanofiller in Energy and Electronics

Because of extraordinary microelectronic chips, high power densities they require conductivity issued as thermal pads, heat spreaders, and thermal interface materials. They are necessary compounds of thermal packaging and are responsible for maintaining a great relationship between metal sinks and chips [133]. The most important characteristic of graphene is modern electrical conductivity.

By inserting graphene into polymer matrix conductive polymer composites are being synthesized. Various polymers specifically polyvinyl chloride, polyvinyl alcohol, polymethyl methacrylate, polyethylene, polyamide, and polystyrene worked as matrices to fabricate graphene composites. They generally demonstrate a non-linear improvement of the electrical conductance as a means of concentration of filler. They can be used to make sensors, solar cells, *etc.*

9. GRAPHENE AEROGELS (GAS)

It is a three-dimensional structure of graphene that produces porous and cheap structures of graphene. Various procedures including chemical-reduction methods [134, 135], hydrothermal methods [136, 137] Resorcinol (R), and Formaldehyde (F) sol-gel chemistry methods [138, 139] were suggested for the preparation of aerogels and monolithic aerogels on the basis of graphene oxide [140]. The chemical reduction method was found to be the most appropriate one since it is an eco-friendly and low-cost method. This method requires a low temperature without the need for a binder. Zhang and co-workers showed high electrical conductance of 102 S/m and a large surface area of 512 m^2g^{-1} was observed in synthesizing graphene aerogels through the chemical-reduction method [135].

10. HYBRID GRAPHENE/MICROFIBER COMPOSITES

An encouraging advance to ready graphene-based composites is by utilization of hybrid fillers that comprises an inorganic material and graphene-based materials. The hybrid combination gives an advantage such as synergistic effects between the fillers or final properties consequential from preservatives. The hybridization process will come across a few drawbacks of filler and progresses its interplay with the host matrix, it depends on its functioning criteria. Multifunctionality plays an important role in manufacturing composites (hybrid) since combined properties of diverse fillers correspond to a substance having a dissimilar set of characteristics as contrasted to characteristics of individually fabricated materials. In addition to this, the value of the ultimate product can be decreased by making use of fine microscale reinforcements like C and fibers of glass having lower levels of graphene where stress moves from micro to nano-scale reinforcement, hence improving the final properties.

Hybridizing graphene materials and the rest inorganic group showed outstanding properties at an atomic scale or nano or synergy can be exposed with complementarities. Hybrid graphene and other-two dimensional structures maintained the electronic properties and specifically, it can alter the bandgap of graphene. The zero-band gap of the unique graphene is unsuccessful in semiconductor materials; it is advantageous to its composites which can help in the manufacturing of graphene-based electronics. The addition of carbon to h-boron nitride increases electric conductance, which acts as an insulating agent in pristine form. The inductive band gap in graphene can effectively fabricate field-effect transistors of graphene/ boron nitride films having 190-2000 $cm^2V^{-1}s^{-1}$ carrier mobilities. Cross-linking of graphene *via* chemicals and h-boron nitride composites could efficiently store energy. The (h-BN) Supercapacitors of graphene or boron nitride have B240 F/g specific capacitance. The bond formed between graphene and individual fillers affects the resulting physical and chemical characteristics of synthesized reinforced composites. To improve the efficiency amongst fillers various procedures were followed like solution blending, functionalization *via* chemicals, mechanical mixing, alternate growth, and many more. Aggregation of graphene takes place because of higher specific surface area and van der Waals interaction. The ornament of graphene is preferred since it avoids aggregation with other nanomaterials. Graphene's reactive sites got improved because of its functionalization which further improved its bond formation capability with the rest of the nanomaterials and finally led to the advancement of less significant covalent bonding between the matrix of host and hybrid-filler.

The functioning of multi-walled carbon nanotubes established effective bonds between multi-graphene platelets and fillers. The two-dimensional filler *i.e.*, multi-graphene platelets and one-dimensional filler *i.e.*, multiwalled carbon nanotubes combine to form a three-dimensional structure having a higher surface area. Thermal conductance of functionalized multiwalled carbon nanotube showed enlargement of >50% in comparison to pristine hybrid filler. Lin and co-workers accounted for the synthesis of Si/reduced GO hybrids through the electrostatic mechanism and then it is mechanical blending with BUNA-S (styrene-butadiene rubber) matrix. Great experimental distribution has been seen in hybrid materials in comparison to individual fillers having great matrix interaction.

11. GRAPHENE BIOACTIVE COMPOSITES

Various bio-sensors which depend on graphene for different sensing purposes consist of separate criteria for electrochemical and optical signals. Because of its quick response, elevated sensitivity, excellent electrochemical properties, straightforward operation, and low-cost graphene is used as a probable contender for biomolecule discovery. Zhou and co-workers examined the electrical and chemical activities of the graphene-modified electrode towards H_2O_2. In comparison to the exposed electrode, noteworthy enlargement in the rate of electron transmission of graphene-based electrodes has been obtained. Shao and co-workers accounted for superior electro-catalytic progress of (N-doped graphene) N-graphene towards reduction of H_2O_2 in comparison to graphene, responsible for the occurrence of N- containing groups, O- functional groups, and structural defects. GO possess fluorescence over a broad array of wavelength *i.e.*, 200 -1200 nm and it successfully quenched the light-emitting property of other fluorescing dyes.

12. CONCLUDING REMARKS

This 2D material which seemed to be of least research in past decades had now become a boon to society since it provides unique and reliable solutions to globally challenged environmental problems. In this chapter, we have briefly discussed its properties and application areas which makes it more content for scientists to depend on. The distinct and captivating properties of graphene make it unique in a way that could lead to substantial advancement in the near future. Its properties make it one of the most reliable materials in various applicable areas which include sensing, tissue engineering, electronics, *etc.* Although many interdisciplinary areas still seek its contribution in different ways which need to be discovered which will further accelerate a better understanding of graphene-based materials. Its properties have paced new ways for scientists working for the betterment of the global environmental progressions.

CONSENT FOR PUBLICATION

Not applicable.

CONFLICT OF INTEREST

The author declares no conflict of interest, financial or otherwise.

ACKNOWLEDGEMENTS

The authors would like to acknowledge Banasthali Vidyapith for providing an internet facility for the completion of this chapter. One of the authors Dr. Nirmala Kumari Jangid is thankful to Dr. Pinki Bala Punjabi and all editors for allowing her to write this chapter.

LIST OF ABBREVIATIONS

AFM	Atomic/Scanning force microscopy
BUNA-S	Styrene-butadiene rubber
CCl_4	Carbon tetrachloride
CM	Chemical enhancement mechanism
CNT	Carbon nanotubes
CPCs	Conductive polymer composites

CTAB	Cetyltrimethylammonium bromide
CVD	Chemical vapor deposition
DNA	Deoxyribonucleic acid
EM	Electromagnetic enhancement mechanism
FET	Field-effect transistor
GO	Graphene oxide
HOMO	Highest occupied molecular orbital
IUPAC	International Union of Pure and Applied Chemistry
IR	Infra-red radiations
LA mode phonons	Longitudinal acoustic phonons
LO	Longitudinal optical phonons
LUMO	Lowest unoccupied molecular orbital
MGP	Multi graphene platelet
MF	Melamine foam
MWCNTs	Multi-walled carbon nanotubes
PC	Percolation threshold
PEG	Polyethylene glycol
PYBS	1- pyrenebutanoic acid succinimidyl ester
PTCDA	Perylenetetracarboxylic dianhydride
SDS	Sodium dodecyl sulfate
SERS	Surface Enhanced Raman Spectroscopy
STM	Scanning tunneling microscopy
TA mode phonons	Transverse acoustic phonons
TEM	Transmission electron microscopy
THF	Tetrahydrofuran
TO	Transverse optical phonons
TPa	Terapascal
QHE	Quantum Hall effect
RNA	Ribonucleic acid
ZA mode phonons	Out of plane acoustic phonons
ZO	Out of plane optical phonons
2D	Two-dimensional

REFERENCES

[1] Geim, A.K. Graphene: Status and Prospects. *Science,* **2009**, *324*(5934), 1530-1534. [http://dx.doi.org/10.1126/science.1158877] [PMID: 19541989]

[2] Whitener, K.E., Jr; Sheehan, P.E. Graphene synthesis. *Diamond Related Materials,* **2014**, *46*, 25-34. [http://dx.doi.org/10.1016/j.diamond.2014.04.006]

[3] Choi, W.; Lahiri, I.; Seelaboyina, R.; Kan, Y.S. Synthesis of Graphene and Its Applications: A Review. *Crit. Rev.,* **2010**, 52-71.

[4] Wallace, P.R. The band theory of graphite. *Phys. Rev.,* **1947**, *71*(9), 622-634. [http://dx.doi.org/10.1103/PhysRev.71.622]

[5] Peierls, R.E. Quelques proprietes typiques des corpses solides. *Annales de l'Institut Henri Poincaré,* **1935**, *5*, 177-222.

[6] Landau, L.D. Zur theorie der phasenumwandlungen II, Physikalische Zeitschri under Sowjetunion. **1937**, 26-35.

[7] Evans, J.W.; Thiel, P.A.; Bartelt, M.C. Morphological evolution during epitaxial thin film growth: formation of 2D islands and 3D mounds, Surface Science Reports. **2006**, 1-28.

[8] Novoselov, K.S.; Geim, A.K.; Morozov, S.V.; Jiang, D.; Zhang, Y.; Dubonos, S.V.; Grigorieva, I.V.; Firsov, A.A. Electric field effect in atomically thin carbon films. *Science,* **2004**, *306*(5696), 666-669. [http://dx.doi.org/10.1126/science.1102896] [PMID: 15499015]

[9] Iijima, S. Helical microtubules of graphitic carbon. *Nature,* **1991**, *354*(6348), 56-58. [http://dx.doi.org/10.1038/354056a0]

[10] Kroto, H.W.; Heath, J.R.; O'Brien, S.C.; Curl, R.F.; Smalley, R.E. C60: Buckminsterfullerene. *Nature,* **1985**, *318*(6042), 162-163. [http://dx.doi.org/10.1038/318162a0]

[11] Lincoln Vogel, F. The electrical conductivity of graphite intercalated with superacid fluorides: experiments with antimony pentafluoride. *J. Mater. Sci.,* **1977**, *12*(5), 982-986. [http://dx.doi.org/10.1007/BF00540981]

[12] Novoselov, K.S.; Geim, A.K.; Morozov, S.V.; Jiang, D.; Zhang, Y.; Dubonos, S.V.; Grigorieva, I.V.; Firsov, A.A. Electric field effect in atomically thin carbon films. *Science,* **2004**, *306*(5696), 666-669. [http://dx.doi.org/10.1126/science.1102896] [PMID: 15499015]

[13] Novoselov, K.S.; Geim, A.K.; Morozov, S.V.; Jiang, D.; Katsnelson, M.I.; Grigorieva, I.V.; Dubonos, S.V.; Firsov, A.A. Two-dimensional gas of massless Dirac fermions in graphene. *Nature,* **2005**, *438*(7065), 197-200. [http://dx.doi.org/10.1038/nature04233] [PMID: 16281030]

[14] Zhang, Y.; Tan, Y.W.; Stormer, H.L.; Kim, P. Experimental observation of the quantum Hall effect and Berry's phase in graphene. *Nature,* **2005**, *438*(7065), 201-204. [http://dx.doi.org/10.1038/nature04235] [PMID: 16281031]

[15] Balandin, A.A.; Ghosh, S.; Bao, W.; Calizo, I.; Teweldebrhan, D.; Miao, F.; Lau, C.N. Superior thermal conductivity of single-layer graphene. *Nano Lett.,* **2008**, *8*(3), 902-907. [http://dx.doi.org/10.1021/nl0731872] [PMID: 18284217]

[16] Nair, R.R.; Blake, P.; Grigorenko, A.N.; Novoselov, K.S.; Booth, T.J.; Stauber, T.; Peres, N.M.R.; Geim, A.K. Fine structure constant defines visual transparency of graphene. *Science,* **2008**, *320*(5881), 1308. [http://dx.doi.org/10.1126/science.1156965] [PMID: 18388259]

[17] Li, X.; Wang, X.; Zhang, L.; Lee, S.; Dai, H. Chemically derived, ultrasmooth graphene nanoribbon semiconductors. *Science,* **2008**, *319*(5867), 1229-1232. [http://dx.doi.org/10.1126/science.1150878] [PMID: 18218865]

[18] Ritter, K.A.; Lyding, J.W. The influence of edge structure on the electronic properties of graphene quantum dots and nanoribbons. *Nat. Mater.*, **2009**, *8*(3), 235-242. [http://dx.doi.org/10.1038/nmat2378] [PMID: 19219032]

[19] Cai, J.; Ruffieux, P.; Jaafar, R.; Bieri, M.; Braun, T.; Blankenburg, S.; Muoth, M.; Seitsonen, A.P.; Saleh, M.; Feng, X.; Müllen, K.; Fasel, R. Atomically precise bottom-up fabrication of graphene nanoribbons. *Nature,* **2010**, *466*(7305), 470-473. [http://dx.doi.org/10.1038/nature09211] [PMID: 20651687]

[20] Ansari, S.; Giannelis, E.P. Functionalized graphene sheet-Poly(vinylidene fluoride) conductive nanocomposites. *J. Polym. Sci., B, Polym. Phys.,* **2009**, *47*(9), 888-897. [http://dx.doi.org/10.1002/polb.21695]

[21] Ramanathan, T.; Abdala, A.A.; Stankovich, S.; Dikin, D.A.; Herrera-Alonso, M.; Piner, R.D.; Adamson, D.H.; Schniepp, H.C.; Chen, X.; Ruoff, R.S.; Nguyen, S.T.; Aksay, I.A.; Prud'Homme, R.K.; Brinson, L.C. Functionalized graphene sheets for polymer nanocomposites. *Nat. Nanotechnol.,* **2008**, *3*(6), 327-331. [http://dx.doi.org/10.1038/nnano.2008.96]

[22] Stankovich, S.; Dikin, D.A.; Dommett, G.H.B.; Kohlhaas, K.M.; Zimney, E.J.; Stach, E.A.; Piner, R.D.; Nguyen, S.T.; Ruoff, R.S. Graphene-based composite materials. *Nature,* **2006**, *442*(7100), 282-286. [http://dx.doi.org/10.1038/nature04969] [PMID: 16855586]

[23] Fan, H.; Wang, L.; Zhao, K.; Li, N.; Shi, Z.; Ge, Z.; Jin, Z. Fabrication, mechanical properties, and biocompatibility of graphene-reinforced chitosan composites. *Biomacromolecules,* **2010**, *11*(9), 2345-2351. [http://dx.doi.org/10.1021/bm100470q] [PMID: 20687549]

[24] Zhang, K.; Zhang, L.L.; Zhao, X.S.; Wu, J. Graphene/Polyaniline Nanofiber Composites as Supercapacitor Electrodes. *Chem. Mater.,* **2010**, *22*(4), 1392-1401. [http://dx.doi.org/10.1021/cm902876u]

[25] Zhao, X.; Zhang, Q.; Chen, D.; Lu, P. Enhanced Mechanical Properties of Graphene-Based Poly(vinyl alcohol) Composites. *Macromolecules,* **2010**, *43*(5), 2357-2363. [http://dx.doi.org/10.1021/ma902862u]

[26] Kuila, T.; Bose, S.; Hong, C.E.; Uddin, M.E.; Khanra, P.; Kim, N.H.; Lee, J.H. Preparation of functionalized graphene/linear low density polyethylene composites by a solution mixing method. *Carbon,* **2011**, *49*(3), 1033-1037. [http://dx.doi.org/10.1016/j.carbon.2010.10.031]

[27] Wu, J.; Becerril, H.A.; Bao, Z.; Liu, Z.; Chen, Y.; Peumans, P. Organic solar cells with solution-processed graphene transparent electrodes. *Appl. Phys. Lett.,* **2008**, *92*(26), 263302. [http://dx.doi.org/10.1063/1.2924771]

[28] Huang, J.; Wang, X.; deMello, A.J.; deMello, J.C.; Bradley, D.D.C. Efficient flexible polymer light emitting diodes with conducting polymer anodes. *J. Mater. Chem.,* **2007**, *17*(33), 3551-3554. [http://dx.doi.org/10.1039/b705918n]

[29] Park, H.; Rowehl, J.A.; Kim, K.K.; Bulovic, V.; Kong, J. Doped graphene electrodes for organic solar cells. *Nanotechnology,* **2010**, *21*(50), 505204. [http://dx.doi.org/10.1088/0957-4484/21/50/505204] [PMID: 21098945]

[30] Li, D.; Müller, M.B.; Gilje, S.; Kaner, R.B.; Wallace, G.G. Processable aqueous dispersions of graphene nanosheets. *Nat. Nanotechnol.,* **2008**, *3*(2), 101-105. [http://dx.doi.org/10.1038/nnano.2007.451] [PMID: 18654470]

[31] Shan, C.; Yang, H.; Han, D.; Zhang, Q.; Ivaska, A.; Niu, L. Water-soluble graphene covalently functionalized by biocompatible poly-L-lysine. *Langmuir,* **2009**, *25*(20), 12030-12033. [http://dx.doi.org/10.1021/la903265p] [PMID: 19769343]

[32] McNaught, A.D.; Wilkinson, A. *Compendium of Chemical Terminology*; Blackwell Scientific Publications: Oxford, **1997**.

[33] Boehm, H.P.; Setton, R.; Stumpp, E. Nomenclature and terminology of graphite intercalation compounds (IUPAC Recommendations 1994). *Pure Appl. Chem.*, **1994**, *66*(9), 1893-1901. [http://dx.doi.org/10.1351/pac199466091893]

[34] Mouras, S.; Hamm, A.; Djurado, D.; Cousseins, J.C. Synthesis of first-stage graphite intercalation compounds with fluorides. *Revue de Chimie Minerale,* **1987**, *24*, 572-582.

[35] Bourzac, K. *Graphene wines nobel prize*; MIT Technology Review, **2010**.

[36] Narita, A.; Chen, Z.; Chen, Q.; Müllen, K. Solution and on-surface synthesis of structurally defined graphene nanoribbons as a new family of semiconductors. *Chem. Sci. (Camb.),* **2019**, *10*(4), 964-975. [http://dx.doi.org/10.1039/C8SC03780A] [PMID: 30774890]

[37] Clair, S.; de Oteyza, D.G. Controlling a Chemical Coupling Reaction on a Surface: Tools and Strategies for On-Surface Synthesis. *Chem. Rev.,* **2019**, *119*(7), 4717-4776. [http://dx.doi.org/10.1021/acs.chemrev.8b00601] [PMID: 30875199]

[38] *Graphene and doped graphene elaborated by pulsed laser deposition*; HAL Open Sciences, **2021**.

[39] Li, D.; Müller, M.B.; Gilje, S.; Kaner, R.B.; Wallace, G.G. Processable aqueous dispersions of graphene nanosheets. *Nat. Nanotechnol.,* **2008**, *3*(2), 101-105. [http://dx.doi.org/10.1038/nnano.2007.451] [PMID: 18654470]

[40] Si, Y.; Samulski, E.T. Synthesis of water soluble graphene. *Nano Lett.,* **2008**, *8*(6), 1679-1682. [http://dx.doi.org/10.1021/nl080604h] [PMID: 18498200]

[41] Smith, A.T.; LaChance, A.M.; Zeng, S.; Liu, B.; Sun, L. Synthesis, properties, and applications of graphene oxide/reduced graphene oxide and their nanocomposites. *Nano Materials Science,* **2019**, *1*(1), 31-47. [http://dx.doi.org/10.1016/j.nanoms.2019.02.004]

[42] Wang, Y.; Shi, Z.; Huang, Y.; Ma, Y.; Wang, C.; Chen, M.; Chen, Y. Supercapacitor devices based on graphene materials. *J. Phys. Chem. C,* **2009**, *113*(30), 13103-13107. [http://dx.doi.org/10.1021/jp902214f]

[43] Tomori, H.; Kanda, A.; Goto, H.; Ootuka, Y.; Tsukagoshi, K.; Moriyama, S.; Watanabe, E.; Tsuya, D. Introducing Nonuniform Strain to Graphene Using Dielectric Nanopillars. *Appl. Phys. Express,* **2011**, 4-7. [http://dx.doi.org/10.1143/APEX.4.075102]

[44] Pop, E.; Varshney, V.; Roy, A.K. Thermal properties of graphene: Fundamentals and applications. *MRS Bull.,* **2012**, *37*(12), 1273-1281. [http://dx.doi.org/10.1557/mrs.2012.203]

[45] Nair, R.R.; Wu, H.A.; Jayaram, P.N.; Grigorieva, I.V.; Geim, A.K. Unimpeded permeation of water through helium-leak-tight graphene-based membranes. *Science,* **2012**, *335*(6067), 442-444. [http://dx.doi.org/10.1126/science.1211694] [PMID: 22282806]

[46] Frank, I.W.; Tanenbaum, D.M.; van der Zande, A.M.; McEuen, P.L. Microelectronics and nanometer structures. J. *J. Vac. Sci. Technol. B,* **2007**, *25*, 2558-2561. [http://dx.doi.org/10.1116/1.2789446]

[47] Roos, W.H. How to perform a nanoindentation experiment on a virus. *Methods Mol. Biol.,* **2011**, *783*, 251-264. [http://dx.doi.org/10.1007/978-1-61779-282-3_14] [PMID: 21909893]

[48] Zou, J.H.; Ye, Z.Q. Phonon thermal properties of graphene from molecular dynamics using different potentials. *J. Chem. Phys,* **2016**, *145*, 134705.

[49] Singh, V.; Joung, D.; Zhai, L.; Das, S.; Khondaker, S.I.; Seal, S. Graphene based materials: Past,

present and future. *Prog. Mater. Sci.*, **2011**, *56*(8), 1178-1271.
[http://dx.doi.org/10.1016/j.pmatsci.2011.03.003]

[50] Chen, S.; Wu, Q.; Mishra, C. Thermal Properties of Isotopically Engineered Graphene. *Nat. Mater.*, **2012**, *11*, 203-207.
[http://dx.doi.org/10.1038/nmat3207] [PMID: 22231598]

[51] Sang, M.; Shin, J.; Kim, K.; Yu, K. Electronic and Thermal Properties of Graphene and Recent Advances in Graphene Based Electronics Applications. *Nanomaterials (Basel)*, **2019**, *9*(3), 374.
[http://dx.doi.org/10.3390/nano9030374] [PMID: 30841599]

[52] Nair, R.R.; Blake, P.; Grigorenko, A.N.; Novoselov, K.S.; Booth, T.J.; Stauber, T.; Peres, N.M.R.; Geim, A.K. Fine structure constant defines visual transparency of graphene. *Science*, **2008**, *320*(5881), 1308.
[http://dx.doi.org/10.1126/science.1156965] [PMID: 18388259]

[53] Kravets, V.G.; Grigorenko, A.N.; Nair, R.R.; Blake, P.; Anissimova, S.; Novoselov, K.S.; Geim, A.K. Spectroscopic ellipsometry of graphene and an exciton-shifted van Hove peak in absorption. *Phys. Rev. B Condens. Matter Mater. Phys.*, **2010**, *81*(15), 155413.
[http://dx.doi.org/10.1103/PhysRevB.81.155413]

[54] Feng Wang, ; Hamdi, M. Strictly non-blocking conditions for the central-stage buffered Clos-network. *IEEE Commun. Lett.*, **2008**, *12*(3), 206-208.
[http://dx.doi.org/10.1109/LCOMM.2008.071752]

[55] Rana, F.; George, P.A.; Strait, J.H.; Dawlaty, J.; Shivaraman, S.; Chandrashekhar, M.; Spencer, M.G. Carrier recombination and generation rates for intravalley and intervalley phonon scattering in graphene. *Phys. Rev. B Condens. Matter Mater. Phys.*, **2009**, *79*(11), 115447.
[http://dx.doi.org/10.1103/PhysRevB.79.115447]

[56] Park, S.; Ruoff, R.S. Chemical methods for the production of graphenes. *Nat. Nanotechnol.*, **2009**, *4*(4), 217-224.
[http://dx.doi.org/10.1038/nnano.2009.58] [PMID: 19350030]

[57] Elias, D.C.; Nair, R.R.; Mohiuddin, T.M.G.; Morozov, S.V.; Blake, P.; Halsall, M.P.; Ferrari, A.C.; Boukhvalov, D.W.; Katsnelson, M.I.; Geim, A.K.; Novoselov, K.S. Control of graphene's properties by reversible hydrogenation: evidence for graphane. *Science*, **2009**, *323*(5914), 610-613.
[http://dx.doi.org/10.1126/science.1167130] [PMID: 19179524]

[58] Bonaccorso, F.; Sun, Z.; Hasan, T.; Ferrari, A.C. Graphene photonics and optoelectronics. *Nat. Photonics*, **2010**, *4*(9), 611-622.
[http://dx.doi.org/10.1038/nphoton.2010.186]

[59] Ma, X.; Tao, H.; Yang, K.; Feng, L.; Cheng, L.; Shi, X.; Li, Y.; Guo, L.; Liu, Z. A functionalized graphene oxide-iron oxide nanocomposite for magnetically targeted drug delivery, photothermal therapy, and magnetic resonance imaging. *Nano Res.*, **2012**, *5*(3), 199-212.
[http://dx.doi.org/10.1007/s12274-012-0200-y]

[60] Shen, A.J.; Li, D.L.; Cai, X.J.; Dong, C.Y.; Dong, H.Q.; Wen, H.Y.; Dai, G.H.; Wang, P.J.; Li, Y.Y. Multifunctional nanocomposite based on graphene oxide for *in vitro* hepatocarcinoma diagnosis and treatment. *J. Biomed. Mater. Res. A*, **2012**, *100*, 9.
[http://dx.doi.org/10.1002/jbm.a.34148] [PMID: 22623284]

[61] Sanchez, V.C.; Jachak, A.; Hurt, R.H.; Kane, A.B. Biological interactions of graphene-family nanomaterials: an interdisciplinary review. *Chem. Res. Toxicol.*, **2012**, *25*(1), 15-34.
[http://dx.doi.org/10.1021/tx200339h] [PMID: 21954945]

[62] Bianco, A. Graphene: safe or toxic? The two faces of the medal. *Angew. Chem. Int. Ed.*, **2013**, *52*(19), 4986-4997.
[http://dx.doi.org/10.1002/anie.201209099] [PMID: 23580235]

[63] Ren, H.; Wang, C.; Zhang, J.; Zhou, X.; Xu, D.; Zheng, J.; Guo, S.; Zhang, J. DNA cleavage system of

nanosized graphene oxide sheets and copper ions. *ACS Nano,* **2010**, *4*(12), 7169-7174. [http://dx.doi.org/10.1021/nn101696r] [PMID: 21082807]

[64] Lu, C.H.; Zhu, C.L.; Li, J.; Liu, J.J.; Chen, X.; Yang, H.H. Using graphene to protect DNA from cleavage during cellular delivery. *Chem. Commun. (Camb.),* **2010**, *46*(18), 3116-3118. [http://dx.doi.org/10.1039/b926893f] [PMID: 20424750]

[65] Wang, Y.; Li, Z.; Hu, D.; Lin, C.T.; Li, J.; Lin, Y. Aptamer/graphene oxide nanocomplex for *in situ* molecular probing in living cells. *J. Am. Chem. Soc.,* **2010**, *132*(27), 9274-9276. [http://dx.doi.org/10.1021/ja103169v] [PMID: 20565095]

[66] Xu, Y.; Wu, Q.; Sun, Y.; Bai, H.; Shi, G. Three-dimensional self-assembly of graphene oxide and DNA into multifunctional hydrogels. *ACS Nano,* **2010**, *4*(12), 7358-7362. [http://dx.doi.org/10.1021/nn1027104] [PMID: 21080682]

[67] Wu, M.; Kempaiah, R.; Huang, P.J.J.; Maheshwari, V.; Liu, J. Adsorption and desorption of DNA on graphene oxide studied by fluorescently labeled oligonucleotides. *Langmuir,* **2011**, *27*(6), 2731-2738. [http://dx.doi.org/10.1021/la1037926] [PMID: 21302946]

[68] Titov, A.V.; Král, P.; Pearson, R. Sandwiched graphene--membrane superstructures. *ACS Nano,* **2010**, *4*(1), 229-234. [http://dx.doi.org/10.1021/nn9015778] [PMID: 20025267]

[69] Rao, C.N.R.; Subrahmanyam, K.S.; Ramakrishna Matte, H.S.S.; Govindaraj, A. Graphene: synthesis, functionalization, and properties. *Mod. Phys. Lett. B,* **2011**, *25*(7), 427-451. [http://dx.doi.org/10.1142/S0217984911025961]

[70] Niyogi, S.; Bekyarova, E.; Itkis, M.E.; McWilliams, J.L.; Hamon, M.A.; Haddon, R.C. Solution properties of graphite and graphene. *J. Am. Chem. Soc.,* **2006**, *128*(24), 7720-7721. [http://dx.doi.org/10.1021/ja060680r] [PMID: 16771469]

[71] Si, Y.; Samulski, E.T. Exfoliated Graphene Separated by Platinum Nanoparticles. *Chem. Mater.,* **2008**, *20*(21), 6792-6797. [http://dx.doi.org/10.1021/cm801356a]

[72] Ghosh, A.; Rao, K.V.; George, S.J.; Rao, C.N.R. Noncovalent functionalization, exfoliation, and solubilization of graphene in water by employing a fluorescent coronene carboxylate. *Chemistry,* **2010**, *16*(9), 2700-2704. [http://dx.doi.org/10.1002/chem.200902828] [PMID: 20108284]

[73] Subrahmanyam, K.S.; Vivekchand, S.R.C.; Govindaraj, A.; Rao, C.N.R. A study of graphenes prepared by different methods: characterization, properties and solubilization. *J. Mater. Chem.,* **2008**, *18*(13), 1517-1523. [http://dx.doi.org/10.1039/b716536f]

[74] Liu, N.; Luo, F.; Wu, H.; Liu, Y.; Zhang, C.; Chen, J. One-step ionic-liquid-assisted electrochemical synthesis of ionic-liquid-functionalized graphene sheets directly from graphite. *Adv. Funct. Mater.,* **2008**, *18*(10), 1518-1525. [http://dx.doi.org/10.1002/adfm.200700797]

[75] Ghosh, A.; Subrahmanyam, K.S.; Krishna, K.S.; Datta, S.; Govindaraj, A.; Pati, S.K.; Rao, C.N.R. Uptake of H_2 and CO_2 by Graphene. *J. Phys. Chem. C,* **2008**, *112*(40), 15704-15707. [http://dx.doi.org/10.1021/jp805802w]

[76] Xu, Y.; Zhao, L.; Bai, H.; Hong, W.; Li, C.; Shi, G. Chemically converted graphene induced molecular flattening of 5,10,15,20-tetrakis(1-methyl-4-pyridinio)porphyrin and its application for optical detection of cadmium(II) ions. *J. Am. Chem. Soc.,* **2009**, *131*(37), 13490-13497. [http://dx.doi.org/10.1021/ja905032g] [PMID: 19711892]

[77] Ramakrishna Matte, H.S.S.; Subrahmanyam, K.S.; Venkata Rao, K.; George, S.J.; Rao, C.N.R. Quenching of fluorescence of aromatic molecules by graphene due to electron transfer. *Chem. Phys. Lett.,* **2011**, *506*(4-6), 260-264.

[http://dx.doi.org/10.1016/j.cplett.2011.03.031]

[78] Fleischmann, M.; Hendra, P.J.; McQuillan, A.J. Raman spectra of pyridine adsorbed at a silver electrode. *Chem. Phys. Lett.*, **1974**, *26*(2), 163-166. [http://dx.doi.org/10.1016/0009-2614(74)85388-1]

[79] Ling, X.; Xie, L.; Fang, Y.; Xu, H.; Zhang, H.; Kong, J.; Dresselhaus, M.S.; Zhang, J.; Liu, Z. Can graphene be used as a substrate for Raman enhancement? *Nano Lett.*, **2010**, *10*(2), 553-561. [http://dx.doi.org/10.1021/nl903414x] [PMID: 20039694]

[80] Manna, A.K.; Pati, S.K. Tuning the electronic structure of graphene by molecular charge transfer: a computational study. *Chem. Asian J.*, **2009**, *4*(6), 855-860. [http://dx.doi.org/10.1002/asia.200800486] [PMID: 19322803]

[81] Xia, J.; Chen, F.; Li, J.; Tao, N. Measurement of the quantum capacitance of graphene. *Nat. Nanotechnol.*, **2009**, *4*(8), 505-509. [http://dx.doi.org/10.1038/nnano.2009.177] [PMID: 19662012]

[82] Shekhirev, M.; Lipatov, A.; Torres, A.; Vorobeva, N.S.; Harkleroad, A.; Lashkov, A.; Sysoev, V.; Sinitskii, A. Highly Selective Gas Sensors Based on Graphene Nanoribbons Grown by Chemical Vapor Deposition. *ACS Appl. Mater. Interfaces*, **2020**, *12*(6), 7392-7402. [http://dx.doi.org/10.1021/acsami.9b13946] [PMID: 32011111]

[83] aYousefi, N.; Sun, X.; Lin, X.; Shen, X.; Jia, J.; Zhang, B.; Tang, B.; Chan, M.; Kim, J.K. Highly aligned graphene/polymer nanocomposites with excellent dielectric properties for high-performance electromagnetic interference shielding. *Adv. Mater.*, **2014**, *26*(31), 5480-5487. [http://dx.doi.org/10.1002/adma.201305293] [PMID: 24715671] bTang, L.; He, M.; Na, X.; Guan, X.; Zhang, R.; Zhang, J.; Gu, J. Functionalized glass fibers cloth/spherical BN fillers/epoxy laminated composites with excellent thermal conductivities and electrical insulation properties. *Composites Communications*, **2019**, *16*, 5-10. [http://dx.doi.org/10.1016/j.coco.2019.08.007]

[84] Rohini, R.; Bose, S. Electromagnetic wave suppressors derived from crosslinked polymer composites containing functional particles: Potential and key challenges. *Nano-Structures & Nano-Objects*, **2017**, *12*, 130-146. [http://dx.doi.org/10.1016/j.nanoso.2017.09.016]

[85] Maidhily, M.; Rajalakshmi, N.; Dhathathreyan, K.S. Electrochemical impedance diagnosis of micro porous layer in polymer electrolyte membrane fuel cell electrodes. *Int. J. Hydrogen Energy*, **2011**, *36*(19), 12352-12360. [http://dx.doi.org/10.1016/j.ijhydene.2011.06.084]

[86] Subrahmanyam, K.S.; Kumar, P.; Maitra, U.; Govindaraj, A.; Hembram, K.P.S.S.; Waghmare, U.V.; Rao, C.N.R. Chemical storage of hydrogen in few-layer graphene. *Proc. Natl. Acad. Sci. USA*, **2011**, *108*(7), 2674-2677. [http://dx.doi.org/10.1073/pnas.1019542108] [PMID: 21282617]

[87] Park, S.; An, J.; Suk, J.W.; Ruoff, R.S. Graphene-Based Actuators. *Small*, **2010**, *6*(2), 210-212. [http://dx.doi.org/10.1002/smll.200901877] [PMID: 19924742]

[88] Xie, X.; Qu, L.; Zhou, C.; Li, Y.; Zhu, J.; Bai, H.; Shi, G.; Dai, L. An asymmetrically surface-modified graphene film electrochemical actuator. *ACS Nano*, **2010**, *4*(10), 6050-6054. [http://dx.doi.org/10.1021/nn101563x] [PMID: 20828146]

[89] Wu, C.; Feng, J.; Peng, L.; Ni, Y.; Liang, H.; He, L.; Xie, Y. Large-area graphene realizing ultrasensitive photothermal actuator with high transparency: new prototype robotic motions under infrared-light stimuli. *J. Mater. Chem.*, **2011**, *21*(46), 18584-18591. [http://dx.doi.org/10.1039/c1jm13311j]

[90] Nagamuthu, S.; Ryu, K.S. Synthesis of Ag/NiO honeycomb structured nanoarrays as the electrode material for high-performance asymmetric supercapacitor devices, Sci. *Sci. Rep.*, **2019**, *9*(1), 4864. [http://dx.doi.org/10.1038/s41598-019-41446-0] [PMID: 30890740]

[91] Borenstein, A.; Hanna, O.; Attias, R.; Luski, S.; Brousse, T.; Aurbach, D. Carbon-based composite materials for supercapacitor electrodes: a review. *J. Mater. Chem. A Mater. Energy Sustain.*, **2017**, *5*(25), 12653-12672.
[http://dx.doi.org/10.1039/C7TA00863E]

[92] Li, H.; Wu, X.; Zhou, J.; Liu, Y.; Huang, M.; Xing, W.; Yan, Z.; Zhuo, S. Enhanced supercapacitive performance of $MnCO_3$@ rGO in an electrolyte with KI as Additive. *ChemElectroChem*, **2019**, *6*(2), 316-319.
[http://dx.doi.org/10.1002/celc.201801290]

[93] Talapatra, S.; Kar, S.; Pal, S.K.; Vajtai, R.; Ci, L.; Victor, P.; Shaijumon, M.M.; Kaur, S.; Nalamasu, O.; Ajayan, P.M. Direct growth of aligned carbon nanotubes on bulk metals. *Nat. Nanotechnol.*, **2006**, *1*(2), 112-116.
[http://dx.doi.org/10.1038/nnano.2006.56] [PMID: 18654161]

[94] Vivekchand, S.R.C.; Rout, C.S.; Subrahmanyam, K.S.; Govindaraj, A.; Rao, C.N.R. Graphene-based electrochemical supercapacitors. *J. Chem. Sci.*, **2008**, *120*(1), 9-13, 9-13.
[http://dx.doi.org/10.1007/s12039-008-0002-7]

[95] Stoller, M.D.; Park, S.; Zhu, Y.; An, J.; Ruoff, R.S. Graphene-Based Ultracapacitors. *Nano Lett.*, **2008**, *8*(10), 3498-3502.
[http://dx.doi.org/10.1021/nl802558y] [PMID: 18788793]

[96] Wang, D.W.; Li, F.; Zhao, J.; Ren, W.; Chen, Z.G.; Tan, J.; Wu, Z.S.; Gentle, I.; Lu, G.Q.; Cheng, H.M. Fabrication of graphene/polyaniline composite paper *via in situ* anodic electro polymerization for high-performance flexible electrode. *ACS Nano*, **2009**, *3*(7), 1745-1752.
[http://dx.doi.org/10.1021/nn900297m] [PMID: 19489559]

[97] Brodie, B.C. On the Atomic Weight of Graphite. *Philos. Trans. R. Soc. Lond.*, **1895**, *149*, 249-259.
[http://dx.doi.org/10.1098/rstl.1859]

[98] Wang, Y.; Shi, Z.; Fang, J.; Xu, H.; Ma, X.; Yin, J. Direct exfoliation of graphene in methanesulfonic acid and facile synthesis of graphene/polybenzimidazole nanocomposites. *J. Mater. Chem.*, **2011**, *21*(2), 505-512.
[http://dx.doi.org/10.1039/C0JM02376K]

[99] Wang, X.; Fulvio, P.F.; Baker, G.A.; Veith, G.M.; Unocic, R.R.; Mahurin, S.M.; Chi, M.; Dai, S. Direct exfoliation of natural graphite into micrometre size few layers graphene sheets using ionic liquids. *Chem. Commun. (Camb.)*, **2010**, *46*(25), 4487-4489.
[http://dx.doi.org/10.1039/c0cc00799d] [PMID: 20485780]

[100] Stankovich, S.; Dikin, D.A.; Piner, R.D.; Kohlhaas, K.A.; Kleinhammes, A.; Jia, Y.; Wu, Y.; Nguyen, S.T.; Ruoff, R.S. Synthesis of graphene-based nanosheets *via* chemical reduction of exfoliated graphite oxide. *Carbon*, **2007**, *45*(7), 1558-1565.
[http://dx.doi.org/10.1016/j.carbon.2007.02.034]

[101] Zheng, Q.B.; Gudarzi, M.M.; Wang, S.J.; Geng, Y.; Li, Z.; Kim, J.K. Improved electrical and optical characteristics of transparent graphene thin films produced by acid and doping treatments. *Carbon*, **2011**, *49*(9), 2905-2916.
[http://dx.doi.org/10.1016/j.carbon.2011.02.064]

[102] Jung, I.; Vaupel, M.; Pelton, M.; Piner, R.; Dikin, D.A.; Stankovich, S.; An, J.; Ruoff, R.S. Characterization of thermally reduced graphene oxide by imaging ellipsometry. *J. Phys. Chem. C*, **2008**, *112*(23), 8499-8506.
[http://dx.doi.org/10.1021/jp802173m]

[103] Xue, F.; Lu, Y.; Qi, X.; Yang, J.; Wang, Y. Melamine foam-templated graphene nanoplatelet framework toward phase change materials with multiple energy conversion abilities. *Chem. Eng. J.*, **2019**, *365*, 20-29.
[http://dx.doi.org/10.1016/j.cej.2019.02.023]

[104] Hong, J-Y.; Jang, J. Highly stable, concentrated dispersions of graphene oxide sheets and their electro-responsive characteristics. *Soft Matter,* **2012**, *8*(28), 7348-7350. [http://dx.doi.org/10.1039/c2sm25865j]

[105] Shin, K.Y.; Lee, S.; Hong, S.; Jang, J. Graphene size control *via* a mechanochemical method and electroresponsive properties. *ACS Appl. Mater. Interfaces,* **2014**, *6*(8), 5531-5537. [http://dx.doi.org/10.1021/am405930k] [PMID: 24665987]

[106] Chua, C.K.; Pumera, M. Chemical reduction of graphene oxide: a synthetic chemistry viewpoint. *Chem. Soc. Rev.,* **2014**, *43*(1), 291-312. [http://dx.doi.org/10.1039/C3CS60303B] [PMID: 24121318]

[107] Wang, Z.; Liu, J.; Wang, W.; Chen, H.; Liu, Z.; Yu, Q.; Zeng, H.; Sun, L. Aqueous phase preparation of graphene with low defect density and adjustable layers. *Chem. Commun. (Camb.),* **2013**, *49*(92), 10835-10837. [http://dx.doi.org/10.1039/c3cc46809g] [PMID: 24121934]

[108] Wang, X.Y.; Narita, A.; Müllen, K. Precision synthesis *versus* bulk-scale fabrication of graphenes. *Nat. Rev. Chem.,* **2017**, 2.

[109] Brodie, B.C. *On the Atomic Weight of Graphite*; The Royal Society, **1859**, pp. 249-259.

[110] Staudenmaier, L. Verfahren zur Darstellung der Graphitsäure. *Ber. Dtsch. Chem. Ges.,* **1898**, *31*(2), 1481-1487. [http://dx.doi.org/10.1002/cber.18980310237]

[111] Hummers, W.S., Jr; Offeman, R.E. Preparation of graphitic oxide. *J. Am. Chem. Soc.,* **1958**, *80*(6), 1339-1339. [http://dx.doi.org/10.1021/ja01539a017]

[112] Pendolino, F.; Armata, N. *Graphene Oxide in Environmental Remediation Process*; Springer, **2017**. [http://dx.doi.org/10.1007/978-3-319-60429-9]

[113] Huang, N.J.; Zang, J.; Zhang, G.D.; Guan, L.Z.; Li, S.N.; Zhao, L.; Tang, L.C. Efficient interfacial interaction for improving mechanical properties of polydimethylsiloxane nanocomposites filled with low content of graphene oxide nanoribbons. *RSC Advances,* **2017**, *7*(36), 22045-22053. [http://dx.doi.org/10.1039/C7RA02439H]

[114] Kovtyukhova, N.I.; Ollivier, P.J.; Martin, B.R.; Mallouk, T.E.; Chizhik, S.A.; Buzaneva, E.V.; Gorchinskiy, A.D. Layer-by-layer assembly of ultrathin composite films from micron-sized graphite oxide sheets and polycations. *Chem. Mater.,* **1999**, *11*(3), 771-778. [http://dx.doi.org/10.1021/cm981085u]

[115] Kuilla, T.; Bhadra, S.; Yao, D.; Kim, N.H.; Bose, S.; Lee, J.H. Recent advances in graphene based polymer composites. *Prog. Polym. Sci.,* **2010**, *35*(11), 1350-1375. [http://dx.doi.org/10.1016/j.progpolymsci.2010.07.005]

[116] Pei, S.; Cheng, H.M. The reduction of graphene oxide. *Carbon,* **2012**, *50*(9), 3210-3228. [http://dx.doi.org/10.1016/j.carbon.2011.11.010]

[117] Yoo, B.M.; Shin, H.J.; Yoon, H.W.; Park, H.B. Graphene and graphene oxide and their uses in barrier polymers. *J. Appl. Polym. Sci.,* **2014**, *131*, 1. [http://dx.doi.org/10.1002/app.39628]

[118] Sun, L.; Fugetsu, B. Mass production of graphene oxide from expanded graphite. *Mater. Lett.,* **2013**, *109*, 207-210. [http://dx.doi.org/10.1016/j.matlet.2013.07.072]

[119] Jalili, R.; Aboutalebi, S.H.; Esrafilzadeh, D.; Shepherd, R.L.; Chen, J.; Aminorroaya-Yamini, S.; Konstantinov, K.; Minett, A.I.; Razal, J.M.; Wallace, G.G. Chen, J.; AminorroayaYamini, S.; Konstantinov, K.; Minett, A.I., Razal, J.M.; Wallace, G.G. Scalable one-step wet-spinning of graphene fibers and yarns from liquid crystalline dispersions of graphene oxide: towards multifunctional

textiles. *Adv. Funct. Mater.*, **2013**, *23*(43), 5345-5354.
[http://dx.doi.org/10.1002/adfm.201300765]

[120] Marcano, D.C.; Kosynkin, D.V.; Berlin, J.M.; Sinitskii, A.; Sun, Z.; Slesarev, A.; Alemany, L.B.; Lu, W.; Tour, J.M. Improved synthesis of graphene oxide. *ACS Nano,* **2010**, *4*(8), 4806-4814.
[http://dx.doi.org/10.1021/nn1006368] [PMID: 20731455]

[121] Zhang, J.; Liu, Q.; Ruan, Y.; Lin, S.; Wang, K.; Lu, H. Monolithic crystalline swelling of graphite oxide: a bridge to ultra-large graphene oxide with high scalability. *Chem. Mater.*, **2018**, *30*(6), 1888-1897.
[http://dx.doi.org/10.1021/acs.chemmater.7b04458]

[122] Dong, L.; Chen, Z.; Zhao, X.; Ma, J.; Lin, S.; Li, M.; Bao, Y.; Chu, L.; Leng, K.; Lu, H.; Loh, K.P. A non-dispersion strategy for large-scale production of ultra-high concentration graphene slurries in water. *Nat. Commun,* **2018**, 1-76.

[123] Li, L.; Wu, G.; Yang, G.; Peng, J.; Zhao, J.; Zhu, J.J. Focusing on luminescent graphene quantum dots: current status and future perspectives. *Nanoscale,* **2013**, *5*(10), 4015-4039.
[http://dx.doi.org/10.1039/c3nr33849e] [PMID: 23579482]

[124] Cheng, C.; Li, S.; Thomas, A.; Kotov, N.A.; Haag, R. Functional graphene nanomaterials-based architectures: bio interactions, fabrications, and emerging biological applications. *Chem. Rev.*, **2017**, *117*(3), 1826-1914.
[http://dx.doi.org/10.1021/acs.chemrev.6b00520] [PMID: 28075573]

[125] Melo, J.P.; Ríos, P.L.; Povea, P.; Morales-Verdejo, C.; Camarada, M.B. Graphene oxide quantum dots as the support for the synthesis of gold nanoparticles and their applications as new catalysts for the decomposition of composite solid propellants. *ACS Omega,* **2018**, *3*(7), 7278-7287.
[http://dx.doi.org/10.1021/acsomega.8b00837] [PMID: 31458888]

[126] Tian, P.; Tang, L.; Teng, K.S.; Lau, S.P. Graphene quantum dots from chemistry to applications. *Mater. Today Chem.*, **2018**, *10*, 221-258.
[http://dx.doi.org/10.1016/j.mtchem.2018.09.007]

[127] Lu, Q.; Wu, C.; Liu, D.; Wang, H.; Su, W.; Li, H.; Zhang, Y.; Yao, S. A facile and simple method for synthesis of graphene oxide quantum dots from black carbon. *Green Chem.*, **2017**, *19*(4), 900-904.
[http://dx.doi.org/10.1039/C6GC03092K]

[128] Wang, W.; Wang, Z.; Liu, J.; Peng, Y.; Yu, X.; Wang, W.; Zhang, Z.; Sun, L. One-pot facile synthesis of graphene quantum dots from rice husks for Fe3þ sensing. *Ind. Eng. Chem. Res.*, **2018**, *57*(28), 9144-9150.
[http://dx.doi.org/10.1021/acs.iecr.8b00913]

[129] Bauhofer, W.; Kovacs, J.Z. A review and analysis of electrical percolation in carbon nanotube polymer composites. *Compos. Sci. Technol.*, **2009**, *69*(10), 1486-1498.
[http://dx.doi.org/10.1016/j.compscitech.2008.06.018]

[130] Wu, C.; Huang, X.; Wang, G.; Lv, L.; Chen, G.; Li, G.; Jiang, P. Highly Conductive Nanocomposites with Three Dimensional, Compactly Interconnected Graphene Networks *via* a Self-Assembly Process. *Adv. Funct. Mater.*, **2013**, *23*, 506513.

[131] Chen, Z.; Xu, C.; Ma, C.; Ren, W.; Cheng, H.M. Lightweight and flexible graphene foam composites for high-performance electromagnetic interference shielding. *Adv. Mater.*, **2013**, *25*(9), 1296-1300.
[http://dx.doi.org/10.1002/adma.201204196] [PMID: 23300002]

[132] Wang, X.; Zhi, L.; Müllen, K. Transparent, conductive graphene electrodes for dye-sensitized solar cells. *Nano Lett.*, **2008**, *8*(1), 323-327.
[http://dx.doi.org/10.1021/nl072838r] [PMID: 18069877]

[133] Slack, G.A. Anisotropic thermal conductivity of pyrolytic graphite. *Phys. Rev.*, **1962**, *127*(3), 694-701.
[http://dx.doi.org/10.1103/PhysRev.127.694]

[134] Chen, W.; Yan, L. *In situ* self-assembly of mild chemical reduction graphene for three-dimensional

architectures. *Nanoscale,* **2011**, *3*(8), 3132-3137.
[http://dx.doi.org/10.1039/c1nr10355e] [PMID: 21698339]

[135] Zhang, X.; Sui, Z.; Xu, B.; Yue, S.; Luo, Y.; Zhan, W.; Liu, B. Mechanically strong and highly conductive graphene aerogel and its use as electrodes for electrochemical power sources. *J. Mater. Chem.,* **2011**, *21*(18), 6494-6497.
[http://dx.doi.org/10.1039/c1jm10239g]

[136] Xu, Y.; Sheng, K.; Li, C.; Shi, G. Self-assembled graphene hydrogel *via* a one-step hydrothermal process. *ACS Nano,* **2010**, *4*(7), 4324-4330.
[http://dx.doi.org/10.1021/nn101187z] [PMID: 20590149]

[137] Nguyen, S.T.; Nguyen, H.T.; Rinaldi, A.; Nguyen, N.P.V.; Fan, Z.; Duong, H.M. Morphology control and thermal stability of binderless-graphene aerogels from graphite for energy storage applications. *Colloids Surf. A Physicochem. Eng. Asp.,* **2012**, *414*, 352-358.
[http://dx.doi.org/10.1016/j.colsurfa.2012.08.048]

[138] Worsley, M.A.; Pauzauskie, P.J.; Kucheyev, S.O.; Zaug, J.M.; Hamza, A.V.; Satcher, J.H., Jr; Baumann, T.F. Properties of single-walled carbon nanotube-based aerogels as a function of nanotube loading. *Acta Mater.,* **2009**, *57*(17), 5131-5136.
[http://dx.doi.org/10.1016/j.actamat.2009.07.012]

[139] Worsley, M.A.; Olson, T.Y.; Lee, J.R.I.; Willey, T.M.; Nielsen, M.H.; Roberts, S.K.; Pauzauskie, P.J.; Biener, J.; Satcher, J.H., Jr; Baumann, T.F. High surface area, sp2- cross-linked three-dimensional graphene monoliths. *J. Phys. Chem. Lett.,* **2011**, *2*(8), 921-925.
[http://dx.doi.org/10.1021/jz200223x] [PMID: 26295629]

[140] Longo, S.; Mauro, M.; Daniel, C.; Musto, P.; Guerra, G. Rayleigh scattering by graphene-oxide in syndiotactic polystyrene aerogels. *Carbon,* **2014**, *77*, 896-905.
[http://dx.doi.org/10.1016/j.carbon.2014.06.003]

CHAPTER 3

Preparation Methods for Graphene and its Derivatives

Sharoni Gupta[1,*], **Pinki Bala Punjabi**[1] and **Rakshit Ameta**[2]

[1] *Department of Chemistry, University College of Science, Mohanlal Sukhadia University, Udaipur-313001, Rajasthan, India*

[2] *Department of Chemistry, J.R.N. Rajasthan Vidyapeeth University, Udaipur-313001, Rajasthan, India*

Abstract: Over the past few decades, graphene and its derivatives have carved a niche for themselves in material science. These carbon nanomaterials exhibit a broad range of applications owing to their enchanting features like high specific surface area, chemical inertness, astonishing electrical and thermal properties, elevated intrinsic mobility, inimitable optical properties, and huge mechanical strength. Considering the ubiquitous applications of graphene in different industries, diverse top-down and bottom-up methods have been developed. This chapter outlines the various methods used for the synthesis of graphene and graphene-based derivatives, such as exfoliation, unrolling or unzipping of carbon nanotubes, electric arc discharge method, laser ablation technique, oxidative exfoliation-reduction of graphene oxide, chemical vapour deposition, epitaxial growth, template synthesis, pyrolysis, substrate-free synthesis, total organic synthesis, and biological methods, highlighting the advantages of these methods. Upcoming challenges concerning the commercial synthesis of graphene have also been addressed in the concluding part.

Keywords: Bottom-up Strategies, Graphene, Graphene Oxide, Reduced Graphene Oxide, Top-down Methods.

INTRODUCTION

Graphene, a 2-D carbon allotrope, is one of the thinnest and strongest materials known. Graphene has a honeycomb lattice structure comprising a monolayer of sp^2 hybridized carbon atoms assembled as hexagonal rings [1]. In 1859, Brodie, while examining the structure of graphite for the first time, observed graphene and

* **Corresponding author Sharoni Gupta:** Department of Chemistry, University College of Science, Mohanlal Sukhadia University, Udaipur-313001, Rajasthan, India; E-mail:sharoni290490@gmail.com

coined the name "graphone" [2]. Later, Boehm *et al.* were the first to use the term 'graphene' [3]. For a long time since its first discovery, very little work has been reported on graphene. Then in 2004, Novoselov, Geim, and their research group isolated graphene from graphite with a simple exfoliation method [4]. The reinstating of graphene by Geim and Novoselov, who were awarded the Noble Prize in 2010 for their work on graphene [5], paved the way for "Graphene Science". Ever since then, a plethora of methods for the synthesis of graphene and its applications in various areas have been explored by researchers worldwide.

The incredible properties namely, enlarged specific surface area (up to 2630 m^2 g^{-1}), enormous intrinsic mobility (200,000 cm^2 v^{-1} s^{-1}), stupendous thermal conductivity (<5000W m^{-1} K^{-1}), extremely large Young's modulus (< 1.0 TPa), exceptional optical transmittance (< 97.7%), outstanding electrical conductivity, and tremendous chemical stability, displayed by graphene arise from its distinctive structure and arrangement of carbon framework [6 - 9].

The arrangement of mono graphitic lamellae of carbon particles in hexagonal planar sheet stacked in one or few layers forms bi-dimensional graphene. It has been reported that graphene lamella has a thickness of about 0.35 nm, comparable to 1/200,000th of the diameter of a strand of a human's hair [10]. The bi-layer and few-layer graphenes consist of two and three to ten layers of graphitic sheets, respectively. Graphene with twenty to thirty sheets is identified as multilayer or thick graphene or nanocrystalline thin graphite [11]. The monolayer graphene has a wavy sheet without any stacking. While, in the bi or few-layered graphene, the graphitic sheets are arranged in AA (hexagonal) or AB (Bernal) or ABC (rhombohedral) stacking [12]. Also, the few-layered graphene with no appreciable stacking order is commonly termed turbostratic [13].

The sp^2 bonding of carbon atoms in graphene exhibits three σ bonds (in-plane) formed by hybridization of s, p_x, and p_y orbitals and a π bond formed by p_z orbital perpendicular to the plane [10, 14]. The σ bonds form a firm backbone of hexagonal geometry, whereas the out-of-plane π bonds result in inter-lamellar interaction between graphene layers [10]. The existence of free electrons in the π bands is thus responsible for the exclusive electrical and optical properties of graphene [15]. Furthermore, the electrical conductivity of graphene is governed by the type of edges. Generally, armchair or zigzag edges exist in graphene. The variety and density of edges can generate metallic or semiconducting behaviour in graphene [16].

In spite of the desirable features of graphene, the lack of industrial-scale production is a major hindrance to its applicability in diverse fields. In such a scenario, the surfacing of graphene oxide and reduced graphene oxide has been a relief for material chemists and industrial workers. This is because preparative protocols for GO and rGO are quite simple, require inexpensive starting materials, and result in mass production of these materials [17].

The oxidized form of graphene, *viz.* graphene oxide (GO), is structurally similar to graphene but with the presence of oxygen functionalities [17]. The precise structure of GO has been a highly controversial subject owing to its amorphous, berthollide character and hygroscopic nature [18]. Despite its debatable structure, several models have been proposed to elucidate the structural architecture of GO. Hofmann and Holst were the first to put forth a model suggesting that the epoxy groups exist at the basal planes of the graphite layer, with an overall molecular formula of C_2O [19]. Next, Ruess anticipated a model which elucidated that the carbon sheets were puckered instead of planar and the oxygen-containing groups like hydroxyl and ether-like oxygen formed bridges between the carbon atoms at positions 1 and 3 of the graphite framework [20]. Thereafter, Clause and co-workers [21], to explain the acidic nature of GO, further illustrated an enol- and keto-type model. In 1969, Scholz and Boehm [22] presented a novel arrangement of GO, which encompassed grooved carbon layers bearing hydroxyl and carbonyl functional groups. Nakajima and his research group [23, 24] also proposed a model elucidating the structural attributes of GO. According to them, GO comprises two carbon layers anchored to each other by sp^3 hybridized carbon-carbon bonds at right angles to the layers with the presence of relative numbers of carbonyl and hydroxyl groups based on the hydration scale. In 1997, Lerf and his colleagues [25] suggested the most widely accepted model for GO. As per their model, GO consists of haphazardly dispersed planar aromatic sections comprising un-oxidized benzene rings and crumpled zones of six-membered alicyclic rings with C=C, C−OH and ether groups at 1 and 2 positions [26]. Finally, Szabó *et al.* presented the most recent structural model for GO [27]. The model revealed that the carbon framework exhibits two kinds of zones, one with trans-linked cyclohexane chairs and the other one bearing ribbons of flat hexagonal rings with C=C bonds as well as functionalities, including tertiary OH, 1,3-ether, ketone, quinone, and phenol. Overall, it is clear that the structure of GO comprises graphitic regions, disordered oxidized areas, and regions with defects (which arise due to holes *via* the release of CO and CO_2 during the belligerent oxidation and exfoliation of the sheet in GO) [28].

Additionally, the partial reduction of graphene oxide restores the carbon framework of a graphene layer with the presence of a few oxygen functional groups. This form is known as reduced graphene oxide (rGO). The atomic structure of rGO was studied by Navarro *et al.* [29], and it was explicated that rGO sheets include large areas of undamaged graphene along with considerable regions rich in defects, namely, pentagonal-hexagonal pairs and planar, quasi amorphous sp^2 bonded carbon clusters. It was reported that the defects occur due to imperfections in the crystallinity of graphene layers after GO reduction. Furthermore, through molecular dynamics simulations, it was revealed that carbonyl and ether groups are quite stable and hinder the complete reduction of GO [30]. Therefore, at low temperatures only, hydroxyl groups desorb without affecting the basal planes of graphene. While the more stable epoxy groups leave the GO surface at high temperatures, resulting in profound distortions in the graphene network. Thus, reinstating the extended π-conjugated network in rGO with some topological defects leads to superior electrical properties of rGO compared to GO. It is also noteworthy that owing to the existence of residual functional groups in rGO, the flow of charge carriers is disrupted *via* sp^2 networks, and thus, electric flow principally occurs by hopping, unlike graphene, where charge carriers flow ballistically [31 - 34]. Another feature of GO and rGO is hydrophilicity, which ascribes to the inclusion of functional groups that renders them soluble in water, organic solvents, and various other polar media [35].

Consequently, by fine-tuning the structural morphology, edges, number of sheets, defects, and functionalities in graphene or graphene oxide/reduced graphene oxide, one can realize desirable characteristic properties of graphene for diverse applications. To achieve this, one can adopt a cost-effective, reliable, simple, high-yielding, time-efficient, and scalable synthetic strategy for graphene synthesis. Considering the crucial inter-relationship between synthesis, property, and application of graphene, the chapter intends to highlight diverse top-down and bottom-up strategies available for designing graphene and its derivatives.

TOP-DOWN SYNTHESIS

Top-down methods are based on the breakdown of graphitic compounds into mono- or few-layer graphene [13]. The top-down approaches for graphene and its derivatives include exfoliation, unrolling or unzipping carbon nanotubes, electric arc discharge method, laser ablation technique, and oxidative exfoliation-reduction of graphene oxide.

Exfoliation Method

Exfoliation is the most cost-effective for the production of graphene from graphite. The process involves the un-stacking of graphite layers by increasing the inter-planar spacing *via* the breakdown of van der Waals forces and causing poor bonding in the direction orthogonal to the planes [36]. The exfoliation method is of two types: mechanical and chemical.

Mechanical Exfoliation

Mechanical exfoliation is also referred to as the "scotch tape" method. This is because in this technique, an adhesive tape is put on a graphite sample, and then the sample flakes are peeled off. The process involves preferential cleaving of flakes that stick to the tape along the crystal plane [37]. Furthermore, few- and single-layered graphene are obtained by pressing a fresh tape against the graphite flakes that adhere to the first piece of tape. Thereafter, the two tapes are peeled apart, slicing graphite into thinner sheets. By repeating the process several times, very thin graphite sheets are obtained [38]. In mechanical exfoliation, milling, stirring, ultrasonication or fluid dynamic techniques are employed to pull apart graphite layers [39].

For instance, Chen and co-workers have reported the synthesis of mono- and few-layer graphene using a three-roll mill machine and a polymer adhesive [40]. A three-roll mill with three adjacent cylindrical rolls (80 mm in diameter), which rotated at the same velocity, was employed, as shown in Fig. (**1**). The first (feed) and third (apron) cylindrical rolls moved in the same direction while the central roll rotated in the opposite direction. The mechanical force of the mill causes the S-shaped movement of the graphite sample in the rolls, causing continuous exfoliation of graphite. To prepare the polymer adhesive, poly vinyl chloride (PVC) was dispersed in dioctyl phthalate and magnetically stirred for half an hour. Then, the adhesive and graphite were placed in the mill. As the rolls rotated, the graphite sample gradually spread on the adhesive, followed by homogenous dispersion on the rolls and exfoliation of the graphite to form graphene. The process of exfoliation was performed for 12 h at ambient temperature. Finally, pure graphene was obtained by adding alcohol to impure graphene in order to remove dioctyl phthalate, followed by combustion in the muffle furnace at 500 °C for 3 h to burn PVC. The method offered a large-scale synthesis of homogeneous graphene sheets of a thickness of about 1.13-1.41 nm.

The use of acoustic cavitations for mechanical exfoliation of graphite flakes is also well known. The dispersion of graphite in suitable solvents like water, ethanol, DMF, *etc.* followed by ultrasonication leads to the breakdown of graphite sheets. The mechanism of ultrasonic exfoliation basically involves distribution of cavitation bubbles around the graphite flakes. The violent collapse of these bubbles generates shock waves which produce stress throughout the graphite material leading to splitting of graphite sheets. Additionally, the unbalanced lateral compressive stress may also separate two adjacent graphite layers by shear effect. The ultrasonic method offers low cost industrial scale production of graphene but sometimes it is deemed unfit because ultrasonic exfoliation in some cases results in fractures in graphene sheet and other undesirable defects.

However, by carefully controlling the reaction parameters like temperature, sonication power, solvent, and sonication time once can overcome this issue as well.

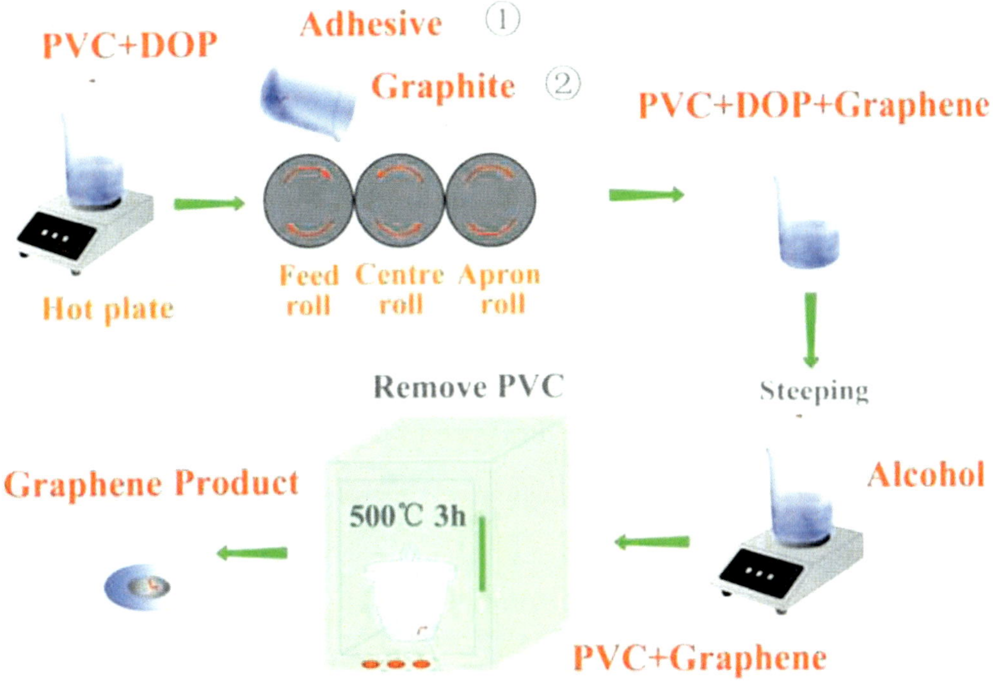

Fig. (1). The use of a three-roll mill machine for the synthesis of graphene by mechanical exfoliation. Reprinted with permission from [40]. Copyright 2012 The Royal Society of Chemistry.

Lately, Noroozi *et al.* demonstrated an ultrasonic exfoliation of graphite under environmentally benign conditions [41]. A mixture of polyvinyl pyrrolidone (PVP), hand soap-deionized water and graphite was stirred followed by sonication for 3 h. The reaction mixture was then kept still for 24 h leading to settling down of the large graphite aggregates at the bottom. A uniform dark suspension was obtained which was left undisturbed for five months. After that the suspension was washed with deionized water and centrifuged at 4,000 rpm for some hours which resulted in the formation of a gel like substance. The gel was dried overnight in a vacuum oven at 80°C followed by annealing at 1000°C for 5 min, to eliminate any residual PVP from the material. At last, pure few layered graphene sheets were obtained as black powder. Thus, graphene sheets were afforded by a low cost and simple sonication based exfoliation method.

Ball-milling of graphite for production of graphene has also gained much attention. Typically, two mechanisms are known for delamellation of graphite *via* ball milling [39]. In the first approach, shear force exerted for exfoliation is employed to produce large sized graphene flakes. The second approach makes use of the collisions or vertical forces applied by the balls in rolling actions to produce small flakes. Recently, Balasubramanyan and his research group have prepared graphene flakes using ball-milling technique for exfoliation [42]. A planetary ball-mill with 14 stainless steel balls was used for this purpose. In this method, sucrose and graphite were milled at 70 rpm for 30 h to obtain graphene-sucrose mixture. Thereafter, the mixture was calcinated for 3 h at 400 °C to obtain pure graphene. The method was high yielding and afforded graphene with fewer defects.

Nowadays, fluid dynamics methods are also employed for mechanical exfoliation. In this method, the graphite flakes flow with the liquid and are exfoliated frequently at different points [43]. An exfoliation method based on fluid dynamics was reported by Tran and his colleagues [44]. In this method, the Taylor-Couette flow reactor (Fig. **2**) consisting of two coaxial cylinders was used with a mixing vessel connected to a circulating chiller for prevention of heating up of the solvent.

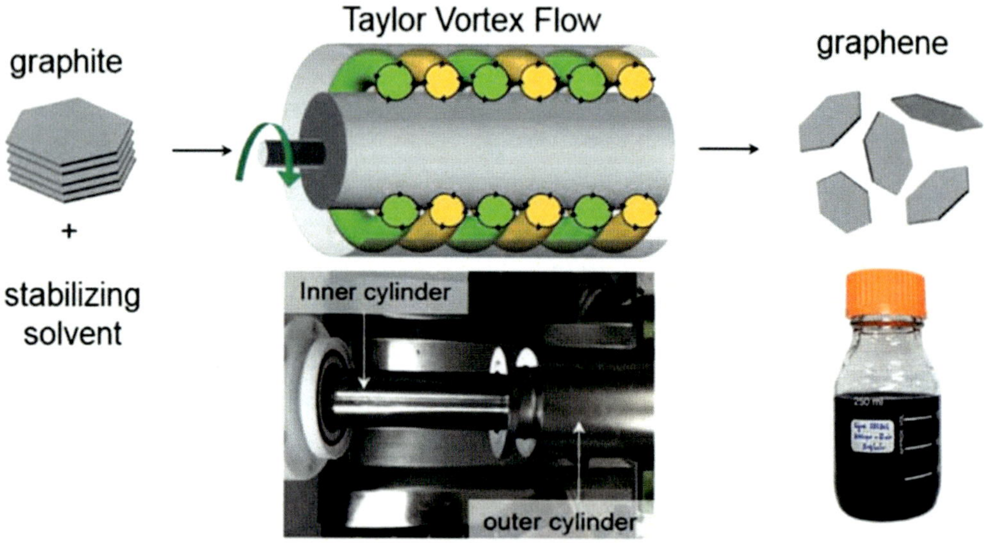

Fig. (2). Synthesis of graphene by using Taylor vortex flow. Reprinted with permission from [44]. Copyright 2016 The Royal Society of Chemistry.

The graphite powder and stabilizing solvents (N-Methyl-2-pyrrolidone and dimethylformamide) were injected into the space between the two cylinders of the reactor. Then, as the inner cylinder of the reactor rotated at a speed of 500 to 3000 rpm for 10 to 120 min., shear mixing of graphite took place. The mixing resulted in the exfoliation of graphite and the resultant dispersion was centrifuged for half an hour at 1000 rpm to discard un-exfoliated graphite particles. Finally, exfoliated graphene was collected from the supernatant. The advantage of fluid dynamic method is that it affords high quality of graphene in short time.

Chemical Exfoliation

In chemical exfoliation, chemical agents are used to ablate graphite sheets. The chemicals first reduce van der Waals forces by increasing inter-layer spacing by intercalating themselves between the graphene layers. Then, thermal/high pressure treatment leads to breakdown of the intercalated graphite to form graphene [45].

Recently, a scalable fabrication of few-layer graphene by chemical exfoliation of graphite flakes has been demonstrated by Liu and his research group [46]. The process involved mixing of peroxyacetic acid with sulphuric acid followed by the addition of graphite to the binary system. The resultant mixture was stirred at room temperature to form a turbid liquid. During stirring, the decomposition of peroxyacetic acid took place accompanying oxygen evolution. Thereafter, the mixture was allowed to stand at ambient temperature and exfoliated graphene was obtained after 4 h. Pure graphene was obtained by filtration and washing with deionized water followed by air drying. In this method sulphuric acid intercalates between the graphite layers while the oxygen gas evolved by decomposition of peroxyacetic acid causes an increase in pressure that results in an increase in inter-layer distance to such an extent that van der Waals forces are conquered resulting in peeling off graphite layers to form graphene. This novel method produced graphene with an average of five layers.

In another chemical exfoliation method, microwave irradiation was used for rapid graphene synthesis [47]. The method involved two steps. In the first step, graphite intercalated with perchloric acid was prepared by mixing graphite with acid followed by 2 min. stirring under ambient conditions. Then, acetic anhydride was added to the mixture and the reaction mixture after stirring was subjected to microwaves for 30 s at 700 W resulting in $HClO_4$-intercalated graphite. In the second step, the intercalated graphite was suspended in deionized water and irradiated by microwaves for 80 s. This process led to large expansion of intercalated graphite's volume resulting in the formation of porous graphene aggregates. The aggregates were then subjected to ultrasonication for 5 min. at 100 W leading to the formation of exfoliated few-layer graphene.

It was reported that the use of perchloric acid with acetic anhydride increased its oxidation of graphite and thereby ensured intercalation of $HClO_4$ in graphite. Further, in the second step, the microwaves facilitated the decomposition of intercalated graphite due to a rapid increase in temperature. Thus, the use of microwaves in combination with sonication led to 100% exfoliated graphene. It was also reported that the few-layer garphene so produced had an average layer number of less than five and an average surface area of 256.8 μm^2.

In recent years, electrochemical exfoliation of graphite in aqueous solutions of inorganic substances and ionic liquids has gained a lot of significance over the years. Generally, the electrochemical exfoliation of graphite involves cleavage of intercalated graphite in solutions of salts or inorganic precursors undergoing insertion into graphite under electric current. The electrochemical exfoliation offers advantages such as better control of oxidation, intercalation and exfoliation of graphite sheets, hence generating good yields of less defective, high quality graphene [48].

An electrochemical exfoliation process for graphene preparation has been reported by Htwe *et al.* [49]. The method was based on the use of the traditional binary electrode system comprising graphite rod as a working electrode and platinum rod as a counter electrode for electrochemical processes. The two electrodes were immersed in an electrolyte solution containing sulphuric acid and phosphoric acid in one study and in another one, ammonium sulphate in distilled water was used as an electrolyte. A 10 V direct current bias voltage was applied until complete exhaustion of the graphitic anode. After electrochemical exfoliation, graphene is sonicated for about 15 to 60 minutes at ambient temperature and washed with distilled water *via* vacuum filtration. After vacuum drying at 80°C, pure graphene is obtained. In this exfoliation process, the inorganic compounds/ions from electrolyte solution are intercalated between graphite sheets *via* covalent or ionic bonds. Next, the graphite undergoes destruction due to electric potential and graphite sheets separated to form graphene. It was further reported that a comparative study of electrolytes and sonication time showed that the sulphuric acid electrolyte coupled with 45 min. of ultrasonication resulted in the most stable and least defective graphene with a few layers. Overall, from the above examples, it can be inferred that both mechanical and chemical exfoliation methods are cost-efficient and lead to high grade graphene in good yields.

Unzipping/Unrolling of Carbon Nanotubes

The unrolling of single-walled or multi-walled carbon nanotubes (SWCNT or MWCNT) is comparatively a new technique for graphene synthesis. Unzipping of

CNT results in the unfolding of a rolled cylindrical structure into planar graphitic sheets of single or few layers. By and large, a high strain rate of about 10^{8-10}/s is the perquisite for breakdown of the C–C bond, leading to unrolling of CNT [50]. Laser irradiation, plasma etching, oxidizing agents, *etc.* are commonly used for the unzipping of SWCNT and MWCNT into mono- and few-layer graphene nanoribbons [51 - 53]. Graphene nanoribbons are basically quasi-monodimensional materials with armchair or zigzag edges [54].

In a recent study, nitrogen doped carbon nanotubes were unzipped to prepare graphene nanoribbons *via* a two-step process [55]. Firstly, an electrochemical process was carried out using 1 M H_2SO_4 solution as electrolyte for a three-electrode arrangement consisting of the nitrogen doped CNT array, platinum mesh, and mercury/mercury (I) sulfate (MMS) as the working, counter, and reference electrode, respectively. A potential of 0.8 V was applied at the working electrode for electrochemical unzipping of N-doped CNT. The unzipped CNTs were then washed thoroughly with deionized water to get rid of impurities. In the next step, the unzipped CNTs were suspended in N-methyl-2-pyrrolidone and subjected to sonication in an ice bath at different powers ranging from 40 W to 200 W (Fig. **3**). Finally, pure graphene nanoribbons were obtained by ultracentrifugation at 40000 rpm to discard the residual N-doped CNTs and other impurities. The method offered an efficient protocol for unzipping of CNTs, leading to graphene nanoribbons of 2 μm length.

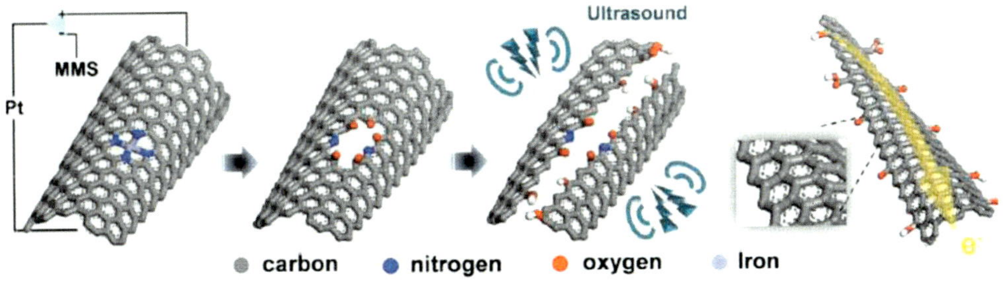

Fig. (3). Two-step synthesis of graphene nanoribbons *via* **(i)** Electrochemical unzipping **(ii)** Ultrasonication. Reprinted with permission from [55]. Copyright © 2019, American Chemical Society.

Plasma etching technique for graphene synthesis has been reported by Jiao *et al.* [56]. In this technique, the MWCNTs were dispersed in 1% aqueous Tween 20 solution and sonicated for 5 min. followed by centrifugation for 10 min. at 16,400*g* to eliminate the aggregates. The MWCNT dispersion was then mounted onto a pretreated Si substrate. Thereafter, the sample was calcined for 10 min. at 350 °C to get rid of the Tween 20. Next, a poly(methyl methacrylate) (PMMA) solution was spin-coated on MWCNTs at 3,000 rounds per minute for 60 s and then scorched at 170 °C on a hot plate for 2 h. This was followed by peeling off the PMMA-MWCNT film in 1 M KOH solution at 80 °C [57]. Later, the film was cleaned with water and deposited onto a Si substrate by heating at 80 °C for 10 min in an oven. A10 W Ar plasma was employed by the workers to etch the PMMA-MWCNT film by applying a pressure of 40 mTorr. Subsequent to the etching process, the PMMA film was adhered to 3-aminopropyltriethoxysilane (APTES)-treated SiO_2/Si substrate with a preassembled Pt/W marker array. Finally, the PMMA was eliminated by acetone vapour followed by calcination for 10 min. at 300 °C to obtain graphene nanoribbons [57]. The graphene nanoribbons synthesized by this method were obtained with smooth edges and a slender width of 10-20 nm. This process has an advantage that it allows better control over the structure and quality of graphene.

Lately, Li and his research group have demonstrated longitudinal unrolling of carbon nanotubes (Fig. **4**) *via* intercalation of MWCNTs with KNO_3 and H_2SO_4 [58]. The advantage of this method is its scalable and mild protocol for the synthesis of graphene nanoribbons in excellent yields using less amount of H_2SO_4. The typical process involved magnetical stirring of MWCNTs with KNO_3 and H_2SO_4 for 2 h at 300 rpm. Thereafter, $KMnO_4$ was added gradually to the solution followed by stirring for another 2 h at ambient temperature. Then the unrolling reaction was carried out for 2 h in a water bath by slowly raising temperature from 30 °C to 70 °C. After completion of the reaction, the product was purified by sonication and centrifugation to yield pure graphene nanoribbons [59]. It was also revealed that the nanoribbons were single-layer to few-layer graphene with an average height of 2.1 nm and width of 101.8 nm.

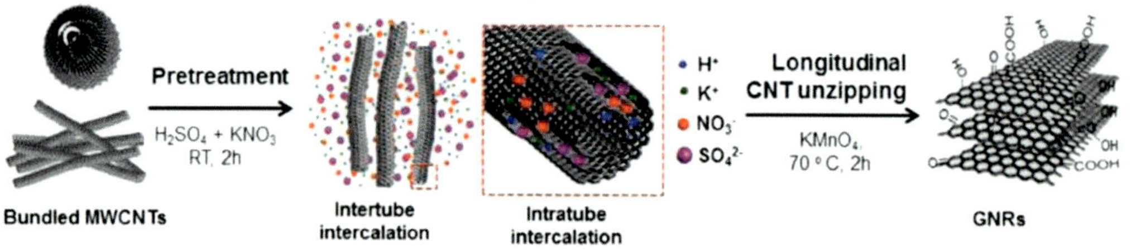

Fig. (4). Graphene nanoribbons are synthesized *via* stepwise longitudinal unzipping of CNTs. Reprinted with permission from [58]. Copyright © 2016, The Author(s).

Electric Arc Discharge

In a characteristic electric arc-discharge method, graphitic rods as a negative cathode and a positive anode are placed at a distance of few millimeters from each other and high current is applied to the electrodes to decompose the graphite rods to generate carbon nanostructures which pile up over the cathode or around the chamber walls [60]. The electric arc method for graphene synthesis has been investigated by several researchers. The advantage of electric arc method is that it results in the production of highly crystalline and pure graphene [61]. Recently a gram scale synthesis of graphene has been established by Wu *et al.* [62]. The method was based on alternating current arc discharge method. An AC arc-discharge was performed in a water-cooled stainless steel chamber containing a mixture of nitrogen and hydrogen as buffer gas. Two graphite electrodes at a distance of 1 mm were used in the chamber. An AC of 150 A was applied at 400 Torr pressure. The electric discharge led to the consumption of two graphite electrodes in a short time of 4 min to 5 min and the carbon products were deposited on the inner and top wall of the chamber. The products were collected and annealed in air for an hour at 400 °C to obtain pure few-layer graphene. The method presented several advantages such as short reaction time, production of gram quantities of graphene and generation of high quality graphene sheets.

In another method, graphene and N-doped graphene were prepared *via* arc discharge by utilizing anodic carbon fillers *viz.* graphite, graphene oxide (GO) and polyaniline (PANi) [63]. In this method, an arc discharge apparatus was employed with graphite cathode and hollowed graphite anode positioned horizontally. Graphite powder, GO and PANi were homogeneously mixed and used as anodic carbon filler. For preparation of N-doped graphene, the arc chamber was first maintained at a pressure of 5×10^{-2} Torr, then filled with He, H_2, and NH_3 gases. The chamber pressure was then sustained at 550 Torr. For graphene, He and H_2 gases were used. A voltage of 30 V and current of 150 A were applied. The discharge resulted in the collection of graphene and N-doped graphene materials at the upper part of chamber. The graphene and N-doped graphene flakes produced by this method exhibited high crystallinity and large number of active edges that displayed super-capacitance.

Laser Ablation

Pulsed laser ablation or pulsed laser deposition is a reductive technique consisting of the production of micro or nanosized materials *via* removal of small portions of bulk substrates using a focused pulsed laser beam. In a typical laser ablation method, a pulsed beam is focused on graphite containing catalytic material in the presence of inert gas and high temperature resulting in ablation of carbon particles [64].

The main advantage of laser ablation techniques is that it yields highly porous graphene materials. A femtosecond laser ablation process for synthesis of extremely porous graphene has been demonstrated by Russo and co-workers [65]. In this approach, a highly oriented pyrolytic graphite (HOPG) was used as a target, placed at the bottom of a quartz cell filled with water and ablated by a femtosecond laser with a pulse duration of 35 fs. The HOPG was ablated continuously for 20 min. to produce highly porous graphene floating in the air-water interface. It was further revealed that a six-layered spongy graphene with an irregular 3D porous structure with an average pore size of 15-20 nm was generated by this method.

A mechanism for graphene synthesis was also proposed by researchers. According to them, the laser beam simultaneously initiates decomposition of water, generates reactive molecules O_2, H_2, and H_2O_2 and exfoliates graphite, causing un-stacking of graphite sheets. Later, oxidation of graphene sheets takes place due to O_2 and H_2O_2 species producing several layered graphene oxide (GO) with sp^3 hybridized carbon atoms due to oxygen bonded atoms. Then, similar to the coal gasification mechanism, the sp^3 carbon atoms of the GO react with oxidized species. Subsequently, the formation of CO and CO_2 molecules occurs, leaving behind carbon vacancies and creating pores of nanodimension. This finally results in the production of porous graphene which floats at the water-air interface from where it is readily collected. Thus, this method affords large-scale synthesis of graphene in a short time and with high porosity.

Very recently, a novel single step liquid phase laser ablation technique has been used by Juvaid and his research group to produce a bilayer graphene [66]. In this method, a commercially graphite bar was suspended in deionized water (ablation medium) and ablated with a KrF excimer laser beam of 248 nm wavelength. The laser irradiation generated plasma which caused exfoliation of graphene films in water. Then, after half an hour of laser exfoliation, the Si/SiO_2 substrate was immersed in deionized water to collect the graphene films and later subjected to a natural drying process. Consequently, this cost-effective method produced bilayer graphene films with a high crystalline nature. Thus, it can be inferred that the laser beams produce high thermal energy which helps in overcoming the van der Waal bonds between the hexagonal sheets of graphite and leads to the formation of graphene.

Oxidative Exfoliation-Reduction of Graphene Oxide

Graphene oxide (GO) is generally prepared by oxidation of graphite using oxidants like $KClO_3$, $KMnO_4$, *etc.* in acidic media (HCl, H_2SO_4 and HNO_3, *etc.*). The first ever synthesis of graphene oxide *via* intercalation of graphite flakes was reported by Schafhacutl in 1840. Since then, various methods have been employed to synthesize graphene oxide including the Brodie's method [2], Staudenmaier method [67], Hofmann method [19, 68], Hummers method [69] and also other modified and improved methods [70, 71]. The reduction of graphene oxide leads to the production of graphene or reduced graphene oxide. In the following sections, oxidation and reduction methods have been briefly discussed.

Graphene Oxide Synthesis

Diverse methods used for oxidation of graphite to form graphene oxide (Fig. **5**) are as follows:

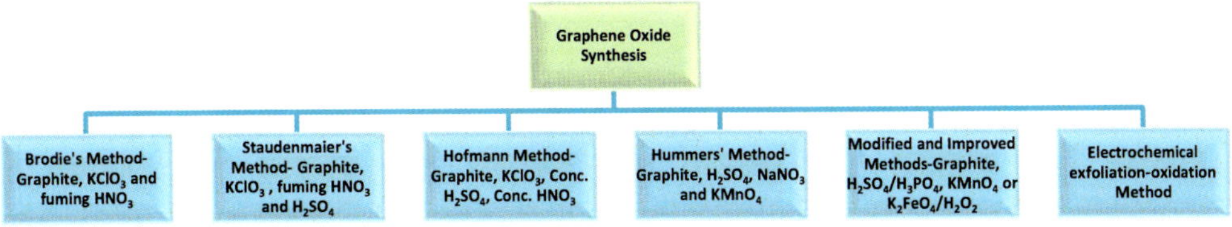

Fig. (5). Diverse methods for synthesis of graphene oxide.

a) Brodie's Method- Graphene oxide was fabricated by Brodie in 1859 using potassium chlorate and fuming nitric acid as oxidants. The synthesized product was a mixture of graphene and graphene oxide and soluble in water. The composition of the product as determined by elemental analysis was found to be 61.04% C, 1.85% H, 37.11% O. The shortcomings of this method were lengthy reaction time and evolution of hazardous gases.

b) Staudenmaier's Method- Staudenmaier's method (1898) was an upgrade to Brodie's method. This method used sulphuric acid in addition to fuming nitric acid and multiple aliquots of potassium chlorate solution, thereby escalating the acidity of the reaction mixture. This led to production of an extensively oxidized graphene oxide. However, this process was also time consuming and released toxic gases, hence it was not so suitable.

c) Hofmann's Method- In 1937, Hofmann and his research group demonstrated the use of concentrated nitric acid along with concentrated H_2SO_4 and $KClO_3$ to prepare graphene oxide.

d) Hummmers' Method- Hummers and Offeman in 1958 elucidated a novel method for the oxidation of graphite. The process was based on stirring of graphite powder with a combination of concentrated H_2SO_4, $NaNO_3$ and $KMnO_4$. When the stirring was complete, the reaction mixture was diluted with water and hydrogen peroxide and filtered. The product was then washed and dried to obtain pure graphene oxide. Although the method yielded highly oxidized graphene oxide in good yields, its major flaws were the explosive reaction process and the evolution of poisonous NO_2 and N_2O_4 gases.

e) Modified and Improved Methods- Over the past few years, improved and modified methods have been developed to conquer the drawbacks of the above mentioned routes and enhance the product yield. In 2010, Marcano and his co-workers designed an improved method for GO synthesis [72]. In this method, they substituted sodium nitrate with a 9:1 mixture of H_2SO_4/H_3PO_4 and raised the concentration of $KMnO_4$. The highlight of this method was that a greater yield of hydrophilic graphene oxide was produced without the release of toxic nitrogenous gases. Recently, an improved Hummers method was demonstrated by Chen *et al.* [73] for GO synthesis using $KMnO_4$ and concentrated sulfuric acid. The advantage of this method was that $NaNO_3$ was not used, which prevented the evolution of hazardous gases Further, very recently an improved $NaNO_3$-free Hummers method was devised by Yu *et al.* employing $KMnO_4$, K_2FeO_4 and concentrated H_2SO_4 as oxidants and boric acid as a stabilizing agent [74]. This protocol was extremely efficacious and environmentally-friendly.

Another non-explosive improved method for GO synthesis was demonstrated by Ranjan *et al.* [75]. The method was a modification of Marcano's method. In this method, graphite flakes and $KMnO_4$ (taken in 1:6 ratio) were mixed using mortar and pestle for 5 min. and kept at 5 °C (the pre-cooling protocol). Next, a solution of sulphuric acid and phosphoric acid (taken in the ratio 9:1) was prepared separately and again pre-cooled at 5 °C. Both the solutions were mixed. At this step, no explosive reaction occurred and temperature also remained constant at room temperature. The suspension was then heated at 65 °C and was continuously stirred for 24 h. Later, the brown colored suspension was cooled down by adding 400 ml of deionized ice water. Then, H_2O_2 (7 mL) was added and stirred to obtain a golden yellow solution indicating the formation of graphene oxide. To the solution, distilled water was added and the precipitate was allowed to settle down. The precipitate was then washed with deionized water, HCl and ethanol, centrifuged and the suspension was decanted to obtain a gel-like substance. Finally, the gel was degassed for 24 h in a desiccator to obtain GO. Thus, the modified approach was cost-effective, free from toxic gas emission as well as non-explosive in nature, generating high purity GO.

Similarly, a modified Hummers approach using a mixture of sulfuric acid, phosphoric acid and nitric acid as intercalating sources and potassium permanganate and hydrogen peroxide as oxidants has been investigated by Al-Gaashani and co-workers [76]. This chemical oxidation method was free from poisonous gas expulsion, a low temperature process and reliable producing high quality of graphene.

f) Electrochemical Exfoliation-Oxidation Method- In a characteristic electrochemical exfoliation-oxidation method, two electrode system with graphite electrode as working or anodic electrode is used with an oxidizing electrolyte [77]. The basic approach is exfoliation of graphite by intercalation followed by oxidation to produce GO.

Of late, a high yielding electrochemical exfoliation procedure for GO preparation has been reported by Wang and his research group [78]. A two electrode system with graphite foil as a working electrode and Pt foil as a counter electrode placed at a distance of 3.0 cm was set-up. A low voltage of 4.0 V from a DC power source was applied and a mixture of H_2SO_4 (1 mol L^{-1}) and $KMnO_4$ (0.4 g) was used as electrolyte. The exfoliation was carried out for an hour. Thereafter, the electrolyte (retreated by sonication at 200 W of power for 30 min.) caused oxidation of the exfoliated graphene. The process continued for 10 hours. The GO was purified by washing with deionized water to remove the electrolyte ions and collected by vacuum filtration using a 0.22 mm microfiltration membrane. Finally, pure GO was obtained by drying at 60 °C for about 12 h. In this process, H_2SO_4 caused intercalation and exfoliation of graphite to afford graphene while $KMnO_4$ resulted in the oxidation of graphene to form hydrophilic graphene oxide. This method generated a 90% yield of GO.

A similar electrochemical oxidation process to produce graphene oxide was reported by Pei *et al.* using graphite intercalation compound paper as anode and Pt as cathode with a 5 V DC power supply, and diluted H_2SO_4 solution as an electrolyte [79]. Overall, the method presented a benign and efficient protocol for GO synthesis (Fig. **6**).

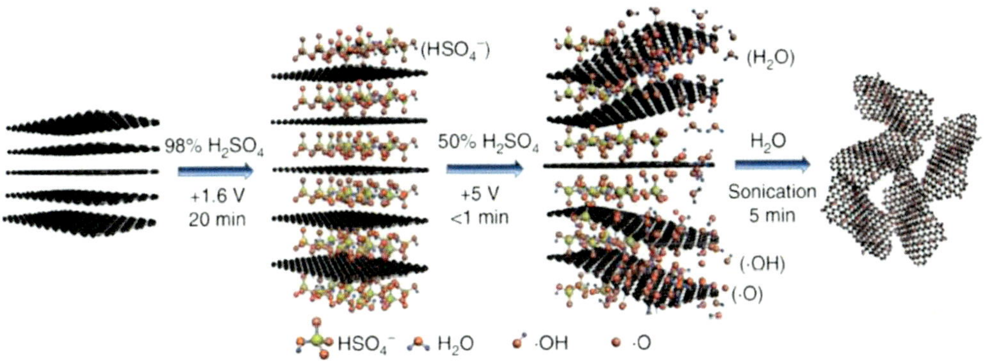

Fig. (6). Synthesis of graphene oxide *via* electrochemical oxidation process. Reprinted with permission from [79]. Copyright © The Author(s) 2018.

Graphene Oxide Reduction

The reduction of graphene oxide to reduced graphene oxide (rGO) or graphene can be accomplished by diverse methods such as chemical reduction, thermal annealing, photo or microwave reduction methods, electrochemical and solvothermal reduction (Fig. **7**). The reduction process removes oxygen functionalities from GO and re-establishes the aromatic framework of graphene.

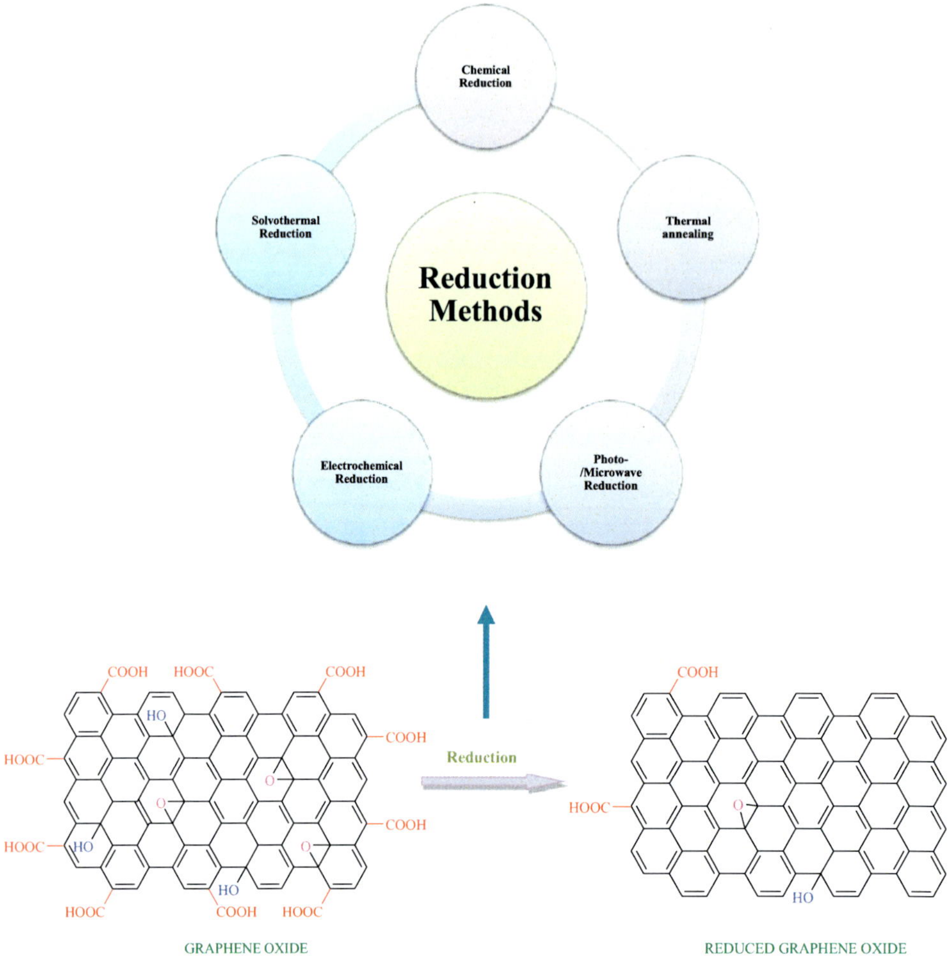

Fig. (7). Reduction of graphene oxide by diverse methods.

a) Chemical Reduction- The most commonly used method for reduction of graphene oxide is chemical reduction. The reduction of GO is basically carried out using chemical reductants in water. A variety of reducing agents such as hydrazine [80], lithium triethylborohydride [81] and elemental sulphur [82], L-ascorbic acid [83], glycine [84], starch [85], vanomycin [86], polydopamine [87], algal extracts [88], bacterium [89], artemisinin [90], green tea extract [91], *Aloe vera* [92], *etc.* have been used to prepare rGO from GO. Despite the scalability of this method, chemical reduction of GO suffers a setback because the rGO so obtained exhibits reasonably poor surface area and electronic conductivity [93].

b) Thermal Annealing- Annealing is a process which is usually carried out in vacuum [94], air [95], in inert [96] or reducing environment [97]. The rapid heating of GO at elevated temperature, not only exfoliates it but also removes oxygen functionality. Further, the thermal annealing introduces morphological defects and damages the graphene sheets due to the release of carbon dioxide [98] which enhances the electronic properties of the reduced form. Additionally, it has been proven that the more elevated the annealing temperature, the better is the reduction and the greater is the surface area of rGO [99].

c) Microwave and Photo-Reduction- The microwave reduction method is quite advantageous because using this method one can simultaneously carry out reduction as well as doping of GO along with fabrication of hybrid composites in short reaction time [100 - 102]. In an instance, it was reported that microwave reduction of graphene oxide powders produced rGO within just 3 minutes [103]. Recently, photo reduction of GO has also emerged as a neat and straightforward technique using solar light [104], ultraviolet light [105], and visible irradiation [106]. These methods lead to supreme quality rGO. Besides, by employing laser beam, the reduction is accomplished in femto-seconds [107].

d) Electrochemical Reduction- The prime highlight of this method is that GO reduction of GO can be achieved in aqueous buffer solution at ambient temperature without using hazardous chemical agents. In this method, the reduction process is performed in an electrochemical cell and reduction of GO occurs *via* the exchange of electrons between the electrodes and GO [108 - 110]. This method produces rGO with a high carbon to oxygen ratio and excellent electrical properties. However, the only concern is scalability that needs to be improved [111].

e) Solvothermal Reduction- In this approach, GO is dispersed in a solvent and reduction is carried out under supercritical conditions which arise due to heating in a sealed vessel [112, 113]. Moreover, solvothermal reduction aided by ultrasonication leads to rapid formation of rGO [114]. Supercritical water is a commonly used green solvent in this technique and results in elimination of functionalities in GO thereby restoring its aromatic framework [115]. The solvothermal route using water as solvent is also branded as hydrothermal reduction [116, 117].

In addition to the above methods, multi-step reductions are also well known for reduction of GO [118, 119]. The multi-step reduction is an extremely capable process which is known to generate high quality rGO. This method is nothing but an amalgamation of the above-mentioned methods.

BOTTOM-UP SYNTHESIS

The bottom-up approach deals with assembly of graphene from carbon precursors, *i.e.* step by step, building up graphene from carbon particles [120]. Chemical vapour deposition (CVD), epitaxial growth, template synthesis, pyrolysis, substrate free synthesis, and total organic synthesis are commonly used bottom-up strategies.

Chemical Vapour Deposition

Chemical vapour deposition (CVD) technique involves decomposition of volatile precursors on a substrate at elevated temperature in the presence of vacuum to prepare nanomaterials [121]. It is an extensively used method for industrial scale preparation of graphene. In this method, carbonaceous precursors undergo pyrolytic decomposition on crystalline substrate resulting in the creation of adsorption sites on the substrate's surface and thereby promoting the growth of high quality homogeneous thin films of graphene [122]. Generally, metals are used as substrates as well as catalysts, whereas carbonaceous precursors include hydrocarbons or biomass materials [123 - 125]. Recently, silicon carbide [126], boron nitride [127], quartz [128], silica [129], *etc.* have also been used as substrates for CVD. The greatest advantage of using CVD is that it produces high quality graphene with less defects, exceedingly cross-linked structure and higher surface area [130]. However, the method also suffers from some shortcomings like high production cost and, need for purification to remove catalysts which can be overcome by optimizing the process to low temperature and pressure [131].

Singh and his research group in their recent work have investigated nickel-catalyzed thermal chemical vapour decomposition of acetylene on steel (GCr15) for graphene synthesis [132]. The method typically involved the use of GCr15 alloy steel discs of 24 mm diameter. The discs were mechanically ground and immersed in acetone and isopropanol, respectively, followed by sonication. Thereafter, the surfaces of steel discs were electroplated with a nickel layer of 25 μm thickness to assist graphene deposition. Acetylene was used as a carbon source with hydrogen gas as a reducing medium. The discs were mounted in a ceramic boat and kept inside the quartz in the centre of the furnace. Then, a vacuum of 1.2×10-2 Torr was created inside the furnace and maintained. Next, the discs were heated at room temperature and at various growth temperatures ranging from 650 °C to 950 °C in 30 min. followed by annealation for 40 min under H_2 gas flow (20 sccm) at the growth temperature. Then, acetylene (6 sccm) was made to flow through the sample for 20 min. along with H_2 (20 sccm). At elevated temperature, the decomposed carbon atoms were absorbed by the nickel and acetylene flow was stopped after reaction completion. Then the cooling process in a hydrogen-rich environment resulted in the formation of graphene due to carbon precipitation. Further, acetylene flow rates (6, 8, and, 10 sccm, growth temperature (850 °C and 950 °C) and reaction time (10 and 20 min.) were also optimized to afford the highest quality graphene. Overall, the study showed that the highest quality graphene over nickel was obtained at a growth temperature of 850 °C for a reaction time of 10 min. under 6 sccm acetylene gas flow rate. This is because at 850 °C, the graphene formed exhibits higher graphitization, and in a short reaction time, the number of graphene layers formed is less and at reduced flow of acetylene, graphene with low defects is formed [133].

In another study, graphene was prepared using a plasma enhanced CVD method. Munoz *et al.* for the first time have demonstrated growth of graphene on silicon dioxide *via* plasma CVD at a low temperature [134]. In this approach, the ASTEX AX 4500 ECR system was used as a microwave power source, a two zone chamber and a two-stage pumping system were also employed. Acetylene and hydrogen gas mixtures and silicon wafers (111) with native oxide as substrate were used in this process. The method involved two steps-(i) growth of high quality graphitic seeds with controlled treatment and density and (ii) promotion of edge growth leading to graphene film formation. In this method, hydrogen flow was adjusted at 50 sccm while acetylene flow was tuned in the range from 0.9 to 0.34 sccm for nucleation and from 0.8 to 0.24 sccm for growth of graphene films. Plasma was generated by applying a power of 200 W and a temperature ranging from 550°C to 650°C was set-up to carry out CVD. After completion of nucleation and growth steps, annealing of graphene films at 650°C was performed in vacuum for 2 h to produce supreme quality graphene films. This method resulted in the fabrication of a monolayer and a few layer graphene films with domain sizes ranging from 50 nm to >300 nm. Additionally, this strategy offered advantages including low temperature process, avoidance of film transfer and scalability.

Epitaxial Growth

The term epitaxial has been derived from two Greek words, "epi" (meaning above) and "taxis" (meaning arrangement), respectively [135]. In epitaxial growth technique, a crystalline film grows over a single crystalline substrate. There are usually two kinds of growth *viz.* homo-epitaxial (when film and substrate are of the same material) and hetero-epitaxial (when film and substrate are of different materials) [136]. For graphene synthesis, a mono crystal of silicon carbide is used as a substrate for epitaxial growth. This technique is usually performed at elevated temperatures under ultra-high vacuum (UHV) [137]. At high temperatures, silicon sublimes, leaving behind carbon atoms which undergo sp^2 hybridization and arrange to form aromatic hexagonal ring network, stimulating graphitization leading to the formation of graphene [138]. This method offers commercial synthesis of graphene but it has certain drawbacks which include expensive protocols and production of graphene lacking uniformity [139]. However, in recent years, modified epitaxial growth methods have been developed to overcome these issues. Of late, Lalmi and his research group have demonstrated a large scale trilayer graphene synthesis *via* epitaxial growth on 6H-SiC (0001) [140]. The typical process included hydrogen etching of SiC at 1550°C to produce 6H-SiC (0001). Thereafter, 6H-SiC (0001) was heated at 1000°C in a semi-ultra high vacuum chamber and later heated at 1550°C in an atmosphere of argon, causing graphitization. An active graphene layer attached to the substrate was obtained and later cooled down for further processing. The method yielded flower-shaped and wrinkled three-layered graphene (Fig. **8**). The unique morphology of graphene is ascribed to partial dislocations between ABA and ABC stacking in trilayered graphene.

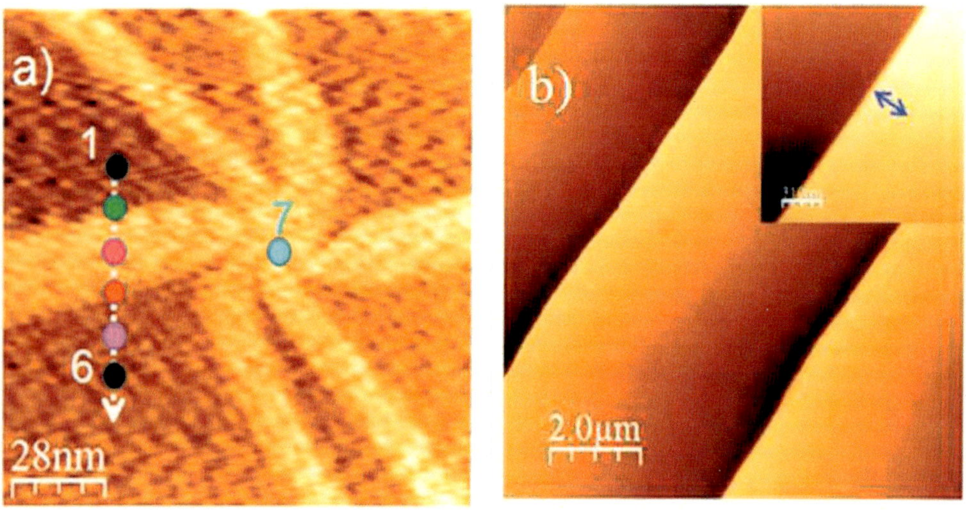

Fig. (8). STM images showing **(a)** flower-shaped domain of graphene **(b)** three-layered graphene. Reprinted with permission from [140] Copyright 2014, The Author(s).

In recent years, other than silicon carbide, boron nitride, metals like copper, nickel, *etc.* have also been used as substrates for epitaxial growth of graphene. For instance, Xu *et al.* reported epitaxial synthesis of graphene on commercially available Cu foil [141]. The method involved placing Cu foil on a quartz substrate and mounting it on a CVD furnace and annealing at 1030°C under the flow of Ar (500 sccm) to yield a Cu (111) crystal. The single Cu crystal (111) was then constantly moved past the furnace at a speed of 2.5 cm min^{-1}. For graphene growth, the CVD furnace was heated to 1030°C under the flow of Ar at 500 sccm and H_2 at 10 sccm. Then methane is made to flow at 1-5 sccm through the CVD reactor. Graphitization leading to growth of graphene film was observed and upon completion of growth, the system was allowed to cool under the atmosphere of Ar (500 sccm) and H_2 (10 sccm). The method afforded ultra high quality graphene film in just 20 min. by epitaxial growth of graphene islands over a single crystal of Cu (111). These islands combine on the Cu surface to form a graphene film displaying exceptional crystallinity. On the whole, the technique presented a rapid and high yielding protocol for ultra-crystalline graphene film synthesis.

Template Synthesis

In template synthesis, an organic or inorganic material is employed as a template to express the growth of novel materials [142]. For graphene fabrication, carbon precursors are used along with pre-synthesized templates. This method is known as a hard-template synthesis. In hard template synthesis, a template works as a mold for replication of carbon materials, and no considerable chemical interactions occur between the template and the carbon precursor [143]. There is another approach in which molecular self-assembly of templates takes place *via* hydrogen bonding, hydrophobic/hydrophilic interactions, ion pairing, or dative bonding. This type of template synthesis involves chemical interactions between templates and carbon and is known as soft-template synthesis [144]. The synthesis of graphene *via* template synthesis using metal oxides [145], micelles [146], polymers [147], *etc.* has been reported to produce graphene with a well ordered structure and high quality.

Yang and co-workers have reported a simple soft-hard template method for the synthesis of graphene [148]. In this method, cetyltrimethyl ammonium bromide (CTAB) was employed as a soft template, silica (SiO_2) obtained from tetramethoxysilane (TMOS) was used as a hard template, and pyrene as a source of carbon (Fig. **9**). The process began with hydrolysis of TMOS in water under acidic conditions at pH=3 at ambient temperature of 2 h. Then, CTAB solution and pyrene were added to it and the reaction mixture was stirred for a few minutes. The homogenous sol so obtained was spin coated on quartz and silicon substrates. Subsequently, the films were thermally treated to vapourize the solvent and to further induce solidification of the composite. Later, the SiO_2/CTAB/pyrene composite was ground and annealed at 900°C for 2 h under N_2 atmosphere to obtain graphene loaded on silica substrate (SiO_2/graphene/SiO_2 composite). The silica was removed by treatment of the composite with HF (10% aqueous solution) for 48 h to obtain pure graphene. The graphene films obtained by template self assembly synthesis were of high quality and had thickness of the order of 1 nm and a size of about 5 μm.

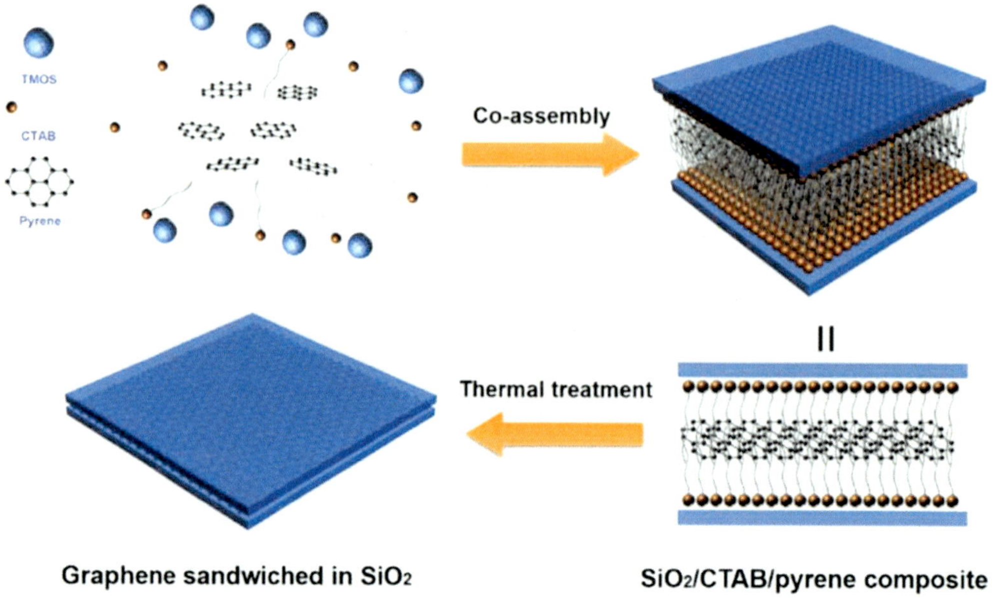

Fig. (9). Template synthesis of graphene film. Reprinted with permission from [148]. Copyright 2014, The Author(s).

In another method, micelle template based synthesis was used to fabricate N-doped mesoporous graphene [149]. Huang *et al.* first mixed 50 mL of GO suspension and 150 mL of hydrazine hydrate solution at ambient temperature *via* vigorous stirring for 12 h. Afterwards, 20 mL of F127 micellar solution was added to the above suspension and the reaction mixture was stirred for 2 h. Later, HCl (14 mL) was added and the mixture was briskly stirred for 24 h. The precipitate was collected by vacuum filtration and sintered at 350°C for 5 h in argon atmosphere and for another 5 h at 900°C to generate highly mesoporous graphene. In order to obtain N-doped graphene, 50 mg of graphene and 250 mg of cyanamide were mixed for 15 min. followed by sintering at 800°C for 3 h in argon atmosphere. Thus, this method was quite efficient and afforded graphene with homogenously distributed mesopores, a large pore volume and a huge surface area.

Pyrolysis

Pyrolysis is a high temperature treatment involving thermal decomposition of a carbon source leading to the formation of C atoms that rearrange to form a graphitic carbon framework, leading to the formation of graphene. The method is inexpensive and produces graphene with superb purity but with few defects [150]. A gram scale synthesis of multi layer graphene was fabricated by Hong *et al.* [151]. This method involved pyrolysis of poly(methyl methacrylate) (PMMA) on nickel microparticles in the restricted microzones created by organophilic montmorillonite (OMT). The typical process involved a pyrolysis apparatus with a quartz tube. The PMMA sample was placed in a quartz tube and placed inside a furnace fitted with a thermocouple under N_2 atmosphere. The carbonization was carried out at 1000°C for 8 min. and the product was obtained as black powder by cooling the reactor under N_2 flow. The product was further purified by etching the impure graphene powder with HF (10 wt.%) and HNO_3 (30 wt.%) under ultrasonication at ambient temperature followed by filtration, washing and drying at 80°C. Initially, PMMA sample was obtained by mixing methyl methacrylate, OMT and benzoyl peroxide under magnetic stirring followed by heating at 85°C under N_2 atmosphere. A thick paste was formed to which Ni particles were added and polymerization was carried out for 48 h at 60°C. Thus, wrinkled graphene sheets with a thickness of 1-4 nm and with few defects were achieved by this method. It was further reported that the addition of OMT restricts the pyrolytic carbon derivatives obtained by PMMA decomposition in microzones and improves yields of graphene produced catalytically by Ni particles in the microzones. Thus, the method offers inexpensive and scalable production of graphene.

Recently, the spray pyrolysis method has been used for the synthesis of graphene. In spray pyrolysis, a precursor solution is sprayed on a heated surface to produce aerosol droplets which undergo reaction and decompose to form particles and thin film [152]. Zou and his research group have demonstrated production of graphene nanosheets by spray pyrolysis of precursor's *viz.* iron carbonyl and pyridine [153]. In this process, the precursors (carbonyl and pyridine) were introduced into a quartz reactor at 5 mL min^{-1} flow rate. Also, N_2 was made to flow through the reactor at 600 L h^{-1}. The resultant sample was collected and mixed with excess ammonium chloride followed by heat treatment at 800°C for 30 min. in a quartz tube to remove the residual precursors. Finally, pure graphene nanosheets of good quality were obtained by washing, filtration and drying.

Substrate Free Synthesis

Substrate free synthesis is a comparatively newly developed strategy where graphene is synthesized *via* gas phase reactions in the absence of substrates, solvents and catalysts. In general, this method includes four strategies including carbonization, arc discharge, combustion, and atmospheric plasma [154]. The advantages of this method comprise scalability and production of high quality graphene with a tunable number of layers, low content of oxygen, and enhanced purity [154].

A substrate free, argon-plasma induced synthesis of graphene sheets was reported by Dato and his research group [155]. The researchers carried out the graphene synthesis in an atmospheric pressure microwave plasma reactor (Fig. **10a**) equipped with a quartz tube for conduction of argon gas stream (1.71 L/min.) *via* a microwave guide. This stream generates argon plasma. Further, with the aid of an alumina tube placed concentrically within the quartz tube, aerosols comprising argon gas (2 L/min.) and ethanol droplets (4×10^{-4} L/min.) were injected straightforwardly into the argon plasma. The ethanol droplets resided within the plasma for a very short time (10^{-1} s) during which these droplets evapourated and dissociated within the plasma forming a solid carbon material. Subsequently, the reaction product *i.e.* graphene after moving through the plasma was quickly cooled and collected on nylon filter membranes. It was reported that the rate of graphene sheets collected on filters was 2 mg/ min, for an input of carbon present in the ethanol of 164 mg/ min. Further, the graphene sheets collected on the filters were dispersed in methanol (Fig. **10b**) and sonicated for 5 min. to obtain a black homogenous suspension. The graphene sheets are composed of mono- and bi-layers as characterized by transmission electron microscopy (Fig. **10c**). The highlight of this method was continuous production of atomically thin graphene sheets.

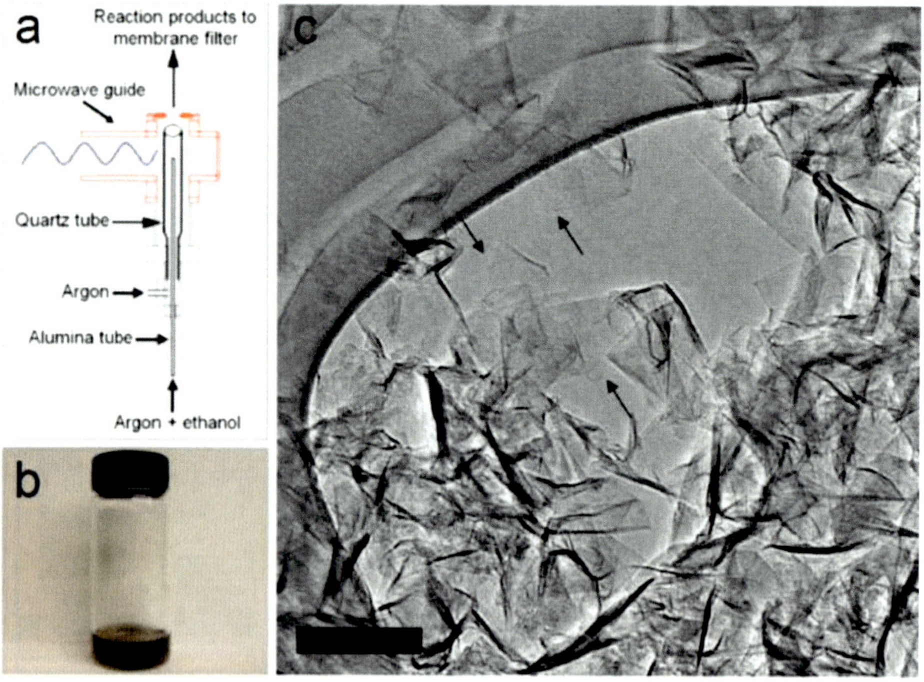

Fig. (10). **(a)** The microwave plasma reactor employed for graphene synthesis. **(b)** Image showing graphene dispersed in methanol. **(c)** TEM image of graphene sheets. Reprinted with permission from [155] Copyright © 2008, American Chemical Society.

In another instance, a multi-layer graphene sheet with a large specific surface area was produced by employing substrate free gas phase synthesis. Liu and Zhang have demonstrated a substrate free fabrication of graphene in a CVD reactor comprising quartz tube placed inside a horizontal furnace [156]. In the first step, camphor was put inside the tube and argon gas flowed through it at atmospheric pressure. The camphor sublimed on heating at 180°C and the volatilized camphor was pyrolyzed at 800-850°C in the furnace. The reaction was accomplished in 40 min., producing a graphene film which adhered to the wall of the chamber. The film was later scraped off and collected in the quartz tube. A graphene film of a large surface area of about 1 cm×1 cm was obtained by this method. Other advantages of this method include scalability and inexpensive production of graphene.

Total Organic Synthesis

The synthesis of graphene from polycyclic aromatic hydrocarbons (PAHs) constitutes total organic synthesis of graphene. PAHs are defined as 2D graphene fragments composed of all sp^2 hybridized carbons and can be readily modified by using appropriate aliphatic chains to tune the solubility of the product [157, 158]. In this method, suitable PAHs are chosen and used to prepare good quality graphene at high yields *via* a facile reaction pathway. Several reactions such as Suzuki coupling, Yamamoto coupling, Diels-Alder's cylcoaddition, *etc.* have been employed for synthesis of nano-graphene materials [159]. Recently, Vo and his research group have demonstrated a total organic synthesis of N-doped graphene nano ribbons [160]. The process involved several steps. In the first step, phenanthrene-9,10-dione was mixed with 98% sulfuric acid and N-bromosuccinimide was added to the mixture. The reaction mixture was then stirred vigorously at ambient temperature for 8 h. Then, ice-cold water was added to the reaction mixture and filtered to obtain orange colored 2,7-dibromophenanthrene-9,10-dione. In the next step, 5,10-dibromo-1,3-diphe-yl-2H-cyclopenta[l]phenanthren-2-one was synthesized by addition of 2,7-dibromophenanthrene-9,10-dione and 1,3-diphenylacetone methanol under stirring. Then, the reaction mixture was refluxed for 2h after addition of 0.3 M KOH solution drop-wise to the reaction mixture. After filtration, a green colored product was obtained. Next, 5-(phenylethynyl)pyrimidine was prepared by reaction between triethylamine, toluene, 5-bromopyrimidine, and triphenylphosphine in the presence of copper iodide. After degassing the suspension for 15 min., bis(triphenylphosphine)palladium (II) chloride was added. The reaction mixture was again constantly degassed for 15 min. followed by heating at 60 °C and stirring for 30 min. Later, phenylacetylene was added to the reaction mixture and heated at 65-70 °C for 18 h. After filtration and column chromatography, the product was isolated.

In the next step, a reaction between 5,10- dibromo-1,3-diphenyl-2H-cyclo-penta[l]phenanthren-2-one, 5-(phenylethynyl)pyrimidine and diphenyl ether was carried out by refluxing for 24 h. this led to formation of 5-(6,11-dibromo-1,-,4-triphenyltriphenylen-2-yl)pyrimidine. Subsequently, bis(cyclooctadiene)nickel was added to a mixture of 2,2'-bipyridyl and 1,5-cyclooctandiene in dried N,N-Dimethylformamide as solvent and heated at 60 °C for 30 min. The degassed solution of 5-(6,11-dibromo-1,3,4-triphenyltriphenylen-2-yl)pyrimidine in toluene was added to the reaction mixture. The reaction mixture was then heated to 80 °C for 3 days and then cooled followed by the addition of methanol/water mixture (50mL/10mL). Thereafter, the slurry was filtered, washed with methanol, concentrated HCl, sodium hydroxide/methanol solution, water, and hexane, followed by drying. Finally, a polymer product was obtained. In the last step, the polymer and dichloromethane were sonicated. Then, iron (III) chloride dissolved in nitromethane was added to the above suspension. Next, nitrogen gas was passed through the reaction mixture for 15- 24 h under stirring condition. Then ethanol was added and the reaction mixture was filtered. The reaction product was then washed with concentrated HCl, water, and methanol/sodium hydroxide solution to obtain a black solid. This solid was dried at 50 °C for 24 h to yield a black powder of N-doped graphene nanoribbons. Thus, this method produces large quantities of high quality N-doped graphene *via* a strategy based on Yamamoto coupling of N-doped precursors and cyclodehydrogenation through Scholl reaction.

A similar scheme was reported by Dichtel *et al.* who used poly(phenylene ethynylene) to produce graphene nanoribbons [161]. The process was based on Asao-Yamamoto alkyne benzannulation [162] promoted by Cu(II) and Brønsted acid that proceeds *via* a benzopyrylium intermediate to generate precursors for nanographene which are further oxidized by Scholl reaction to produce nano graphene ribbons. In this process, diynes undergo benzannulation to form a polyaromatic hydrocarbon intermediate which undergoes oxidation to form nanographene derivatives. The process was quite efficient and resulted in the production of good quality graphene nanoribbons.

Biosynthesis of Graphene and its Derivatives

Besides the aforementioned methods, biosynthesis has also emerged as a sustainable strategy for of preparation of graphene and its derivatives. The biomolecules derived from living organisms or cells or plant extracts serve as reducing, capping as well as stabilizing agents for the synthesis of graphene materials thereby eliminating the use of hazardous chemical agents [163]. Recently, Khanam and Hasan have reported a benign biosynthesis of graphene using extracts of *Allium cepa* (onion) as non-toxic reducing agents [164]. In this method GO was synthesized by modified Hummers method and then reduced by extracts of *Allium cepa* to produce graphene. The onion skins were first peeled off; the scales were chopped up and then blended to produce juice. Then, about 10 mL of the onion juice was added to GO and the suspension was sonicated for 30 min. Thereafter, the suspension of GO in juice was stirred for 6 h at ambient temperature and the color of dispersion changed from brown to black. This confirmed the transformation of GO to graphene. The mixture was then filtered and the impure graphene was washed with ethanol and water followed by drying to yield pure graphene. Thus, this method afforded a green route for efficacious and economically compatible production of graphene. It was further reported that the graphene so obtained was thin, crumpled and multilayer with a thickness of the order of 140 nm. In another report, green tea leaf extracts containing polyphenols which can serve as reducing and stabilizing agents were used for biofabrication of reduced graphene oxide.

The microbial reduction of graphene oxide is also an important biological method for production of rGO or graphene. For instance, Gurunathan and co-workers have reported reduction of GO using *Escherichia coli* biomass [165]. The *E. coli* bacteria were grown on LB medium (Luria-Bertani broth comprising 10 g tryptone, 5 g yeast extract, and 10 g/L NaCl) in an Erlenmeyer flask. The flask was incubated in a rotatory shaker agitating at a speed of 120 rpm/min for 21 h at 37°C. The bacterial culture after incubation was centrifuged at 10,000 rpm/min. then, the bacterial biomass was employed for reduction of GO. Next, to a GO dispersion, about 200 mg of *E. coli* biomass was added and the mixture was stirred for 72 h at 37°C. The reduction of GO produced a black suspension that was sonicated for 5 min. followed by centrifugation for 10 min. at 10,000 rpm/min to eliminate bacterial biomass as supernatant liquid. Thus, reduced graphene oxide was obtained as black pellets. This method resulted in highly efficient, large scale production water dispersible reduced graphene oxide by eliminating the use of toxic chemical reductants. Therefore, *E. coli* based reduction of GO presented an eco-friendly protocol. Similarly, an inexpensive and environmentally-benign synthesis of microbially reduced graphene oxide was also reported by Wang and his research group using *Shewanella oneidensis* MR-1 cell culture. Consequently, it is quite obvious that the bioproduction of graphene or reduced graphene oxide has gained great attention over the past few years owing to the simplicity, eco-compatibility, cost-effectiveness and sustainability of these methods.

CONCLUDING REMARKS

The superlative and desirable characteristics of graphene and its derivatives, including large specific surface area, high chemical stability, extraordinary electrical and thermal properties, high intrinsic mobility, unique optical properties and enormous mechanical strength, have rendered them as the most useful carbonaceous nanomaterials. Considering the wide scope of these materials, several researchers from time to time have immersed themselves in developing simple, sustainable and efficacious synthetic strategies for graphene synthesis. In this chapter, we have summarized some of the most widely employed methods for fabrication of graphene and its derivatives. Several methods including exfoliation, unrolling or unzipping of carbon nanotubes, electric arc discharge method, laser ablation technique, oxidative exfoliation-reduction of graphene oxide, chemical vapour deposition, epitaxial growth, template synthesis, pyrolysis, substrate free synthesis, total organic synthesis, and biological methods have been used for efficient synthesis of high quality graphene and its derivatives. However, despite the rapidly expanding research in the field of graphene chemistry and availability of a large spectrum of methods, there are still a few challenges that need to be addressed for industrial scale synthesis of graphene. For instance, most of these methods largely yield few layer graphene. Scalable synthesis of monolayer graphene is still an area that needs to be worked upon. From economic perspective, designing of inexpensive modus operandi is the need of the hour for commercial graphene production. Also, eco-friendly graphene synthesis which minimizes the release of toxic gases, makes use of benign solvents and precursors, energy intensive protocols, and avoids further processing should be encouraged so at to reduce adverse impact on our environment. Overall, it can be said that in order to realize the dream of industrialization and commercialization of graphene and its derivatives, fabrication of scalable, reproducible, economically compatible and benign synthetic methods is of utmost significance. On a positive note, it is worth mentioning that the pace with which the graphene research is growing, it is quite certain that in near future a plethora of improved methodologies will be accessible for manufacturing of graphene with promising applications in our day to day life.

LIST OF ABBREVIATIONS

2-D	Two-dimensional
3-D	Three-dimensional
AC	Alternating current
APTES	3-Aminopropyl triethoxysilane
CNT	Carbon nanotubes
CTAB	Cetyltrimethyl ammonium bromide
CVD	Chemical vapour deposition
DC	Direct current

DMF	Dimethyl formamide
GO	Graphene oxide
HOPG	Highly oriented pyrolytic graphite
LB	Luria-Bertani
MMS	Mercury/Mercury (I) sulfate
MWCNT	Multi-walled carbon nanotubes
OMP	Organophilic montmorillonite
PAC	Polycyclic Aromatic Hydrocarbon
PANi	Polyaniline
PMMA	Polymethyl methacryalte
PVC	Polyvinyl chloride
PVP	Polyvinyl pyrrolidine
rGO	Reduced graphene oxide
SWCNT	Single-walled carbon nanotubes
TMOS	Tetramethoxysilane
UHV	Ultra high vacuum

CONSENT FOR PUBLICATION

Not applicable.

CONFLICT OF INTEREST

The author declares no conflict of interest, financial or otherwise.

ACKNOWLEDGEMENTS

The authors are grateful to the esteemed publishers for providing copyright permission to publish figures/diagrams/schemes used in this book chapter.

REFERENCES

[1] Allen, M.J.; Tung, V.C.; Kaner, R.B. Honeycomb carbon: a review of graphene. *Chem. Rev.*, **2010**, *110*(1), 132-145.
[http://dx.doi.org/10.1021/cr900070d] [PMID: 19610631]

[2] Brodie, B.C. XIII. On the atomic weight of graphite. *Philos. Trans. R. Soc. Lond.*, **1859**, *149*, 249-259.
[http://dx.doi.org/10.1098/rstl.1859.0013]

[3] Boehm, H.P.; Setton, R.; Stumpp, E. Nomenclature and terminology of graphite intercalation compounds. *Carbon,* **1986**, *24*(2), 241-245.
[http://dx.doi.org/10.1016/0008-6223(86)90126-0]

[4] Novoselov, K.S.; Geim, A.K.; Morozov, S.V.; Jiang, D.; Zhang, Y.; Dubonos, S.V.; Grigorieva, I.V.; Firsov, A.A. Electric field effect in atomically thin carbon films. *Science,* **2004**, *306*(5696), 666-669.

[http://dx.doi.org/10.1126/science.1102896] [PMID: 15499015]

[5] Gerstner, E. Nobel Prize 2010. *Andre Geim & Konstantin Novoselov. Nature Phys.,* **2020**, *6*, 836.

[6] Zhu, Y.; Murali, S.; Cai, W.; Li, X.; Suk, J.W.; Potts, J.R.; Ruoff, R.S. Graphene and graphene oxide: synthesis, properties, and applications. *Adv. Mater.*, **2010**, *22*(35), 3906-3924. [http://dx.doi.org/10.1002/adma.201001068] [PMID: 20706983]

[7] Bharech, S.; Kumar, R.J. A review on the properties and applications of graphene. *Mat. Sci. Mech. Eng.,* **2015**, *2*, 70-73.

[8] Singh, R.K.; Kumar, R.; Singh, D.P. Graphene oxide: strategies for synthesis, reduction and frontier applications. *RSC Advances,* **2016**, *6*(69), 64993-65011. [http://dx.doi.org/10.1039/C6RA07626B]

[9] Olenych, I.B.; Aksimentyeva, O.I.; Monastyrskii, L.S.; Horbenko, Y.Y.; Partyka, M.V. Electrical and photoelectrical properties of reduced graphene oxide-porous silicon nanostructures. *Nanoscale Res. Lett.,* **2017**, *12*(1), 272. [http://dx.doi.org/10.1186/s11671-017-2043-7] [PMID: 28410550]

[10] Zhen, Z.; Zhu, H. Structure and properties of graphene. In: *Graphene: Fabrication, Characterizations, Properties and Applications*; Zhen, Z.; Zhu, H.; Xie, D.; Fang, Y., Eds.; Academic Press: Cambridge, **2018**. [http://dx.doi.org/10.1016/B978-0-12-812651-6.00001-X]

[11] Zhang, Y.; Small, J.P.; Pontius, W.V.; Kim, P. Fabrication and electric-field-dependent transport measurements of mesoscopic graphite devices. *Appl. Phys. Lett.,* **2005**, *86*(7), 073104. [http://dx.doi.org/10.1063/1.1862334]

[12] Yang, G.; Li, L.; Lee, W.B.; Ng, M.C. Structure of graphene and its disorders: a review. *Sci. Technol. Adv. Mater.,* **2018**, *19*(1), 613-648. [http://dx.doi.org/10.1080/14686996.2018.1494493] [PMID: 30181789]

[13] Edwards, R.S.; Coleman, K.S. Graphene synthesis: relationship to applications. *Nanoscale,* **2013**, *5*(1), 38-51. [http://dx.doi.org/10.1039/C2NR32629A] [PMID: 23160190]

[14] Tiwari, S.K.; Mishra, R.K.; Ha, S.K.; Huczko, A. Evolution of graphene oxide and graphene: from imagination to industrialization. *ChemNanoMat,* **2018**, *4*(7), 598-620. [http://dx.doi.org/10.1002/cnma.201800089]

[15] Sang, M.; Shin, J.; Kim, K.; Yu, K. Electronic and thermal properties of graphene and recent advances in graphene based electronics applications. *Nanomaterials (Basel),* **2019**, *9*(3), 374. [http://dx.doi.org/10.3390/nano9030374] [PMID: 30841599]

[16] Jia, X.; Campos-Delgado, J.; Terrones, M.; Meunier, V.; Dresselhaus, M.S. Graphene edges: a review of their fabrication and characterization. *Nanoscale,* **2011**, *3*(1), 86-95. [http://dx.doi.org/10.1039/C0NR00600A] [PMID: 21103548]

[17] Chen, D.; Feng, H.; Li, J. Graphene oxide: preparation, functionalization, and electrochemical applications. *Chem. Rev.,* **2012**, *112*(11), 6027-6053. [http://dx.doi.org/10.1021/cr300115g] [PMID: 22889102]

[18] Dreyer, D.R.; Park, S.; Bielawski, C.W.; Ruoff, R.S. The chemistry of graphene oxide. *Chem. Soc. Rev.,* **2010**, *39*(1), 228-240. [http://dx.doi.org/10.1039/B917103G] [PMID: 20023850]

[19] Hofmann, U.; Holst, R. Über die Säurenatur und die Methylierung von Graphitoxyd. *Ber. Dtsch. Chem. Ges. B,* **1939**, *72*(4), 754-771. [http://dx.doi.org/10.1002/cber.19390720417]

[20] Ruess, G. Uber das Graphitoxyhydroxyd (Graphitoxyd). *Monatsh. Chem.,* **1947**, *76*(3-5), 381-417. [http://dx.doi.org/10.1007/BF00898987]

[21] Clause, A.; Plass, R.; Boehm, H-P.; Hofmann, U. Untersuchungen zur Struktur des Graphitoxyds. *Z. Anorg. Allg. Chem.*, **1957**, *291*(5-6), 205-220. [http://dx.doi.org/10.1002/zaac.19572910502]

[22] Scholz, W.; Boehm, H.P. Untersuchungen am Graphitoxid. VI. Betrachtungen zur Struktur des Graphitoxids. *Z. Anorg. Allg. Chem.*, **1969**, *369*(3-6), 327-340. [http://dx.doi.org/10.1002/zaac.19693690322]

[23] Nakajima, T.; Mabuchi, A.; Hagiwara, R. A new structure model of graphite oxide. *Carbon,* **1988**, *26*(3), 357-361. [http://dx.doi.org/10.1016/0008-6223(88)90227-8]

[24] Nakajima, T.; Matsuo, Y. Formation process and structure of graphite oxide. *Carbon,* **1994**, *32*(3), 469-475. [http://dx.doi.org/10.1016/0008-6223(94)90168-6]

[25] Lerf, A.; He, H.; Riedl, T.; Forster, M.; Klinowski, J. 13C and 1H MAS NMR studies of graphite oxide and its chemically modified derivatives. *Solid State Ion.*, **1997**, *101-103*, 857-862. [http://dx.doi.org/10.1016/S0167-2738(97)00319-6]

[26] Lerf, A.; Buchsteiner, A.; Pieper, J.; Schöttl, S.; Dekany, I.; Szabo, T.; Boehm, H.P. Hydration behavior and dynamics of water molecules in graphite oxide. *J. Phys. Chem. Solids,* **2006**, *67*(5-6), 1106-1110. [http://dx.doi.org/10.1016/j.jpcs.2006.01.031]

[27] Szabó, T.; Tombácz, E.; Illés, E.; Dékány, I. Enhanced acidity and pH-dependent surface charge characterization of successively oxidized graphite oxides. *Carbon,* **2006**, *44*(3), 537-545. [http://dx.doi.org/10.1016/j.carbon.2005.08.005]

[28] Eigler, S.; Hirsch, A. Chemistry with graphene and graphene oxide-challenges for synthetic chemists. *Angew. Chem. Int. Ed.*, **2014**, *53*(30), 7720-7738. [http://dx.doi.org/10.1002/anie.201402780] [PMID: 24962439]

[29] Gómez-Navarro, C.; Weitz, R.T.; Bittner, A.M.; Scolari, M.; Mews, A.; Burghard, M.; Kern, K. Electronic transport properties of individual chemically reduced graphene oxide sheets. *Nano Lett.*, **2007**, *7*(11), 3499-3503. [http://dx.doi.org/10.1021/nl072090c] [PMID: 17944526]

[30] Bagri, A.; Mattevi, C.; Acik, M.; Chabal, Y.J.; Chhowalla, M.; Shenoy, V.B. Structural evolution during the reduction of chemically derived graphene oxide. *Nat. Chem.*, **2010**, *2*(7), 581-587. [http://dx.doi.org/10.1038/nchem.686] [PMID: 20571578]

[31] Becerril, H.A.; Mao, J.; Liu, Z.; Stoltenberg, R.M.; Bao, Z.; Chen, Y. Evaluation of solution-processed reduced graphene oxide films as transparent conductors. *ACS Nano,* **2008**, 463-470. [http://dx.doi.org/10.1021/nn700375n]

[32] Mattevi, C.; Eda, G.; Agnoli, S.; Miller, S.; Mkhoyan, K.A.; Celik, O.; Mastrogiovanni, D.; Granozzi, G.; Garfunkel, E.; Chhowalla, M. Evolution of Electrical, Chemical, and Structural Properties of Transparent and Conducting Chemically Derived Graphene Thin Films. *Adv. Funct. Mater.*, **2009**, *19*(16), 2577-2583. [http://dx.doi.org/10.1002/adfm.200900166]

[33] Castro Neto, A.H.; Guinea, F.; Peres, N.M.R.; Novoselov, K.S.; Geim, A.K. The electronic properties of graphene. *Rev. Mod. Phys.*, **2009**, *81*(1), 109-162. [http://dx.doi.org/10.1103/RevModPhys.81.109]

[34] Du, X.; Skachko, I.; Barker, A.; Andrei, E.Y. Approaching ballistic transport in suspended graphene. *Nat. Nanotechnol.*, **2008**, *3*(8), 491-495. [http://dx.doi.org/10.1038/nnano.2008.199] [PMID: 18685637]

[35] Huang, X.; Yin, Z.; Wu, S.; Qi, X.; He, Q.; Zhang, Q.; Yan, Q.; Boey, F.; Zhang, H. Graphene-based materials: synthesis, characterization, properties, and applications. *Small,* **2011**, *7*(14), 1876-1902.

[http://dx.doi.org/10.1002/smll.201002009] [PMID: 21630440]

[36] Kumar, R.; Sahoo, S.; Joanni, E.; Singh, R.K.; Tan, W.K.; Kar, K.K.; Matsuda, A. Recent progress in the synthesis of graphene and derived materials for next generation electrodes of high performance lithium ion batteries. *Pror. Energy Combust. Sci.*, **2019**, *75*, 100786. [http://dx.doi.org/10.1016/j.pecs.2019.100786]

[37] Dresselhaus, M.S.; Dresselhaus, G. Intercalation compounds of graphite. *Adv. Phys.*, **2002**, *51*(1), 1-186. [http://dx.doi.org/10.1080/00018730110113644]

[38] Whitener, K.E., Jr; Sheehan, P.E. Graphene synthesis. *Diamond Related Materials*, **2014**, *46*, 25-34. [http://dx.doi.org/10.1016/j.diamond.2014.04.006]

[39] Yi, M.; Shen, Z. A review on mechanical exfoliation for the scalable production of graphene. *J. Mater. Chem. A Mater. Energy Sustain.*, **2015**, *3*(22), 11700-11715. [http://dx.doi.org/10.1039/C5TA00252D]

[40] Chen, J.; Duan, M.; Chen, G. Continuous mechanical exfoliation of graphene sheets *via* three-roll mill. *J. Mater. Chem.*, **2012**, *22*(37), 19625-19628. [http://dx.doi.org/10.1039/c2jm33740a]

[41] Noroozi, M.; Zakaria, A.; Radiman, S.; Abdul Wahab, Z. Environmental synthesis of few layers graphene sheets using ultrasonic exfoliation with enhanced electrical and thermal properties. *PLoS One*, **2016**, *11*(4), e0152699. [http://dx.doi.org/10.1371/journal.pone.0152699] [PMID: 27064575]

[42] Balasubramanyan, S.; Sasidharan, S.; Poovathinthodiyil, R.; Ramakrishnan, R.M.; Narayanan, B.N. Sucrose-mediated mechanical exfoliation of graphite: a green method for the large scale production of graphene and its application in catalytic reduction of 4-nitrophenol. *New J. Chem.*, **2017**, *41*(20), 11969-11978. [http://dx.doi.org/10.1039/C7NJ01900A]

[43] Yi, M.; Shen, Z.; Zhu, J. A fluid dynamics route for producing graphene and its analogues. *Chin. Sci. Bull.*, **2014**, *59*(16), 1794-1799. [http://dx.doi.org/10.1007/s11434-014-0303-9]

[44] Tran, T.S.; Park, S.J.; Yoo, S.S.; Lee, T.R.; Kim, T. High shear-induced exfoliation of graphite into high quality graphene by Taylor–Couette flow. *RSC Advances*, **2016**, *6*(15), 12003-12008. [http://dx.doi.org/10.1039/C5RA22273G]

[45] Bhuyan, M.S.A.; Uddin, M.N.; Islam, M.M.; Bipasha, F.A.; Hossain, S.S. Synthesis of graphene. *Int. Nano Lett.*, **2016**, *6*(2), 65-83. [http://dx.doi.org/10.1007/s40089-015-0176-1]

[46] Liu, M.; Zhang, X.; Wu, W.; Liu, T.; Liu, Y.; Guo, B.; Zhang, R. One-step chemical exfoliation of graphite to ~100% few-layer graphene with high quality and large size at ambient temperature. *J. Mater. Chem. C Mater. Opt. Electron. Devices*, **2019**, *355*, 181-185.

[47] Wu, W.; Liu, M.; Gu, Y.; Guo, B.; Ma, H.; Wang, P.; Wang, X.; Zhang, R. Fast chemical exfoliation of graphite to few-layer graphene with high quality and large size *via* a two-step microwave-assisted process. *Chem. Eng. J.*, **2020**, *381*, 122592. [http://dx.doi.org/10.1016/j.cej.2019.122592]

[48] Bakunin, E.S.; Obraztsova, E.Y.; Rukhov, A.V. Modern methods for synthesis of few-layer graphene structures by electrochemical exfoliation of graphite. *Inorg. Mater. Appl. Res.*, **2019**, *10*(2), 249-255. [http://dx.doi.org/10.1134/S2075113319020047]

[49] Htwe, Y.Z.N.; Chow, W.S.; Suda, Y.; Thant, A.A.; Mariatti, M. Effect of electrolytes and sonication times on the formation of graphene using an electrochemical exfoliation process. *Appl. Surf. Sci.*, **2019**, *469*, 951-961. [http://dx.doi.org/10.1016/j.apsusc.2018.11.029]

[50] Tiwary, C.S.; Javvaji, B.; Kumar, C.; Mahapatra, D.R.; Ozden, S.; Ajayan, P.M.; Chattopadhyay, K. Chemical-free graphene by unzipping carbon nanotubes using cryo-milling. *Carbon,* **2015**, *89*, 217-224.
[http://dx.doi.org/10.1016/j.carbon.2015.03.036]

[51] Kumar, P.; Yamijala, S.S.R.K.C.; Pati, S.K. Optical unzipping of carbon nanotubes in liquid media. *J. Phys. Chem. C,* **2016**, *120*(30), 16985-16993.
[http://dx.doi.org/10.1021/acs.jpcc.6b02524]

[52] Thomas, R.; Hembram, K.P.S.S.; Kumar, B.V.M.; Rao, G.M. High density oxidative plasma unzipping of multiwall carbon nanotubes. *RSC Advances,* **2017**, *7*(76), 48268-48274.
[http://dx.doi.org/10.1039/C7RA04318J]

[53] Dimiev, A.M.; Khannanov, A.; Vakhitov, I.; Kiiamov, A.; Shukhina, K.; Tour, J.M. Revisiting the mechanism of oxidative unzipping of multiwall carbon nanotubes to graphene nanoribbons. *ACS Nano,* **2018**, *12*(4), 3985-3993.
[http://dx.doi.org/10.1021/acsnano.8b01617] [PMID: 29578700]

[54] Xie, L.; Wang, H.; Jin, C.; Wang, X.; Jiao, L.; Suenaga, K.; Dai, H. Graphene nanoribbons from unzipped carbon nanotubes: atomic structures, Raman spectroscopy, and electrical properties. *J. Am. Chem. Soc.,* **2011**, *133*(27), 10394-10397.
[http://dx.doi.org/10.1021/ja203860a] [PMID: 21678963]

[55] Lee, H.J.; Lim, J.; Cho, S.Y.; Kim, H.; Lee, C.; Lee, G.Y.; Sasikala, S.P.; Yun, T.; Choi, D.S.; Jeong, M.S.; Jung, H.T.; Hong, S.; Kim, S.O. Intact crystalline semiconducting graphene nanoribbons from unzipping nitrogen-doped carbon nanotubes. *ACS Appl. Mater. Interfaces,* **2019**, *11*(41), 38006-38015.
[http://dx.doi.org/10.1021/acsami.9b08876] [PMID: 31544452]

[56] Jiao, L.; Zhang, L.; Wang, X.; Diankov, G.; Dai, H. Narrow graphene nanoribbons from carbon nanotubes. *Nature,* **2009**, *458*(7240), 877-880.
[http://dx.doi.org/10.1038/nature07919] [PMID: 19370031]

[57] Jiao, L.; Fan, B.; Xian, X.; Wu, Z.; Zhang, J.; Liu, Z. Creation of nanostructures with poly(methyl methacrylate)-mediated nanotransfer printing. *J. Am. Chem. Soc.,* **2008**, *130*(38), 12612-12613.
[http://dx.doi.org/10.1021/ja805070b] [PMID: 18763767]

[58] Li, Y.S.; Liao, J.L.; Wang, S.Y.; Chiang, W.H. Intercalation-assisted longitudinal unzipping of carbon nanotubes for green and scalable synthesis of graphene nanoribbons. *Sci. Rep.,* **2016**, *6*(1), 22755.
[http://dx.doi.org/10.1038/srep22755] [PMID: 26948486]

[59] Kosynkin, D.V.; Higginbotham, A.L.; Sinitskii, A.; Lomeda, J.R.; Dimiev, A.; Price, B.K.; Tour, J.M. Longitudinal unzipping of carbon nanotubes to form graphene nanoribbons. *Nature,* **2009**, *458*(7240), 872-876.
[http://dx.doi.org/10.1038/nature07872] [PMID: 19370030]

[60] Yang, L.; Zhang, L.; Webster, T.J. Carbon nanostructures for orthopedic medical applications. *Nanomedicine (Lond.),* **2011**, *6*(7), 1231-1244.
[http://dx.doi.org/10.2217/nnm.11.107] [PMID: 21929458]

[61] Agudosi, E.S.; Abdullah, E.C.; Numan, A.; Mubarak, N.M.; Khalid, M.; Omar, N. A review of the graphene synthesis routes and its applications in electrochemical energy storage. *Crit. Rev. Solid State Mater. Sci.,* **2020**, *45*(5), 339-377.
[http://dx.doi.org/10.1080/10408436.2019.1632793]

[62] Wu, X.; Liu, Y.; Yang, H.; Shi, Z. Large-scale synthesis of high-quality graphene sheets by an improved alternating current arc-discharge method. *RSC Advances,* **2016**, *6*(95), 93119-93124.
[http://dx.doi.org/10.1039/C6RA22273K]

[63] Pham, T.V.; Kim, J.G.; Jung, J.Y.; Kim, J.H.; Cho, H.; Seo, T.H.; Lee, H.; Kim, N.D.; Kim, M.J. High areal capacitance of n-doped graphene synthesized by arc discharge. *Adv. Funct. Mater.,* **2019**, *29*(48), 1905511.

[http://dx.doi.org/10.1002/adfm.201905511]

[64] Shojaei, T.R.; Azhari, S. *Emerging Applications of Nanoparticles and Architecture Nanostructures*; Barhoum, A.; Makhlouf, A.S.H., Eds.; Elsevier: Amsterdam, **2018**, pp. 501-531. [http://dx.doi.org/10.1016/B978-0-323-51254-1.00016-6]

[65] Russo, P.; Hu, A.; Compagnini, G.; Duley, W.W.; Zhou, N.Y. Femtosecond laser ablation of highly oriented pyrolytic graphite: a green route for large-scale production of porous graphene and graphene quantum dots. *Nanoscale,* **2014**, *6*(4), 2381-2389. [http://dx.doi.org/10.1039/C3NR05572H] [PMID: 24435549]

[66] Juvaid, M.M.; Kumar, D.; Ramachandra Rao, M.S. Realization of good quality bilayer graphene by single step laser ablation process. *Mater. Res. Bull.,* **2020**, *126*, 110840. [http://dx.doi.org/10.1016/j.materresbull.2020.110840]

[67] Staudenmaier, L. Verfahren zur Darstellung der Graphitsäure. *Ber. Dtsch. Chem. Ges.,* **1898**, *31*(2), 1481-1487. [http://dx.doi.org/10.1002/cber.18980310237]

[68] Hofmann, U.; König, E. Untersuchungen über Graphitoxyd. *Z. Anorg. Chem.,* **1937**, *234*(4), 311-336. [http://dx.doi.org/10.1002/zaac.19372340405]

[69] Hummers, W.S., Jr; Offeman, R.E. Preparation of graphitic oxide. *J. Am. Chem. Soc.,* **1958**, *80*(6), 1339. [http://dx.doi.org/10.1021/ja01539a017]

[70] Cao, N.; Zhang, Y. Study of reduced graphene oxide preparation by Hummers' method and related characterization. *J. Nanomater.,* **2015**, *2015*, 1-5. [http://dx.doi.org/10.1155/2015/168125]

[71] Alam, S.N.; Sharma, N.; Kumar, L. Synthesis of graphene oxide (GO) by modified Hummers method and its thermal reduction to obtain reduced graphene Oxide (rGO). *Graphene (Irvine, Calif.),* **2017**, *6*(1), 1-18. [http://dx.doi.org/10.4236/graphene.2017.61001]

[72] Marcano, D.C.; Kosynkin, D.V.; Berlin, J.M.; Sinitskii, A.; Sun, Z.; Slesarev, A.; Alemany, L.B.; Lu, W.; Tour, J.M. Improved synthesis of graphene oxide. *ACS Nano,* **2010**, *4*(8), 4806-4814. [http://dx.doi.org/10.1021/nn1006368] [PMID: 20731455]

[73] Chen, J.; Yao, B.; Li, C.; Shi, G. An improved Hummers method for eco-friendly synthesis of graphene oxide. *Carbon,* **2013**, *64*, 225-229. [http://dx.doi.org/10.1016/j.carbon.2013.07.055]

[74] Yu, H.; Zhang, B.; Bulin, C.; Li, R.; Xing, R. High-efficient synthesis of graphene oxide based on improved Hummers method. *Sci. Rep.,* **2016**, *6*(1), 36143. [http://dx.doi.org/10.1038/srep36143] [PMID: 27808164]

[75] Ranjan, P.; Agrawal, S.; Sinha, A.; Rao, T.R.; Balakrishnan, J.; Thakur, A.D. A low-cost non-explosive synthesis of graphene oxide for scalable applications. *Sci. Rep.,* **2018**, *8*(1), 12007. [http://dx.doi.org/10.1038/s41598-018-30613-4] [PMID: 30104689]

[76] Al-Gaashani, R.; Najjar, A.; Zakaria, Y.; Mansour, S.; Atieh, M.A. XPS and structural studies of high quality graphene oxide and reduced graphene oxide prepared by different chemical oxidation methods. *Ceram. Int.,* **2019**, *45*(11), 14439-14448. [http://dx.doi.org/10.1016/j.ceramint.2019.04.165]

[77] Liu, F.; Wang, C.; Sui, X.; Riaz, M.A.; Xu, M.; Wei, L.; Chen, Y. Synthesis of graphene materials by electrochemical exfoliation: Recent progress and future potential. *Carbon Energy,* **2019**, *1*(2), 173-199. [http://dx.doi.org/10.1002/cey2.14]

[78] Wang, H.; Fu, Q.; Pan, C. Green mass synthesis of graphene oxide and its MnO_2 composite for high performance supercapacitor. *Electrochim. Acta,* **2019**, *312*, 11-21.

[http://dx.doi.org/10.1016/j.electacta.2019.04.178]

[79] Pei, S.; Wei, Q.; Huang, K.; Cheng, H.M.; Ren, W. Green synthesis of graphene oxide by seconds timescale water electrolytic oxidation. *Nat. Commun.*, **2018**, *9*(1), 145. [http://dx.doi.org/10.1038/s41467-017-02479-z] [PMID: 29321501]

[80] Chua, C.K.; Pumera, M. The reduction of graphene oxide with hydrazine: elucidating its reductive capability based on a reaction-model approach. *Chem. Commun. (Camb.),* **2016**, *52*(1), 72-75. [http://dx.doi.org/10.1039/C5CC08170J] [PMID: 26525927]

[81] Xu, G.; Malmström, J.; Edmonds, N.; Broderick, N.; Travas-Sejdic, J.; Jin, J. Investigation of the reduction of graphene oxide by lithium triethylborohydride. *J. Nanomater.*, **2016**.

[82] Nam, K.H.; Kim, K.; Kim, S.G.; Lee, H.S.; Jung, H.; Yu, J.; Jang, S.G.; Ku, B-C.; Moon, B.; You, N-H. Sustainable production of reduced graphene oxide using elemental sulfur for multifunctional composites. *Compos., Part B Eng.*, **2019**, *176*, 107236. [http://dx.doi.org/10.1016/j.compositesb.2019.107236]

[83] Go, S.H.; Kim, H.; Yu, J.; You, N.H.; Ku, B.C.; Kim, Y.K. Synergistic effect of UV and l-ascorbic acid on the reduction of graphene oxide: Reduction kinetics and quantum chemical simulations. *Solid State Sci.*, **2018**, *84*, 120-125. [http://dx.doi.org/10.1016/j.solidstatesciences.2018.09.001]

[84] Khandelwal, M.; Kumar, A. Electrochemical behavior of glycine-mediated N-doped reduced graphene oxide. *New J. Chem.*, **2017**, *41*(16), 8333-8340. [http://dx.doi.org/10.1039/C7NJ01662J]

[85] Narayanan, K.B.; Kim, H.D.; Han, S.S. Biocompatibility and hemocompatibility of hydrothermally derived reduced graphene oxide using soluble starch as a reducing agent. *Colloids Surf. B Biointerfaces,* **2020**, *185*, 110579. [http://dx.doi.org/10.1016/j.colsurfb.2019.110579] [PMID: 31689675]

[86] Xu, L.Q.; Liao, Y.B.; Li, N.N.; Li, Y.J.; Zhang, J.Y.; Wang, Y.B.; Hu, X.F.; Li, C.M. Vancomycin-assisted green synthesis of reduced graphene oxide for antimicrobial applications. *J. Colloid Interface Sci.*, **2018**, *514*, 733-739. [http://dx.doi.org/10.1016/j.jcis.2018.01.014] [PMID: 29316529]

[87] Luo, F.; Wu, K.; Shi, J.; Du, X.; Li, X.; Yang, L.; Lu, M. Green reduction of graphene oxide by polydopamine to a construct flexible film: superior flame retardancy and high thermal conductivity. *J. Mater. Chem. A Mater. Energy Sustain.*, **2017**, *5*(35), 18542-18550. [http://dx.doi.org/10.1039/C7TA04740A]

[88] Ahmad, S.; Ahmad, A.; Khan, S.; Ahmad, S.; Khan, I.; Zada, S.; Fu, P. Algal extracts based biogenic synthesis of reduced graphene oxides (rGO) with enhanced heavy metals adsorption capability. *J. Ind. Eng. Chem.*, **2019**, *72*, 117-124. [http://dx.doi.org/10.1016/j.jiec.2018.12.009]

[89] Kalathil, S.; Katuri, K.P.; Alazmi, A.S.; Pedireddy, S.; Kornienko, N.; Costa, P.M.F.J.; Saikaly, P.E. Bioinspired synthesis of reduced graphene oxide-wrapped *Geobacter sulfurreducens* as a hybrid electrocatalyst for efficient oxygen evolution reaction. *Chem. Mater.*, **2019**, *31*(10), 3686-3693. [http://dx.doi.org/10.1021/acs.chemmater.9b00394]

[90] Hou, D.; Liu, Q.; Wang, X.; Quan, Y.; Qiao, Z.; Yu, L.; Ding, S. Facile synthesis of graphene *via* reduction of graphene oxide by artemisinin in ethanol. *Journal of Materiomics,* **2018**, *4*(3), 256-265. [http://dx.doi.org/10.1016/j.jmat.2018.01.002]

[91] Vatandost, E.; Ghorbani-HasanSaraei, A.; Chekin, F.; Naghizadeh Raeisi, S.; Shahidi, S.A. Green tea extract assisted green synthesis of reduced graphene oxide: Application for highly sensitive electrochemical detection of sunset yellow in food products. *Food Chem. X,* **2020**, *6*, 100085. [http://dx.doi.org/10.1016/j.fochx.2020.100085] [PMID: 32577617]

[92] Bhattacharya, G.; Sas, S.; Wadhwa, S.; Mathur, A.; McLaughlin, J.; Roy, S.S. *Aloe vera* assisted facile

green synthesis of reduced graphene oxide for electrochemical and dye removal applications. *RSC Advances,* **2017**, *7*(43), 26680-26688. [http://dx.doi.org/10.1039/C7RA02828H]

[93] Reduced Graphene Oxide: Properties, Applications and Production Methods. *AZoNano,* https://www.azonano.com/article.aspx?ArticleID=4041

[94] Silipigni, L.; Salvato, G.; Di Marco, G.; Fazio, B.; Torrisi, A.; Cutroneo, M.; Torrisi, L. Band-like transport in high vacuum thermal reduced graphene oxide films. *Vacuum,* **2019**, *165*, 254-261. [http://dx.doi.org/10.1016/j.vacuum.2019.04.025]

[95] Tian, S.; Sun, J.; Yang, S.; He, P.; Ding, S.; Ding, G.; Xie, X. Facile thermal annealing of graphite oxide in air for graphene with a higher C/O ratio. *RSC Advances,* **2015**, *5*(85), 69854-69860. [http://dx.doi.org/10.1039/C5RA09388K]

[96] Sofer, Z.; Jankovský, O.; Šimek, P.; Sedmidubský, D.; Šturala, J.; Kosina, J.; Mikšová, R.; Macková, A.; Mikulics, M.; Pumera, M. Insight into the mechanism of the thermal reduction of graphite oxide: deuterium-labeled graphite oxide is the key. *ACS Nano,* **2015**, *9*(5), 5478-5485. [http://dx.doi.org/10.1021/acsnano.5b01463] [PMID: 25894311]

[97] Kumar, R.; Kumar, R.M.; Lahiri, D.; Lahiri, I. Thermally reduced graphene oxide film on soda lime glass as transparent conducting electrode. *Surf. Coat. Tech.,* **2017**, *309*, 931-937. [http://dx.doi.org/10.1016/j.surfcoat.2016.10.060]

[98] Schniepp, H.C.; Li, J.L.; McAllister, M.J.; Sai, H.; Herrera-Alonso, M.; Adamson, D.H.; Prud'homme, R.K.; Car, R.; Saville, D.A.; Aksay, I.A. Functionalized single graphene sheets derived from splitting graphite oxide. *J. Phys. Chem. B,* **2006**, *110*(17), 8535-8539. [http://dx.doi.org/10.1021/jp060936f] [PMID: 16640401]

[99] Li, X.; Wang, H.; Robinson, J.T.; Sanchez, H.; Diankov, G.; Dai, H. Simultaneous nitrogen doping and reduction of graphene oxide. *J. Am. Chem. Soc.,* **2009**, *131*(43), 15939-15944. [http://dx.doi.org/10.1021/ja907098f] [PMID: 19817436]

[100] Agusu, L.; Ahmad, L.O.; Anggara, D.; Alimin, ; Mitsudo, S.; Fujii, Y.; Kikuchi, H. Alimin, Mitsudo, S.; Fujii, Y.; Kikuchi, H. Microwave hydrothermal synthesis of reduced graphene oxide: effects of microwave power and irradiation time. *J. Phys. Conf. Ser.,* **2018**, *1011*, 012012. [http://dx.doi.org/10.1088/1742-6596/1011/1/012012]

[101] Umrao, S.; Gupta, T.K.; Kumar, S.; Singh, V.K.; Sultania, M.K.; Jung, J.H.; Oh, I.K.; Srivastava, A. Microwave-assisted synthesis of boron and nitrogen co-doped reduced graphene oxide for the protection of electromagnetic radiation in Ku-band. *ACS Appl. Mater. Interfaces,* **2015**, *7*(35), 19831-19842. [http://dx.doi.org/10.1021/acsami.5b05890] [PMID: 26287816]

[102] Wu, J.; Zhao, J.; Vaidhyanathan, B.; Zhang, H.; Anshuman, A.; Nare, A.; Saremi-Yarahmadi, S. Rapid microwave-assisted bulk production of high-quality reduced graphene oxide for lithium ion batteries. *Materialia (Oxf.),* **2020**, *13*, 100833. [http://dx.doi.org/10.1016/j.mtla.2020.100833]

[103] Iskandar, F.; Hikmah, U.; Stavila, E.; Aimon, A.H. Microwave-assisted reduction method under nitrogen atmosphere for synthesis and electrical conductivity improvement of reduced graphene oxide (rGO). *RSC Advances,* **2017**, *7*(83), 52391-52397. [http://dx.doi.org/10.1039/C7RA10013B]

[104] Mohandoss, M.; Gupta, S.S.; Nelleri, A.; Pradeep, T.; Maliyekkal, S.M. Solar mediated reduction of graphene oxide. *RSC Advances,* **2017**, *7*(2), 957-963. [http://dx.doi.org/10.1039/C6RA24696F]

[105] He, X.; Zhang, S.; Pan, H.; Chen, J.; Xu, J. Horizontally aggregation of monolayer reduced graphene oxide under deep UV irradiation in solution. *Nanoscale Res. Lett.,* **2019**, *14*(1), 117. [http://dx.doi.org/10.1186/s11671-019-2940-z] [PMID: 30941577]

[106] Hasani, A.; Dehsari, H.S.; Zarandi, A.A.; Salehi, A.; Taromi, F.A.; Kazeroni, H. Visible light-assisted photoreduction of graphene oxide using CdS nanoparticles and gas sensing properties. *J. Nanomater,* **2015**.

[107] Low, M.J.; Lee, H.; Lim, C.H.J.; Suchand Sandeep, C.S.; Murukeshan, V.M.; Kim, S.W.; Kim, Y.J. Laser-induced reduced-graphene-oxide micro-optics patterned by femtosecond laser direct writing. *Appl. Surf. Sci.,* **2020**, *526*, 146647. [http://dx.doi.org/10.1016/j.apsusc.2020.146647]

[108] Feng, X.; Chen, W.; Yan, L. Electrochemical reduction of bulk graphene oxide materials. *RSC Advances,* **2016**, *6*(83), 80106-80113. [http://dx.doi.org/10.1039/C6RA17469H]

[109] Tong, H.; Zhu, J.; Chen, J.; Han, Y.; Yang, S.; Ding, B.; Zhang, X. Electrochemical reduction of graphene oxide and its electrochemical capacitive performance. *J. Solid State Electrochem.,* **2013**, *17*(11), 2857-2863. [http://dx.doi.org/10.1007/s10008-013-2195-z]

[110] Nia, P.M.; Abouzari-Lotf, E.; Arshad, A.; Nasef, M.M.; Ripin, A. Tunable electrochemical approach for reduction of graphene oxide: Taguchi-assisted chemical and structural optimization. *J. Electrochem. Soc.,* **2018**, *165*(9), E429-E438. [http://dx.doi.org/10.1149/2.0031810jes]

[111] Ray, S.C. *Applications of Graphene and Graphene-Oxide Based Nanomaterials,* 1st ed; Elsevier Science: Amsterdam, **2015**.

[112] Dubin, S.; Gilje, S.; Wang, K.; Tung, V.C.; Cha, K.; Hall, A.S.; Farrar, J.; Varshneya, R.; Yang, Y.; Kaner, R.B. A one-step, solvothermal reduction method for producing reduced graphene oxide dispersions in organic solvents. *ACS Nano,* **2010**, *4*(7), 3845-3852. [http://dx.doi.org/10.1021/nn100511a] [PMID: 20586422]

[113] Kim, S.; Choi, K.; Park, S. Solvothermal reduction of graphene oxide in dimethylformamide. *Solid State Sci.,* **2016**, *61*, 40-43. [http://dx.doi.org/10.1016/j.solidstatesciences.2016.07.013]

[114] Mikhaylov, P.A.; Levin, I.S.; Lubenchenko, A.V.; Bondarenko, G.N.; Kulichihin, V.G. Ultrasonication-assisted ultrafast solvothermal reduction of graphene oxide. *IOP Conf. Ser.: Mater. Sci. Eng.,* 525 012039. **2019**. [http://dx.doi.org/10.1088/1757-899X/525/1/012039]

[115] Torres, D.; Arcelus-Arrillaga, P.; Millan, M.; Pinilla, J.; Suelves, I. Enhanced reduction of few-layer graphene oxide *via* supercritical water gasification of glycerol. *Nanomaterials (Basel),* **2017**, *7*(12), 447. [http://dx.doi.org/10.3390/nano7120447] [PMID: 29240720]

[116] Zheng, X.; Peng, Y.; Yang, Y.; Chen, J.; Tian, H.; Cui, X.; Zheng, W. Hydrothermal reduction of graphene oxide; effect on surface-enhanced Raman scattering. *J. Raman Spectrosc.,* **2017**, *48*(1), 97-103. [http://dx.doi.org/10.1002/jrs.4998]

[117] Hayes, W.I.; Joseph, P.; Mughal, M.Z.; Papakonstantinou, P. Production of reduced graphene oxide *via* hydrothermal reduction in an aqueous sulphuric acid suspension and its electrochemical behaviour. *J. Solid State Electrochem.,* **2015**, *19*(2), 361-380. [http://dx.doi.org/10.1007/s10008-014-2560-6]

[118] Wei, M.; Qiao, L.; Zhang, H.; Karakalos, S.; Ma, K.; Fu, Z.; Swihart, M.T.; Wu, G. Engineering reduced graphene oxides with enhanced electrochemical properties through multiple-step reductions. *Electrochim. Acta,* **2017**, *258*, 735-743. [http://dx.doi.org/10.1016/j.electacta.2017.11.120]

[119] Lin, Y.; Hou, G.; Bi, S.; Su, X.; Li, H. Synthesis of reduced graphene oxide paper for EMI shielding

by a multi-step process. *Funct. Mater. Lett. (Singap.),* **2020**, *13*(5), 2051024. [http://dx.doi.org/10.1142/S1793604720510248]

[120] Suraya Abdul Rashid, Tongling Tan. Recent Development on the Synthesis Techniques and Properties of Graphene Derivatives. In: *Synthesis, Technology and Applications of Carbon Nanomaterials*; Rashid, S.A.; Othman, R.N.I.R.; Hussein, M.Z., Eds. Elsevier: Amsterdam, **2019**; pp. 77-107. [http://dx.doi.org/10.1016/B978-0-12-815757-2.00004-8]

[121] Martin, P.M. *Handbook of Deposition Technologies for Films and Coatings,* 3rd ed; William Andrew Publishing: New York, **2010**.

[122] Reinke, M.; Kuzminykh, Y.; Hoffmann, P. Low temperature chemical vapour deposition using atomic layer deposition chemistry. *Chem. Mater.,* **2015**, *27*(5), 1604-1611. [http://dx.doi.org/10.1021/cm504216p]

[123] Fujita, J.; Hiyama, T.; Hirukawa, A.; Kondo, T.; Nakamura, J.; Ito, S.; Araki, R.; Ito, Y.; Takeguchi, M.; Pai, W.W. Near room temperature chemical vapor deposition of graphene with diluted methane and molten gallium catalyst. *Sci. Rep.,* **2017**, *7*(1), 12371. [http://dx.doi.org/10.1038/s41598-017-12380-w] [PMID: 28959046]

[124] Jeong, H.; Hwang, W.T.; Song, Y.; Kim, J.K.; Kim, Y.; Hihath, J.; Chung, S.; Lee, T. Highly uniform monolayer graphene synthesis *via* a facile pretreatment of copper catalyst substrates using an ammonium persulfate solution. *RSC Advances,* **2019**, *9*(36), 20871-20878. [http://dx.doi.org/10.1039/C9RA02689D] [PMID: 35515571]

[125] Seo, D.H.; Pineda, S.; Fang, J.; Gozukara, Y.; Yick, S.; Bendavid, A.; Lam, S.K.H.; Murdock, A.T.; Murphy, A.B.; Han, Z.J.; Ostrikov, K. Single-step ambient-air synthesis of graphene from renewable precursors as electrochemical genosensor. *Nat. Commun.,* **2017**, *8*(1), 14217. [http://dx.doi.org/10.1038/ncomms14217] [PMID: 28134336]

[126] Strupinski, W.; Grodecki, K.; Wysmolek, A.; Stepniewski, R.; Szkopek, T.; Gaskell, P.E.; Grüneis, A.; Haberer, D.; Bozek, R.; Krupka, J.; Baranowski, J.M. Graphene epitaxy by chemical vapor deposition on SiC. *Nano Lett.,* **2011**, *11*(4), 1786-1791. [http://dx.doi.org/10.1021/nl200390e] [PMID: 21438581]

[127] Tang, S.; Wang, H.; Wang, H.S.; Sun, Q.; Zhang, X.; Cong, C.; Xie, H.; Liu, X.; Zhou, X.; Huang, F.; Chen, X.; Yu, T.; Ding, F.; Xie, X.; Jiang, M. Silane-catalysed fast growth of large single-crystalline graphene on hexagonal boron nitride. *Nat. Commun.,* **2015**, *6*(1), 6499. [http://dx.doi.org/10.1038/ncomms7499] [PMID: 25757864]

[128] Hwang, J.; Kim, M.; Cha, H.Y.; Spencer, M.G.; Lee, J.W. Metal free growth of graphene on quartz substrate using chemical vapor deposition (CVD). *J. Nanosci. Nanotechnol.,* **2014**, *14*(4), 2979-2983. [http://dx.doi.org/10.1166/jnn.2014.8583] [PMID: 24734720]

[129] Zhou, L.; Wei, S.; Ge, C.; Zhao, C.; Guo, B.; Zhang, J.; Zhao, J. Ultrafast growth of uniform multi-layer graphene films directly on silicon dioxide substrates. *Nanomaterials (Basel),* **2019**, *9*(7), 964. [http://dx.doi.org/10.3390/nano9070964] [PMID: 31266221]

[130] Min, B.H.; Kim, D.W.; Kim, K.H.; Choi, H.O.; Jang, S.W.; Jung, H.T. Bulk scale growth of CVD graphene on Ni nanowire foams for a highly dense and elastic 3D conducting electrode. *Carbon,* **2014**, *80*, 446-452. [http://dx.doi.org/10.1016/j.carbon.2014.08.084]

[131] Lee, X.J.; Hiew, B.Y.Z.; Lai, K.C.; Lee, L.Y.; Gan, S.; Thangalazhy-Gopakumar, S.; Rigby, S. Review on graphene and its derivatives: Synthesis methods and potential industrial implementation. *J. Taiwan Inst. Chem. Eng.,* **2019**, *98*, 163-180. [http://dx.doi.org/10.1016/j.jtice.2018.10.028]

[132] Singh, S.; Chen, X.; Zhang, C.; Gautam, R.K.; Tyagi, R.; Luo, J. Nickel-catalyzed direct growth of graphene on bearing steel (GCr15) by thermal chemical vapor deposition and its tribological behavior. *Appl. Surf. Sci.,* **2020**, *502*, 144135. [http://dx.doi.org/10.1016/j.apsusc.2019.144135]

[133] Regmi, M.; Chisholm, M.F.; Eres, G. The effect of growth parameters on the intrinsic properties of large-area single layer graphene grown by chemical vapor deposition on Cu. *Carbon,* **2012**, *50*(1), 134-141.
[http://dx.doi.org/10.1016/j.carbon.2011.07.063]

[134] Muñoz, R.; Martínez, L.; López-Elvira, E.; Munuera, C.; Huttel, Y.; García-Hernández, M. Direct synthesis of graphene on silicon oxide by low temperature plasma enhanced chemical vapor deposition. *Nanoscale,* **2018**, *10*(26), 12779-12787.
[http://dx.doi.org/10.1039/C8NR03210F] [PMID: 29946620]

[135] Sze, S.M.; Ng, K.K. *Physics of Semiconductor Devices,* 3rd ed; John Wiley & Sons: New York, **1981**.

[136] Capper, P.; Irvine, S.; Joyce, T. Epitaxial Crystal Growth: Methods and Materials. In: *Springer Handbook of Electronic and Photonic Materials*; Kasap, S.; Capper, P., Eds.; Springer: Cham, New York, **2017**.
[http://dx.doi.org/10.1007/978-3-319-48933-9_14]

[137] Briggs, N.; Gebeyehu, Z.M.; Vera, A.; Zhao, T.; Wang, K.; De La Fuente Duran, A.; Bersch, B.; Bowen, T.; Knappenberger, K.L., Jr; Robinson, J.A. Epitaxial graphene/silicon carbide intercalation: a minireview on graphene modulation and unique 2D materials. *Nanoscale,* **2019**, *11*(33), 15440-15447.
[http://dx.doi.org/10.1039/C9NR03721G] [PMID: 31393495]

[138] Mishra, N.; Boeckl, J.; Motta, N.; Iacopi, F. Graphene growth on silicon carbide: A review (Phys. Status Solidi A 9/2016). *Phys. Status Solidi., A Appl. Mater. Sci.,* **2016**, *213*(9), 2269-2289.
[http://dx.doi.org/10.1002/pssa.201670657]

[139] Riedl, C.; Coletti, C.; Starke, U. Structural and electronic properties of epitaxial graphene on SiC(0 0 0 1): a review of growth, characterization, transfer doping and hydrogen intercalation. *J. Phys. D Appl. Phys.,* **2010**, *43*(37), 374009.
[http://dx.doi.org/10.1088/0022-3727/43/37/374009]

[140] Lalmi, B.; Girard, J.C.; Pallecchi, E.; Silly, M.; David, C.; Latil, S.; Sirotti, F.; Ouerghi, A. Flower-shaped domains and wrinkles in trilayer epitaxial graphene on silicon carbide. *Sci. Rep.,* **2015**, *4*(1), 4066.
[http://dx.doi.org/10.1038/srep04066] [PMID: 24513669]

[141] Xu, X.; Zhang, Z.; Dong, J.; Yi, D.; Niu, J.; Wu, M.; Lin, L.; Yin, R.; Li, M.; Zhou, J.; Wang, S.; Sun, J.; Duan, X.; Gao, P.; Jiang, Y.; Wu, X.; Peng, H.; Ruoff, R.S.; Liu, Z.; Yu, D.; Wang, E.; Ding, F.; Liu, K. Ultrafast epitaxial growth of metre-sized single-crystal graphene on industrial Cu foil. *Sci. Bull. (Beijing),* **2017**, *62*(15), 1074-1080.
[http://dx.doi.org/10.1016/j.scib.2017.07.005]

[142] Wade, T.L.; Wegrowe, J.E. Template synthesis of nanomaterials. *Eur. Phys. J. Appl. Phys.,* **2005**, *29*(1), 3-22.
[http://dx.doi.org/10.1051/epjap:2005001]

[143] Lu, A.H.; Schuth, F. Nanocasting: A versatile strategy for creating nanostructured porous materials. *Adv. Mater,* **2006**, 179-1805.

[144] Liang, C.; Li, Z.; Dai, S. Mesoporous carbon materials: synthesis and modification. *Angew. Chem. Int. Ed.,* **2008**, *47*(20), 3696-3717.
[http://dx.doi.org/10.1002/anie.200702046] [PMID: 18350530]

[145] Lei, L.; Wu, Z.; Liu, H.; Qin, Z.; Chen, C.; Luo, L.; Wang, G.; Fan, W.; Wang, J. A facile method for the synthesis of graphene-like 2D metal oxides and their excellent catalytic application in the hydrogenation of nitroarenes. *J. Mater. Chem. A Mater. Energy Sustain.,* **2018**, *6*(21), 9948-9961.
[http://dx.doi.org/10.1039/C8TA02338G]

[146] Huang, X.; Sun, B.; Su, D.; Zhao, D.; Wang, G. Soft-template synthesis of 3D porous graphene foams with tunable architectures for lithium–O_2 batteries and oil adsorption applications. *J. Mater. Chem. A Mater. Energy Sustain.,* **2014**, *2*(21), 7973-7979.

[http://dx.doi.org/10.1039/C4TA00829D]

[147] von Kugelgen, S.; Piskun, I.; Griffin, J.H.; Eckdahl, C.T.; Jarenwattananon, N.N.; Fischer, F.R. Templated synthesis of end-functionalized graphene nanoribbons through living ring-opening alkyne metathesis polymerization. *J. Am. Chem. Soc.,* **2019**, *141*(28), 11050-11058. [http://dx.doi.org/10.1021/jacs.9b01805] [PMID: 31264864]

[148] Yang, Y.; Liu, R.; Wu, J.; Jiang, X.; Cao, P.; Hu, X.; Pan, T.; Qiu, C.; Yang, J.; Song, Y.; Wu, D.; Su, Y. Bottom-up fabrication of graphene on silicon/silica substrate *via* a facile soft-hard template approach. *Sci. Rep.,* **2015**, *5*(1), 13480. [http://dx.doi.org/10.1038/srep13480] [PMID: 26311022]

[149] Huang, X.; Zhao, Y.; Ao, Z.; Wang, G. Micelle-template synthesis of nitrogen-doped mesoporous graphene as an efficient metal-free electrocatalyst for hydrogen production. *Sci. Rep.,* **2015**, *4*(1), 7557. [http://dx.doi.org/10.1038/srep07557] [PMID: 25523276]

[150] Wang, F.; Wang, F.; Hong, R.; Lv, X.; Zheng, Y.; Chen, H. High-purity few-layer graphene from plasma pyrolysis of methane as conductive additive for $LiFePO_4$ lithium ion battery. *J. Mater. Res. Technol.,* **2020**, *9*(5), 10004-10015. [http://dx.doi.org/10.1016/j.jmrt.2020.06.072]

[151] Hong, N.; Yang, W.; Bao, C.; Jiang, S.; Song, L.; Hu, Y. Facile synthesis of graphene by pyrolysis of poly(methyl methacrylate) on nickel particles in the confined microzones. *Mater. Res. Bull.,* **2012**, *47*(12), 4082-4088. [http://dx.doi.org/10.1016/j.materresbull.2012.08.049]

[152] Perednis, D.; Gauckler, L.J. Thin film deposition using spray pyrolysis. *J. Electroceram.,* **2005**, *14*(2), 103-111. [http://dx.doi.org/10.1007/s10832-005-0870-x]

[153] Zou, B.; Wang, X.X.; Huang, X.X.; Wang, J.N. Continuous synthesis of graphene sheets by spray pyrolysis and their use as catalysts for fuel cells. *Chem. Commun. (Camb.),* **2015**, *51*(4), 741-744. [http://dx.doi.org/10.1039/C4CC08197H] [PMID: 25421428]

[154] Sun, Y.; Zhang, J. Strategies for scalable gas-phase preparation of free-standing graphene. *CCS Chem.,* **2020**, *2*, 1058-1077.

[155] Dato, A.; Radmilovic, V.; Lee, Z.; Phillips, J.; Frenklach, M. Substrate-free gas-phase synthesis of graphene sheets. *Nano Lett.,* **2008**, *8*(7), 2012-2016. [http://dx.doi.org/10.1021/nl8011566] [PMID: 18529034]

[156] Liu, F.; Zhang, Y. Substrate-free synthesis of large area, continuous multi-layer graphene film. *Carbon,* **2010**, *48*(9), 2394-2400. [http://dx.doi.org/10.1016/j.carbon.2010.02.033]

[157] Choi, W.; Lee, J. *Graphene: Synthesis and applications*; CRC Press: Ohio, **2016**. [http://dx.doi.org/10.1201/b11259]

[158] Zheng, Q.; Kim, J-K. Synthesis, structure, and properties of graphene and graphene oxide. In: *Graphene for transparent conductors: synthesis, properties and applications*; Springer: New York, **2015**; pp. 29-94. [http://dx.doi.org/10.1007/978-1-4939-2769-2_2]

[159] Narita, A.; Wang, X.Y.; Feng, X.; Müllen, K. New advances in nanographene chemistry. *Chem. Soc. Rev.,* **2015**, *44*(18), 6616-6643. [http://dx.doi.org/10.1039/C5CS00183H] [PMID: 26186682]

[160] Vo, T.H.; Perera, U.G.E.; Shekhirev, M.; Mehdi Pour, M.; Kunkel, D.A.; Lu, H.; Gruverman, A.; Sutter, E.; Cotlet, M.; Nykypanchuk, D.; Zahl, P.; Enders, A.; Sinitskii, A.; Sutter, P. Axel Enders, Sinitskii, A.; Sutter, P. Nitrogen-doping induced self-assembly of graphene nanoribbon-based two-dimensional and three-dimensional metamaterials. *Nano Lett.,* **2015**, *15*(9), 5770-5777.

[http://dx.doi.org/10.1021/acs.nanolett.5b01723] [PMID: 26258628]

[161] Dichtel, W.R.; Arslan, H.; Uribe-Romo, F.J. Graphene nanoribbons derived from poly(phenylene ethynylene) polymer, methods of making same, and uses thereof. **2017**.

[162] Asao, N.; Nogami, T.; Lee, S.; Yamamoto, Y. Lewis acid-catalyzed benzannulation *via* unprecedented [4+2] cycloaddition of o-alkynyl(oxo)benzenes and enynals with alkynes. *J. Am. Chem. Soc.,* **2003**, *125*(36), 10921-10925. [http://dx.doi.org/10.1021/ja036927r] [PMID: 12952473]

[163] Sharma, D.; Kanchi, S.; Bisetty, K. Biogenic synthesis of nanoparticles: A review. *Arab. J. Chem.,* **2019**, *12*(8), 3576-3600. [http://dx.doi.org/10.1016/j.arabjc.2015.11.002]

[164] Noorunnisa Khanam, P.; Hasan, A. Biosynthesis and characterization of graphene by using non-toxic reducing agent from *Allium Cepa* extract: Anti-bacterial properties. *Int. J. Biol. Macromol.,* **2019**, *126*, 151-158. [http://dx.doi.org/10.1016/j.ijbiomac.2018.12.213] [PMID: 30584937]

[165] Gurunathan, S.; Han, J.W.; Eppakayala, V.; Kim, J.H. Microbial reduction of graphene oxide by *Escherichia coli*: A green chemistry approach. *Colloids Surf. B Biointerfaces,* **2013**, *102*, 772-777. [http://dx.doi.org/10.1016/j.colsurfb.2012.09.011] [PMID: 23107955]

CHAPTER 4

Characterization Techniques for Graphene-Based Materials

Vickramjeet Singh[1] and **Ramesh L. Gardas**[2,*]

[1] *Department of Chemistry, Dr B R Ambedkar National Institute of Technology, Jalandhar-144011, Punjab, India*

[2] *Department of Chemistry, Indian Institute of Technology Madras, Chennai - 600 036, India*

Abstract: Graphene bearing 2D (dimensional) layer of carbon atoms bonded in sp^2 hybridized state are only 1 atomic-scale thick. However, the graphene can be extended along the horizontal dimension. The alternate double bonds leading to perfect conjugation with sp^2 hybridization are exhibited in the hexagonal structure (*honeycomb*) of graphene. Theoretically and experimentally, the thicknesses of graphene have been determined and are in the nano-meter range. The extraordinary mechanical and electrical properties exhibited by such a 2D material have inspired scientists for device fabrication methodologies that can shift the synthesis from lab scale to large scale. It is considered the strongest material on earth, almost 100 times stronger (*i.e.*, strength) than the best steel. Since graphene is only 1 atomic-scale thick and transparent, the characterization of graphene is complex but essential. The thickness down to one atomic layer in graphene can be identified by the light interference causing color contrast. Thus, optical microscopy-based methods enable the identification of graphene or its derivatives; on the other hand, Raman spectroscopy, which is sensitive to molecular bonding and geometric structure, is commonly employed for the quality determination of graphene-based materials. In this chapter, various characterization techniques are discussed, enabling the characterization of graphene and graphene-based materials (GBMs).

Keywords: Characterization, Fourier-transform infra-red Spectroscopy (FT-IR), Graphene, Graphene Sayers, Graphene Oxide, Optical Microscopy, Raman Spectroscopy, Reduced Graphene Oxide, Scanning Tunneling Microscope (STM), Transmission Electron Microscope (TEM), X-ray Shotoelectron Spectroscopy (XPS).

* **Corresponding author Ramesh L. Gardas:** Department of Chemistry, Indian Institute of Technology Madras, Chennai - 600 036, India; E-mail:gardas@iitm.ac.in

Pinki Bala Punjabi, Rakshit Ameta & Sharoni Gupta (Eds.)

1. INTRODUCTION

Graphene is a 2D nano crystallite material composed of a single carbon atom layer packed in a hexagonal arrangement (honeycomb) with alternating single and double bonds [1, 2]. The sp^2 hybridized state of carbon atoms has π-bonds (out-o--plane) responsible for conduction (delocalized π-electrons) and weak interaction between graphene layers [1, 2]. The strongest bond in this carbon allotrope is the in-plane C-C bond [2, 3]. The graphene is thus a 2D semiconductor with zero band gaps or nil band gap energy, *i.e.*, having a minor overlap among the conduction and valence bands [2, 4]. While maintaining the sp^2 hybridization, graphene sheets can be rolled up to make materials with dimensions as 3 D (graphite), 1 D (nanotubes), and 0D (fullerenes) [5]. Since graphene is 1 atomic layer thick, the first report of graphene discovery (structural analysis) was still based on optical microscopy [4, 6]. The ease of analysis by the visualization using an optical microscope is possible for graphene synthesized by mechanical peeling and transferred on a few layers of thick SiO_2 (onto a silicon surface) [6]; however, when deposited on cleaned Si surfaces, this visualization of graphene is lost [6]. The graphene-based materials (GBMs) can be single or multi-layered, graphene oxide, graphene reduced oxide, graphene ribbons, and graphene nano sheets [1, 6, 7]. Graphene with various layers or GBMs can be characterized by a number of techniques, and in this chapter following techniques are being discussed:

1. Optical Microscopic techniques
2. Raman Spectroscopy
3. FT-IR Spectroscopy
4. X-ray Photoelectron Spectroscopy
5. Thermogravimetric Analysis

2. OPTICAL METHODS

As mentioned above, optical methods can be used to characterise graphene synthesized by mechanical peeling unless the graphene is pasted onto a clean Si substrate [6]. The graphene layers onto a SiO_2 substrate can cause light interferences, so the color and the contrast depend upon graphene layers [3, 6, 8, 9]. Besides SiO_2-based surfaces, other substrates for supporting graphene layers that can be visibility distinct are polymethylmethacrylate and silicon nitrite (Si_3N_4) [3, 6, 8, 9]. Mechanical exfoliation may form flakes of various thicknesses; the image of graphite on SiO_2 substrate (oxide thickness of 500 nm) is shown in Fig. (**1**) [8]. Besides the optical image, the SEM (scanning electron microscope) and AFM (atomic force microscope) images are also shown in Fig. **1 b** and **c**, respectively. The picture shows the regions having varied thicknesses [8]. The rectangular box (broken lines) in these images indicates the region with a few monolayer thicknesses. The images can be resolved to get a better picture with an electron microscope, such as SEM with better electron detectors [8].

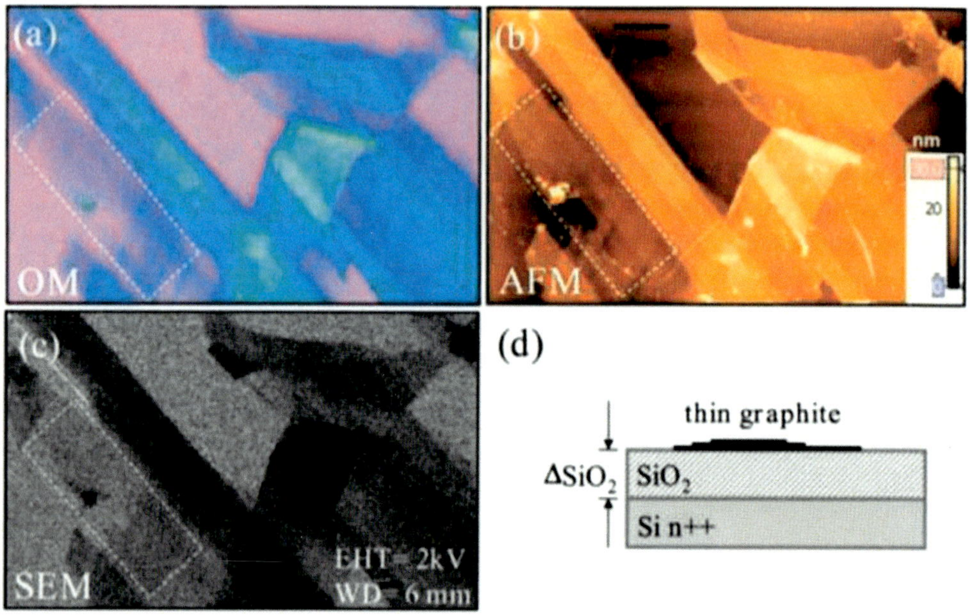

Fig. (1). The image of graphite flake having regions with different thicknesses, as obtained from **(a)** optical microscope (OM), **b)** AFM, **c)** SEM. The graphite was deposited on a Si substrate with a SiO_2 thickness of about 500 nm. Reprinted (adapted) with permission from ref [8]. Copyright (2007) American Chemical Society.

The AFM analysis can directly identify the layer number of graphene, but this technique can damage the crystal lattice while performing the measurement, and the true thickness cannot be estimated [10]. Ni *et al.* [10] developed a method based on contrasting graphene and the substrate onto which graphene layers were transferred. The light reflection makes graphene visible from ten layers down to one layer. Furthermore, experimental data were correlated with theoretical analysis and a good agreement was reported [10]. An analytical expression for determining the number of graphene layers was also provided [10]. Blake *et al.* [3] represented the optical microscopic images of graphene transferred onto the SiO_2 layer. It was demonstrated that the optically visualized contrast depends upon the opacity of graphene and the optical path, that is, the dependence on the SiO_2 layer as well as light wavelength [3]. The images of graphene obtained under normal light illumination on SiO_2 layers (200 nm and 300 nm) are shown in Fig. (**2**) [3]. It was demonstrated from the images that contrast could be employed as a tool to quantitatively define the number of graphene layers deposited on a suitable substrate (in this case, a Si wafer) [3]. Moreover, the color contrast at different wavelengths of light and SiO_2 thickness is also shown in Fig. (**3**) [3]. The Fresnel theory was applied to show that by using appropriate filters, the contrast could be enhanced for various SiO_2 thicknesses [3].

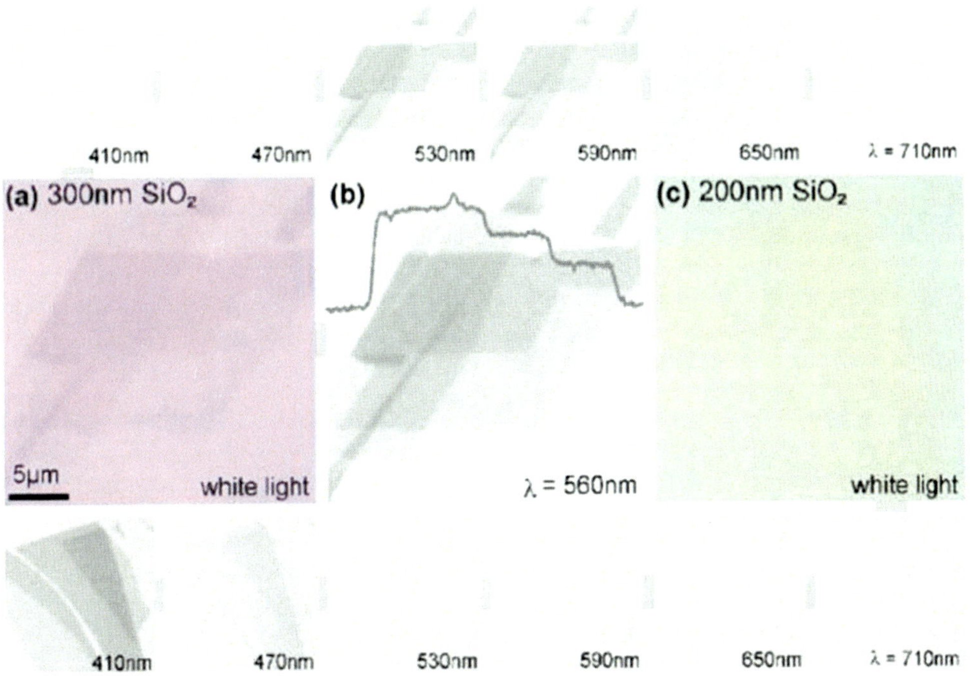

Fig. (2). The images of graphene crystallites **(a)** deposited onto Si wafer (SiO_2 thickness was 300 nm) obtained using normal white light, **(b)** deposited onto Si wafer (SiO_2 thickness was 300 nm) obtained using green light and **(c)** deposited onto SiO_2 (thickness 200 nm) obtained using normal white light, here in image **(c)** graphene is not visible. The images shown above and below those of a, b and c were taken by illumination through various narrow-band filters [3]. Reproduced from ref [3], with the permission of AIP Publishing.

Mechanically exfoliated graphene deposited on SiO_2 (300 nm thickness) was also observed using an optical microscope along with a scanning force microscopy (SFM) micrograph for the selected flakes, as reported previously [11]. The color and contrast were qualitatively correlated to the intensity of light (white to blue to violet light). The number of layers with an uncertainty of ± 1 layer estimated from SFM analysis; the uncertainty arises due to surface roughness and the difference in height between the first layer and the substrate [11]. A methodology was developed by Jia *et al.* [12] by which graphene domains/boundaries can be optically identified on a surface other than SiO_2 layers. The copper oxide enhances the interference, leading to color contrast, thereby making the visible detection of graphene with an optical microscope possible [12] (see Fig. **4**).

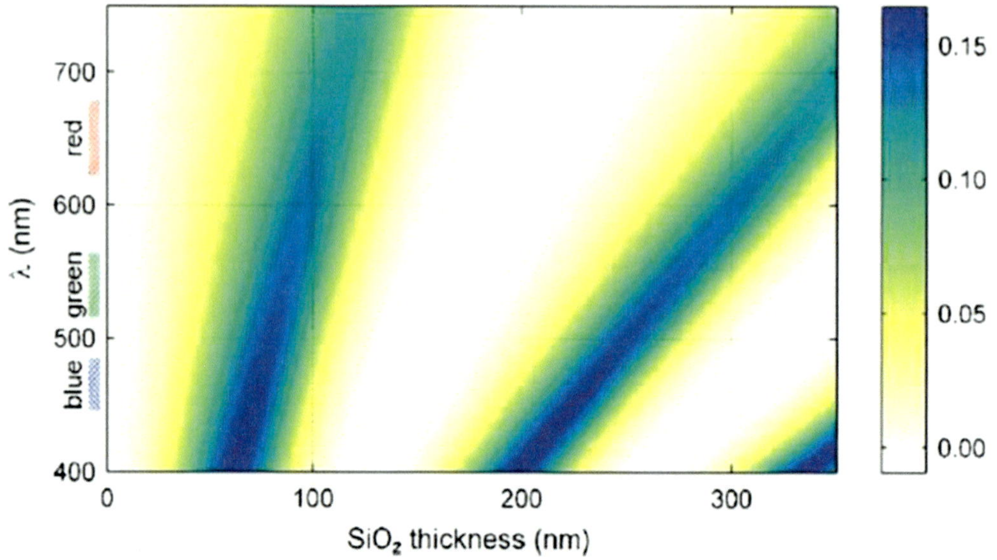

Fig. (3). The plot (colored) depicted by variation in wavelength as well as SiO_2 thickness. The scale in color on the right side depicts the expected contrast [3]. Reproduced from [3], with the permission of AIP Publishing.

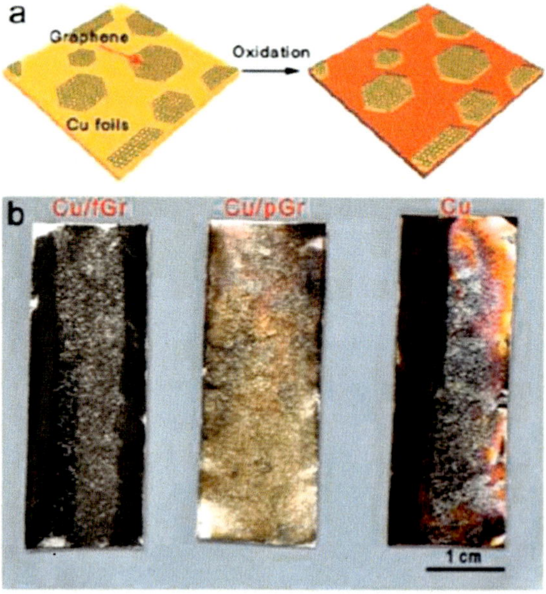

(Fig. 4) contd.....

Fig. (4). The **(a)** schematic illustration of optical characterization and oxidation of graphene domains deposited on Cu substrate **(b)** pictures of Cu substate (foil) having fully grown graphene (represented as Cu/fGr), partially grown graphene (Cu/pGr), Cu without graphene after annealing at 160 °C for 6 min. in air. The optical images of **(c)** Cu/fGr, **(d)** Cu/pGr, and **(e)** Cu foil after the annealing [12]. Reused with permission from ref [12]. under Creative Commons license.

Hong *et al.* [13] have visualized graphene deposited on Cu-foil by the area-specific atomic layer deposition (ALD) method. The optical images were shown for graphene on Cu substrate obtained by annealing at 200 °C and with various cycles of ALD [13]. The underling Cu substrate was oxidized by water vapor, with selective deposition of single-layer graphene. The Al_2O_3 was used as a mask to cover single-layer graphene (SLG) and the area with multi-layer graphene (MLG). The MLG was removed by ion etching, and the copper substrate was thermally oxidized before visualization. The images of graphene on Cu substrate obtained by ALD as a function of time and ALD cycles were reported [13]. Overall, the process is a four-step involving chemical vapor deposition, area selective ALD, treatment with plasma, followed by thermal oxidation [13]. These combinations of surface modifications provide a way to demonstrate the spatial distribution of graphene multi-layers and control the growth and quality evaluation of graphene [13]. The multi-step process was monitored/characterized by TEM, Raman spectroscopy, and SEM [13].

Besides Cu and SiO_2, Ni surfaces have also been used for the growth of graphene multi-layers [14]. The chemical vapor deposition (CVD) method at 900 °C was employed for graphene formation on polycrystalline as well as single Ni (111) crystals [14]. The thermal annealing rate was modulated for the growth of graphene on polycrystalline Ni. The X-ray diffraction (XRD) spectra collected from single as well as polycrystalline Ni with variable annealing rates and the optical images of graphene are shown in Fig. (**5**). Only mono- or bi-layer graphene was formed on a single Ni crystal [*i.e.*, Ni (111)] due to a lack of grain boundaries (interplane) on a single crystal and thus reduction in the nucleation sites [14]. The multi-layer graphene (MLG) was depicted by the dark regions in the optical images, and the light regions depict the bilayer (or monolayer) graphene, and this was confirmed by Raman spectroscopic analysis [14]. The various samples were obtained with slow, medium, or fast annealing rates. It was reported [14] that with an increase in polycrystallinity of Ni, the growth of MLG was enhanced. Bilayer or mono-layer graphene formed on central regions of larger crystalline grains, whereas the enhanced carbon segregation confined to polycrystalline grain boundaries was responsible for MLG formation [14].

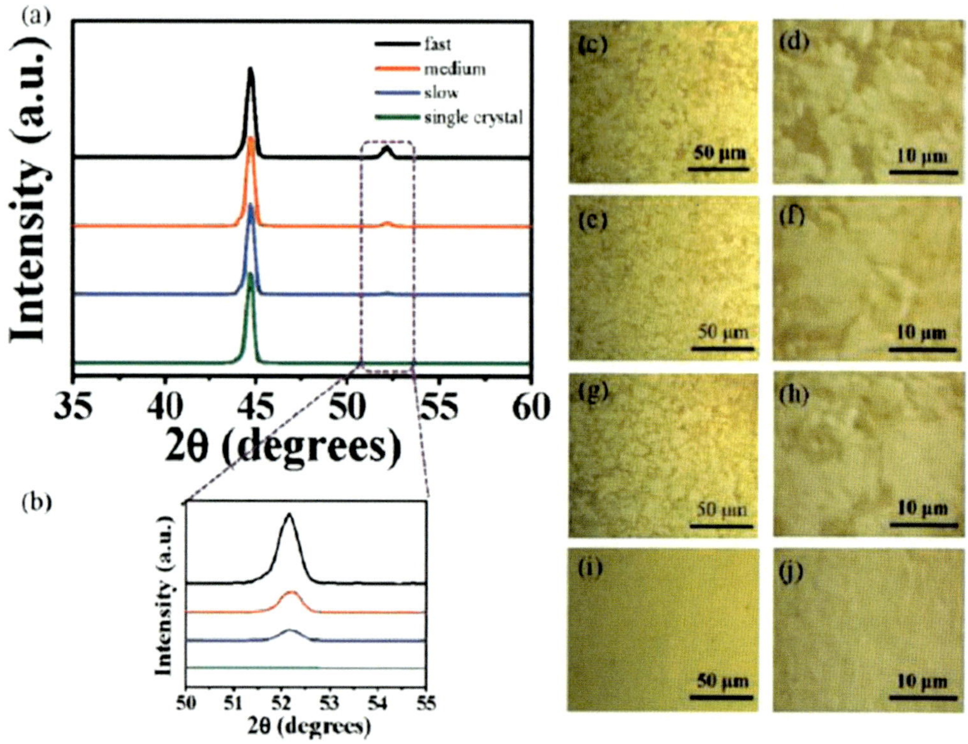

Fig. (5). a) XRD spectra from polycrystalline Ni with variable annealing rates {*i.e.* black (fast), red (medium), and blue (slow)} and XRD spectrum from single-crystal {Ni (111) (green)} after thermal annealing (XRD spectrum is identical for Ni (111) using different annealing rates). **(b)** XRD spectra (Zoomed-in) of peaks at 2θ = 52.16° (that is assigned as Ni (200)). **(c-j)** Optical images after CVD grown graphene on polycrystalline Ni with variable annealing rates {**c, d** (fast); **e, f** (medium); and **g, h** (slow)}; **i,j** optical images after CVD grown graphene on single crystal Ni (111) [14]. The figure has been reprinted (adapted) with permission from ref [14]. Copyright (2010) American Chemical Society.

2.1. Optical Microscopic Analysis

Due to the limited wavelength of visible light, the resolution of an optical microscope is low (~ 200 nm); therefore, to observe the structure at the atomic level after the basic optical microscopic analysis, the most popular tool for the characterization of graphene or graphene derivatives (Go, rGO *etc.*) is based on electron microscopic methods [15]. This way, structural morphology, quality, and crystal structure can be investigated [15]. The commonly used optical microscopic techniques are AFM, TEM or high-resolution electron microscope (HR-TEM), SEM, and scanning tunnelling microscope (STM) [15].

2.2. Transmission Electron Microscope (TEM)

The electrons which are transmitted through a thin sample kept on a TEM grid are analysed in the technique [15 - 17]. The local information at the atomic level about the bulk and surface of a sample is obtained by using TEM. The technique also provides information such as three-dimensional structure, electronic, and chemical properties [16]. In spite of a versatile technique for optical analysis, TEM demonstrates drawbacks such as a small relative area of view as compared to SEM [15, 16]. Very high magnification and a much better resolution down to a group of atoms or even a single atom are provided by HRTEM [15]. The advanced TEM provides various structural and material information as: a) in- real space; or b) reciprocal space or energy space (the spectroscopic mode) [16 - 19].

The C-C bond distance of about 0.142 nm [19] in graphene can be directly resolved by aberration-corrected TEM (AC-TEM), which employs accelerating voltage in the range of (60-80 kV) [16 - 19]. This voltage range is relatively low and avoids beam damage [19]. The better resolution images of graphene and carbon nanotubes were thus recorded by AC-TEM and reported by Suenaga and co-workers [20 - 25]. A lot of new information was provided for various nanomaterials, including graphene [19 - 25]. The image of graphene has been reported by Suenga and Koshino [25]. The TEM-based characterization of carbon-based material is explained in a review by Zhang and Su [16]. The sample preparation is required for TEM analysis; the support film used for materials like carbon nanotubes or graphene can produce a stronger contrast as compared to the sample under analysis [16 - 19]. Thus, various techniques have been employed for sample preparations, such as mechanical exfoliation and transfer of graphene onto TEM grids or CVD on grids [26]. The graphene or GBMs (graphene-based materials) can be synthesized by employing various methods, viz. mechanical exfoliation (top-down), reduction or oxidation of bulk graphite, CVD, and flash Joule heating [2, 26 - 29].

The recent report regarding the scalable synthesis of graphene from cheap sources (tyres, coal, carbon black, waste food) in a short period of time employs a flash joule heating process [30]. The process involves the heating of precursor carbon material above 3000 K for a period of < 100 ms. The heating is performed by means of high voltage electric discharge between two electrodes forming turbostratic graphene [30]. The graphene derived from the process was shown to have single-layer graphene, also graphene showing Moire pattern (interference pattern) [30]. The graphene thus produced does not require a purification step and, in addition to TEM, is also characterized by Raman spectroscopic technique [30]. Ding *et al.* [31] have reported the green synthesis process to obtain graphene *via* a one-pot synthesis process. The precursor used is an alkaline mixture with lignin (called as black liquor) [31]. The graphene, called black liquid converted graphene (BL-G) with thickness between one to three atomic carbon layers, was obtained [31]. The BL-G obtained *via* this green process has a hexagonal lattice structure and thickness similar to reduced graphene sheets fabricated by the conventional chemical process [31]. The TEM and high-resolution TEM (HR-TEM) images of BL-graphene were recorded to analyse synthesized graphene. The BL-G was spin-coated on a flexible substrate with decreased transparencies and enhanced conductivity [31].

The remarkable properties of graphene oxide (GO) have led to extensive research on GO synthesis, characterization, and application [32 - 36]. The monolayer graphite oxide layer is considered as GO, typically fabricated by oxidation of graphite and then stored in dispersed form or as agglomerated (dried) form [36]. Wilson *et al.* [27] have performed the structural analysis of GO by electron microscopy. It was concluded that the hexagonal symmetry was maintained with order and the C-C bond length of the pristine (unmodified) graphene sheet. The GO was used as an underlying film for the study of the structure of macromolecules, and this has become possible as GO was stable under the electron beam and demonstrated transparency in the electron beam [36]. Thus, GO can be used as a TEM grid [36].

However, the structural characterization of GO is challenging owing to its non-stoichiometric structure and the structured dependency on the fabrication process [37, 38]. Moreover, surface contaminants (oxidative debris [39], amorphous airborne hydrocarbons [40]) influence not only the surface properties but also the chemical structure and defect density [38 - 41]. The image in the TEM detector may not be a direct image showing atomic position; rather can be a modification of the wave coming from behind the sample by the optical (electron) system of microscope or modification of incoming electron wave by the under studied sample [42]. In such a case, aberration-corrected TEM is employed [38, 42]. The direct image can be obtained for graphene but still cannot be successful for GO or GO with contaminants [38, 43]. Dave *et al.* [38] demonstrated a method to remove contaminants for a much better atomic resolution of GO. The *in-situ* heating (up to 700 °C) enables the elucidation of GO structure [38], which otherwise inhibits the sample imaging by adsorbed surface contamination [38].

2.3. Scanning Tunnelling Microscope (STM)

The graphene layers stack in a number of arrangements in graphite; the layer arrangement is described in detail by Mayer, Li and Andrei [43]. The atoms marked A in the top layer can reside above the B atoms of the lower layer in such a manner that three stacking modes can occur: AA stacking, AB stacking (Bernal stacking) and twisted stacking (giving Moiré pattern). In the AA stacking, layers are stacked in such a way that all the carbon atoms of one layer are on top of the other carbon atoms; Bernal stacking or AB type stacking is done with a shifted second layer such that A atoms reside on top of atom B of the layer, which is twisted by some degree [43 - 45]. The twisted or rotated stacked layers form the Moiré pattern [43 - 45]. Lucian *et al.* [46] employed home build low-temperature STM to study the graphene layers. They employed CVD at ambient pressure for the growth of graphene layer on polycrystalline films. The graphene layers were mechanically supported by spin-coated polymethyl methacrylate (PMMA) layers; afterward, the graphene layers were transferred onto the TEM grid for analysis. It was demonstrated that the electronic properties (low energy) of twisted graphene layers were managed by Dirac femion [46].

The twisted graphene layers (TGL) (particularly bi layers) have also been analysed with an optical microscope, and it was demonstrated that various coloured regions variable in shape and size are present in a given sample [47]. The red-, blue-, and yellow-colored domains are shown in Fig. (**6**) [47]. The methodology for bilayer fabrication is depicted in Fig. (**6a**) and Fig. (**6b**) represents the coloured domains. Briefly, the fabrication involves the growth of graphene films onto Cu support by CVD at ambient pressure, then transferring the layers onto Si/SiO_2 substrate by a wet process using PMMA protection layer and etching process [47]. The process was repeated to grow a graphene bilayer onto an already transferred graphene layer supported on SiO_2/Si [47]. The distinct colored regions occur without the use of any filters; as shown in Fig. (**6b**), the graphene layers were annealed and chemically treated (solvent cleaning) to modify the extent of the color domains, but there is no change in colored regions suggesting that the colors do not produce by impurities or residues [47]. The colored regions are observed in the bilayer but not by the neighbouring single layer [47].

The scanning tunnelling spectroscopy provides tunnelling conductance as a function of the quantity considered proportional to the density (local) of states (DOS). This quantity is bias voltage [48, 49]. The tunnelling spectrum obtained in zero magnetic field and low temperature is obtained for graphene flakes and graphite. A V-shaped curved is obtained for graphene DOS in the proximity of the Fermi level and reaches a minimum (at zero bias voltage) at Dirac point [48, 49]. The graphene flakes obtained after mechanical exfoliation of highly oriented pyrolitic–graphite (HOPG) are obtained as loosely bound graphene after exfoliation on graphite surface [48]. The STM images of the graphene layers were recorded as well. It was concluded that graphite as a substrate for the growth of graphene could be used for studying the intrinsic properties of charge carriers in graphene [48].

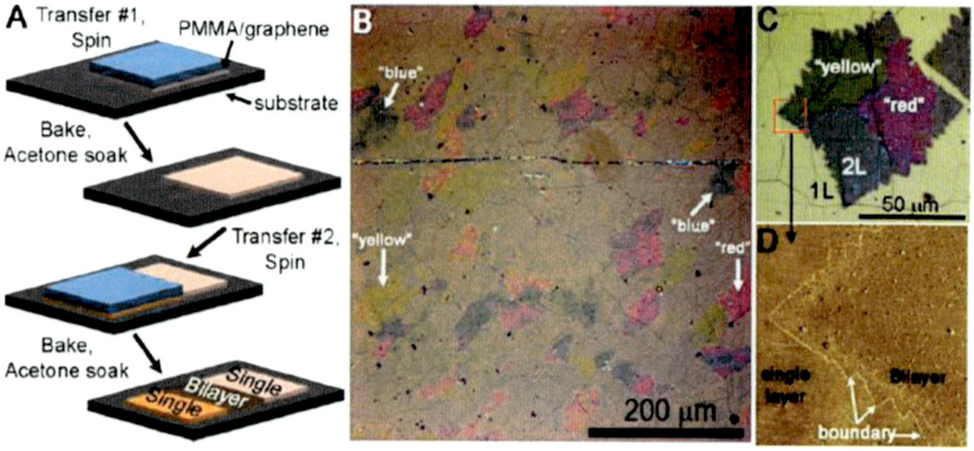

Fig. (6). The fabrication as well as characterization of stacked graphene films. **(a)** Schematic process for bilayer graphene preparation starting from single-layer CVD graphene films. **(b)** Images (Optical microscope) of a twisted bilayer graphene (TBG) film. The colored domains {"red", "blue" and "yellow") are depicted. **(c)** Sample showing top (second) graphene layer is only a partial layer, forming single-(1L) and bi-layer (2L) regions. **(d)** AFM images (15x15 μm^2) depicting the boundary between the single-layer and bilayer region highlighted by the red box in figure **(c)** [47]. Reprinted (adapted) with permission from ref [47]. Copyright (2013) American Chemical Society.

3. RAMAN SPECTROSCOPY

The identification of graphene or graphene-based materials has also been performed extensively by Raman spectroscopic technique [50, 51]. Simultaneous study of vibrational and electronic properties of graphene-based materials (GBM) is obtained from Raman spectroscopy [50, 51]. It is a reliable, rapid, non-destructive, sensitive analytical technique that provides high resolution and better precision [51 - 53]. The technique is based on the Raman Effect involving the Stokes process (absorption) and emission of phonons (anti-Stokes process) [50, 51]. The technique is sensitive to molecular structure and bonding in the molecule. The Raman bands obtained for graphite are influenced by the defect, flakes, layer of flakes, functionalization, doping, *etc* [53, 54]. Since the band gap is negligible in graphene, the incident wavelength resonates information regarding electronic and atomic levels [53, 55]. The position of the peaks, intensity and shape (of Raman band) varies with flake size, degree of doping, oxidation, number of layers, *etc* [51 - 55]. Thus, Raman spectroscopy quantitatively provides information that is mostly in agreement with microscopic analysis data [53 - 55].

The technique, which is based on Raman Effect, involves the phenomena in which light of wavelength (750-850 nm) excites the molecules in the sample. The incident light interacts with the sample, causing light to be reflected with different frequencies (or wave numbers) [53 - 55]. The wave number change occurs due to molecular rotations and vibrations in the sample [55, 56]. The modulated frequency provides information regarding the molecular structure of the sample [55, 56]. In the case of graphene with a negligible band gap, the incident wavelength resonates, providing atomic and electronic information. Thus, peak shape, intensity, and position vary with oxidation, disorder, doping, layers and flakes of graphene [54, 55].

The absorption or emission of phonons that are observed in Raman spectroscopy occurs owing to the interaction between incident radiation (light) and vibrations causing energy shift. The shift causes a change in wave number of emission or absorption [50, 57]. The energy difference between absorption and emission is known as the Raman shift [58]. When an electron scattered by phonon leads to first-order Raman scattering or the scattering can be of higher-order (second or more), involving scattering by two or more than two phonons [58, 59] and these provide a lot of information regarding the phonon coupling in the studied material [58]. Thus, Raman spectroscopy is effectively used to evaluate carbon nanostructures ranging from amorphous carbon, and carbon nanotubes (CNTs) to graphene or GBMs [58]. Jorio *et al.* [60] have shown the Raman spectra of various carbon-based materials. The similarities among the spectra, as well as differences between the spectra, help to study various carbon allotropes [15].

For understanding the characterization process for graphene or graphene-based materials, the graphene with honeycomb structure (Fig. **7a**) having sp^2 hybridization [60 - 64] of carbon and the reciprocal space unit cell depicting the first Brillouin zone (phonon dispersions of graphene, Fig. (**7b**) are shown in Fig. (**7**). The monolayer graphene showing hexagonal symmetry belongs to D^1_{6h} (space group *P6/mmm*) [60, 61]. Graphene having six normal modes (2 are double degenerate) and 2 atoms per unit cell at the Brillouin zone (Γ) center with (Γ) = $A_{2u} + B_{2g} + E_{1u} + E_{2g}$ [50, 62]. The B_{2g} phonon is Raman and IR inactive, while E_{2g} phonons are active, showing Raman spectra [50]. In the case of graphene, the phonon dispersion (six phonon branches shown in Fig. (**7b**) from the non-equivalent 2 carbon atoms present in the unit cell [60, 63]. The characteristic peaks in Raman are of two types called the G (~ 1582 cm^{-1}) and the G' bands (~ 2670 cm^{-1}) [58]. The in-plane sp^2 C-C vibration of carbon atoms (C-C in-plane vibration) leads to the band called as G band (first-order Raman mode) [50, 51, 58, 60]. The G band is related to doubly degenerate phonon modes, namely iTO (in-plane transverse optic) and LO at the Brillouin zone center Γ point, with the E_{2g} symmetry [51, 58, 60]. The LO stands for longitudinal optic [51, 64]. The iTO and iLO phonons are causing the Raman active bands in the graphene [64]. The second-order Raman scattering *i.e.*, G' band involves inelastic inter-valley scattering with iTO phonon close to the K point [51, 64].

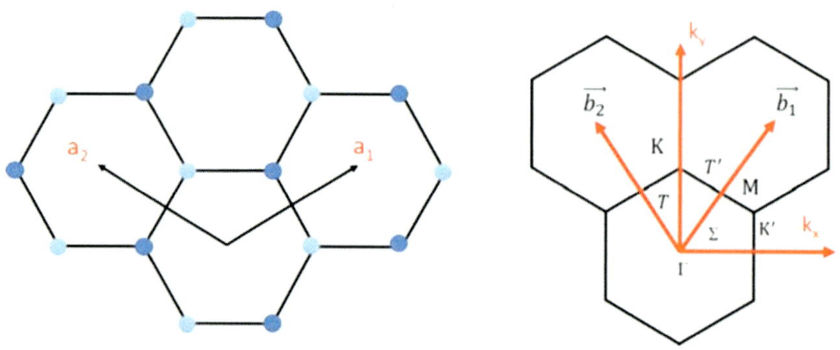

(a) (b)

Fig. (7). **(a)** Monolayer graphene from a top view showing the unit vectors a_1 and a_2 and inequivalent atoms A and B; **(b)** reciprocal space unit cell depicting the first Brillouin zone (BZ) with high-symmetry points and lines. Reciprocal space coordinate axes are K_x and K_y (primitive vectors) [64].

The G' band is also known as 2 D mode [50]. The G, G', and D peak positions are used to study graphene layer or doping in graphene [50]. The GO obtained from oxidation of graphite having sp^3 hybridized carbon is characterized based on G and D bands [50]. The D and D' peaks are Raman bands (second-order); here D band is inter-valley scattering with iTO phonon and defect; in other words, it arises due to disordered C atoms [50, 51, 53]. Whereas D' band is intra-valley double resonance scattering phenomena with iLO phonon and the defect [50, 51]. So, the G band occurs due to bond stretching of carbon sp_2 atoms both in ring and chain compounds, whereas the D peak appeared owing to the breathing mode of the sp^2 hybridized atoms in the ring [55, 65, 66]. Since D mode is known to occur by disorder or defects in the graphene-based materials, the ratio of D band to G band intensities (I_D/I_G) denotes the disorder/defect level in graphene synthesis [53, 67]. The 2D band is a G' band but has a frequency twice that of the D band [51]. The 2D band is present in both graphite and graphene [53]. The D and G' band peaks change as the excitation wavelength of the laser changes [51, 53, 60].

The prominent G' (second-order) band ~ 2700 cm^{-1} is produced by two phonons which is the iTO (in-plane transversal optical) phonon close to the K point in the Brillouin zone [60, 63]. This phonon displays strong electron-phonon coupling and produces a prominent G' band [60, 63]. In graphene, other phonons can produce similar scattering processes, causing various smaller peaks to appear in Raman spectra [60, 68]. Tuinstra and Koenig [66] performed the Raman spectroscopic analysis of single-crystal graphite and other graphite materials. The peaks at ~1580 cm^{-1} are assigned to E_{2g} and at ~1350 cm^{-1}, to A_{1g} breathing mode

[66]. The former high-frequency peak and the mode at Brillouin Zone (BZ) boundary K are known as G and D peaks, respectively [66, 69, 70]. The Raman spectrum of graphene (crystalline) on SiO_2/Si substrate is shown in Fig. (**8a**); the excitation laser with 514.4 nm was employed [64, 68]. The spectrum has two Raman bands at 1580 cm^{-1} (G band) and another one at 2700 cm^{-1}{the G' band or termed at 2D band} [64, 68, 71]. The number of graphene layer influences the intensity of the 2D band, which is approximately four times as compared to the G band in the case of crystalline graphene single-layer atoms [10, 64, 68, 72 - 75]. The number of layers in graphene increases, causing variation in an electronic environment, and this change occurs due to interactions among adjacent layers [53]. This consequently results in the broadening of 2D peaks and decreasing of the peak intensity [53], as shown in Fig. (**8b**). After a few layers (five), graphene and graphite Raman spectra are indistinguishable [53, 68]. The D band is weak, and the presence of defects in graphene nanosheets makes the D band appear in the spectrum [64]. The intense D peak is observed for amorphous carbon isotopes, which are absent in graphitic material [64].

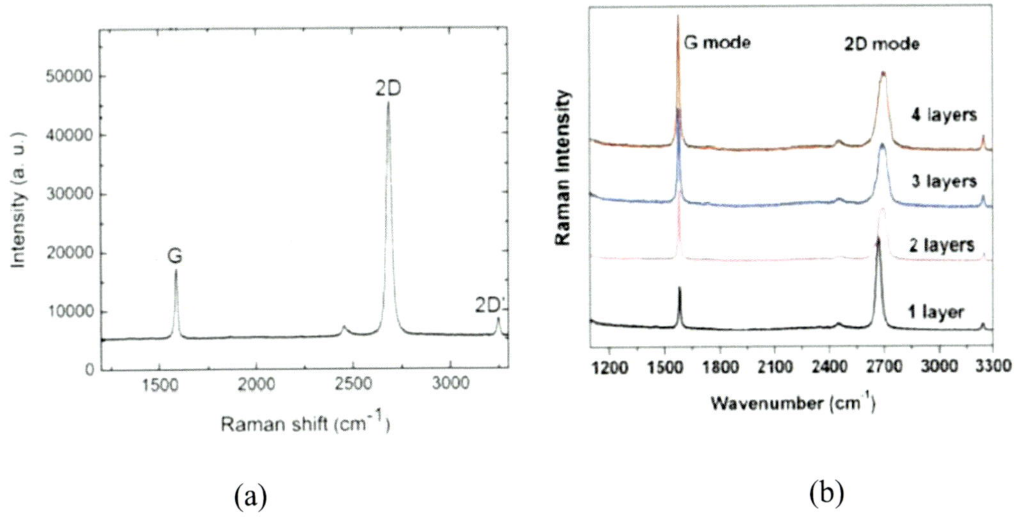

Fig. (8). The Raman spectra of **a)** crystalline graphite, **(b)** few layer graphene. Reproduced/reprinted (adapted) from refs [10, 64].

Ni *et al.* [10] employed a 532 nm excitation source for recording Raman spectra of graphene having different staked layers [10]. The sharp 2D peak is observed for single-layer graphene, which is distinguishable from the 2D band of multi-layer graphene. But in the case of graphene with 2 to 4 layers, the difference in 2D peak is not noticeable, but a visible difference is observed in G band intensity [10]. The appearance or disappearance of D peaks corresponds to the defects or imperfections in the hexagonal sp^2 carbon network [60]. The Raman spectra are shown for amorphous carbon materials with or without high-temperature treatments, with D band peaks appearing in the spectra having different peak widths and intensities [60]. With few defects in the structure will exhibit D, and in the case of crystalline ordered graphene or graphite, the D band was missing [60]. Luong *et al.* [30] fabricated the high-quality graphene by the flash Joule heating (FJH) process and observed that the I_{2D}/I_G is greater than 10. The D band with extremely low intensity is observed, indicating a low defect concentration [30]. The ratio I_{2D}/I_G was as high as 17 for graphene obtained from carbon black by the FJH process and was reported to be the highest value for any form of graphene [30]. This is due to the extremely high temperature achieved for graphene fabrication by the flash process [30].

Cancado *et al.* [70] studied the Raman spectra of single-layer graphene samples with various defect densities. The defect was created by the bombardment of the sample by Ar^+ ions. The Raman spectra of various SLG bombarded with different doses of Ar^+ ions are shown in Fig. (**9**). Thus, the measure of defects (point-like) in graphene was evaluated from the I_D/I_G ratio by Raman spectroscopy using different excitation laser energies [70].

The one-dimensional defects (or the amount of border) with respect to the crystalline area give the degree of disorder in a nano-crystallite and the measurement of L_a [70]. In the case of zero-dimension point-like defects, as in graphene, L_d is the distance between defects [70, 75]. Various methods can quantify the measure of the amount of disorder, *i.e.*, L_d, by using Raman spectroscopic analysis [70, 75]. Thus, the influence of L_d on peak frequency, width, integrated area and intensity has been studied [70, 76, 77]. In Fig. (**9a**), the peaks at 1580 cm^{-1} and 1350 cm^{-1} correspond to G and D bands, respectively [70]. The strongly dispersive intervalley double resonance process having excitation energy due to Kohn Anomaly at K gives rise to the D peak [70]. The intervalley process can also occur by connecting the two points which belong to the same cone around the K or K' [78]. Thus, giving a peak at ~1620 cm^{-1} in a defective sample is obtained using a laser at 514.15 nm [70]. The second-order 2D peak appears as a single peak in graphene but breaks into four peaks in the case of BLG (bilayer graphene), indicating the electron band structure evolution [68, 70, 79]. Again, the second-order peak of D' known as 2D' and the 2D peak do not require any defects for their activation and are found in the Raman spectra [70]. However, the peak corresponding to ~ 2940 cm^{-1} (D + D') requires defect activation, and this peak is a combination of phonons having dissimilar momenta at K and Γ [70]. In Fig. (**9b**), the plot of I_D/I_G *vs.* L_D is represented [70]. The ratio increases with a decrease in L_D and reaches maxima at $L_D \sim 3$ nm and reaches a minimum (zero) at $L_D < 3$ nm [70]. Thus, Raman spectroscopy has been employed for the analysis of defects in graphene and related materials [80]. The detailed application of Raman

Spectroscopy is provided in the reference [51]. Kumar and Pattammattel [53] highlighted a few limitations and opportunities for errors in employing Raman spectroscopy; these are summarised previously in the literature [53]:

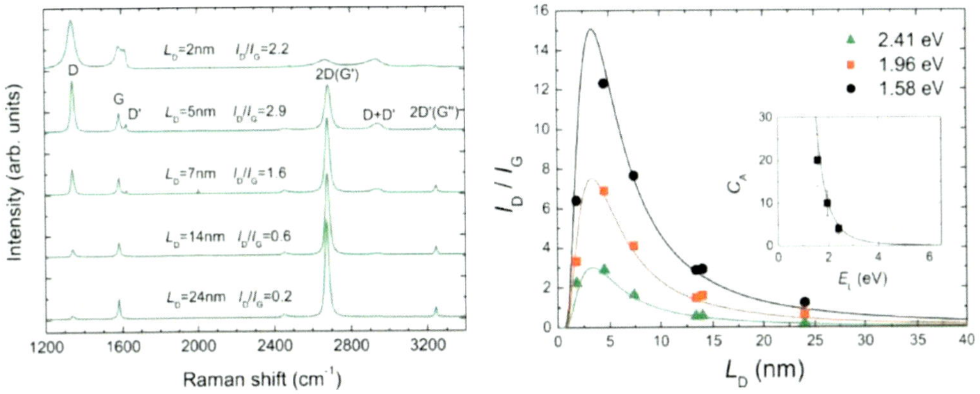

(a) (b)

Fig. (9). (a) Raman spectra of ion bombarded SLG with different L_D (measured at laser excitation energy, E_L = 2.41 eV and wavelength, λ_L = 514.5 nm). The L_D (distance between defects) is independently measured as per ref [70, 75]. The I_D/I_G values are shown for each spectrum. The notation, such as in the parentheses [2D(G′)], indicates two commonly used notations for the same peak (2D and G′); and **b)** I_D/I_G for all SLGs and E_L [70, 75]. The inset figure is the plot of C_A *vs.* E_L. Reprinted (adapted) with permission from ref [70]. Copyright (2011) American Chemical Society.

A) The sampling for Raman analysis is crucial; for instance, the spectroscopic analysis of exfoliated graphene samples can estimate the graphene layer. The transfer of exfoliated samples on glass slides can lead to the rest of the sheets, thus providing a rough evaluation of the number of layers [53]. The samples (films) fabricated from various stock solutions under dilute conditions can provide a better way of sample analysis. However, they may encounter problems related signal-to-noise ratio [53].

B) The particle size estimation (formed by exfoliation) from Raman analysis should be complemented with absorption spectroscopy and microscopic studies [53].

C) During spectroscopic analysis, samples can undergo photoreduction, as in the case of graphene oxide, influencing the conclusions drawn [53]. By reducing laser excitation energy (laser power) or by defocusing, the damage caused to samples can be minimized, and the same can be judged by optical analysis using an optical microscope [53].

4. FT-IR SPECTROSCOPY

FT-IR spectroscopic analysis is one of the efficient approaches for the detection of functional groups in the molecule [53]. The modification or functionalization of graphene or graphene oxide can be analysed by IR spectroscopy along with other spectroscopic techniques such as NMR, and XPS [6, 53, 81]. The 2D hexagonal arrangement of sp^2 hybridized carbons atoms is present in graphene [82, 83]. In the FTIR spectrum, no significant peak due to any functional group is observed [82, 83]. But in the case of graphene oxide, IR spectra provide information regarding -COOH, -OH, C-O and C=O functional groups, as shown in Fig. (**10**) [53, 84]. Fig. (**10**) shows spectra for crystalline graphite, GO and rGO [84]. The in-plane stretching vibration of C=C in graphite appears at ~1500–1600 cm^{-1} [81, 84]. In the case of GO, -OH peak appears at 3162 cm^{-1}, C=O peak at 1710 $cm^{-1,}$ and for alkoxy and epoxy groups, the peaks are at 1050 cm^{-1} and 1220 cm^{-1}, respectively [53, 84]. The functionalization of graphene can be performed by a number of reagents, including amino acids, carboxylic acids, polymers, phosphates, *etc* [73, 84 - 87].

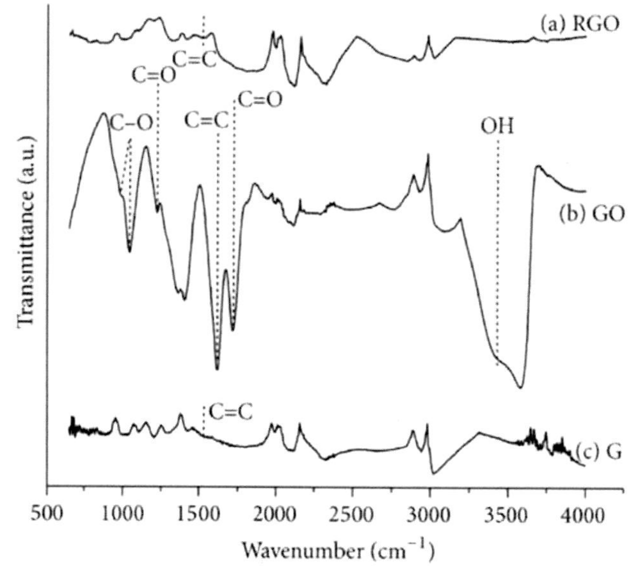

Fig. (10). The figure represents the IR spectra of **(a)** (RGO), **(b)** graphene oxide (GO), and **(c)** pure graphite (G). Reproduced with permission from ref [84]. under Creative Commons Attribution License.

Mittal *et al.* [88] performed thermal oxidation of graphene (a few layer) in the presence of oxygen, and the oxidized graphene was analysed by FT-IR spectroscopic technique. The attached oxygen groups onto the surface of oxidized graphene (called porous graphene) were confirmed by the peaks at 3418 cm^{-1} for -OH; similarly, the peaks at 1726 cm^{-1}, 1096 cm^{-1}, and 915 cm^{-1} ascribed to -COO-,

-C-O-, and -C-O-C- groups, respectively [88]. Kuila *et al.* [89] employed a green approach to reducing for the reduction of GO to rGO, by using carrot root (endophytic micro-organisms in the carrot root). The partial reduction of GO to graphene was confirmed by X-ray photoelectron and FT-IR [89]. The C=O stretching peak and -O-H deformation peaks at 1728 cm^{-1} and 1395 cm^{-1}, respectively, decrease significantly after carrot-assisted reduction of GO [89]. The intensities of epoxy and alkoxy C-O stretching peaks were unchanged after reduction, but the peaks' position was significantly altered [89], owing to a partial reduction of GO. New peaks due to carotene present in carrots appear at 2928 cm^{-1} and 2851 cm^{-1} [89].

Acik *et al.* [81] performed *in situ* transmission IR absorption analysis of GO during vacuum thermal annealing (temperature 60-850 °C). In this work, Acik *et al.* [81] have reported that the peak assignments (identification of chemical composition) in IR of GO are difficult and challenging due to conjugation and proximity effects. Modified Hummer's method was used for the fabrication of GO and analysed by the peaks at 3000-3600 cm^{-1} of hydroxyls C-OH (-COOH, H_2O contribution), ketones C=O at ~ 1750-1850 cm^{-1}, carboxyls (COOH) at ~ 1600-1750 cm^{-1}, in-plane C=C stretching at ~ 1500-1600 cm^{-1}, C-O-C epoxide peaks at ~1280-1330 cm^{-1} and 800-900 cm^{-1}, along with weak contributions by all these species in the region 900-1500 cm^{-1} [81]. The possible IR peak assignments of functional groups needed for the characterization of graphene-based materials are reviewed by Tucureanu *et al.* [82]. Since the IR analysis is complementary to other spectroscopic analyses, *e.g.*, XPS, NMR, *etc.*, the characterization of GBMs by XPS analysis is discussed in the next section.

5. X-RAY PHOTOELECTRON SPECTROSCOPY (XPS)

The characterization of surface functionalities is done using X-ray photoelectron spectroscopy (XPS). This is a quantitative technique for the evaluation of surface chemistry, the chemical and electronic state of elements, and also the elemental composition [43, 90 - 92]. The X-ray beam irradiates the material of interest; the surface generates a spectrum of electrons with characteristic kinetic energy [92]. The number and kinetic energy of released electrons are measured in XPS [93]. Thus, XPS peaks with characteristic binding energies are obtained by the various elements exposed to the X-ray beam [92]. The peaks provide information regarding electronic configuration within atoms [92]. The technique is based on the photoelectric effect [93], and for obtaining XPS spectra, the sample is irradiated with X-ray beams [91 - 93]. The law of energy can be used to understand the working principle of the XPS technique as [91 - 93]:

$$E_p = E_{ke} + \varphi + E_{bind} \quad (1)$$

In the above eq. (1), E_p, E_{ke}, φ, and E_{bind}, represent the energy of the incident X-ray radiation (*i.e.*, of incident photons), the kinetic energy of the photon emitted core electrons, the part of energy absorbed (work function), and ionization energy (binding energy) in eV, respectively [91 - 93]. So, the binding energy is calculated from the above equation [91, 93]. The photons of incident radiations eject core-level electrons from the atoms, and the kinetic energy of electrons depends upon the binding energy in the atoms and the elements from which electrons are ejected [91, 93]. The ejected electrons are collected from the uppermost 3 nm to 7 nm part of the material under study [91, 93].

XPS has been employed to characterize GBMs, GO, and rGO [53, 91, 93]. The representative binding energy of C1s, depending upon the chemical environment, is detected around 285 eV [53, 90]. The various C1s components of graphene with binding energy are shown in Table **1** [53].

Table 1. The binding energy (eV) of C1s components in graphene [92 - 97].

State	C1s Binding Energy (eV)
C-C	284.9-285.6
C=C	284.3-284.6
C=O	287.9-288.3
C-OH	285.4-286.7
C-O-C	285.6-286.9
COOR/COOH	288.3-289.4
O=C-O	288.5
O-C-O	285.6-286.9
π-π*	290.5-291.2
C-H	285.3
C=O	289
C-N (sp^3 N)	283.7
C=N	286.8

As shown in Table **1**, the common chemical state of carbon along with oxygen is used to identify or distinguish graphene from GO or rGO, heteroatom doped graphene [92 - 97]. The GO is rich in elemental oxygen, so in XPS analysis strong oxygen-bearing C1s peaks are observed [92]. Among the oxygen-bearing groups in GO, such as C-O, C=O, O-C=O, the most dominating XPS peak is C-O [92,

98]. Yu *et al.* [98] have reported improved Hummer's method for fabrication of GO in a high yield. The fabricated GO was characterized by various techniques of spectroscopic (FTIR, UV, XPS, Raman) [98]. XPS spectra for the fabricated GOs are shown in Fig. (**11**), in which the GO2 is the GO fabricated by their methodology, and GO1 is fabricated by improved Hummer's method reported by Kovtyukhova *et al.* [99]. Few XPS spectra indicate a strong peak of C=C [92 - 97, 99], and some report a strong main peak of the C-O bond [92, 98].

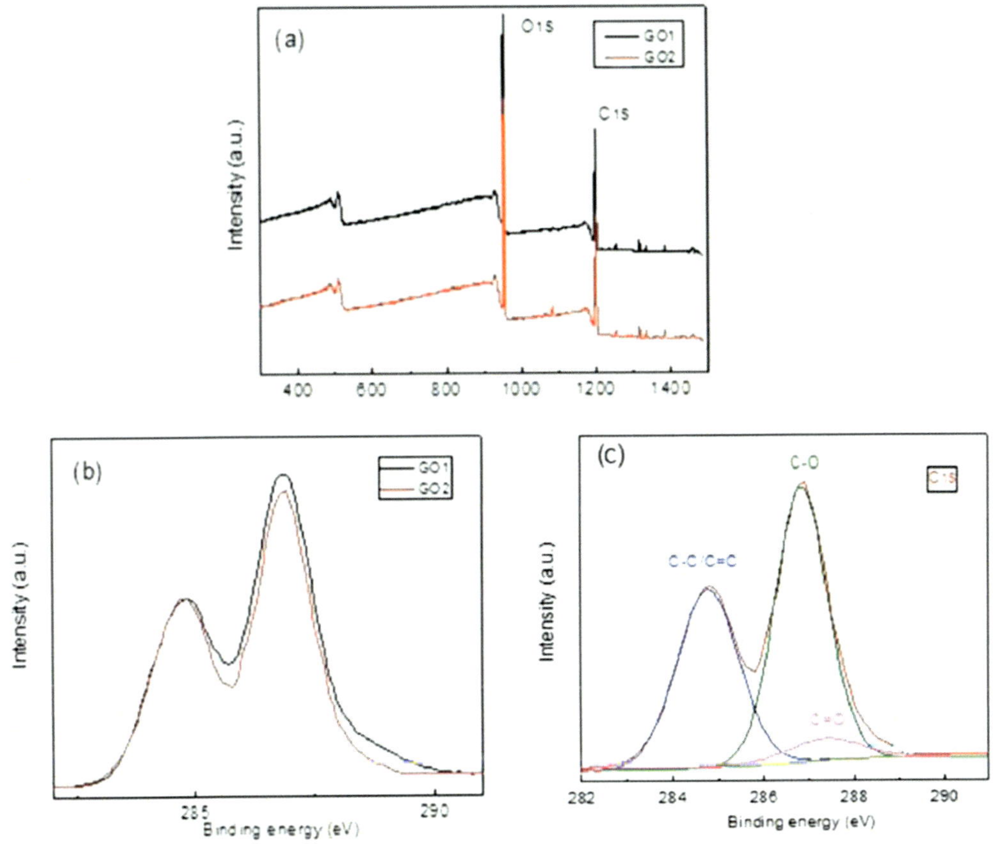

Fig. (11). The XPS of GO samples showing **(a)** GO1 and GO2, **(b)** C1s of GO1 and GO2 normalized to C=C peak, **(c)** deconvoluted C1s of GO2. Reused with permission from ref [98]. under Creative Commons CC BY license.

Ganguly *et al.* [100] reported *in situ* temperature-dependent XPS analysis of GO upon heat treatment process and observed that the unstable groups, such as the edge-plane COOH and C=O were stable and difficult to remove [100]. The C=O peak at 200 °C appears at ~ 288.3 eV, and the C-OH peak remains even at 1000 °C [100]. This peak indicates that thermally pyrolysis of the C-OH group is not complete even at very high temperature and due to intercalating position of the C-OH group in the interlayer galleries, which is between intact conjugated regions [100, 101].

The C1s components in GO samples are a) carbon atoms of graphitic material with a peak at 284.5 eV, b) carbon atoms of alcohols (tertiary) and epoxides at 286. 5 eV, and c) the carboxylic acid groups at 289.2 eV [93]. The signal of ketones is placed between 287.7 to 288.1 eV [93, 102]. However, the interval has been placed either into 286.5 to 287.5 eV (along with alcohol and epoxides) or 288.5–289.2 eV interval (along with carboxyl groups) [93]. The deconvoluting integral C1s spectra are reported in various ways, for instance, the spectral deconvolutions of C-O-C and C-OH peaks [93, 100, 103]. Nonetheless, the overall oxidation level of GO or rGO can be interpreted with confidence using XPS analysis [93].

Siokuo *et al.* [104] have adopted two different methodologies to grow atomically clean graphene layers on Cu substrates. One of the methodologies is the stepwise annealing up to a temperature of 750 °C and the second process is the treatment using mild Ar^+ sputtering under UHV conditions [104]. The XPS spectra of C1s of single-layer graphene grown on Cu foil treated at different temperatures. The XPS analysis in this report [104] has a much stronger C-OH peak. The sample of pristine graphene has peaks at 284.5 eV of C=C bonds, 285.6 eV for sp^3 hybridized carbon and peaks for C–OH and O–C–O bonds at 286.7 eV and 288.3 eV, respectively. The intensity of peaks corresponding to O-C-O and C-OH bonds indicates the presence of amorphous and oxidized carbon contamination [92, 104].

Rabchinksii *et al.* [105] have fabricated aminated graphene by simultaneous reduction and amination of GO under mild conditions. This facile and scalable fabrication involves treating GO with hydrobromic acid and ammonia solution [105]. The aminated graphene bears a low content of various other nitrogen species (graphitic N and pyridines), and this was confirmed from XAS (X-ray absorption studies) and XPS analysis [105]. The XPS spectra of brominated GO, rGO-Br, aminated graphene oxide, rGO-Am and precursor GO are shown in Fig. (**12**) [105].

The C1s peak in GO appears at ~284.7 eV, and O1s peaks in GO appear at ~ 532.6 eV, demonstrating no impurities were present [105]. The presence of bromine in the brominated GO demonstrates the Br3p doublet peaks at ~ 182eV and 189 eV, for the Br3d band at ~ 69 eV was confirmed from XPS analysis [105]. The absence of these Br-related peaks and the presence of N1s peak at ~ 400 eV indicated the amination of rGO-Br; furthermore, the incorporation of nitrogen functionalities had a concentration of about 5.5 at. % [105]. The peaks appearing at 285.1 eV, 284.7 eV, and 283.9 eV correspond to sp^2 C=C domains, carbon atoms having bond distortion due to the presence of functional groups at the neighbouring C-C atoms vacancy defects in graphene lattice [105].

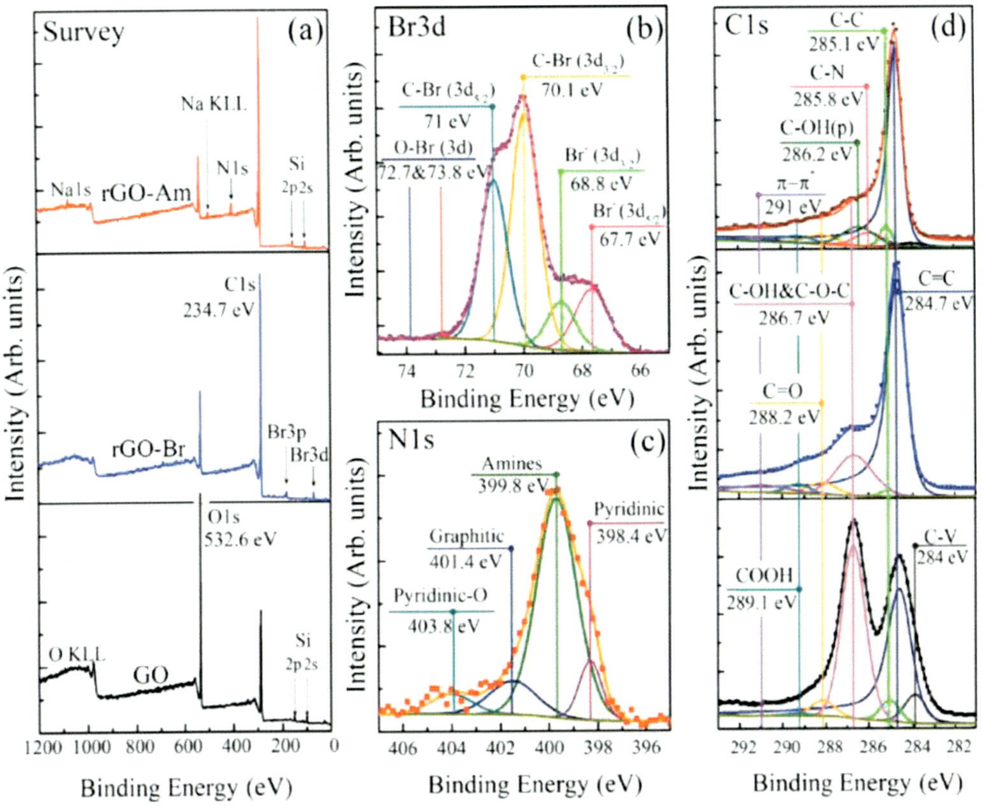

Fig. (12). (a) XPS survey spectra of the pristine GO, brominated rGO-Br and rGO-Am. **(b)** High-resolution XPS spectrum of rGO-Br (Br3d). **(c)** N1s XPS (high-resolution) of rGO-Am. **(d)** C1s spectra (high-resolution) of the initial GO and modified rGOs. For clarity, C1s spectrum and their fits are vertically offset from the fitting components [105]. Reused from ref [105]. under Creative Commons CC BY license.

Besides XPS, extended X-ray absorption fine structure (EXAFS) is also utilized as a powerful technique for structural characterization (bond length and coordination number) of GBMs [106 - 108]. The local structure around the element of interest can be evaluated using EXAFS; thus, metal coordination or metal absorption can be studied to evaluate the potential applications [106 - 108]. GBMs are employed for a number of applications related to environmental remediation, catalysis, energy, water desalination, *etc* [106 - 108]. For contamination absorption (*e.g.*, heavy metal) or the use of GBMs as a catalyst, the local structure around an element of interest can be studied using EXAFS [106 - 108].

The strength of X-ray absorbed by material as a function of energy provided by the X-ray radiations is measured [108]. The absorption of heavy metals [Pb(II), Cd(II), and U(VI)] using multi-layered graphene oxide (MLGO) was studied recently using XAFS [107]. Inner-sphere and the outer-sphere mechanisms were concluded to be the crucial mechanisms of heavy metal absorption onto MLGO [107]. EXAFS and Near-edge X-ray absorption fine-structure have also been used to study the structure of GBMs [107 - 110].

6. THERMOGRAVIMETRIC ANALYSIS (TGA)

Various techniques discussed above are used for the characterization of graphene or graphene-based materials, typically probing properties only of a small sample area. These localized analytical techniques can also characterize the industrially produced graphene materials; however, it has been shown that graphene material produced on a large scale may not be graphene [111, 112] but instead can be a few layers of graphene (FLG) having non-exfoliated graphitic materials or impurities [111, 112]. The FLG, as described by ISO (International Organization for Standardization) [113], is a 2-D material with 3-10 stacked graphene layers [111]. Recently, the TGA analysis of GBMs has been performed to demonstrate the analytical application of TGA for characterization by recording and analyzing TGA/DTA thermograms [111, 112]. TGA, a thermoanalytical technique, measures the mass change with respect to temperature rise. The mass change can occur due to oxidation, decomposition, moisture loss, or loss of volatiles [114]. In this regard, the TGA-DTG of materials, such as graphite, FLG, GO, and rGO was recorded, and the decomposition with temperature was analyzed in terms of various characteristic parameters, such as DTG (derivative thermogravimetry) peak-range and -shape, maximum mass decomposition rate (T_{max}) and thermal stability. These parameters closely related to intrinsic physical and chemical structures were studied to identify graphene-based materials [111, 112] undoubtedly. The authors [111, 112] designed a hypothesis of using TGA as an analytical method, and this hypothesis was verified by recording TGA/DTA of controlled, verified samples of FLG, GO and rGO. These samples were also characterized by XPS, SEM, TEM, particle size distribution (PSD) and XRD. The TGA/DTA characteristics of these samples were grouped into analytical parameters. These parameters (DTG peak range and shape, T_{max} and thermal stability) were further used for the identification of GBMs [111, 112].

The TGA-DTA signatures of GBMs involved the division of thermographs into three distinct temperature regions [111]. Region I (RGI): water evaporation region (temperature, $T < 100\ °C$); region II (RGII): decomposition of oxygen-bearing groups (T =100°-360°) and region III (RGIII): carbon combustion (T = 360°-

1000°). Three DTG peaks were reported [111] for GO samples in each region, with T_{max} values of 221.2 ± 1.2 °C in region II and 555.5 ± 5.0 °C in region III. For rGO, only one DTG peak with a different position and T_{max} was observed in region III. The T_{max} values in RIII for rGO, FLG, and graphite were observed at 536.5 ± 5.0 °C, 703.9 ± 0.5 °C, and 871.3 ± 0.1 °C, respectively [111, 112]. It was reported that the T_{max} values are highly distinct for the studied GBMs [111, 112]. The T_{max} values were different for these GBMs, and the values vary with their chemical structures and carbon structures. The highest T_{max} was reported for graphite, followed by FLG, rGO and GO with lower values, resulting in significant differences in carbon decomposition temperatures. The graphite decomposes at higher temperatures than FLG, rGO and GO. Graphite, which contains many stacked graphene layers, is more thermodynamically stable owing to additional van der Waals interactions within the layers [111]. Thus, graphite requires higher temperature treatment for its decomposition due to the strength provided due to coexisting covalent bonding and the van der Waals interactions. The three regions in thermographs of graphite, graphene, and GO are shown in Fig. (**13**) [112].

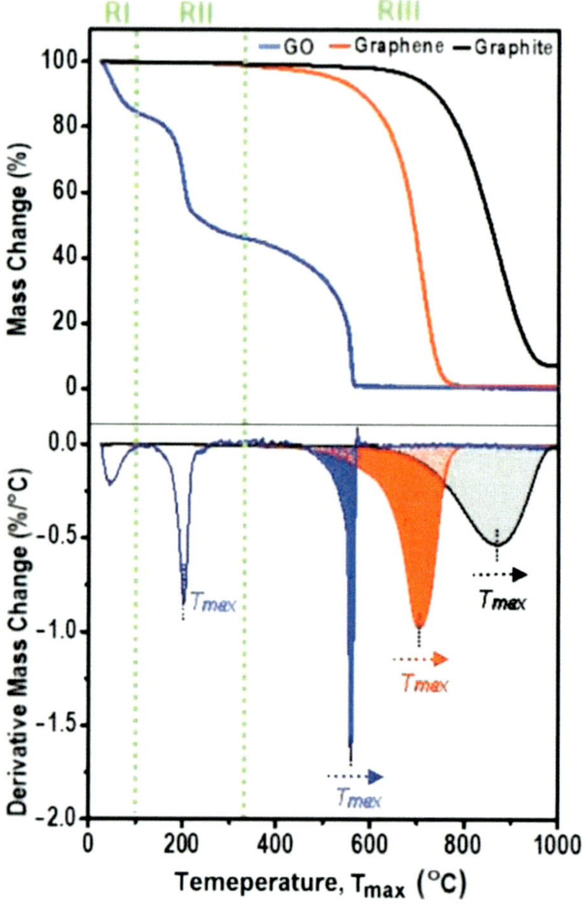

Fig. (13). The representative TGA-DTA thermographs along with T_{max} values of graphene, graphite, and GO [112].

The oxidation of graphite sp^2 hybridized carbon atoms forming GO leads to the conversion of the majority of sp^2 hybridized C-atoms to non-graphitic sp^3 hybridized carbon atoms. Additionally, the defects' density also increases due to extensive oxidation during the said conversion. The presence of defects and oxygenated groups leads to a reduction of GO strength, and GO decomposes at a lower temperature [111, 112]. The reduced-GO shows similar behavior in RGIII to that of GO, indicating that the sp^3 carbon framework has not recovered to the graphitic structure. The distinguished TGA/DTA behavior of GO and rGO was reported in RGII; here, no DTG peak was observed as most of the oxygen functional groups present in GO were removed during the reduction process when rGO was formed. It is worth mentioning that the authors have compared the results with available literature data and the data collected from more than hundreds of samples from other laboratories [106]. Thus, DTG peaks and values of T_{max} can be used to distinguish and identify graphite, FLG, GO and rGO. For qualitative analysis, several samples (commercially available as well as synthesized in lab) were analyzed using TGA/DTA, and the results were compared with verified GBM samples. The TGA/DTA results used for quantitative and qualitative analysis of GBMs were also in good agreement with other commonly used characterization techniques such as Raman spectroscopy, TEM, AFM, XRD, SEM and XPS [111, 112].

Wang *et al.* [115] facilely fabricated and then characterized graphene nanosheets. The synthesis was performed by oxidation, ultrasonic exfoliation, and chemical reduction. The nanosheets were characterized using XRD, HRTEM, TEM, and selected area electron diffraction (SAED). The TGA analysis of graphene nanosheets, GO, and pristine graphite powders was also performed in dry air to characterize and study the thermal behavior [115]. GO shows two-step mass loss with the rise of temperatures, one at 200 °C due to removing oxygenated groups and the other at a higher temperature of 500 °C because of carbon oxidation [115]. The pristine graphite losses mass at a much higher temperature of 650 °C, which was combustion to carbon dioxide. The mass loss in the case of less stable graphene nanosheet (as compared to graphite bulk powder) was reported at 250 °C. Similarly, Shtein *et al.* [116] have also characterized graphene-nanoplatelets by thermogravimetry in addition to Raman spectroscopy, electron microscopy, AFM and XPS analysis.

Recently, GO and rGO were used as fillers for the synthesis of polymer composite with poly(vinyl alcohol) as the polymer matrix [117]. The composite was studied for IR shielding applications. The filler GO and rGO were characterized along with TGA analysis. The TGA/DTA was performed to check the thermal stability of GBMs [117]. Three peaks were observed for GO, the first peak corresponding to water loss, the second peak at 200 °C due to the decomposition of GO as oxygen groups were removed and at 500 °C, beyond which the GO was completely decomposed. The GO was reduced chemically (using citric acid or hydrazine) and thermally (at 200 °C or 1000 °C) GO, and the rGO was again analyzed by TGA/DTA. The analysis indicated that a strong reducing agent or higher temperature treatment was able to lower the oxygen content in rGO significantly [117], as supported by decomposition trends in thermographs. Thus TGA/DTA analysis has been used for the characterization of GBMs. The degree of oxidation during GO formation can also be evaluated facilely using TGA [106]. In oxidized graphene samples, the mass loss of the functional groups can be studied quantitatively by TGA, and when combined with mass spectra or FTIR, more accurate information can be obtained [106, 111, 112].

7. CONCLUDING REMARKS

The affinity of hydrocarbons toward graphene leads to hydrocarbon contamination [53]. The hydrocarbons can come from vacuum grease, solvents, oil pumps *etc* [53]. This contamination is a challenge for the characterization of graphene by XPS analysis. Furthermore, oxygen-rich functionalities are liable to get reduced in the presence of X-ray beams, so this issue needs to be considered while recording XPS spectra [53, 93]. Additionally, the sample analysis is performed under a vacuum compared to the ambient atmosphere required by other techniques, such as Raman analysis [53, 93].

The characterization techniques in this chapter provided crucial information regarding atomic level as well as surface, chemical, and electronic properties of graphene-based materials. Optical microscopic techniques can make the visible distinction between graphene layer-by-layer or stacking. The information regarding 3D structural integration along with simultaneous vibrational and electronic properties can be evaluated. For practical application of GBMs, which depends upon controlling surface contamination, tuning surface properties, gram-scale fabrication process, and modulating electronic properties can be possible by targeted fabrication/modification approach. In this regard, the characterization of the material, both raw and processed (synthesized), is highly desirable. Thus, a brief discussion is made in this chapter regarding some of the important GBMs characterization techniques.

LIST OF ABBREVIATIONS

AC-TEM	Aberration-corrected transmission electron microscope
AFM	Atomic force microscope
ALD	Atomic layer deposition
BLG	Bilayer graphene
BZ	Brillouin zone
CNTs	Carbon nanotubes
CVD	Chemical vapor deposition
DOS	Density of states
DTG	Derivative thermogravimetry
FJH	Flash Joule heating
FT-IR	Fourier-transform infra-red spectroscopy
GBMs	Graphene-based materials
GO	Graphene oxide

HOPG	Highly oriented pyrolitic graphite
HR-TEM	High-resolution electron microscope
ISO	International Organization for Standardization
LO	Longitudinal optic
MLG	Multi-layer graphene
NMR	Nuclear magnetic resonance
OM	Optical Microscope
PMMA	Polymethyl methacrylate
PSD	Particle size distribution
rGO	Reduced graphene oxide
SAED	Selected area electron diffraction
SFM	Scanning force microscopy
SLG	Single layer graphene
STM	Scanning tunneling microscope
TEM	Transmission electron microscope
TGA	Thermogravimetric analysis
TGL	Twisted graphene layers
XPS	X-ray photoelectron spectroscopy
XRD	X-ray diffraction

CONSENT FOR PUBLICATION

Not applicable.

CONFLICT OF INTEREST

The author declares no conflict of interest, financial or otherwise.

ACKNOWLEDGEMENTS

Declared none.

REFERENCES

[1] Lee, H.C.; Liu, W.W.; Chai, S.P.; Mohamed, A.R.; Aziz, A.; Khe, C.S.; Hidayah, N.M.S.; Hashim, U. Review of the synthesis, transfer, characterization and growth mechanisms of single and multilayer graphene. *RSC Advances,* **2017**, *7*(26), 15644-15693. [http://dx.doi.org/10.1039/C7RA00392G]

[2] Huang, X.; Yin, Z.; Wu, S.; Qi, X.; He, Q.; Zhang, Q.; Yan, Q.; Boey, F.; Zhang, H. Graphene-based materials: synthesis, characterization, properties, and applications. *Small,* **2011**, *7*(14), 1876-1902. [http://dx.doi.org/10.1002/smll.201002009] [PMID: 21630440]

[3] Blake, P.; Hill, E.W.; Castro Neto, A.H.; Novoselov, K.S.; Jiang, D.; Yang, R.; Booth, T.J.; Geim, A.K. Making graphene visible. *Appl. Phys. Lett.*, **2007**, *91*(6), 063124. [http://dx.doi.org/10.1063/1.2768624]

[4] Novoselov, K.S.; Geim, A.K.; Morozov, S.V.; Jiang, D.; Zhang, Y.; Dubonos, S.V.; Grigorieva, I.V.; Firsov, A.A. Electric field effect in atomically thin carbon films. *Science,* **2004**, *306*(5696), 666-669. [http://dx.doi.org/10.1126/science.1102896] [PMID: 15499015]

[5] Arco, L.G.D.; Zhang, Y.; Zhou, C. Large scale graphene by chemical vapor deposition: synthesis, characterization and applications. In: *Graphene - synthesis, characterization, properties and applications*; Jian Ru Gong, Intech Open, **2011**; pp. 161-184. [http://dx.doi.org/10.5772/22358]

[6] Chen, Q.; Zhu, H. Structural characterizations of graphene. In: *Graphene Fabrication, Characterizations, Properties and Applications*; Zhu, H.; Xu, Z.; Xie, D.; Fang, Y., Eds.; Academic Press, **2018**; pp. 13-26.

[7] Kumar, V.; Lee, Y.S.; Shin, J.W.; Kim, K.H.; Kukkar, D.; Fai Tsang, Y. Potential applications of graphene-based nanomaterials as adsorbent for removal of volatile organic compounds. *Environ. Int.*, **2020**, *135*, 105356. [http://dx.doi.org/10.1016/j.envint.2019.105356] [PMID: 31881425]

[8] Roddaro, S.; Pingue, P.; Piazza, V.; Pellegrini, V.; Beltram, F. The optical visibility of graphene: interference colors of ultrathin graphite on $SiO_{(2)}$. *Nano Lett.*, **2007**, *7*(9), 2707-2710. [http://dx.doi.org/10.1021/nl071158l] [PMID: 17665963]

[9] Cheng, Y.; Song, Y.; Zhao, D.; Zhang, X.; Yin, S.; Wang, P.; Wang, M.; Xia, Y.; Maruyama, S.; Zhao, P.; Wang, H. Direct identification of multilayer graphene stacks on copper by optical microscopy. *Chem. Mater.*, **2016**, *28*(7), 2165-2171. [http://dx.doi.org/10.1021/acs.chemmater.6b00053]

[10] Ni, Z.H.; Wang, H.M.; Kasim, J.; Fan, H.M.; Yu, T.; Wu, Y.H.; Feng, Y.P.; Shen, Z.X. Graphene thickness determination using reflection and contrast spectroscopy. *Nano Lett.*, **2007**, *7*(9), 2758-2763. [http://dx.doi.org/10.1021/nl071254m] [PMID: 17655269]

[11] Graf, D.; Molitor, F.; Ensslin, K.; Stampfer, C.; Jungen, A.; Hierold, C.; Wirtz, L. Spatially resolved Raman spectroscopy of single- and few-layer graphene. *Nano Lett.*, **2007**, *7*(2), 238-242. [http://dx.doi.org/10.1021/nl061702a] [PMID: 17297984]

[12] Jia, C.; Jiang, J.; Gan, L.; Guo, X. Direct optical characterization of graphene growth and domains on growth substrates. *Sci. Rep.*, **2012**, *2*(1), 707. [http://dx.doi.org/10.1038/srep00707] [PMID: 23050091]

[13] Hong, K.P.; Lee, K.H.; Nam, J.; Kim, K.S.; Kim, S.H.; Kim, B.; Kim, K.; Park, J.S.; Lee, J.Y.; Jeong, T.; Song, Y.J.; Hwang, J.Y.; Choi, J.B.; Cho, S.; Kim, K.S.; Kim, H. Visualization of CVD-grown graphene on Cu film using area-selective ALD for quality management. *Appl. Surf. Sci.*, **2019**, *496*, 143614. [http://dx.doi.org/10.1016/j.apsusc.2019.143614]

[14] Zhang, Y.; Gomez, L.; Ishikawa, F.N.; Madaria, A.; Ryu, K.; Wang, C.; Badmaev, A.; Zhou, C. Comparison of graphene growth on single-crystalline and polycrystalline Ni by chemical vapor deposition. *J. Phys. Chem. Lett.*, **2010**, *1*(20), 3101-3107. [http://dx.doi.org/10.1021/jz1011466]

[15] Badruddoza, A.Z.M.; Chepyala, R.; Ashraf, K. State-of-the-Art characterization methods for graphene and its derivatives. In: *Micro and nano technologies, biomedical applications of graphene and 2D nanomaterials*; Nurunnabi, M.; McCarthy, J.R., Eds.; Elsevier, **2019**; pp. 43-86. [http://dx.doi.org/10.1016/B978-0-12-815889-0.00003-9]

[16] Zhang, B.; Su, D.S. Transmission electron microscopy and the science of carbon nanomaterials. *Small,* **2013**, 1-8.

[PMID: 23913822]

[17] Yang, J.C.; Small, M.W.; Grieshaber, R.V.; Nuzzo, R.G. Recent developments and applications of electron microscopy to heterogeneous catalysis. *Chem. Soc. Rev.*, **2012**, *41*(24), 8179-8194. [http://dx.doi.org/10.1039/c2cs35371g] [PMID: 23120754]

[18] Gai, P.L. Developments of electron microscopy methods in the study of catalysts. *Curr. Opin. Solid State Mater. Sci.*, **2001**, *5*(5), 371-380. [http://dx.doi.org/10.1016/S1359-0286(01)00036-5]

[19] Harris, P. J. F. Transmission electron microscopy of carbon: A brief history. **2018**, 1-17.

[20] Guan, L.; Suenaga, K.; Iijima, S. Smallest carbon nanotube assigned with atomic resolution accuracy. *Nano Lett.*, **2008**, *8*(2), 459-462. [http://dx.doi.org/10.1021/nl072396j] [PMID: 18186659]

[21] Suenaga, K.; Wakabayashi, H.; Koshino, M.; Sato, Y.; Urita, K.; Iijima, S. Imaging active topological defects in carbon nanotubes. *Nat. Nanotechnol.*, **2007**, *2*(6), 358-360. [http://dx.doi.org/10.1038/nnano.2007.141] [PMID: 18654307]

[22] Sato, Y.; Suenaga, K.; Okubo, S.; Okazaki, T.; Iijima, S. Structures of D5d-C80 and Ih-Er3N@C80 fullerenes and their rotation inside carbon nanotubes demonstrated by aberration-corrected electron microscopy. *Nano Lett.*, **2007**, *7*(12), 3704-3708. [http://dx.doi.org/10.1021/nl0720152]

[23] Liu, Z.; Suenaga, K.; Harris, P.J.F.; Iijima, S. Open and closed edges of graphene layers. *Phys. Rev. Lett.*, **2009**, *102*(1), 015501. [http://dx.doi.org/10.1103/PhysRevLett.102.015501] [PMID: 19257205]

[24] Sato, Y.; Yanagi, K.; Miyata, Y.; Suenaga, K.; Kataura, H.; Iijima, S. Chiral-angle distribution for separated single-walled carbon nanotubes. *Nano Lett.*, **2008**, *8*(10), 3151-3154. [http://dx.doi.org/10.1021/nl801364g] [PMID: 18729412]

[25] Suenaga, K.; Koshino, M. Atom-by-atom spectroscopy at graphene edge. *Nature,* **2010**, *468*(7327), 1088-1090. [http://dx.doi.org/10.1038/nature09664] [PMID: 21160475]

[26] Meyer, J.C. Transmission electron microscopy (TEM) of graphene. Graphene, Woodhead Publishing, **2014**; pp. 101-123.

[27] Wilson, N.R.; Pandey, P.A.; Beanland, R.; Young, R.J.; Kinloch, I.A.; Gong, L.; Liu, Z.; Suenaga, K.; Rourke, J.P.; York, S.J.; Sloan, J. Graphene oxide: structural analysis and application as a highly transparent support for electron microscopy. *ACS Nano,* **2009**, *3*(9), 2547-2556. [http://dx.doi.org/10.1021/nn900694t] [PMID: 19689122]

[28] Dong, L.X.; Chen, Q. Properties, synthesis, and characterization of graphene. *Front. Mater. Sci. China,* **2010**, *4*(1), 45-51. [http://dx.doi.org/10.1007/s11706-010-0014-3]

[29] Seekaew, Y.; Arayawut, O.; Timsorn, K.; Wongchoosuk, C. Synthesis, characterization, and applications of graphene and derivatives. In: *Carbon-Based Nanofillers and Their Rubber Nanocomposites*; Elsevier, **2019**.

[30] Luong, D.X.; Bets, K.V.; Algozeeb, W.A.; Stanford, M.G.; Kittrell, C.; Chen, W.; Salvatierra, R.V.; Ren, M.; McHugh, E.A.; Advincula, P.A.; Wang, Z.; Bhatt, M.; Guo, H.; Mancevski, V.; Shahsavari, R.; Yakobson, B.I.; Tour, J.M. Gram-scale bottom-up flash graphene synthesis. *Nature,* **2020**, *577*(7792), 647-651. [http://dx.doi.org/10.1038/s41586-020-1938-0] [PMID: 31988511]

[31] Ding, Z.; Yuan, T.; Wen, J.; Cao, X.; Sun, S.; Xiao, L.P.; Shi, Q.; Wang, X.; Sun, R. Green synthesis of chemical converted graphene sheets derived from pulping black liquor. *Carbon,* **2020**, *158*, 690-697. [http://dx.doi.org/10.1016/j.carbon.2019.11.041]

[32] Dreyer, D.R.; Park, S.; Bielawski, C.W.; Ruoff, R.S. The chemistry of graphene oxide. *Chem. Soc. Rev.*, **2010**, *39*(1), 228-240.
[http://dx.doi.org/10.1039/B917103G] [PMID: 20023850]

[33] Zhu, J.; Li, B.; Xu, M.; Liu, R.; Xia, T.; Zhang, Z.; Xu, Y.; Liu, S. Graphene oxide promotes cancer metastasis through associating with plasma membrane to promote TGF-β signaling-dependent epithelial–mesenchymal transition. *ACS Nano,* **2020**, *14*(1), 818-827.
[http://dx.doi.org/10.1021/acsnano.9b07891] [PMID: 31877027]

[34] Georgakilas, V.; Tiwari, J.N.; Kemp, K.C.; Perman, J.A.; Bourlinos, A.B.; Kim, K.S.; Zboril, R. Noncovalent functionalization of graphene and graphene oxide for energy materials, biosensing, catalytic, and biomedical applications. *Chem. Rev.*, **2016**, *116*(9), 5464-5519.
[http://dx.doi.org/10.1021/acs.chemrev.5b00620] [PMID: 27033639]

[35] Chen, D.; Feng, H.; Li, J. Graphene oxide: preparation, functionalization, and electrochemical applications. *Chem. Rev.*, **2012**, *112*(11), 6027-6053.
[http://dx.doi.org/10.1021/cr300115g] [PMID: 22889102]

[36] Tian, Y.; Yu, Z.; Cao, L.; Zhang, X.L.; Sun, C.; Wang, D.W. Graphene oxide: An emerging electromaterial for energy storage and conversion. *J. Energy.Chem.*, **2021**, *55*, 323-344.
[http://dx.doi.org/10.1016/j.jechem.2020.07.006]

[37] Dave, S.H.; Gong, C.; Robertson, A.W.; Warner, J.H.; Grossman, J.C. Correction to chemistry and structure of graphene oxide *via* direct imaging. *ACS Nano,* **2017**, *11*(1), 1121-1121.
[http://dx.doi.org/10.1021/acsnano.6b07705] [PMID: 28026164]

[38] Dave, S.H.; Gong, C.; Robertson, A.W.; Warner, J.H.; Grossman, J.C. Chemistry and structure of graphene oxide *via* direct imaging. *ACS Nano,* **2016**, *10*(8), 7515-7522.
[http://dx.doi.org/10.1021/acsnano.6b02391] [PMID: 27397115]

[39] Rourke, J.P.; Pandey, P.A.; Moore, J.J.; Bates, M.; Kinloch, I.A.; Young, R.J.; Wilson, N.R. The real graphene oxide revealed: stripping the oxidative debris from the graphene-like sheets. *Angew. Chem. Int. Ed.*, **2011**, *50*(14), 3173-3177.
[http://dx.doi.org/10.1002/anie.201007520] [PMID: 21432951]

[40] Gómez-Navarro, C.; Meyer, J.C.; Sundaram, R.S.; Chuvilin, A.; Kurasch, S.; Burghard, M.; Kern, K.; Kaiser, U. Atomic structure of reduced graphene oxide. *Nano Lett.*, **2010**, *10*(4), 1144-1148.
[http://dx.doi.org/10.1021/nl9031617] [PMID: 20199057]

[41] Li, Z.; Wang, Y.; Kozbial, A.; Shenoy, G.; Zhou, F.; McGinley, R.; Ireland, P.; Morganstein, B.; Kunkel, A.; Surwade, S.P.; Li, L.; Liu, H. Effect of airborne contaminants on the wettability of supported graphene and graphite. *Nat. Mater.*, **2013**, *12*(10), 925-931.
[http://dx.doi.org/10.1038/nmat3709] [PMID: 23872731]

[42] Skákalová, V.; Kaiser, A.B., Eds. Meyer. J. C. Transmission electron microscopy (TEM) of graphene. In: *Graphene*; Woodhead Publishing, **2014**; pp. 101-123.
[http://dx.doi.org/10.1533/9780857099334.2.101]

[43] Luican-Mayer, A.; Li, G.; Andrei, E.Y. Atomic scale characterization of mismatched graphene layers. *J. Electron Spectrosc. Relat. Phenom.*, **2017**, *219*, 92-98.
[http://dx.doi.org/10.1016/j.elspec.2017.01.005]

[44] Bottari, G.; Herranz, M.Á.; Wibmer, L.; Volland, M.; Rodríguez-Pérez, L.; Guldi, D.M.; Hirsch, A.; Martín, N.; D'Souza, F.; Torres, T. Chemical functionalization and characterization of graphene-based materials. *Chem. Soc. Rev.*, **2017**, *46*(15), 4464-4500.
[http://dx.doi.org/10.1039/C7CS00229G] [PMID: 28702571]

[45] Andrei, E.Y.; Li, G.; Du, X. Electronic properties of graphene: a perspective from scanning tunneling microscopy and magnetotransport. *Rep. Prog. Phys.*, **2012**, *75*(5), 056501.
[http://dx.doi.org/10.1088/0034-4885/75/5/056501] [PMID: 22790587]

[46] Luican, A.; Li, G.; Reina, A.; Kong, J.; Nair, R.R.; Novoselov, K.S.; Geim, A.K.; Andrei, E.Y. Single-

layer behavior and its breakdown in twisted graphene layers. *Phys. Rev. Lett.*, **2011**, *106*(12), 126802. [http://dx.doi.org/10.1103/PhysRevLett.106.126802] [PMID: 21517338]

[47] Robinson, J.T.; Schmucker, S.W.; Diaconescu, C.B.; Long, J.P.; Culbertson, J.C.; Ohta, T.; Friedman, A.L.; Beechem, T.E. Electronic hybridization of large-area stacked graphene films. *ACS Nano,* **2013**, *7*(1), 637-644.
[http://dx.doi.org/10.1021/nn304834p] [PMID: 23240977]

[48] Luican, A.; Li, G.; Andrei, E.Y. Scanning tunneling microscopy and spectroscopy of graphene layers on graphite. *Solid State Commun.*, **2009**, *149*(27-28), 1151-1156.
[http://dx.doi.org/10.1016/j.ssc.2009.02.059]

[49] Vázquez De Parga, A.L.; Miranda, R. Scanning tunneling microscopy (STM) of graphene. In: *Graphene: Properties, Preparation, Characterisation and Devices*; Skákalová, V.; Kaiser, A.B., Eds.; Woodhead Publishing, **2014**; pp. 124-155.
[http://dx.doi.org/10.1533/9780857099334.2.124]

[50] Nanda, S.S.; Kim, M.J.; Yeom, K.S.; An, S.S.A.; Ju, H.; Yi, D.K. Raman spectrum of graphene with its versatile future perspectives. *Trends Analyt. Chem.*, **2016**, *80*, 125-131.
[http://dx.doi.org/10.1016/j.trac.2016.02.024]

[51] Wu, J.B.; Lin, M.L.; Cong, X.; Liu, H.N.; Tan, P.H. Raman spectroscopy of graphene-based materials and its applications in related devices. *Chem. Soc. Rev.*, **2018**, *47*(5), 1822-1873.
[http://dx.doi.org/10.1039/C6CS00915H] [PMID: 29368764]

[52] Hulman, M. Raman spectroscopy of graphene. In: *Graphene: Properties, Preparation, Characterization and Devices*; Skákalová, V.; Kaiser, A.B., Eds.; Woodhead Publishing, **2014**.
[http://dx.doi.org/10.1533/9780857099334.2.156]

[53] Kumar, C.V.; Pattammattel, A. Characterization techniques for graphene. In: *Introduction to Graphene, Chemical and Biochemical Applications*; Kumar, C.V.; Pattammattel, A., Eds.; Elsevier, **2017**; pp. 45-74.

[54] Ferrari, A.C.; Basko, D.M. Raman spectroscopy as a versatile tool for studying the properties of graphene. *Nat. Nanotechnol.*, **2013**, *8*(4), 235-246.
[http://dx.doi.org/10.1038/nnano.2013.46] [PMID: 23552117]

[55] Ferrari, A.C. Raman spectroscopy of graphene and graphite: Disorder, electron–phonon coupling, doping and nonadiabatic effects. *Solid State Commun.*, **2007**, *143*(1-2), 47-57.
[http://dx.doi.org/10.1016/j.ssc.2007.03.052]

[56] Synetos, A.; Tousoulis, D. Invasive imaging techniques. In: *Coronary Artery Disease, from Biology to clinical practice*; Tousoulis, D., Ed.; Academic Press, **2018**, pp. 359-376.
[http://dx.doi.org/10.1016/B978-0-12-811908-2.00018-0]

[57] Cardona, M. *Light Scattering in Solids II*; Springer: Berlin, **1982**, Vol. 50, pp. 19-176.
[http://dx.doi.org/10.1007/3-540-11380-0_14]

[58] Wu, J.; Xie, L. Structural quantification for graphene and related two–dimensional materials by Raman spectroscopy. *Anal. Chem.*, **2019**, *91*(1), 468-481.
[http://dx.doi.org/10.1021/acs.analchem.8b04991] [PMID: 30408419]

[59] Loudon, R. The Raman effect in crystals. *Adv. Phys.*, **2001**, *50*(7), 813-864.
[http://dx.doi.org/10.1080/00018730110101395]

[60] Jorio, A.; Souza Filho, A.G. Raman studies of carbon nanostructures. *Annu. Rev. Mater. Res.*, **2016**, *46*(1), 357-382.
[http://dx.doi.org/10.1146/annurev-matsci-070115-032140]

[61] Malard, L.M.; Guimarães, M.H.D.; Mafra, D.L.; Mazzoni, M.S.C.; Jorio, A. Group-theory analysis of electrons and phonons in N layer graphene systems. *Phys. Rev. B Condens. Matter Mater. Phys.*, **2009**, *79*(12), 125426.
[http://dx.doi.org/10.1103/PhysRevB.79.125426]

[62] Nemanich, R.J.; Lucovsky, G.; Solin, S.A. Infrared active optical vibrations of graphite. *Solid State Commun.,* **1977**, *23*(2), 117-120.
[http://dx.doi.org/10.1016/0038-1098(77)90663-9]

[63] Jorio, A.; Dresselhaus, M.; Saito, R.; Dresselhaus, G. *Raman spectroscopy in graphene related systems*; Weinheim, Ger: Wiley-VCH, **2011**.
[http://dx.doi.org/10.1002/9783527632695]

[64] Lan, Y.; Zondode, M.; Deng, H.; Yan, J.A.; Ndaw, M.; Lisfi, A.; Wang, C.; Pan, Y.L. Basic Concepts and recent advances of crystallographic orientation determination of graphene by Raman spectroscopy. *Crystals (Basel),* **2018**, *8*(10), 375.
[http://dx.doi.org/10.3390/cryst8100375]

[65] Ferrari, A.C.; Robertson, J. Interpretation of Raman spectra of disordered and amorphous carbon. *Phys. Rev. B Condens. Matter,* **2000**, *61*(20), 14095-14107.
[http://dx.doi.org/10.1103/PhysRevB.61.14095]

[66] Tuinstra, F.; Koenig, J.L. Raman Spectrum of Graphite. *J. Chem. Phys.,* **1970**, *53*(3), 1126-1130.
[http://dx.doi.org/10.1063/1.1674108]

[67] Childres, I.; Jauregui, L.A.; Park, W.; Cao, H.; Chen, Y.P. Raman spectroscopy of graphene and related materials. In: *New Developments in Photon and Materials Research*; Jang, J.I., Ed.; Nova Science Publishers: Hauppauge, NY, **2013**.

[68] Ferrari, A.C.; Meyer, J.C.; Scardaci, V.; Casiraghi, C.; Lazzeri, M.; Mauri, F.; Piscanec, S.; Jiang, D.; Novoselov, K.S.; Roth, S.; Geim, A.K. Raman spectrum of graphene and graphene layers. *Phys. Rev. Lett.,* **2006**, *97*(18), 187401.
[http://dx.doi.org/10.1103/PhysRevLett.97.187401] [PMID: 17155573]

[69] Geim, A.K.; Novoselov, K.S. The rise of graphene. *Nat. Mater.,* **2007**, *6*(3), 183-191.
[http://dx.doi.org/10.1038/nmat1849] [PMID: 17330084]

[70] Cançado, L.G.; Jorio, A.; Ferreira, E.H.M.; Stavale, F.; Achete, C.A.; Capaz, R.B.; Moutinho, M.V.O.; Lombardo, A.; Kulmala, T.S.; Ferrari, A.C. Quantifying defects in graphene *via* Raman spectroscopy at different excitation energies. *Nano Lett.,* **2011**, *11*(8), 3190-3196.
[http://dx.doi.org/10.1021/nl201432g] [PMID: 21696186]

[71] Vidano, R.P.; Fischbach, D.B.; Willis, L.J.; Loehr, T.M. Observation of Raman band shifting with excitation wavelength for carbons and graphites. *Solid State Commun.,* **1981**, *39*(2), 341-344.
[http://dx.doi.org/10.1016/0038-1098(81)90686-4]

[72] Gupta, A.; Chen, G.; Joshi, P.; Tadigadapa, S.; Eklund, P.C. Raman scattering from high-frequency phonons in supported n-graphene layer films. *Nano Lett.,* **2006**, *6*(12), 2667-2673.
[http://dx.doi.org/10.1021/nl061420a] [PMID: 17163685]

[73] Park, J.S.; Cho, S.M.; Kim, W.J.; Park, J.; Yoo, P.J. Fabrication of graphene thin films based on layer-by-layer self-assembly of functionalized graphene nanosheets. *ACS Appl. Mater. Interfaces,* **2011**, *3*(2), 360-368.
[http://dx.doi.org/10.1021/am100977p] [PMID: 21207942]

[74] Liu, Y.; Liu, Z.; Lew, W.S.; Wang, Q.J. Temperature dependence of the electrical transport properties in few-layer graphene interconnects. *Nanoscale Res. Lett.,* **2013**, *8*(1), 335.
[http://dx.doi.org/10.1186/1556-276X-8-335] [PMID: 23885802]

[75] Lucchese, M.M.; Stavale, F.; Ferreira, E.H.M.; Vilani, C.; Moutinho, M.V.O.; Capaz, R.B.; Achete, C.A.; Jorio, A. Quantifying ion-induced defects and Raman relaxation length in graphene. *Carbon,* **2010**, *48*(5), 1592-1597.
[http://dx.doi.org/10.1016/j.carbon.2009.12.057]

[76] Jorio, A.; Lucchese, M.M.; Stavale, F.; Martins Ferreira, E.H.; Moutinho, M.V.O.; Capaz, R.B.; Achete, C.A. Raman study of ion-induced defects in *N* -layer graphene. *J. Phys. Condens. Matter,* **2010**, *22*(33), 334204-334208.

[http://dx.doi.org/10.1088/0953-8984/22/33/334204] [PMID: 21386494]

[77] Martins Ferreira, E.H.; Moutinho, M.V.O.; Stavale, F.; Lucchese, M.M.; Capaz, R.B.; Achete, C.A.; Jorio, A. Evolution of the Raman spectra from single-, few-, and many-layer graphene with increasing disorder. *Phys. Rev. B Condens. Matter Mater. Phys.*, **2010**, *82*(12), 125429-125437. [http://dx.doi.org/10.1103/PhysRevB.82.125429]

[78] Saito, R.; Jorio, A.; Souza Filho, A.G.; Dresselhaus, G.; Dresselhaus, M.S.; Pimenta, M.A. Probing phonon dispersion relations of graphite by double resonance Raman scattering. *Phys. Rev. Lett.*, **2001**, *88*(2), 027401-027404. [http://dx.doi.org/10.1103/PhysRevLett.88.027401] [PMID: 11801034]

[79] Cançado, L.G.; Reina, A.; Kong, J.; Dresselhaus, M.S. Geometrical approach for the study of G ′ band in the Raman spectrum of monolayer graphene, bilayer graphene, and bulk graphite. *Phys. Rev. B Condens. Matter Mater. Phys.*, **2008**, *77*(24), 245408-245416. [http://dx.doi.org/10.1103/PhysRevB.77.245408]

[80] Jorio, A.; Cançado, L.G. Disorder and Defects in Two-Dimensional Materials Probed by Raman Spectroscopy. In: *Raman Spectroscopy of Two-Dimensional Materials. Springer Series in Materials Science*; Tan, P.H., Ed.; Springer: Singapore, **2019**; Vol. 276, pp. 99-110. [http://dx.doi.org/10.1007/978-981-13-1828-3_5]

[81] Acik, M.; Lee, G.; Mattevi, C.; Pirkle, A.; Wallace, R.M.; Chhowalla, M.; Cho, K.; Chabal, Y. The Role of oxygen during thermal reduction of graphene oxide studied by infrared absorption spectroscopy. *J. Phys. Chem. C,* **2011**, *115*(40), 19761-19781. [http://dx.doi.org/10.1021/jp2052618]

[82] Ţucureanu, V.; Matei, A.; Avram, A. M. FTIR spectroscopy for carbon family study, critical reviews in analytical chemistry. *Crit. Rev. Anal. Chem,* **2016**, *46*(6), 502-20.

[83] Sangermano, M.; Marchi, S.; Valentini, L.; Bon, S.B.; Fabbri, P. Transparent and conductive graphene oxide-polyethylenglycol diacrylate coatings obtained by photopolymerization. *Macromol. Mater. Eng.*, **2011**, *296*(5), 401-407. [http://dx.doi.org/10.1002/mame.201000372]

[84] Ramos-Galicia, L.; Mendez, L.N.; Martínez-Hernández, A.L.; Espindola-Gonzalez, A.; Galindo-Esquivel, I.R.; Fuentes-Ramirez, R.; Velasco-Santos, C. Improved performance of an epoxy matrix as a result of combining graphene oxide and reduced graphene. *Int. J. Polym. Sci.*, **2013**, *2013*, 1-7. [http://dx.doi.org/10.1155/2013/493147]

[85] Layek, R.K.; Nandi, A.K. A review on synthesis and properties of polymer functionalized graphene. *Polymer (Guildf.)*, **2013**, *54*(19), 5087-5103. [http://dx.doi.org/10.1016/j.polymer.2013.06.027]

[86] Yu, S.; Liu, J.; Zhu, W.; Hu, Z.T., Lim, T.T.; Yan, X. Facile room-temperature synthesis of carboxylated graphene oxide-copper sulfide nanocomposite with high photodegradation and disinfection activities under solar light irradiation. *Sci. Rep.*, **2015**, *5*(1), 16369. [http://dx.doi.org/10.1038/srep16369] [PMID: 26553709]

[87] Mallakpour, S.; Abdolmaleki, A.; Borandeh, S. Covalently functionalized graphene sheets with biocompatible natural amino acids. *Appl. Surf. Sci.*, **2014**, *307*, 533-542. [http://dx.doi.org/10.1016/j.apsusc.2014.04.070]

[88] Mittal, G.; Rhee, K.Y.; Park, S.J.; Hui, D. Generation of the pores on graphene surface and their reinforcement effects on the thermal and mechanical properties of chitosan-based composites. *Compos., Part B Eng.*, **2017**, *114*, 348-355. [http://dx.doi.org/10.1016/j.compositesb.2017.02.018]

[89] Kuila, T.; Bose, S.; Khanra, P.; Mishra, A.K.; Kim, N.H.; Lee, J.H. A green approach for the reduction of graphene oxide by wild carrot root. *Carbon,* **2012**, *50*(3), 914-921. [http://dx.doi.org/10.1016/j.carbon.2011.09.053]

[90] Penkov, O.V. Preparation and characterization of graphene. In: *Tribology of Graphene*; Elsevier, **2020**.

[91] Lefebvre, J.; Galli, F.; Bianchi, C.L.; Patience, G.S.; Boffito, D.C. Experimental methods in chemical engineering: X-ray photoelectron spectroscopy-XPS. *Can. J. Chem. Eng.*, **2019**, *97*(10), 2588-2593. [http://dx.doi.org/10.1002/cjce.23530]

[92] Chen, X.; Wang, X.; Fang, D. A review on C1s XPS-spectra for some kinds of carbon materials. *Fuller. Nanotub. Carbon Nanostruct.*, **2020**, *28*(12), 1048-1058. [http://dx.doi.org/10.1080/1536383X.2020.1794851]

[93] Eigler, S.; Dimiev, A.M. *Characterization techniques in graphene oxide fundamentals and applications*; Eigler, S.; Dimiev, A.M., Eds.; John Wiley & Sons, Inc., **2017**, pp. 85-118.

[94] Scardamaglia, M.; Aleman, B.; Amati, M.; Ewels, C.; Pochet, P.; Reckinger, N.; Colomer, J.F.; Skaltsas, T.; Tagmatarchis, N.; Snyders, R.; Gregoratti, L.; Bittencourt, C. Nitrogen implantation of suspended graphene flakes: Annealing effects and selectivity of sp^2 nitrogen species. *Carbon,* **2014**, *73*, 371-381. [http://dx.doi.org/10.1016/j.carbon.2014.02.078]

[95] Susi, T.; Pichler, T.; Ayala, P. X-ray photoelectron spectroscopy of graphitic carbon nanomaterials doped with heteroatoms. *Beilstein J. Nanotechnol.*, **2015**, *6*, 177-192. [http://dx.doi.org/10.3762/bjnano.6.17] [PMID: 25671162]

[96] Thermoscientific, X.P.S. http://xpssimplified.com/elements/carbon.php

[97] Cooper, A.J.; Wilson, N.R.; Kinloch, I.A.; Dryfe, R.A.W. Single stage electrochemical exfoliation method for the production of few-layer graphene *via* intercalation of tetraalkylammonium cations. *Carbon,* **2014**, *66*, 340-350. [http://dx.doi.org/10.1016/j.carbon.2013.09.009]

[98] Yu, H.; Zhang, B.; Bulin, C.; Li, R.; Xing, R. High-Efficient synthesis of graphene oxide based on improved hummers method. *Sci. Rep.*, **2016**, *6*(1), 36143. [http://dx.doi.org/10.1038/srep36143] [PMID: 27808164]

[99] Kovtyukhova, N.I.; Ollivier, P.J.; Martin, B.R.; Mallouk, T.E.; Chizhik, S.A.; Buzaneva, E.V.; Gorchinskiy, A.D. Layer-by-layer assembly of ultrathin composite films from micron-sized graphite oxide sheets and polycations. *Chem. Mater.*, **1999**, *11*(3), 771-778. [http://dx.doi.org/10.1021/cm981085u]

[100] Ganguly, A.; Sharma, S.; Papakonstantinou, P.; Hamilton, J. Probing the thermal deoxygenation of graphene oxide using high-resolution *in situ* x-ray-based spectroscopies. *J. Phys. Chem. C,* **2011**, *115*(34), 17009-17019. [http://dx.doi.org/10.1021/jp203741y]

[101] Szabó, T.; Berkesi, O.; Forgó, P.; Josepovits, K.; Sanakis, Y.; Petridis, D.; Dékány, I. Evolution of surface functional groups in a series of progressively oxidized graphite oxides. *Chem. Mater.*, **2006**, *18*(11), 2740-2749. [http://dx.doi.org/10.1021/cm060258+]

[102] Hofflund, G.B. Spectroscopic techniques: X-ray photoelectron spectroscopy (XPS), Auger electron spectroscopy (AES) and ion scattering spectroscopy (ISS). In: *Handbook of Surface and Interface Analysis*; Marcel Dekker: New York, **1998**.

[103] Chiu, P.L.; Mastrogiovanni, D.D.T.; Wei, D.; Louis, C.; Jeong, M.; Yu, G.; Saad, P.; Flach, C.R.; Mendelsohn, R.; Garfunkel, E.; He, H. Microwave- and nitronium ion-enabled rapid and direct production of highly conductive low-oxygen graphene. *J. Am. Chem. Soc.*, **2012**, *134*(13), 5850-5856. [http://dx.doi.org/10.1021/ja210725p] [PMID: 22385480]

[104] Siokou, A.; Ravani, F.; Karakalos, S.; Frank, O.; Kalbac, M.; Galiotis, C. Surface refinement and electronic properties of graphene layers grown on copper substrate: An XPS, UPS and EELS study. *Appl. Surf. Sci.*, **2011**, *257*(23), 9785-9790. [http://dx.doi.org/10.1016/j.apsusc.2011.06.017]

[105] Rabchinskii, M.K.; Ryzhkov, S.A.; Kirilenko, D.A.; Ulin, N.V.; Baidakova, M.V.; Shnitov, V.V.; Pavlov, S.I.; Chumakov, R.G.; Stolyarova, D.Y.; Besedina, N.A.; Shvidchenko, A.V.; Potorochin, D.V.; Roth, F.; Smirnov, D.A.; Gudkov, M.V.; Brzhezinskaya, M.; Lebedev, O.I.; Melnikov, V.P.; Brunkov, P.N. From graphene oxide towards aminated graphene: facile synthesis, its structure and electronic properties. *Sci. Rep.*, **2020**, *10*(1), 6902. [http://dx.doi.org/10.1038/s41598-020-63935-3] [PMID: 32327708]

[106] Hu, M.; Yao, Z.; Wang, X. Characterization techniques for graphene-based materials in catalysis. *AIMS Mater. Sci.*, **2017**, *4*(3), 755-788. [http://dx.doi.org/10.3934/matersci.2017.3.755]

[107] Showalter, A.R.; Duster, T.A.; Szymanowski, J.E.S.; Na, C.; Fein, J.B.; Bunker, B.A. An X-ray absorption fine structure spectroscopy study of metal sorption to graphene oxide. *J. Colloid Interface Sci.*, **2017**, *508*, 75-86. [http://dx.doi.org/10.1016/j.jcis.2017.08.040] [PMID: 28822863]

[108] Bunker, G. *Introduction to XAFS: A Practical Guide to X-ray Absorption Fine Structure Spectroscopy*; Cambridge University Press: New York, **2010**. [http://dx.doi.org/10.1017/CBO9780511809194]

[109] Papagno, M.; Fraile Rodríguez, A.; Girit, Ç.Ö.; Meyer, J.C.; Zettl, A.; Pacilé, D.; Polarization-dependent, C. Polarization-dependent C K near-edge X-ray absorption fine-structure of graphene. *Chem. Phys. Lett.*, **2009**, *475*(4-6), 269-271. [http://dx.doi.org/10.1016/j.cplett.2009.05.054]

[110] Wahab, H.; Mertins, H.C.; Timmers, H.; Frankcombe, T.J. X-ray absorption fine structure of carboxyl and other adventitious moieties attached to copper-supported graphene. *Carbon,* **2019**, *141*, 457-466. [http://dx.doi.org/10.1016/j.carbon.2018.09.086]

[111] Farivar, F.; Yap, P.L.; Hassan, K.; Tung, T.T.; Tran, D.N.H.; Pollard, A.J.; Losic, D. Unlocking thermogravimetric analysis (TGA) in the fight against "Fake graphene" materials. *Carbon,* **2021**, *179*, 505-513. [http://dx.doi.org/10.1016/j.carbon.2021.04.064]

[112] Farivar, F.; Yap, P. L.; Karunagaran, R. U.; Losic, D. Thermogravimetric Analysis (TGA) of Graphene materials: effect of particle size of graphene, graphene oxide and graphite on thermal parameters. **2021**, *7*(2), 41. [http://dx.doi.org/10.3390/c7020041]

[113] ISO. Nanotechnologies - Vocabulary - Part 13: Graphene and related two- dimensional (2D) materials. **2017**.

[114] Polini, A.; Yang, F. Physicochemical characterization of nanofiber composites. Ramakrishna, S.; Ramalingam, M. In: *Nanofiber composites for biomedical applications*; Woodhead Publishing, **2017**; pp. 97-115.

[115] Wang, G.; Yang, J.; Park, J.; Gou, X.; Wang, B.; Liu, H.; Yao, J. Facile synthesis and characterization of graphene nanosheets. *J. Phys. Chem. C,* **2008**, *112*(22), 8192-8195. [http://dx.doi.org/10.1021/jp710931h]

[116] Shtein, M.; Pri-Bar, I.; Varenik, M.; Regev, O. Characterization of graphene-nanoplatelets structure *via* thermogravimetry. *Anal. Chem.*, **2015**, *87*(8), 4076-4080. [http://dx.doi.org/10.1021/acs.analchem.5b00228] [PMID: 25796961]

[117] Mazurkiewicz-Pawlicka, M.; Nowak, M.; Malolepszy, A.; Witowski, A.; Wasik, D.; Hu, Y.; Stobinski, L. Graphene oxide with controlled content of oxygen groups as a filler for polymer composites used for infrared radiation shielding. *Nanomaterials (Basel),* **2019**, *10*(1), 32. [http://dx.doi.org/10.3390/nano10010032] [PMID: 31877714]

CHAPTER 5

Functionalization of Graphene and Factors Affecting Catalytic Performance

Gopal Avashthi[1,2,3] and **Man Singh**[1,*]

[1] *School of Chemical Sciences, Central University of Gujarat, Gandhinagar-382030, Gujarat, India*

[2] *Department of Chemistry, Sankalchand Patel University, Visnagar-384315, Gujarat, India*

[3] *School of Sciences, P P Savani University, Kosamba-394125, Surat, Gujarat, India*

Abstract: Carbon and its driven materials have been a foundation of living and non-living systems for centuries due to their amazing experimental expressions in light, temperature, pressure, and pH. Being light-weighted and electronically active with equal energy partitioning in its four orbitals ($2s^1 2px^1 2py^1 2pz^1$), the C atoms have been at the core of natural sciences, providing valuable resources like high-grade wood, cotton, and many others. Thus, carbon-driven materials like diamond, graphite, and graphene ink have been attracting the attention of scientists, researchers, and industries. The chapter reviews recent chemical methodologies for the synthesis and structural investigation of graphene and its derivatives by various analytical techniques that provide information about basic knowledge to understand the role of graphene and graphene-based composites in various qualitative and quantitative applications. Here, several methods have been enlisted for the surface alteration of graphene oxide by a synthetic approach, such as ultrasound, a microwave-assisted synthesis that avoids the use of hazardous chemicals. Also, conventional methods have been discussed, including various types of reactions, such as nucleophilic, electrophilic, condensation, and cycloaddition. This review article highlights the key points to understanding the 2D carbon material for researchers and users to learn about the chemical modification of graphene at the initial stage. This write-up also discusses a brief explanation of various carbon nanomaterials that concern graphene and its oxide forms. We have explained the synthetic value of 2D carbon materials so that it covers a lot of the needs of researchers for synthetic aspects in graphene and allied fields of interest. Currently, such unique experiments are noted as milestones in the field of material synthesis for various applications. So, a review of chemically altered graphene materials reinforced with structural multi-functionalities is highly informative as a ready reckoner for needed information and understanding.

* **Corresponding author Man Singh:** School of Chemical Sciences, Central University of Gujarat, Gandhinagar-382030, Gujarat, India. Tel: +91-079-23260210; Fax: +91-079-23260076. E-mail: mansingh50@hotmail.com

Pinki Bala Punjabi, Rakshit Ameta & Sharoni Gupta (Eds.)

Keywords: Covalent Functionalization, Graphene, Non-covalent Functionalization, Surface Modified Carbon Materials, Utrasound and Microwave-assisted Functionalization.

INTRODUCTION TO CARBON CHEMISTRY

Carbon is the fourth most abundant chemical element in the universe and the second most abundant in the human body by mass [1]. The atomic number of carbon is six and it belongs to the second period of the fourteenth group [IVA group] of the periodic table. A carbon atom has six electrons with $1s^2$, $2s^2$, and $2p^2$ electronic configurations, which infer tetravalency. Two inner shell electrons are spherically symmetric; out of which, 1s orbital is tightly close to the nucleus. The energy difference between 1st and 2nd shells is far from the Fermi energy levels, and remaining four valence electrons ($2s^2$, $2p^2$) have very similar energy to silicon. The energy difference between the 2s and 2p orbitals of carbon is relatively low compared to the bond enthalpy. The tetravalent nature of carbon and its distinctive catenation property can be accessible for hybridization, forming new compounds. Thus, 2s and 2p orbitals hybridize (Fig. **1a,b**). The catenation property of carbon results in stable allotropes of 0D, 1D, 2D, and 3D dimensionality compared to other elements [2].

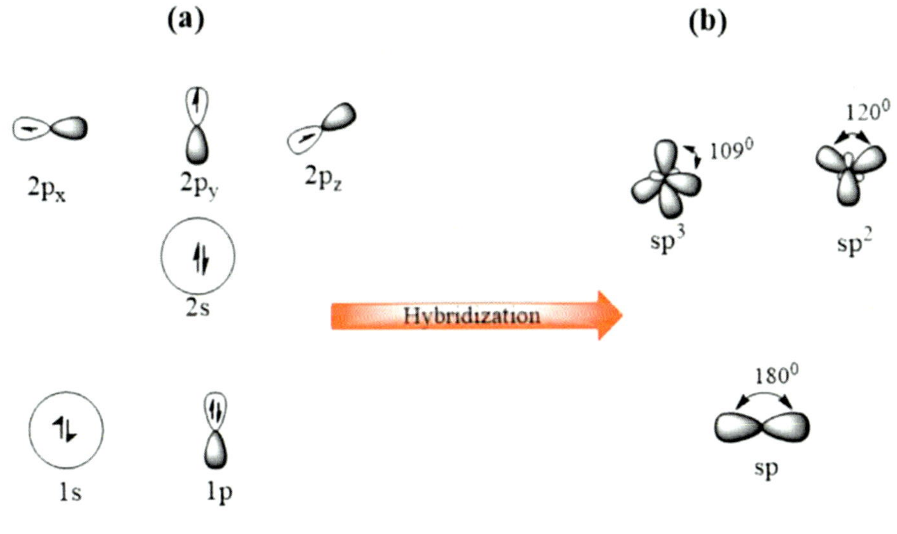

Fig. (1). (**a**) Diagram of atomic orbitals, and (**b**) hybrid orbitals of carbon.

POLYMORPHISM OF CARBON

Carbon has the ability to bind with various elements like H, Si according to optimal hybridization, supporting the synthesis of compounds with different structural properties and applications. Therefore, newly synthesized compounds have different geometrical features, like sp hybridized carbon which is linear in nature, while sp^2 and sp^3 have trigonal planar and tetrahedral structures, respectively [2]. The one p and one s-orbital hybridize as sp while, two p and one s-orbital hybridize as sp^2 carbon, and remaining unhybridized orbitals participate in forming an additional π bond. Therefore, the tetrahedral lattice of the diamond (Greek 'adamant' meaning impossible to tame) belongs to sp^3-hybridized carbon atoms in terms of strong covalent bonding to itself [1, 2]. The wave function of four valence electrons ($2s^2$ and $2p^2$) of carbon, hybridize easily and it adds to the solid-state features of sp^2 hybridized graphite (Gt). The diamond and Gt are two natural allotropes of carbon. The diamond and Gt allotropes have been known since prehistoric times for their distinctive properties such as hardness, thermal conductivity, and lubrication behaviour [3, 4]. In the last few decades, the surprising discoveries of three artificial allotropes of carbon, *i.e.*, 0D fullerene, 1D carbon nanotube, and 2D graphene (Gr) have confirmed the efficiencies of carbon materials for their amazing applications [1 - 8] (Fig. **2a-e**).

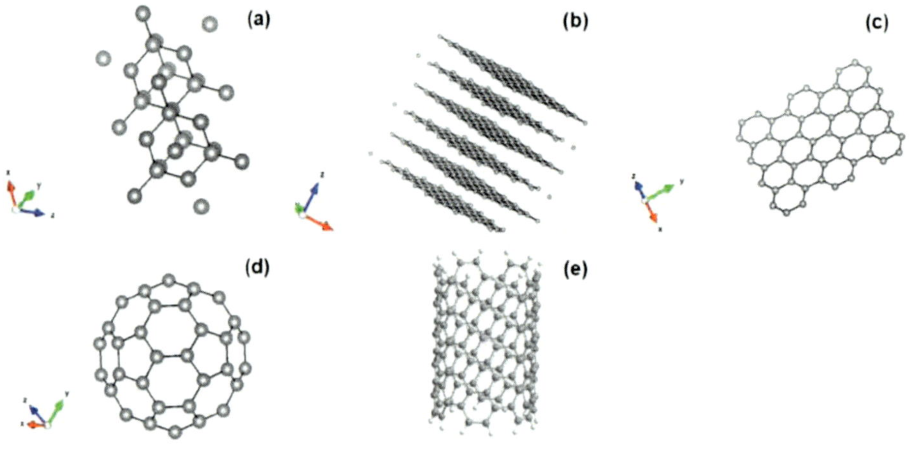

Fig. (2). Carbon allotropes (**a**) Diamond, (**b**) Graphite, (**c**) Graphene, (**d**) Fullerene, and (**e**) Carbon nanotube.

Natural Allotropes

Diamond

Diamond is a sp^3 hybridized natural allotrope of carbon with structurally strong C-C covalent bonds [1, 2]. The crystal structure (Figs. **3a, b**) of the diamond is known for its excellent properties; such as high Young's modulus, hardness (hardest material with a supreme Mohs value of 10), chemical inertness, low electrical conductivity, high thermal conductivity (900–2,320 W/m/K), high X-ray, and high optical scattering ability of visible spectrum 1-1.5 eV [1, 9, 10].

Natural diamond, high-pressure synthetic diamond, low-pressure synthetic diamond, and metastable carbon or diamond-like carbon are the four types of diamonds known to date [8]. Natural diamonds are produced at a high temperature and pressure in volcanic eruptions. Moreover, English chemist H. Davy proved that the burning of diamonds produced only CO_2. Initially, in 1955, the first unquestioned high-pressure synthetic diamond was synthesized in the Institute for High-Pressure Physics of the Soviet Union, the U.S. and Sweden. It was commercialized for various industrial applications such as wear, abrasive, and tools. The preparation of low-pressure synthetic diamond is based on chemical vapour deposition (CVD). A diamond-like carbon is produced by a thin coating at low pressure [8]. The C-C strong covalent bonds in diamond are responsible for the thermal conductivity as well as its valuable applications in microelectronics, optoelectronics, and carbon fibre electronics. Diamond acts as an insulator at RT with a resistivity 10^2 Ω and a wide bandgap (~5.5 eV) [11, 12].

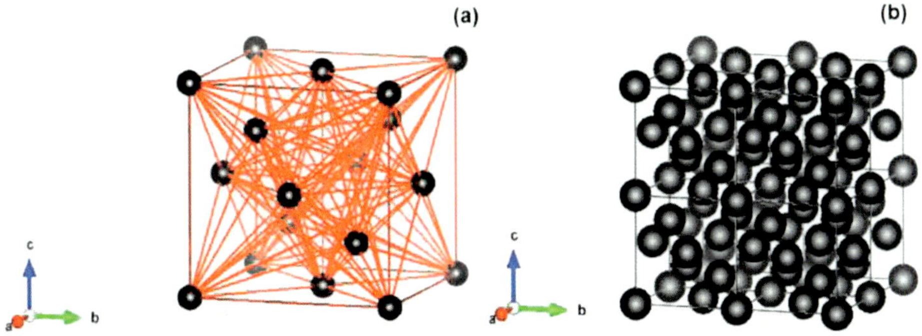

Where a, b, and c are primitive lattice vectors:

Fig. (3). (a) Crystal structure of diamond, and **(b)** fcc lattice with a set of primitive lattice vectors.

The direct lattice (linear combinations) for the diamond is stated to be face-centred cubic (fcc) (Fig. **3b**) with primitive lattice vectors a, b, and c. The dissimilarities of primitive lattice vectors shown in Figs. (**3a** and **b**) are given below:

$$a = (0, a/2, a/2),$$

$$b = (a/2, 0, a/2),$$

$$c = (a/2, a/2, 0),$$

where a, b, and c are the lengths of the edges of the smallest cube in the fcc lattice. The unit cube, or crystallographic unit cell, is the smallest cube of the direct lattice [10]. The CVD-synthesized diamonds have potential applications in various fields such as electronics, optics, mechanics, thermals, chemicals, and others. The supreme hardness of a diamond is useful for cutting-edge applications. In chemical synthesis of nanodiamonds (NDs), the detonation process retains heterogeneous functional moieties such as carbonyl, alcohol, epoxy, ether, and unsaturated double bonds on ND surfaces. The surface functionalization of DNDs to produce highly unstable intermediates such as acyl and ester is accountable for amidation and esterification [12 - 14]. Thus, newly functionalized NDs have various applications in biomedical engineering, biophysics, electronics, optical devices, and mechanical studies. Recently, the NDs have been utilized as drug carriers due to their stable dispersibility subjected to drug and gene delivery and imaging systems [12 - 15].

Graphite

3D flakes of Gt are amorphous and crystalline in nature. The multi-layered planar hexagonal 2D sheets of Gr are stacked by weak Van der Waals forces. In each plane, the carbon atoms are arranged in such a way that each carbon atom is sp^2 hybridized along the hexagon lattice [16] (Fig. **4a,b**). The length of the hexagon lattice is 0.142 nm and the interplanar distance is 1 nm. Due to weak interactions between the sheets, the Gt is reported as the softest natural material (Mohs hardness of 1). Its electrons move over the sheets and, being a soft material, graphite is used as a lubricating agent in industries [1 - 5]. However, the reason for the interaction between the sheets is not entirely understood. Unlike diamond, the delocalized π-electron clouds over the sheets of Gt produce good electrical conductivity [17, 18]. Gt and graphite oxide (GtO) are used in the synthetic transformations as carbocatalysts to avoid hazardous organic reagents and solvents. The Gt acts as a carbocatalyst in microwave, conventional, and ultrasound-assisted synthesis, offering various synthetic transformations such as reduction of substituted nitro functionalities to their respective amines using hydrazine hydrate, cleavage of ethers followed by coupling with an acyl halide, Friedel-Crafts-type substitution, Diels-Alder reaction, cycloaddition reactions, and ene-cyclization. Also, GtO is used as a selective and highly efficient oxidant of thiols and sulphides [19 - 25]. Oxidation of Gt produces GtO which further facilitates ultrasound-assisted exfoliation into graphene oxide (GO) followed by surface functionalization [18, 26 - 31]. Birch reduction of Gt with lithium in liquid ammonia gives a highly reduced exfoliated product which is free from metal. The reduction is due to edge and interior hydrogenation [32]. Gt, GtO, reduced GtO and surface functionalized GO composites are found to be potentially applicable in electrical, optical, mechanical, energy storage devices, capacitors, and biological areas of research and industries [18, 27, 33, 34].

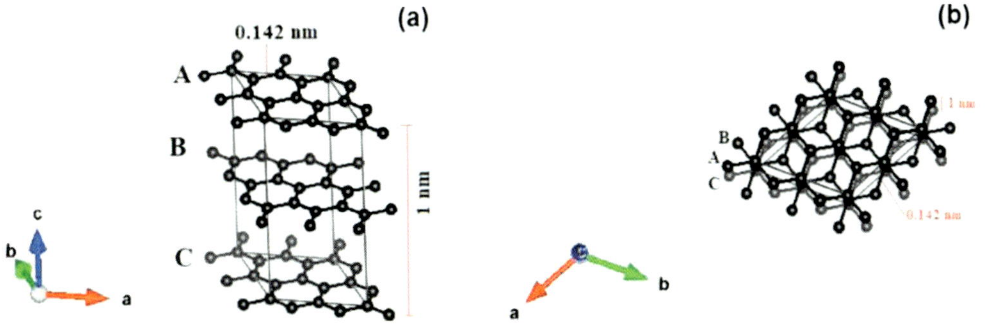

Fig. (4). (**a**) Layers sequencing in hexagonal, and (**b**) Rhombohedral graphite (front view).

Synthetic Allotropes

Fullerene

Fullerene is a 0D synthetic allotrope of carbon with a face-centred cubic crystal lattice. The credit of fullerene's discovery goes to a group of scientists including Richard Smalley, Robert Curl, Harry Kroto, and Heath and O'Brien, at Rice University, Huston, who compared the structure of fullerene with the geodesic dome model of American architect Buckminster fuller. In 1996, Richard Smalley, Robert Curl, and Harry Kroto were awarded the Nobel Prize in Physics [35]. They proposed the icosahedron shape of the molecule possessing 32 faces, 60 apexes (carbon atoms) and 90 ribs within 12 pentagons and 20 hexagon rings to make a spheroid shape of the molecule. The point group symmetry of this molecule is I_h [36]. The chemical stability of the closed-hollow-cell of C_{60} is proved by a large energy gap between HOMO and LUMO [36 - 38]. The Van der Waals diameter of fullerene is 10.1 Å and the nucleus-to-nucleus diameter of C_{60} is 7.09 Å. The C-C bond length between two hexagons is 1.40 Å. It is considered as a double bond and slightly shorter than the C-C bond length of hexagon-pentagon *i.e.* 1.46 Å [36 - 38]. The functionalization of fullerene comprises exohedral and endohedral orientations, depending on the added substituents outside and trapped inside the cages. The chemistry of fullerene covers a wider range of chemical reactions like nucleophilic addition, electrophilic addition, electrocyclic reactions, radical addition, metal doping, and supramolecular complexation [39 - 57]. The exohedral modification of fullerene cages has been done by reacting with sulphur-containing group thiols, thioethers to facilitate the adsorption on the metal surfaces [39] and explored for their applications in various research areas and industries. C_{60} acts like a 2π electron-deficient dienophile and dipolarophile, hence enduring cycloaddition reactions such as [2+2], [3+2], and [4+2] on the 6, 6-ring junctions of the fullerene skeleton, either thermally or photochemically [40, 41]. The reactive divalent species, a carbene, has synthesized new derivatives and enhanced the reactivity of fullerene [42]. In a one-pot reaction, a charged zwitterion intermediate with added nucleophiles with multiple functional groups is fitted onto the fullerene [43]. Fullerene has been named "radical sponge" due to its extraordinary reactivity towards radical species. Therefore, the highly reactive species are very attractive for functionalization [44, 45]. Although fullerene is insoluble in water, treatment with hydrogen peroxide converts it into a water-soluble polyhydroxy functionalized fullerene. Apposite functionalization of fullerene also makes the molecule soluble in water. Hence, water-soluble derivatives interact with DNA, protein, and living cells and find biomedical applications [46, 47]. There is aggregation of fullerenes into colloidal crystals due to fullerene-fullerene and fullerene-water driving forces in an aqueous or biological medium. Most of the photoactivity of nC_{60} is lost during the process, whereas aggregates of hydrophilic functionalized nC_{60} may or may not have reduced photoactivity, depending on the functional characteristics [47, 48]. Fullerene-based Friedel-Crafts arylation, alkylation, and acylation are multistep procedures for functionalization in synthetic chemistry [49]. The hydrogenation reaction is accomplished *via* the Birch reduction. The reaction of fullerene with hydrogen atoms makes hydrofullerenes, similar to the addition of H_2 at the C=C double bond, which results in the adduct of two C-H bonds [50]. The radical derivative of the fullerene is analysed by the most efficient analytical technique, *i.e.*, electron spin resonance (ESR) spectroscopy [51]. Fullerene and its alkali-doped composites such as M_3C_{60} (M is an alkali metal) [37, 52 - 54] and organic ferromagnets have been found as superconductors [54, 55]. The vast array of chemical reactivity promises a growth of various fullerene-based materials with their uses in polymer chemistry [56], biological applications [57, 58], catalysts [59, 60], organic electronics [61], photovoltaic [62], and thin-film transistors [63, 64].

Carbon Nanotube

In the year 1991, Sumio Iijima discovered the multi-walled carbon nanotube (MWCNT) [65], with multiple coaxial tubular units of single-walled carbon nanotube (SWCNT) rolled up by Van der Waals forces [66, 67]. MWCNTs are multi-stacked structures with unique properties [65 - 68]. SWCNT can be synthesized by various methods, such as chemical approach (Fig. **5**) [67], arc discharge [68], laser ablation [68], and CVD [69]. The principle of SWCNT synthesis implies that the hexagonal honeycomb carbon lattice of single-layer Gr gets rolled up into cylindrical helical hexagonal microtubules with a diameter range of 1-2 nm [70, 71]. CNT acts as a nanomechanical resonator in water to enhance the viscosity at low temperatures. It also applies to the mass measurement of biological molecules and various sensing applications due to their lightweight nature, characteristic ultrahigh mechanical strength, and high surface activity [72, 73].

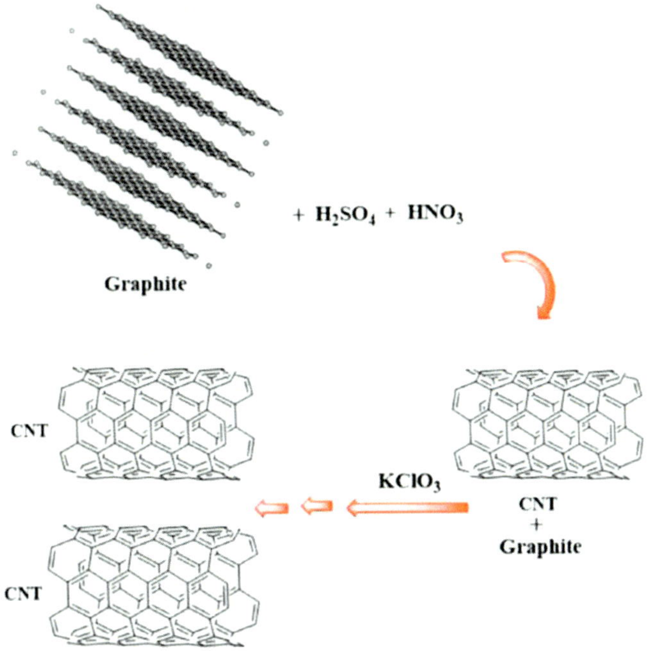

Fig. (5). Graphical representation of CNT synthesis by chemical approach.

The functionalization of CNT with organometallic compounds, metals, organic compounds, and biological molecules helps to manipulate the bandgap and heterojunction energetics [74 - 77]. Transformation of the assembled SWCNT into ultrathin long CNT fibrils has a great potential for applications as conductive wires and probe electrodes in industries [78]. The strength, small dimensions, and electronic, optical, and chemical properties of CNT composite materials open a new direction for innovative applications, including [74 - 77, 79], energy storage electronic and semiconductor devices [80, 81], hydrogen storage media [82, 83], and high-performance hydrogen production devices [84].

Graphene

The Single-layered isolated sheet from Gt, is noted as Gr, which was first discovered through mechanical exfoliation using the "Scotch-tape" method from bulk Gt [85]. Epitaxial CVD [86] is also considered as a suitable method to grow Gr. The first experimental Gr was reported by Hanns-Peter Boehm in 1962, but it was not experimentally proven for so many years [87]. The comprehensive Gr study was started by Novoselov and Geim after getting the exfoliated single layer

Gr in 2004 and both were awarded the Nobel Prize in 2010 for ground-breakthrough research and its extraordinary properties. The atomic-level thin Gr possesses a 2D planar structure, having sp^2-hybridized carbon atoms with high charge mobility, mechanical properties (breaking strength ~40 N/m, Young's modulus~1.0 Tpa), thermal properties (at RT thermal conductivity~3000 $Wm^{-1}K^{-1}$), electronic, quantum hall effect, spin transport, and magnetism [85 - 94]. The electronic behaviour of materials is characterized in terms of the Fermi surface and controlled by external electric and magnetic fields [95 - 100]. However, due to the synergistic impact of nitrogen, N-doped Gr has become a promising material for various applications such as batteries, electronics, sensors, catalysis, and supercapacitors [96 - 100]. Furthermore, the ultrasound and microwave-assisted methods for surface modification of Gr have a wide range of applications [101 - 121]. In synthetic chemistry, unmodified Gt, GtO, and their derivatives catalyze redox reactions, alkylation, cycloaddition, and Friedel-Craft reactions [102 - 112]. The covalent functionalization of Gr could be demonstrated with the help of reactive intermediates such as radicals, nitrenes, carbenes, arynes, and nucleophilic substitutions like amidation or esterification [113 - 116]. The non-covalent approach for the surface modification of carbon-based materials comprises π-π stacking between the Gr surfaces and various biological molecules to synthesize bio-conjugates-proteins, enzyme-drug, DNA-protein complexes, and supra molecules-inorganic metallic complexes [119 - 121]. The impact of surface modification of Gr and graphene oxide (GO) demonstrates excellent electrical, mechanical, and optical properties, which could be further explored for the development of several biomedical devices, industrial metallic goods, conductors, energy storage devices, and transistors [120, 121]. The surface modification of Gr by using a robust chemical approach is the latest thrust area in the field of carbon nanotechnology to explore its several potential applications [102 - 116]. Moreover, CNT and Gr are both carbon-based functional nanostructures, and they have different properties due to the diverse nature of their geometries which are planar in Gr and tubular in CNT.

REFINING OF GRAPHENE OXIDE

Initially, four methods were introduced in GtO synthesis. Each successive method has advancement in the simplicity, degree of oxidation, yields, and product quality. The impurities, *i.e.*, highly flammable salts and acids, are a major problem for GtO synthesis and purification, which makes it a dull procedure in GtO production. Multiple time washing, centrifugation, and filtration are important steps for the purification of GtO. Inorganic salts and acids are both quite water-soluble, but the purification of them is not easy since a gelatinous appearance of dispersed GtO is observed due to an increase in pH during washing with water, which slows down the further purification, centrifugation, and especially filtration processes. Another best procedure for purification involves two-step washing with aqueous HCl and acetone to avoid gelation and accelerate the synthesis of purified GtO. Therefore, GtO is washed with an excess amount of HCl to remove salts and ions effectively, then acetone is used to eliminate the acid to make the pH neutral, which avoids gelation. The GtO is air-dried and allowed to disperse in acetone after 3 to 4 washes with HCl, followed by filtration and washing with more acetone, which removes residual HCl without gelation. In 1974, a similar purification process was described by Nikolaev *et al.* This purification method does not damage the various functionalities present over the surface. Thus, the solid GtO synthesis at a large scale can be achieved by filtration and vacuum drying. The salt content of GtO, washed with HCl and acetone 3 to 4 times, is measured. If the concentration is lower than 0.1 wt.%, then it goes unnoticed by Energy Dispersive X-Ray (EDX). Therefore, the purified GtO is quite stable against ignition. The GtO obtained by this method is utilized as a precursor in graphene synthesis [122].

SURFACE FUNCTIONALIZATION OF GRAPHENE OXIDE FOR STRENGTHENING CATALYTIC PERFORMANCES

Fabrication of Gr is cost-effective as compared to other carbon materials. Therefore, the researchers have focused their attention on exploring this material for various vital applications *via* surface modification of Gr, also known as the functionalization of Gr (f-Gr). Chemical functionalization of Gr avoids agglomeration and aids to sustain the intrinsic properties [122]. GtO is utilized as a starting material for the synthesis of Gr through chemical, thermal and photochemical reduction [123 - 127]. Oxidation of Gt helps to synthesize GtO, possessing various oxygenated functionalities over the surface, *i.e.*, hydroxyl, epoxide, diol, ketone, carboxyl, phenol, lactone and quinone [128 - 134]. The active functional groups expand the intercalative distance, to reduce interlayer Van der Waals forces. These help to improve the solubility and stability in water and organic solvents [135 - 139]. The presence of carbonyl and carboxyl groups at the edges of GtO sheets, makes it strongly hydrophilic and make stable dispersion in water using sound energy [135 - 137]. The different structures of GtO has been explained in the literature [128, 140]. But the stability of Gr is the necessity for chemical reduction of GtO due to rapid irreversible aggregation of the graphene sheet. Hence, the surface functionalization is linked with the composite materials which are synthesized by covalent and non-covalent functionalization [101].

Covalent Functionalization

Oxidation of the Gt produces various functionalities over the Gr surface. The numerous oxygenated functional groups on the Gr surface offer covalent modification. The aim of surface alteration of Gr in organic solvents is the stable dispersion to synthesize new composite materials for various applications. The principle of functionalization is to rehybridize one or more sp^2 carbon atoms into sp^3, which deals with the bandgap engineering by losing the conjugated electrons [2, 141]. The covalent modification of Gr can be achieved in four different ways [142 - 178].

1. Nucleophilic substitution reaction

2. Electrophilic substitution reaction

3. Condensation reaction

4. Addition reaction

Nucleophilic Substitution Reaction

The nucleophilic substitution reaction of GO is based on the accessibility of epoxy and carboxylic acid groups over the surface of dispersed GO in the aqueous and organic solvent phases. It has been reported that the epoxy and hydroxyl functional groups lie above and below, while the carboxyl group is at the edge of the carbon layer. Surface functionalization of GO with amine substituted organic compounds is simple at RT as a result of nucleophilic substitution and condensation reactions. Amide bond formation is one of the most vital reactions in organic chemistry because of its extensive applicability in the synthesis of modern biological and pharmaceutical compounds. The Lone pair electron of amine attacks the epoxy or electrophilic carbon center of the carboxyl group of GO. Since then, this method has been known for the large-scale production of functionalized (f) Gr. The acyclic and aromatic amines, amino acids, amine substituted biomolecules, ionic liquids, and small molecular weight polymeric compounds are functionalized with graphene oxide (f-GO) [101 - 104, 112 - 116, 120, 142 - 151]. Athanasios B. Bourlinos *et al.* explored the primary amines, possessing chemical formulas equivalent to $C_nH_{2n+1}NH_2$ (n = 2, 4, 8, 12, 18) of amino acids and nucleobases for surface modification of GtO and GO at RT [146 - 148]. Thus, the reaction condition depends upon the size of amine-substituted compounds. For a long-chain aliphatic amine, the reaction mixture is heated under reflux for several hours. The surface functionalization enhances the stability and dispersibility of f-GO in water and organic solvents. X-Ray Diffraction (XRD) verified the functionalization of GtO at higher interlayer spacing (d-spacing) due to the amidation of GtO with amine substituted molecules [147]. The reaction between amino acids and GtO occurs in an alkaline solution due to the hydrophilicity of GtO. According to Bourlinos et al., the tilted orientation is more likely to occur due to its hydrophilicity [147]. Fan Zhang *et al.* for the first-time synthesized guanine-based covalently modified Gr. According to Fan Zhang, the amine group of guanines is covalently attached to the GOCl group, which is made *via* reacting COOH with $SOCl_2$. Thus, the synthesized compound is highly soluble in organic solvents with good potential in nanoelectronics [148]. The GO sheets were functionalized with the amine-end of the ionic liquid. The resulting composites were well dispersed in water, DMF, and DMSO without any addition of polymer or surfactant stabilizers [149].

Sang Hwa Shim *et al.* have shown a method to convert the hydrophilic groups of GO into alkyl or alkyl ether *via* a single-step bimolecular substitution reaction with alkyl bromide. Thus, the f-GO is dispersed as exfoliated sheets in various solvents, accomplished by covalent modification. The substitution of f-GO with solvophilic groups like polyethylene terephthalate (PET) on a small scale improves the electronic, mechanical tensile strength, and optical properties [152]. The dispersibility of GtO in organic solvents is potentially applicable to synthesizing polymeric-based composites [120, 153 - 157]. The GO surface was modified by an amidation reaction between carboxylic acid and amine substituted alkylamine [147], and esterification between carboxyl groups and alcohol polyvinyl alcohol [116]. Sungjin Park *et al.* produced a homogenous colloidal suspension of Gr sheets with the addition of polyallylamine to an aqueous suspension of GO sheet. Thus, this type of cross-linked sheet possesses outstanding mechanical stiffness and strength [153]. Wei-Hao Liao *et al.* have studied the fabrication of amine-substituted GO-polyimide composite film with high-performance capability using in-situ polymerization [154]. Zhiyuan Xu *et al.* made a polymeric composite of GO with 6 armed polyethylene glycol (PEG) by facile covalent amidation under mild conditions to make modified GO. GO-PEG was biocompatible in physiological solution and nontoxic to human lung cancer A-549 and human breast cancer MCF-7 cells. It was tested *via* cell feasibility examination [155]. Sunanda Roy *et al.* have established an effective chemical alteration with polyethylneimine (PEI) that converts into modified GO, which can extensively increase the overall strength of the nylon-12 composite. This method overcomes the drawbacks of developing advanced graphene-based polymer composites. Poor dispersibility is caused by factors such as restacking, weak interfacial bonding, and incompatibility with polymeric matrices, all of which suppress the potential action of Gr sheets [156]. GO esterification proceeds through converting the COOH of GO into an acyl chloride reacting with $SOCl_2$. Then, alcohol was substituted with organic molecules such as hydroxypropyl cellulose or chitosan, which further covalently binds to the activated graphene oxide through esterification [116, 157].

Electrophilic Substitution Reaction

The synthesis of f-Gr or f-GO is based on an electrophilic substitution reaction in which a hydrogen atom is replaced by an electrophile [158 - 165]. The high electron density of the Gr induces more reactivity towards the strong electrophile. The substitution reaction was attained *via* grafting of aryldiazonium salt, Friedel–Crafts acylation and hydrogen–lithium exchange methods [158 - 165]. The diazonium salt of p-nitroaniline and 4-bromoaniline or para-nitro substituted aryl compounds have been embedded on the Gr surface [159 - 161]. M. Tour *et al.* have synthesized organic-soluble Gr by electrophilic substitution of aryldiazonium salt on the surfactant wrapped Gr surface [159, 160]. 4-bromoaniline was used for edge-selective functionalization [161]. Thus, chemically functionalized Gr is more dispersible than pristine Gr. The Friedel-Crafts acylation is still the only method for introducing the aryl ketone group onto the Gr surface [164]. A Friedel-Crafts acylation occurs through an intermediate of an acyl anion which is catalyzed in the presence of Lewis acid. Recently, T. Govindaraju *et al.* proposed the Friedel Crafts acylation of an exfoliated single-layer GO with ferrocene at RT. The reaction was achieved by making a solution of ferrocene in dichloromethane, which was adsorbed on acidic alumina by adding trifluoroacetic anhydride at RT [162].

Fang, Wang *et al.* have synthesized covalently f-GO using polystyrene through a three-step mechanism subsequently followed by oxidation of Gt, fixing of the atom through radical initiator and in-situ polymerization to the surface of a reduced Gr sheet [163]. In hydrogen-lithium exchange methods, the metal-substituted aromatic compounds (Ar-M) are more reactive than simple aromatic compounds (Ar-H). The metallated aromatic compounds are more reactive towards the electrophile for the formation of a covalent bond. It increases the length of the aromatic ring. Yan and co-workers proposed the functionalization of Gr using hydrogen-lithium exchange (deprotonation/carbometallation and subsequent electrophilic attack) method [165]. However, towards making a new solid catalyst with trimethylamine, it shows an electrophilic substitution reaction [164, 165]. The f-GO exhibited higher catalytic activity than sodium hydroxide for hydrolysis of ethyl acetate [164]. Jin Suk Chung *et al.* have proposed a new idea for the functionalization of Gr through a solvothermal process by using GO and NMP as precursors [166]. The solvothermal reduction and functionalization occur *via* generated free radicals from NMP and proceed similarly as in the functionalization of Gr by aryldiazonium salts [158 - 161, 166]. The restoration of electronic conjugation of C=C in Gr increases with reaction time. The f-Gr forms a stable dispersion in organic solvents. Thus, the electrical conductivity of f-Gr is more efficient than pristine Gr [166].

Condensation Reaction

The condensation reaction is carried out by combining two compounds (functional groups) into a single molecule with a loss of energy. The condensation reaction of Gr takes place with isocyanate, diisocyanate, and amine through amide and carbamate ester linkages [164, 167 - 172]. Yanfei Xu *et al.* proposed a condensation reaction between 5-4-(aminophenyl)-10,15,20-triphenylporphyrin (TPP) and GO *via* an amide linkage. Modified GO was standardised and optimised for homogeneous dispersion in DMF. Consequently, the addition of the TPP-NH_2 has changed the electronic properties of GO, but no major effect is detected on the TPP-NH_2 part individually. The prevention of agglomeration is significant for processability and applications. Fluorescence quenching and energy loss occur as a result [167]. Zhuang Liu *et al.* have functionalized nanoscale graphene oxide (NGO) sheets (50 nm) by branching with biocompatible polyethylene glycol (PEG) to achieve higher aqueous solubility and stability in physiological solutions of serum. Thus, the unique features of Gr are revealed for drug delivery with water-insoluble aromatic drugs. The solubility of the drug decreases its efficacy. It revealed that PEGylated NGO (NGO-PEG) easily synthesizes complexes with water-insoluble drugs *via* noncovalent Van der Waals interaction. The hybrid NGO-PEG-SN38 composite displays excellent aqueous solubility and holds the high influence of free SN38 dissolved in organic solvents. The toxicity exceeds that of irinotecan (CPT-11, an FDA approved SN38 prodrug for colon cancer treatment) by 2-3 times [168]. Jianfeng Shen *et al.* functionalized GO nanosheets (GOS) with bovine serum albumin (BSA) *via* diimide-activated amidation. The GOS–BSA complex is highly water-soluble. BSA protein was used in this study; it contains 60 amino moieties in lysine and 26 arginine units in guanidino side chains. Infrared spectroscopy reveals the strong bands at 1690 and 1570 cm^{-1} for N-H stretching frequency of amide I and amide II, respectively. The thermogravimetric analysis (TGA) curves presented the lower thermal stability of f-GO compared to GO. AFM and Raman have demonstrated surface-functionalized GO. This method was used to synthesize the next generation of functionalized GO, with applicability for various applications [169].

Yuhua Xue *et al.* synthesized a polyhedral oligomeric silsesquioxane (POSS)-functionalized GO through an amide bond which is highly soluble in various organic solvents including tetrahydrofuran (THF), dichloromethane, chloroform, toluene, hexane, and acetone. Thus, POSS-Gr has developed super-hydrophobicity with a water/air contact angle of 157°. It could be utilized for the construction of liquid marbles. The hybrid POSS-Gr was also used as nanofiller to raise the glass transition and decomposition temperature of polymers [170]. Sasha Stankovich *et al.* used several isocyanate compounds for surface functionalization of GO. Since the isocyanates functionalized graphene oxide (iGO) formed a stable dispersion in polar aprotic solvents, The functionalization of the hydroxyl and carboxyl groups in GO occurs through the formation of amides and carbamate esters, respectively. The efficiency of the functionalization is controlled by the isocyanates' reactivity or reaction time. iGO was not completely dispersed in the water. It was formed a colloidal dispersion in apolar aprotic solvents such as dimethyl sulfoxide (DMSO), hexamethylphosphoramide (HMPA), N-methyl pyrrolidone (NMP), and DMF using sonochemical technique. iGO material does not disperse in less polar solvents, such as THF, acetone, toluene, methylene chloride, or common polar solvents, such as ethanol and methanol. The stable brown dispersion of phenyl iGO in DMF contains no precipitate. Thus, the mechanism is very useful for the preparation of polymeric nanocomposites. Hence, isocyanate and diisocyanate are useful in GO modification through cross-linkage. The functionalization occurs through the formation of a covalent (amide linkage) bond between Ar-COCl (produced by the reaction of GO with $SOCl_2$) and the NH_2 functional group of octadecylamine (ODA) [171]. Mitra Yoonessi *et al.* proposed that the synthesized polyimide nanocomposite be shown to have superior mechanical properties and good compatibility with N-methyl-2-pyrrolidone. Since then, the thermal stability of polyimide nanocomposite has been enhanced by the addition of Gr [172].

Cycloaddition Reaction (Pericyclic Reaction)

The cycloaddition reaction differs from most of the organic reactions in which the formation of cation and anion is not essential. The electrons are moved cyclically, with the bond making and breaking spontaneously. The Cycloaddition reaction of Gr offers a new route for improving the bandgap, dispersibility, and solubility. There are four types of graphene cycloaddition reactions: 1,3 dipolar cycloaddition, [2+2] cycloaddition, Diels Alder reaction, and [2+1] cycloaddition reaction [101, 150, 164, 173 - 178]. The 1,3 dipolar cycloaddition reaction, also known as [3+2] cycloaddition, occurs *via* six electron cycloaddition between 1,3 dipolar and dipolarophile (graphene) to produce five-membered rings of azomethine ylides on Gr (functionalized Gr). Mildred Quintana *et al.* have prepared functionalized Gr by condensation of formaldehyde and protected a-amino acid, followed by a deprotection of the t-butyl carbamate group (Boc group). Meanwhile, functionalized Gr has developed a new method for making nanocomposite materials by using gold nanorods [164, 173]. The azomethine ylides are unstable in most of the cases of functionalization and are made by the decarboxylation of carbonyl species and an N-methylglycine (sarcosine). Other than the edges, 1,3 dipolar reactions also occur at the C=C of the centre of the Gr sheet. Aijian Wang *et al.* proposed two possible routes for covalent functionalization of reduced GO (RGO) with porphyrin moieties by 1,3 dipolar reactions. The first method involves reacting RGO with sarcosine and 4-hydroxybenzaldehyde in dimethylformamide (DMF) at 145 °C, followed by nucleophilic substitution with w 5-[4-(2-bromoethoxy) phenyl]-10,15,2--triphenylporphyrin (TPP 1) in DMF and K_2CO_3 at 80 °C to derivatize RGO [174]. The second method involves making a formyl-inclosing porphyrin 5-[4--2-(4-formylphenoxy) ethyloxy) phenyl]-10,15,20-triphenylporphyrin (TPP 2) that is attached to the RGO through azomethine ylide. The attachment of RGO with porphyrin ensures efficient electron and energy transfer between them and improves the optical and electronic properties compared to the RGO and pristine Gr [174]. [2+2] cycloaddition comprises the four-electron cyclization process with Gr.

Zhong *et al.* have established a new efficient approach for the synthesis of chemically modified Gr by aryne cycloaddition under mild and neutral reaction conditions. Aryne modified Gr is highly thermally stable and well dispersed in water and organic solvents such as 1,2-dichlorobenzene, ethanol, and chloroform [175]. After the photochemical or thermal heat treatment, the Diels-Alder cycloaddition reaction or [4+2] cycloaddition has proceeded in a single step *via* six-membered ring formation. The reaction occurs between conjugated cis diene (electron-rich) and dienophile (electron-deficient). The reaction would take place favourably for dienophiles in the existence of an electron-withdrawing group. The interaction between the diene and dienophile is due to the good symmetrical overlapping between HOMO of the diene and LUMO of the dienophile. At high temperatures, Jeong-Min Seo and Jong-Beom Baek have developed the solvent-free mechanochemical approach for the Diels–Alder reaction of Gt (diene) with dienophile maleic anhydride (MA) or maleimide (MI) in a sealed glass bulb of argon at a high pressure of 160–220 °C. Dispersed in polar solvents, functionalized composites were used in a variety of applications [164, 176]. Hence, well-dispersed graphene function permits it to act as a more reactive dienophile or diene in Diels-Alder's reaction. Such a synthetic method has a high potential for reversible modification of Gr electronic properties *via* bandgap opening under mild conditions [177]. The unique method for the functionalization of Gr nanosheets is based on nitrene chemistry. This scheme is easy and efficient to introduce various functional groups (hydroxyl, amine, carboxyl, and bromine), long alkyl chains, and polymers, *e.g.*, poly (ethylene glycol, polystyrene) by covalent bond and stably stacked on the Gr surface to yield single-layer functionalized Gr from GO in a one-step reaction. Nitrenes are reactive intermediates that can be made from azide species through irradiation or thermolysis. Further, it can be tried for covalent functionalization with epitaxial

Gr and exfoliated Gt by [2+1] cycloaddition [178]. Thus, functionalized composites are confirmed with X-ray photoelectron spectroscopy (XPS), Atomic force microscopy (AFM), Scanning electron microscopy (SEM), EDX, XRD, Transmission electron microscopy (TEM), Brunauer–Emmett–Teller (BET), Fourier transform infrared spectroscopy (FTIR), Elemental analyzer (EA), and Solid-state nuclear magnetic resonance (SSNMR) spectroscopy techniques.

Noncovalent Functionalization

Noncovalent functionalization is an established technique for the surface modification of carbon-based nanomaterials. The noncovalent approach involves the hydrophobic Van der Waals interaction, which is prompted by an electrical interaction between two or more closely placed atoms, - interactions produced by the electronic cloud, and electrostatic force of interaction produced by electrically charged particles [101 - 121, 128, 140, 179 - 188]. Such functionalization requires the physical adsorption of suitable molecules on the Gr surface. Noncovalent functionalization is achieved by polymer wrapping, adsorption of surfactants [139] or small aromatic molecules, and interaction with sulfonated polyaniline [184], porphyrins [186], or biomolecules like proteins, enzyme-drug complexes, DNA-protein complexes, supramoleculars, and inorganic metallic complexes [117 - 119]. Surface modification of Gr and GO demonstrates excellent electrical, mechanical, and optical properties, which could be further explored for the development of several biomedical devices, industrial metallic goods, conductors, energy storage devices, and transistors [120, 121, 128, 140].

Surface modification of Gr using a robust chemical approach is the latest thrust area in the field of carbon nanotechnology to explore its several potential industrial applications [101 - 116]. The wet chemical dispersion of the Gr layer was prompted by multiple noncovalent functionalization in a suitable solvent, NMP. The first analysis was done on surfactants, *i.e.*, sodium cholate, sodium deoxycholate and dodecyl benzenesulfonate, for major alterations in material applications for enhancing efficiencies. In amphiphilic sodium dodecyl benzenesulfonate dispersion, the majority of the material uptake endorsements were observed. The utilisation of sodium deoxycholate leads by far to the uppermost level of individualization [139]. However, the selection of various matrices is an important factor for the incorporation and homogenous distribution of Gr to restore the electrical properties. Hence, adequate solubility of Gr-based materials in various solvents is a necessity for their future experimental applications [180, 181]. The stability of reduced Gr in various solvents is achieved *via* noncovalent functionalization of amine-terminated polymers [182]. The GO reduction is carried out in an aqueous medium by treating with hydrazine monohydrate and subsequently washing for the removal of agglomerated Gr sheets and impurities. The amine-substituted polymer is solubilized in an organic solvent. The aqueous dispersion of RGO is added to the organic polymer solution [183, 184]. The electric biosensor was made by non-covalent functionalization of RGO with conducting polypyrrole, implant copolymer, *i.e.*, poly (styrenesulfonic acid-g-pyrrole) [183]. The sulfonated polyaniline functionalized Gr sheet enhanced the electrochemical activity and stability [184]. The dispersibility of Gr in water is increased by noncovalent functionalization of RGO with aromatic dyes like methylene blue [185] and porphyrin [186]. Its research is useful in the preparation of graphene-based electric conducting sheets [185, 186]. Gr has created a blue fluorescent material by combining coronene derivatives such as the tetrapotassium salt of coronene tetracarboxylic acid. Its liquid crystalline derivatives, ester and imide, exhibit high fluorescence emission in solution and solid phase [187]. Non-covalent functionalization of RGO with sulfanilic acid azocromotrop shows its applicability as a supercapacitor electrode material [188].

GRAPHENE OXIDE FUNCTIONALIZATION WITH AMINE SUBSTITUTED ORGANIC MOLECULES

Substituted organic compounds are highly attractive due to their importance in agrochemicals, pharmaceuticals, device manufacturing, supercapacitors, and nanomaterial synthesis. Amine substituted molecules may be either aromatic or nonaromatic, heterocyclic or non-heterocyclic, or aliphatic. The heterocyclic compounds are considered for their potency in various fields. Heterocyclic molecules are cyclic structures that bear one or two heteroatoms (other than carbon) in the ring. The ring size differs from three-membered onward and is most efficient up to eight members. The number of heterocyclic compounds is still increasing rapidly. Almost all natural drugs, such as papaverine, procaine, emetine, atropine, theophylline, morphine, reserpine, and codeine, are heterocyclic natural drugs. The most commonly known synthetic drugs such as chlorpromazine, antipyrine, metronidazole, methotrexate, captopril, diazepam, barbiturates, and azidothymidine are heterocyclic. Moreover, some pesticides, dyes, and herbicides are also heterocyclic [189 - 192]. The nucleobases of DNA and RNA are nitrogen-based heterocyclic compounds that are directly involved in the recording of genetic code for therapeutic applications [146]. The essential amino acids [115], natural pigments such as haemoglobin [193], chlorophyll [194] and some hormones have a heterocyclic core. These molecules have changed their potency after functionalization with Gr. The whole physiological process of life, such as the provision of energy, metabolism, respiration, hereditary transformation, and nerve impulse, is based on biochemical reactions in which many heterocyclic compounds are associated with vitamins, serotonin, enzymes, coenzymes, adenosine triphosphate [ATP], nucleic acids, and others. According to the literature survey, most of the natural products contain heterocyclic molecules, revealing that nature itself likes heterocyclic compounds for equilibrating their life cycle because of their unique properties. The presence of heteroatoms in the heterocyclic compounds maintains the pH of many synthetic and biochemical reactions. The heterocyclic molecules undergo nucleophilic, electrophilic, condensation, and cycloaddition reactions. Such flexible reactivity is responsible for changing their electronic properties to enhance their efficiency for various applications.

ALTERNATIVES OF CONVENTIONAL METHODS FOR FUNCTIONALIZATION OF GRAPHENE OXIDE

The synthesis needs environmentally benign routes to maintain sustainability. Researchers are trying to develop a new chemical process for minimising the use of hazardous chemicals. The chemist is constantly looking for greener alternatives to traditional synthetic methods. Ultrasound-energy and microwave-irradiation are two of the thrust areas for achieving this goal. To explore new efficient and eco-friendly alternatives for chemical reactions to minimise the waste product by avoiding or reducing the use of conventional volatile organic solvents. Thus, several approaches have been reported, such as reaction under solvent-free (dry media) conditions, by the use of solid-supported reagents, mechanochemical mixing (grinding), and alternative activation and heating methods that utilize microwave and ultrasonic irradiation for synthesis. Therefore, to avoid overheating problems, the sonochemical approach is utilized for the synthesis of surface modified Gr-based nanomaterials [195 - 200].

Ultrasound-Assisted Functionalization: A Sustainable Technology

Ultrasound refers to sound waves (>20 kHz) higher than the upper limit of human hearing (16-18 kHz). The ultrasound frequency is generally kHz) higher than the upper limit of human hearing (16-18 kHz). The ultrasound frequency is generally considered to be between 20 and 100 MHz. The laboratory sonicator for performing chemical reactions generally has a frequency of 20 to 100 kHz, while the higher frequency of ultrasound waves is used in medical applications. The principle of ultrasound waves is based on the generation of acoustic cavitation, which favours a chemical reaction between 20 and 40 kHz. Higher ultrasound frequencies of 5 MHz and above do not produce cavitation, and this frequency range is used in medical imaging. The ultrasound waves are generated by the ultrasonic transducer, which converts the electrical and mechanical energy into sound energy.

The electromechanical transducer is made of quartz and is usually based on the piezoelectric effect. Applying equal and opposite electrical charges to the opposite faces of quartz tends to cause simultaneous expansion and contraction. The speedily reversed charges put up a vibration that produces ultrasonic waves, *i.e.*, a piezoelectric effect. Recently, the piezoelectric transducer of ultrasonic equipment is made of ceramic impregnated barium titanate that transforms 95% of electrical energy into ultrasound energy [195, 196]. The proliferation of sound waves, with a continuous succession of compression and refraction, goes through the liquid medium, which permits the molecules to oscillate about their mean. Hence, the average distance between molecules decreases and increases simultaneously. In the refraction cycle, the average distance and attractive forces between the molecules in the liquid medium may be overwhelmed, causing the formation of bubbles. The bubbles continuously rise within a few cycles by taking some vapours or gases from the liquid medium to gain an equilibrium size which matches the frequency of bubbles to the resonance of applied ultrasound frequency. Therefore, the acoustic field experienced by the bubbles is not stable because of the interference of other bubbles forming and resonating around them. However, the internal forces are too high to confirm the violent collapsing of bubbles, producing a very high confined temperature of nearly 5000 K and a pressure of over 1000 bar. High temperatures and pressure can be involved in chemical reactions.

The term "sonochemistry" is used to explain the effect of sound energy through bond oscillation on chemical reactivity. Among several theories, the 'hot-spot' is the most acceptable for describing sonochemical reactions. It explains the release of energy through cavitation. A cavitation bubble acts as a localised microreactor in an aqueous system, generating thousands of kelvins of temperature and high pressure. The driving force for a chemical reaction under the influence of ultrasonic waves is driven by cavitation bubbles [197]. The broad applications of sonochemistry in synthesis have attracted the attention of many researchers. Interest has spread beyond academic laboratories into the industry and chemical engineering. Various ultrasound-assisted organic reactions, exfoliation of GtO [198] and nanomaterial synthesis, have been conducted in aqueous as well as in non-aqueous media and homogenous liquid mixtures. As a result, the wonderful properties of ultrasonic-irradiation have been highlighted for successful application in organic transformations.

Microwave-Assisted Functionalization: A Green Chemical Transformations

Microwave-assisted heating with a Bunsen burner is a 21st-century asset for its use in the synthesis of organic, nanomaterials, polymers, and inorganic materials. It is the homogeneous heating process that speeds up the reaction rate compared to conventional methods. The confined zone is electromagnetic radiation range of size 1 cm to 1 m, frequency 30 GHz to 300 MHz, and photonic energy 1.24 meV to 1.24 eV. The transition between infrared irradiation and radio frequency is referred to as microwave irradiation [199, 200]. The mechanism of microwave irradiation is the interaction of electromagnetic radiation with dipolar molecules. Hence, the dipole moment directs it to become aligned and change its orientation due to electric field oscillations. Speedily reoriented molecules produce homogenous internal heating to proceed with the chemical reactions. Therefore, the choice of solvent is important for the microwave-assisted synthesis to consider a dipole moment [199, 200]. Surface modification of GO occurs in terms of an amide bond between the carboxyl group of GO and the amine group of chitosan, subsequently followed by reduction with hydrazine monohydrate. Thus, the dispersibility, aqueous solubility, and biocompatibility were enhanced by the grafting of chitosan on the GO sheet [201]. Therefore, a microwave is more valuable to get a quantifiable yield with more precision than conventional heating. The researchers are attempting to advance nanomaterial design and synthesis by monitoring all of these parameters such as temperature, pressure, temperature ramp, and polar solvent selection [199 - 201].

CHEMICAL REDUCTION OF GRAPHENE OXIDE

Most of the research work has been published for the functionalization of Gr using chemical, electrochemical, sonochemical, and microwave methods. GO functionalization has been done through the reduction of epoxy, carboxylic and ketone groups using reducing agents [124, 125]. The most commonly used reducing reagents are hydrazine monohydrate [126], sodium borohydride and their substituted derivatives [202, 203], lithium aluminium hydride [204], alkaline solution (KOH, NaOH) [205], metal iodides [123], hydrogen iodide [206], aluminium foil [207], alcohol [208], thiourea dioxide [209], p-phenylenediamine [210], L-ascorbic acid [211], nascent hydrogen [212], iron/hydrochloric acid [213], bacterial respiration [214] and pyrrole ring [215]. These chemicals are hazardous to the environment and living organisms. Hydrazine monohydrate, sodium borohydride, and lithium aluminium hydride are sensitive and explosive chemicals to moisture. The alkaline solution and organic solvents unbalance the whole aquatic system. Therefore, precautions are needed for Gr synthesis using bulk amounts of a strong reducing reagent. The alternative methods, *i.e.* metal/HCl-based reduction, such as Fe/HCl, are secure compared to explosive chemicals . But still, some problems exist due to the presence of an impurity in the reduced Gr. Recently, an acetic acid and HCl-based alternative method has been proposed, which is comparable to the reduction by hydrazine monohydrate. The elemental analysis of reduced Gr revealed a high C:O ratio vis-à-vis a loss of oxygen (reduction), high conductivity (4600 Sm^{-1}), and specific capacitance (35 Fg^{-1}) due to the attainment of unsaturation [208]. Green approaches such as microwave, ultrasound, and solvent-assisted, flash, thermal, and photochemical-based GO reduction have piqued the interest of researchers for large-scale Gr production. Considering the literature review, the recent progress in Gr research is the enzymatic reduction of GO on a priority basis. For example, Martin Pumera *et al.* have proposed GO reduction by thiourea dioxide [209]. Aravind Dasari *et al.* have reported the dopamine-induced reduction and subsequently functionalization of the GO sheet [210]. Jingyan Zhang and Shouwu Guo *et al.* found the GO reduction using L-ascorbic acid [211], which is very close to that of hydrazine monohydrate [126]. James M. Tour *et al.* have shown bacterial respiration for GO reduction [214].

GRAPHENE-BASED MATERIALS TOWARDS VARIOUS CATALYTIC PERFORMANCES

The single-atom thick, 2D carbon hexagon network of Gr and its derivatives have remarkable applications in the fields of biomedicals [146, 148, 168, 216 - 225], nanoelectronics [226 - 230], energy storage devices [231 - 233] and organic synthesis [21, 110]. The magic material of the 21st century shows good mechanical [234, 235], optical [236 - 239], sensing [98, 121, 240 - 245] and adsorption [179, 246 - 255] properties. The amphiphilic nature of GO is seen from edge to centre in the dispersion of hydrophilic to hydrophobic domains. Therefore, GO acts as a surfactant and sticks to interfaces and lower interfacial energy sites [249]. The use of pure Gr is challenging due to its agglomeration and processing difficulties. So, the modification of Gr has attracted broad research interest. The grafting of polymers with GO increases the biocompatibility and stability in the cellular medium, and it is used as a nanocarrier for hydrophilic and hydrophobic drugs [220, 246]. Also, the GO functionalization with enzymes enhanced its thermostability, which has been extensively used in biotechnological applications to stabilise heat-sensitive enzymes [247]. Recently, the surface modification of GO has been utilized to make nanofiltration membranes for desalination purposes [248]. Still, there is not much research done to search for graphene in biological systems, despite much exertion in the area of CNT for *in vitro* and *in vivo* biological uses. f-GO has biocompatibility with organic molecules such as nucleobases [146, 148] to increase the high aqueous solubility and stability. A new hybrid nanomaterial was established as an exhibiting enzyme for *in vitro* analysis and therapeutic cure of cancer cells [168, 219]. Gr-based composites show a positive impact when imparting antimicrobial [222, 223] and antibacterial [224, 225] properties.

Arghya Paul *et al.* have discovered an injectable and biocompatible hydrogel of polyethylenimine (PEI) functionalized GO to easily deliver vascular endothelial growth factor-165 (VEGF) pro-angiogenic gene for myocardial therapy [221]. Self-assembled 3D Gr or functionalized Gr-based conducting nanostructures are suitable for designing flexible, thermally and electrically stable electronics [226 -

229], supercapacitors [227, 228, 230], energy storage devices [231, 232] and solar energy conversion schemes [233]. Gr-based nanomaterials are appropriate as highly efficient complexes with good mechanical properties such as high strength and modulus, which are prepared by adding a few numbers of fillers [234, 235]. Gr and Gr-based composites have sparked interest in optoelectronic applications such as photodetection [236], the synthesis of low-cost optical photovoltaic devices [237], and fluorescent nanocomposites [238, 239]. The most important application of f-Gr is to act as a sensor to diagnose the disease, which depends upon the large surface area of Gr. f-Gr can be sensitive to a particular disease. The dispersed f-Gr interacts with the biomolecules in the matrix. The more effective nanocomposites of GO are allowed to bond with single-stranded DNA, subsequently added by fluorescent molecules [98, 121, 240 - 245]. Tetiana Kurkina *et al.* established an f-GO-based device as an electrical biosensor for the immunodetection of amyloid-beta peptide [244]. Also, f-GO-based sensors have been developed for humidity and temperature sensing due to their suitable features such as flexibility, transparency, and bulk-scale production [245]. The porous nature of Gr and its derivatives is utilized as an adsorbent for hazardous chemicals, elements, and gases for environmental remediation. Gr-based composites decontaminate naphthalene and its derivatives [179], bisphenol A [250], toxic Cr (VI) [251], copper (II) [252], carbon dioxide [253] and mercury [254] from the ecosystem. The catalytic properties of unmodified Gt are found to be utilized for alkylation [21] and Friedel–Crafts acylation of aromatic compounds [20] to avoid the use of hazardous solvents and reagents. The fluorinated Gr acts as an electrophile, which substitutes amine and alcohol nucleophiles *via* covalent bond formation [255]. The excellent properties of Gr and its derivatives in several research inventions have given fruitful solutions. The synthesis of Gr-based composites by a robust chemical approach is a new and recent thrust area. The covalently modified Gr-based materials are further confirmed in the product with detailed analysis and electrical as well as photochemical applications [256, 257]. Hence The robust chemical approach for the surface functionalization of GO is suitable for synthesizing nanocomposites for biomedical applications [258 - 260].

Biomedical Catalysis

Gr-based modified materials have been extensively studied and reported for biomedical research like cell viability, cancer therapy, tissue implantation, antibacterial activities, antioxidant behaviour and as drug carriers due to their improved solvent processabilities. Furthermore, in the literature review, alginate-based functionalized GO has been selectively obstructed in the MDA-MB-231 cell line adhesion efficiency (human breast cancer cells) while promoting adhesion towards healthy murine fibroblasts (3T3) and human osteosarcoma

(HOS) [261]. There is variation in the adhesion efficiency towards the cancer cell MDA-MB-231. HOS is not similar to healthy cell 3T3 due to tumour cell specificity, but it is driven by topographical signals originated by alginate-GO based substrate. The use of various antibiotic medicines causes resistance against the pathogens, which needs to be modified with GO that endorses the stability with agglomeration reduction. So, the materials are used as drug carriers and antibacterial agents too [262 - 264]. The optimised nanosized GO functionalization with two derivatives, *i.e.*, polyethylene glycol and polyethyleneimine, is a new method for developing a si-RNA (si = small interfering) carrier-based delivery system. It improves the osseointegration implantation, which has been confirmed by cellular cytocompatibility investigation [265]. The modulation of the Gr surface as well as its functionalization have developed various functionalities like hydroxyl groups with surfaces' defects, which show excellent cytocompatibility with better mechanical strength. These facilitate cell differentiation and regeneration in the tissue engineering field [266 - 269]. The combination of highly electrically conductive Gr and titanium dioxide facilitates the separation of electrons and holes in the TiO_2 energy band and prevents recombination when ultrasound is applied. Thus, the TiO_2-graphene-based material is used as a nanosonosensitizer for therapeutic effects against tumours [270]. Therefore, biomedical science is essentially required for chemical modification of graphene to develop a new cost-effective artificial cytocompatible tissue that can be used as an alternative.

Photoluminescent Catalysis

The zero-bandgap has less applicability in optics. So, the oxidation of Gr opens the zero bandgap after generating various oxygen functionalities such as hydroxyl, carbonyl, carboxyl, and aldehyde to focus on the optical properties [271 - 277]. The photoluminescence (PL) of GO is seen in the visible or ultraviolet region, which is more intense than Gr. Two PL bands, one in blue and another in the red region (500-600 nm), are seen. The PL band of the GO is not yet simplified like organic molecules due to the presence of various electronic domains and uncertain functionality on the GO surface. The origin of PL bands is still being debated. The blue band has been observed due to the the π-π^* transition of sp^2 electronic domains at the C-O sp^3 surface [277], while it is also claimed that π-π^* transition has observed the broad long-wavelength (LW) band [278]. The broadband due to defects has caused a variation in the sp^2 domains' bandgap. The pH-dependent tunable PL bands of the GO resemble those of the polycyclic aromatic compound in a basic medium because the carboxylic functionalities are coupled with their neighbouring atoms of Gr. The blue band is intense at pH>8 and allocated for the optical-electronic transition from GO−COO* to GO−COO−. Hence, the LW band has developed an important factor at pH 8. It is ascribed to the LW band for an optical-electronic transition from (GOCOOH)* to (GOCOOH) as an excited carboxyl group protonated into carboxylic acid. Further, the study concludes that the protonation and deprotonation of carboxyl and hydroxyl functionalities of GO were accountable for the dissimilar PL strength with pH of solutions [278, 279]. The COOH and OH are accountable for the blue and broadening the LW band [280, 281]. According to the theory of the red-edge effect, the performance of the blue PL band depends on the excitation wavelength. The red-edge-effect also explains that the solvation dynamics of GO in a polar medium decelerates miserably with the time scale of PL, resulting in a broadening of the red-shift of the PL band [282]. Hence, the surface-modified GO shows change in the PL bandgap as their corresponding PL peaks slightly differ compared to previously reported GO due to the difference in electronic domains at the GO surface. Therefore, the GO-based derived materials have attracted a lot of attention for PL and sensing applications due to their variable electronic structural insight and solubility in various solvents.

Magnetic Catalysis

The magnetic properties of the Gr and its composites are very challenging due to the absence of d and f shell electrons, which are accountable for the ferromagnetism. The theoretical and experimental studies infer that magnetism can be explored only in specified circumstances. Magnetism is attributed by density functional theory (DFT) calculations to the confinement of spin-polarized electronic states in Fermi energy levels. In honeycomb nanostructured, Gr "zig-zag" edges possess magnetic moments while zero magnetic moments are recognised at the "armchair" edge [283]. However, the temperature-dependent magnetic moment is also produced due to the point defects in the Gr lattice. It is found in the form of missing carbon from the continuous lattice or additional doping or functionalization on the surface to create vacancies [283 - 290]. Ion irradiation of atoms is bombarded on the carbon lattice to create defects, and it is susceptible to generating magnetism. The ferromagnetic behaviour at (room temperature) RT in Gr-based materials needs great attention to the inference of the doubts as well as the validity of results along with reproducibility. To get appropriate results, extra care is necessary to protect against impurities and contaminants [291]. Hence, we have initiated GO functionalization by a chemical approach to induce magnetic behaviour at RT. Understanding the thermal behaviour of defected Gr is closely related to magnetism.

Electrocatalysis and Photocatalysis For Energy Applications

Electrocatalysis was initiated by Kobosev and Monblanowa in 1936 [292, 293], to explain the electrochemical reactions which occur at the electrode surface. The catalytic activity of the Gr can be modulated by various dopants like N, S, or O on the surface. The heteroatom dopants affect the electronic charge and spin density, which determine the catalytic activities. The electrocatalytic reactions of heterogeneous Gr have the efficiency to act as an intermediate or a reactant to interact with the electrodes. Also, the catalytic activity depends on the properties of the electrode like surface area, electron density, and thermal stability. To investigate electrocatalyst binding efficacy with electrode surfaces and to investigate new electrocatalytic applications. Pyridinic and quaternary N are the main dopants of chemically modified Gr. Several N-containing species are useful for doping on the Gr surface to explain the oxygen reduction reaction (ORR) [292]. The consumption of fuel cells (fcs) converts chemical energy into electricity. In ORR, the H_2 gas is produced at the anode. It splits into electrons and protons by a hydrogen oxidation reaction (HOR) in which the proton is diffused at the cathode with the motion of electrons throughout the cell to generate electricity. Consequently, the presence of cathodic O_2 reacts with protons and electrons to generate the water, which is generally known as ORR. Another electrocatalytic property of Gr-based materials is termed oxygen evolution reaction (OER), in which renewable natural energy is converted into chemical fuels like hydrocarbons and hydrogen energy (methanol, formic acid, methane, and butane, *etc*). OER is an excellent source of energy storage [293 - 300]. In OER, the water splits into H_2 and hydrocarbons by the reduction of CO_2. The mechanistic path of OER involves the generation of O_2 at the anode, which is important for energy storage devices like batteries, solar cells, and reversible fuel cells [293 - 300]. OER has an essential type of kinetics produced by numerous proton-coupled electron-transfer steps, which are successively made possible by the use of an active electrocatalyst to increase the reaction rate [301 - 304]. The noble metal oxide-based electrocatalysts act most efficiently [305 - 308], but due to the high cost of materials, a lot of transition metals have been applied as alternatives [303, 309, 310]. Thus, cost-effective heteroatoms doped with Gr have recently been reported for an efficient and novel class of electrocatalyst [302, 311]. Also, heterogeneous and homogenous electrocatalysts have attracted attention for OER applications due to their active electrocatalytic centres and simple OER mechanisms [304, 312 - 315].

Due to the availability of limited natural resources like petrol, diesel, coal, and others, scientists have intensified research for H_2 energy production. The accessibility of H_2O is much greater on earth than on water, which is only a workable and clean resource to produce H_2 energy in the future as an excellent source of energy within locally available resources. However, the extreme temperature (2000 °C) is required for water dissociation to produce H_2 and O_2 while avoiding instant recombining capability.The small overpotential is required for H_2O splitting to produce H_2 *via* the cathodic HER pathway of electrocatalytic performance at ambient circumstances [293, 316]. The most effective platinum-based electrocatalyst exhibits potentially effective HER, but its high cost and lack of robustness prevent it from being commercialised [317, 318]. Therefore, scientists are highly motivated to identify alternatives to Pt-based electrocatalysts that are cost-effective, highly competent, and most durable for HER. A lot of investigations have been completed. But some are in continuation in the same direction. Transition metal-based homogenous or heterogeneous materials *e.g.* W, Co, Mo, Ni and their various derivatives like nitrides, phosphides, carbides and sulphides have been discovered, despite their corrosion vulnerability to acidic and basic electrolytes [319 - 323]. The preparation of the membrane electrode by an assembling method in which the powder form of catalyst is dissolvate in solvent with binders, subsequently forming ink on the surface with the support of copper foil and glassy carbon. As a result of the inevitable accumulation of catalyst mass and detachment from its chemical composition during the gas-based progression of HER, the dried membrane electrode has excellent catalytic activity as prepared [324, 325]. The electrocatalyst performance depends on the surface morphology, chemical constituents, pore size, electronic homogeneity or heterogeneity. Also, the architecture applied for the electrode preparation is important for HER applications.

The scientific community has focused on photocatalytic methods for photon-based energy generation in various chemical reactions. This energy solves the major global problems of pollutant remediation like water purification from toxic chemicals and dye degradation [326 - 332]. Simultaneously, it produces solar energy as a fuel substitute [298]. Photocatalysis occurs in the solar cell as a chemical bond that can be utilized in the future for various uses. Thus, photocatalysis is applicable, but there is a lack of efficiency for practical industrial application. Photocatalysis is an energy-storage device-based process. Therefore, a lot of procedures have been applied to enhance the photocatalytic activity of semiconductor-based photocatalysts, like architecture designing of materials [333, 334], doping of homo or heteroatoms [335, 336], metals and synthesis of semiconductor composites [337]. (Hydroxyl radical) attacks on organic pollutants like dyes to breakdown into nontoxic or less hazardous units that are released in gaseous form (CO_2, NO_2, N_2O, and HCO_3) and as H_2O also [327 - 332]. The toxicity of various metal-based photocatalysts forced researchers to develop metal-free carbon-based materials, especially heteroatom doped Gr, constituting carbon, nitrogen, oxygen, sulphur, and boron. So, a heterogeneous gr enhances the photoefficiency and degradation rate of organic pollutants. It makes the environment a benign water resource for the biotic community [329].

In conclusion (Fig. **6**), recent developments in the development of heteroatom-doped Gr-based materials have had a tremendous electrocatalysis application for

HER, OER, ORR, photocatalysis, and sonocatalysis to make energy storage devices: batteries, superconducting materials, and imprinting chips. Therefore, the excellent behaviour of Gr and its derivatives in many research areas has attracted us to initiate this research work. The chemically modified Gr-based materials explore out-of-the-box applications in the 21st century. Not only Gr but also GO and its derivatives have been investigated for structural studies and numerous industrial applications through several experimental conditions.

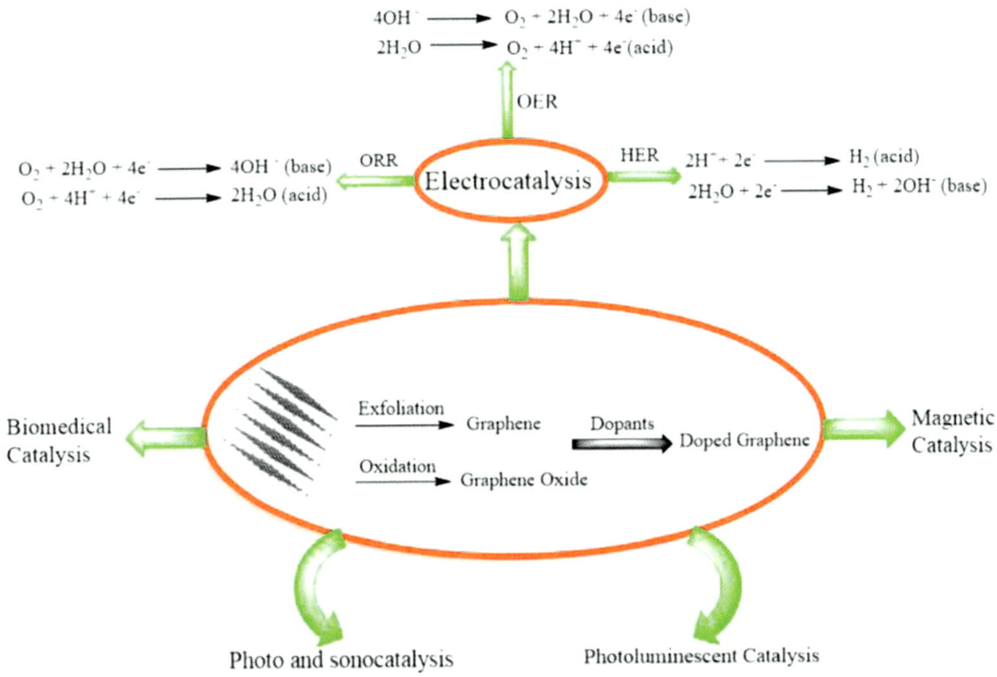

Fig. (6). Graphical representation of various catalytic performance of graphene-based materials.

We have thoroughly reviewed the initial research to explore the advanced understanding of a new dimension of the Gr and its wider derivatives and applications. Reportedly, a lot of studies have been done within the box model with certain *restrictions*. Therefore, our theme, hypothesis, and objectives have been at the cutting edge of research in the area of Gr research for enhancing industrial applications and processes.

PHYSICOCHEMICAL MECHANISM TOWARDS REINFORCING CATALYTIC PERFORMANCES

The Friccohesity is the most authentic physicochemical database to determine the structural stability, transitions, and electronic state with respect to temperature (wrt), electrical field, magnetic field, UV light, and laser. It is found that Gt and GO are the most tunable nanofunctional materials which are structurally reorganised by these external forces. These forces could be studied with the GO sheet. The database with respect to temperature, electric field, and magnetic field, the G (Gibbs free energy) and friccohesity could be measured in certain solvents at variable stoichiometric ratios. The stoichiometric ratio determines the activity of the GO with its compositions where it is on the first order or has certain transitions in structure. This will also define the uniformity of sheet repetition. Currently, these are measured with the D and G characteristic peaks of Gt and GO with Raman spectroscopy.

However, this friccohesity tracks each and every chemical activity of the GO. The polarisation of the solvent could be studied. Normal energy could be studied. There are some intra-GO points where the solvent interacts more. This could generate the transitions in its fluid dynamics in the steady state because the force will be affected due to the different orientation of friccofluids owing to its natural flow. Hence, microfluids could also be studied. The structure of GO has an intersheet binding force which makes it difficult to weaken the intersheet Van der Waals force (ISVF) through the wet reaction procedures. As a result, before proceeding with the Gt oxidation, it could be treated with a suitable nondestructive solvent with mild interaction, allowing dipolar tuning in the ISVF, and the sheet could be slightly free with a higher surface area (HSA). The HSA develops low activation energy and higher activity. Thus, the soaked Gt could be the most economical model. There could be the possibility of reacting systems. Solvents should be explored for use before adding real reacting species. Different solvents take place one by one in the reaction medium. These could be studied to fix their impact on their interaction with ISVF. These forces are the key factors which are responsible for the sheet's togetherness. These forces originated where the Gt sheet was developed initially. At the normal energy scale, mixed solvent could also be used to identify specific solvated molecules dispersed within the sheets or out edges with moderate binding forces. Gt and GO sheets have specific spatial geometry to synchronise and synergize these structures. Thus, the points of functionalizing reagent create Lennard Jone potential distance The capillary sciences explore carbon-based functional nanomaterials (CFN). For checking the stability of the solvent-GO binding dynamics, van der Waals force theory (VFT) could be achieved. Further, the activation energy (Ea) with increased temperatures could be estimated to understand the economics of the GO functionalization. How much heat (q) is needed for functionalization could be studied by monodispersed GO in suitable solvents. Certain fluorescent dyes could also be added to understand the real science of their electronic linkages mutually. The activity coefficient could also depict the electrolyte behaviour of the GO solution in a specific solvent. The capillary action develops a remarkable alignment to have a favourable geometrical orientation to receive maximum photons for photocatalysis of the materials. The friccofluids are different microfluids. Hence, more surface area is obtained by orienting the thinnest flat surface capillary so that it seems monodispersed . However, there could be a maximum frictional force. There is gradient flow where the GO is automatically sparsely dispersed and represents the photon O_2 cellular forces. The area of GO distribution for better electronic activity reflection investigates the various catalytic performances.

CONCLUSION AND PERSPECTIVES

Since inception, carbon nanomaterials have attracted enormous attention due to their extraordinary properties, moldability, and applications. Therefore, along with the natural allotropes of carbon, the artificial allotropes such as Fullerene, CNT, and Graphene have grown extensively in the last few decades to replace metallic devices to store energy. Because of their unique synergistic effects and physical properties such as optical, electrical, thermal, mechanical, piezoelectric, bionics, and supercatalytic, surface-modified carbon materials (SMCM) have become significant in the generation. The synergistic impacts of SMCM have emerged as the new avenues for biomedical application and catalytic activity with green environmental impact. The synthesis of graphene and surface-modified graphene oxide has both unlocked the latest thrust area in material sciences for space applications. The robust chemical approach for the surface modification of GO has been developed *via* direct amidation. Functionalized GO for semiconductor formulation has resolved the challenges to capturing solar energy. The sonochemical approach is employed for the greener surface modification of GO with unique edge intensities to design novel and state-of-the-art superconductor nanomaterials. The mechanism and outcome both lay a foundation for the formation of mechanistic intermediates with quasistatic irreversible and reversible states. For an insight into structural confirmation, the various high-end and sophisticated instrumentation techniques like ultraviolet-visible spectroscopy, FTIR, TGA, XRD AFM, SEM XPS, and HRTEM have been the assets to illustrate the sheet structures of the GO with an eye on ridging eutectic electronics for designing high-intensity CV (cyclic voltameter) sensors. Upgrading carbon footprints through surface plasmon resonance (SPR) and upconversion nanoparticles (UCNPs) out of the newly prepared functionalized graphene oxide (f-GO) could lay a foundation for the multifunctionalities for advanced and thermodynamically and kinetically stable nanoemulsion for stable electronic and heat-conducting nanoparticles for sensing thin-film technology.

CONSENT FOR PUBLICATION

Not applicable.

CONFLICT OF INTEREST

The authors declare no conflict of interest, financial or otherwise.

ACKNOWLEDGEMENTS

Authors are thankful to the Central University of Gujarat, India for providing infrastructural support. Authors acknowledge UGC-DAE CSR Indore for collaborative research scheme [Ref: CSR-IC-BL-79/CRS-196]. Authors are thankful to Principal Collaborators, Dr. Mukul Gupta, UGC-DAE CSR Indore, and Dr. S.N. Jha, Indus-2 RRCAT Indore, DAE India for extending their kind cooperation and support during the Collaborative Research Scheme. Mr. Gopal Avashthi is thankful to UGC-DAE CSR Indore for financial assistance in the form of Project Fellow.

REFERENCES

[1] Ren, Z.; Lan, Y.; Wang, Y. Aligned Carbon Nanotubes: Physics, Concepts, Fabrication and DevicesSpringer: Verlag Berlin Heidelberg, **2013**.

[2] Krueger, A. *Carbon Materials and Nanotechnology*; Wiley-VCH: Verlag GmbH and Co. KGaA, Weinheim, **2010**.

[3] Shenderova, O.A.; Zhirnov, V.V.; Brenner, D.W. Carbon Nanostructures. *Crit. Rev. Solid State Mater. Sci.*, **2002**, *27*(3-4), 227-356. [http://dx.doi.org/10.1080/10408430208500497]

[4] Delhaes, P. Carbon-based solids and materials. In: *ISTE Ltd and John Wiley & Sons*; ISTE Ltd and John Wiley & Sons, Inc.: UK and USA, **2011**.

[5] Serp, P.; Machado, B. *Nanostructured Carbon Materials for Catalysis*; Wiley-VCH: France, **2015**, pp. 1-45.

[6] Titirici, M.M.; White, R.J.; Brun, N.; Budarin, V.L.; Su, D.S.; del Monte, F.; Clark, J.H.; MacLachlan, M.J. Sustainable carbon materials. *Chem. Soc. Rev.*, **2015**, *44*(1), 250-290. [http://dx.doi.org/10.1039/C4CS00232F] [PMID: 25301517]

[7] Martín, N.; Guldi, D.M.; Echegoyen, L. Carbon nanostructures—Introducing the latest web themed issue. *Chem. Commun. (Camb.)*, **2011**, *47*(2), 604-605. [http://dx.doi.org/10.1039/C0CC90122A] [PMID: 21125092]

[8] Miyoshi, K. *Structures and Mechanical Properties of Natural and Synthetic Diamonds*, **1998**.

[9] Liu, Y.; Gu, Z.; Margrave, J.L.; Khabashesku, V.N. Functionalization of nanoscale diamond powder: fluoro-, alkyl-, amino-, and amino acid-nanodiamond derivatives. *Chem. Mater.*, **2004**, *16*(20), 3924-3930. [http://dx.doi.org/10.1021/cm048875q]

[10] Yu, P.Y.; Cardona, M. *Electronic Band Structures: Fundamentals of Semiconductors: Physics and Materials Properties*; Springer: Verlag Berlin Heidelberg, **2010**, pp. 17-106.

[11] Motochi, I.; Makau, N.W.; Amolo, G.O. Metal–semiconductor ohmic contacts: An ab initio Density Functional Theory study of the structural and electronic properties of metal–diamond (111)-(1×1) interfaces. *Diamond Related Materials*, **2012**, *23*, 10-17. [http://dx.doi.org/10.1016/j.diamond.2011.12.021]

[12] Lafosse, A.; Bertin, M.; Azria, R. Electron driven processes in ices: Surface functionalization and

synthesis reactions. *Prog. Surf. Sci.,* **2009**, *84*(5-6), 177-198.
[http://dx.doi.org/10.1016/j.progsurf.2009.03.001]

[13] El Tall, O.; Hou, Y.; Abou-Hamad, E.; Raja, I.U.; Hedhili, M.N.; Peng, W.; Mahfouz, R.; Bakr, O.M.; Beaujuge, P.M. Direct functionalization of nanodiamonds with maleimide. *Chem. Mater.,* **2014**, *26*(9), 2766-2769.
[http://dx.doi.org/10.1021/cm500036x]

[14] Burleson, T.; Yusuf, N.; Stanishevsky, A. Surface modification of nanodiamonds for biomedical application and analysis by infrared spectroscopy. *J. Achiev. Mater. Manuf. Eng.,* **2009**, *37*, 258-263.

[15] Kaur, R.; Badea, I. Nanodiamonds as novel nanomaterials for biomedical applications: drug delivery and imaging systems. *Int. J. Nanomedicine,* **2013**, *8*, 203-220.
[PMID: 23326195]

[16] Lukovits, I. Resonance energy in graphite. *J. Chem. Inf. Comput. Sci.,* **2004**, *44*(5), 1565-1570.
[http://dx.doi.org/10.1021/ci049931a] [PMID: 15446813]

[17] Wallace, P.R. The band theory of graphite. *Phys. Rev.,* **1947**, *71*(9), 622-634.
[http://dx.doi.org/10.1103/PhysRev.71.622]

[18] Sengupta, R.; Bhattacharya, M.; Bandyopadhyay, S.; Bhowmick, A.K. A review on the mechanical and electrical properties of graphite and modified graphite reinforced polymer composites. *Prog. Polym. Sci.,* **2011**, *36*(5), 638-670.
[http://dx.doi.org/10.1016/j.progpolymsci.2010.11.003]

[19] Dreyer, D.R.; Bielawski, C.W. Carbocatalysis: Heterogeneous carbons finding utility in synthetic chemistry. *Chem. Sci. (Camb.),* **2011**, *2*(7), 1233-1240.
[http://dx.doi.org/10.1039/c1sc00035g]

[20] Kodomari, M.; Suzuki, Y.; Yoshida, K. Graphite as an effective catalyst for Friedel–Crafts acylation. *Chem. Commun. (Camb.),* **1997**, (16), 1567-1568.
[http://dx.doi.org/10.1039/a703947f]

[21] Sereda, G.A.; Rajpara, V.B.; Slaba, R.L. The synthetic potential of graphite-catalyzed alkylation. *Tetrahedron,* **2007**, *63*(34), 8351-8357.
[http://dx.doi.org/10.1016/j.tet.2007.05.005]

[22] Dreyer, D.R.; Jia, H.P.; Todd, A.D.; Geng, J.; Bielawski, C.W. Graphite oxide: a selective and highly efficient oxidant of thiols and sulfides. *Org. Biomol. Chem.,* **2011**, *9*(21), 7292-7295.
[http://dx.doi.org/10.1039/c1ob06102j] [PMID: 21909587]

[23] Bawaked, S.; He, Q.; Dummer, N.F.; Carley, A.F.; Knight, D.W.; Bethell, D.; Kiely, C.J.; Hutchings, G.J. Selective oxidation of alkenes using graphite-supported gold-palladium catalysts. *Catal. Sci. Technol.,* **2011**, *1*(5), 747-759.
[http://dx.doi.org/10.1039/c1cy00122a]

[24] Garrigues, B.; Laporte, C.; Laurent, R.; Laporterie, A.; Dubac, J. Microwave-assisted Diels-Alder reaction supported on graphite. *Liebigs Ann.,* **1996**, *1996*(5), 739-741.
[http://dx.doi.org/10.1002/jlac.199619960515]

[25] Vijay Kumar, A.; Rama Rao, K. Recyclable graphite oxide catalyzed Friedel–Crafts addition of indoles to α,β-unsaturated ketones. *Tetrahedron Lett.,* **2011**, *52*(40), 5188-5191.
[http://dx.doi.org/10.1016/j.tetlet.2011.08.002]

[26] Girard-Lauriault, P.L.; Illgen, R.; Ruiz, J.C.; Wertheimer, M.R.; Unger, W.E.S. Surface functionalization of graphite and carbon nanotubes by vacuum-ultraviolet photochemical reactions. *Appl. Surf. Sci.,* **2012**, *258*(22), 8448-8454.
[http://dx.doi.org/10.1016/j.apsusc.2012.03.012]

[27] Sadasivuni, K.K.; Ponnamma, D.; Thomas, S.; Grohens, Y. Evolution from graphite to graphene elastomer composites. *Prog. Polym. Sci.,* **2014**, *39*(4), 749-780.
[http://dx.doi.org/10.1016/j.progpolymsci.2013.08.003]

[28] Boukhvalov, D.W. Oxidation of a graphite surface: The role of water. *J. Phys. Chem.*, **2014**, *118*, 27594-27598.

[29] Dimiev, A.; Kosynkin, D.V.; Alemany, L.B.; Chaguine, P.; Tour, J.M. Pristine graphite oxide. *J. Am. Chem. Soc.*, **2012**, *134*(5), 2815-2822.
[http://dx.doi.org/10.1021/ja211531y] [PMID: 22239610]

[30] Dongil, A.B.; Bachiller-Baeza, B.; Guerrero-Ruiz, A.; Rodríguez-Ramos, I.; Martínez-Alonso, A.; Tascón, J.M.D. Surface chemical modifications induced on high surface area graphite and carbon nanofibers using different oxidation and functionalization treatments. *J. Colloid Interface Sci.*, **2011**, *355*(1), 179-189.
[http://dx.doi.org/10.1016/j.jcis.2010.11.066] [PMID: 21190698]

[31] Beckett, R.J.; Croft, R.C. The structure of graphite oxide. *J. Phys. Chem.*, **1952**, *56*(8), 929-935.
[http://dx.doi.org/10.1021/j150500a001]

[32] Yang, Z.; Sun, Y.; Alemany, L.B.; Narayanan, T.N.; Billups, W.E. Birch reduction of graphite. Edge and interior functionalization by hydrogen. *J. Am. Chem. Soc.*, **2012**, *134*(45), 18689-18694.
[http://dx.doi.org/10.1021/ja3073116] [PMID: 23072630]

[33] Liu, S.; Zeng, T.H.; Hofmann, M.; Burcombe, E.; Wei, J.; Jiang, R.; Kong, J.; Chen, Y. Antibacterial activity of graphite, graphite oxide, graphene oxide, and reduced graphene oxide: membrane and oxidative stress. *ACS Nano,* **2011**, *5*(9), 6971-6980.
[http://dx.doi.org/10.1021/nn202451x] [PMID: 21851105]

[34] Robledo, C.B.; Otero, M.; Luque, G.; Cámara, O.; Barraco, D.; Rojas, M.I.; Leiva, E.P.M. First-principles studies of lithium storage in reduced graphite oxide. *Electrochim. Acta,* **2014**, *140*, 232-237.
[http://dx.doi.org/10.1016/j.electacta.2014.07.013]

[35] Ulloa, E. Fullerenes and their Applications in Science and Technology. *Introduction to Nanotechnology,* **2013**, (Spring), 1-5.

[36] Sheka, E.F.; Razbirin, B.S.; Nelson, D.K. Continuous symmetry of C_{60} fullerene and its derivatives. *J. Phys. Chem. A,* **2011**, *115*(15), 3480-3490.
[http://dx.doi.org/10.1021/jp109128v] [PMID: 21449547]

[37] Haddon, R.C. Electronic structure, conductivity and superconductivity of alkali metal doped (C60). *Acc. Chem. Res.*, **1992**, *25*(3), 127-133.
[http://dx.doi.org/10.1021/ar00015a005]

[38] Lieber, C.M.; Zhang, Z. Physical properties of metal-doped fullerene superconductors. *Solid State Phys.*, **1994**, *48*, 349-384.
[http://dx.doi.org/10.1016/S0081-1947(08)60581-0]

[39] del Carmen Gimenez-Lopez, M.; Räisänen, M.T.; Chamberlain, T.W.; Weber, U.; Lebedeva, M.; Rance, G.A.; Briggs, G.A.D.; Pettifor, D.; Burlakov, V.; Buck, M.; Khlobystov, A.N. Functionalized fullerenes in self-assembled monolayers. *Langmuir,* **2011**, *27*(17), 10977-10985.
[http://dx.doi.org/10.1021/la200654n] [PMID: 21744819]

[40] Gogotsi, Y. Carbon Nanomaterials. *Taylor and Francis, CRC Press, USA,* **2006**.

[41] Hoke, S.H., II; Molstad, J.; Dilettato, D.; Jay, M.J.; Carlson, D.; Kahr, B.; Cooks, R.G. Reaction of fullerenes and benzyne. *J. Org. Chem.*, **1992**, *57*(19), 5069-5071.
[http://dx.doi.org/10.1021/jo00045a012]

[42] Yamada, M.; Akasaka, T.; Nagase, S. Carbene additions to fullerenes. *Chem. Rev.*, **2013**, *113*(9), 7209-7264.
[http://dx.doi.org/10.1021/cr3004955] [PMID: 23773169]

[43] Zhang, W.; Sprafke, J.K.; Ma, M.; Tsui, E.Y.; Sydlik, S.A.; Rutledge, G.C.; Swager, T.M. Modular functionalization of carbon nanotubes and fullerenes. *J. Am. Chem. Soc.*, **2009**, *131*(24), 8446-8454.
[http://dx.doi.org/10.1021/ja810049z] [PMID: 19480427]

[44] Tzirakis, M.D.; Orfanopoulos, M. Radical reactions of fullerenes: from synthetic organic chemistry to materials science and biology. *Chem. Rev.,* **2013,** *113*(7), 5262-5321. [http://dx.doi.org/10.1021/cr300475r] [PMID: 23570603]

[45] Kokubo, K.; Matsubayashi, K.; Tategaki, H.; Takada, H.; Oshima, T. Facile synthesis of highly water-soluble fullerenes more than half-covered by hydroxyl groups. *ACS Nano,* **2008,** *2*(2), 327-333. [http://dx.doi.org/10.1021/nn700151z] [PMID: 19206634]

[46] Nakamura, E.; Isobe, H. Functionalized fullerenes in water. The first 10 years of their chemistry, biology, and nanoscience. *Acc. Chem. Res.,* **2003,** *36*(11), 807-815. [http://dx.doi.org/10.1021/ar030027y] [PMID: 14622027]

[47] Snow, S.D.; Kim, K.C.; Moor, K.J.; Jang, S.S.; Kim, J.H. Functionalized fullerenes in water: a closer look. *Environ. Sci. Technol.,* **2015,** *49*(4), 2147-2155. [http://dx.doi.org/10.1021/es504735h] [PMID: 25632831]

[48] Zhang, G.; Liu, Y.; Liang, D.; Gan, L.; Li, Y. Facile synthesis of isomerically pure fullerenols and formation of spherical aggregates from $C_{60}(OH)_8$. *Angew. Chem. Int. Ed.,* **2010,** *49*(31), 5293-5295. [http://dx.doi.org/10.1002/anie.201001280] [PMID: 20575126]

[49] Iwashita, A.; Matsuo, Y.; Nakamura, E. $AlCl_{(3)}$-mediated mono-, di-, and trihydroarylation of [60]fullerene. *Angew. Chem. Int. Ed.,* **2007,** *46*(19), 3513-3516. [http://dx.doi.org/10.1002/anie.200700062] [PMID: 17385815]

[50] Talyzin, A.V.; Tsybin, Y.O.; Purcell, J.M.; Schaub, T.M.; Shulga, Y.M.; Noréus, D.; Sato, T.; Dzwilewski, A.; Sundqvist, B.; Marshall, A.G. Reaction of hydrogen gas with C_{60} at elevated pressure and temperature: hydrogenation and cage fragmentation. *J. Phys. Chem. A,* **2006,** *110*(27), 8528-8534. [http://dx.doi.org/10.1021/jp0557971] [PMID: 16821837]

[51] Morton, J.R.; Preston, K.F.; Krusic, P.J.; Hill, S.A.; Wasserman, E. ESR studies of the reaction of alkyl radicals with fullerene (C60). *J. Phys. Chem.,* **1992,** *96*(9), 3576-3578. [http://dx.doi.org/10.1021/j100188a006]

[52] Wang, H.H.; Kini, A.M.; Savall, B.M.; Carlson, K.D.; Williams, J.M.; Lathrop, M.W.; Lykke, K.R.; Parker, D.H.; Wurz, P.; Pellin, M.J.; Gruen, D.M.; Welp, U.; Kwok, W-K.; Fleshler, S.; Crabtree, G.W.; Schirber, J.E.; Overmyer, D.L. Superconductivity at 28.6 K in a rubidium-C_{60} fullerene compound, Rb_xC_{60}, synthesized by a solution-phase technique. *Inorg. Chem.,* **1991,** *30*(15), 2962-2963. [http://dx.doi.org/10.1021/ic00015a004]

[53] Glarum, S.H.; Duclos, S.J.; Haddon, R.C. Microwave study of superconductivity in alkali-metal-doped fullerene C60 films. *J. Am. Chem. Soc.,* **1992,** *114*(6), 1996-2001. [http://dx.doi.org/10.1021/ja00032a010]

[54] Komatsu, K.; Takimoto, N.; Murata, Y.; Wan, T.S.M.; Wong, T. Synthesis and properties of dialkyl derivatives of di[60]fullerenylbutadiyne and di[60]fullerenylacetylene: the buckydumbbells. *Tetrahedron Lett.,* **1996,** *37*(34), 6153-6156. [http://dx.doi.org/10.1016/0040-4039(96)01335-4]

[55] Innocenzi, P.; Brusatin, G. Fullerene-based organic-inorganic nanocomposites and their applications. *Chem. Mater.,* **2001,** *13*(10), 3126-3139. [http://dx.doi.org/10.1021/cm0110223]

[56] Giacalone, F.; Martín, N. Fullerene polymers: synthesis and properties. *Chem. Rev.,* **2006,** *106*(12), 5136-5190. [http://dx.doi.org/10.1021/cr068389h] [PMID: 17165684]

[57] Bakry, R.; Vallant, R.M.; Najam-ul-Haq, M.; Rainer, M.; Szabo, Z.; Huck, C.W.; Bonn, G.K. Medicinal applications of fullerenes. *Int. J. Nanomedicine,* **2007,** *2*(4), 639-649. [PMID: 18203430]

[58] Bosi, S.; Da Ros, T.; Spalluto, G.; Prato, M. Fullerene derivatives: an attractive tool for biological applications. *Eur. J. Med. Chem.,* **2003,** *38*(11-12), 913-923.

[http://dx.doi.org/10.1016/j.ejmech.2003.09.005] [PMID: 14642323]

[59] Tohji, K.; Paul, A.; Moro, L.; Malhotra, R.; Lorents, D.C.; Ruoff, R.S. Selective and high-yield synthesis of higher fullerenes. *J. Phys. Chem.*, **1995**, *99*(50), 17785-17788. [http://dx.doi.org/10.1021/j100050a018]

[60] Gan, L.; Huang, S.; Zhang, X.; Zhang, A.; Cheng, B.; Cheng, H.; Li, X.; Shang, G. Fullerenes as a tert-butylperoxy radical trap, metal catalyzed reaction of tert-butyl hydroperoxide with fullerenes, and formation of the first fullerene mixed peroxides C(60)(O)(OO(t)Bu)(4) and C(70)(OO(Bu)(10). *J. Am. Chem. Soc.*, **2002**, *124*(45), 13384-13385. [http://dx.doi.org/10.1021/ja027714p] [PMID: 12418881]

[61] Bendikov, M.; Wudl, F.; Perepichka, D.F. Tetrathiafulvalenes, oligoacenenes, and their buckminsterfullerene derivatives: the brick and mortar of organic electronics. *Chem. Rev.*, **2004**, *104*(11), 4891-4946. [http://dx.doi.org/10.1021/cr030666m] [PMID: 15535637]

[62] Imahori, H.; Fukuzumi, S. Porphyrin- and fullerene-baed molecular photovoltaic devices. *Adv. Funct. Mater.*, **2004**, *14*(6), 525-536. [http://dx.doi.org/10.1002/adfm.200305172]

[63] Anthopoulos, T.D.; Singh, B.; Marjanovic, N.; Sariciftci, N.S.; Montaigne Ramil, A.; Sitter, H.; Cölle, M.; de Leeuw, D.M. High performance n-channel organic field-effect transistors and ring oscillators based on C_{60} fullerene films. *Appl. Phys. Lett.*, **2006**, *89*(21), 213504-3. [http://dx.doi.org/10.1063/1.2387892]

[64] Haddon, R.C.; Perel, A.S.; Morris, R.C.; Palstra, T.T.M.; Hebard, A.F.; Fleming, R.M. C_{60} thin film transistors. *Appl. Phys. Lett.*, **1995**, *67*(1), 121-123. [http://dx.doi.org/10.1063/1.115503]

[65] Iijima, S. Helical microtubules of graphitic carbon. *Nature,* **1991**, *354*(6348), 56-58. [http://dx.doi.org/10.1038/354056a0]

[66] Ajayan, P.M.; Charlier, J.C.; Rinzler, A.G. Carbon nanotubes: From macromolecules to nanotechnology. *Proc. Natl. Acad. Sci. USA,* **1999**, *96*(25), 14199-14200. [http://dx.doi.org/10.1073/pnas.96.25.14199] [PMID: 10588679]

[67] Lee, D.W.; Seo, J.W Preparation of carbon nanotubes from graphite powder at room temperature. **2010**.

[68] Journet, C.; Maser, W.K.; Bernier, P.; Loiseau, A.; de la Chapelle, M.L.; Lefrant, S.; Deniard, P.; Lee, R.; Fischer, J.E. Large-scale production of single-walled carbon nanotubes by the electric-arc technique. *Nature,* **1997**, *388*(6644), 756-758. [http://dx.doi.org/10.1038/41972]

[69] Cassell, A.M.; Raymakers, J.A.; Kong, J.; Dai, H. Large scale CVD synthesis of single-walled carbon nanotubes. *J. Phys. Chem. B,* **1999**, *103*(31), 6484-6492. [http://dx.doi.org/10.1021/jp990957s]

[70] Odom, T.W.; Huang, J.L.; Kim, P.; Lieber, C.M. Atomic structure and electronic properties of single-walled carbon nanotubes. *Nature,* **1998**, *391*(6662), 62-64. [http://dx.doi.org/10.1038/34145]

[71] Tersoff, J.; Ruoff, R.S. Structural properties of a carbon-nanotube crystal. *Phys. Rev. Lett.*, **1994**, *73*(5), 676-679. [http://dx.doi.org/10.1103/PhysRevLett.73.676] [PMID: 10057509]

[72] Aykol, M.; Hou, B.; Dhall, R.; Chang, S.W.; Branham, W.; Qiu, J.; Cronin, S.B. Clamping instability and van der Waals forces in carbon nanotube mechanical resonators. *Nano Lett.*, **2014**, *14*(5), 2426-2430. [http://dx.doi.org/10.1021/nl500096p] [PMID: 24758201]

[73] Sawano, S.; Arie, T.; Akita, S. Carbon nanotube resonator in liquid. *Nano Lett.*, **2010**, *10*(9), 3395-

3398.
[http://dx.doi.org/10.1021/nl101292b] [PMID: 20681632]

[74] Wong, S.S.; Joselevich, E.; Woolley, A.T.; Cheung, C.L.; Lieber, C.M. Covalently functionalized nanotubes as nanometre- sized probes in chemistry and biology. *Nature,* **1998**, *394*(6688), 52-55. [http://dx.doi.org/10.1038/27873] [PMID: 9665127]

[75] Blake, R.; Gun'ko, Y.K.; Coleman, J.; Cadek, M.; Fonseca, A.; Nagy, J.B.; Blau, W.J. A generic organometallic approach toward ultra-strong carbon nanotube polymer composites. *J. Am. Chem. Soc.,* **2004**, *126*(33), 10226-10227.
[http://dx.doi.org/10.1021/ja0474805] [PMID: 15315418]

[76] Biercuk, M.J.; Llaguno, M.C.; Radosavljevic, M.; Hyun, J.K.; Johnson, A.T.; Fischer, J.E. Carbon nanotube composites for thermal management. *Appl. Phys. Lett.,* **2002**, *80*(15), 2767-2769. [http://dx.doi.org/10.1063/1.1469696]

[77] An, W.; Turner, C.H. Linking carbon and boron-nitride nanotubes: Heterojunction energetics and band gap tuning. *J. Phys. Chem. Lett.,* **2010**, *1*(15), 2269-2273.
[http://dx.doi.org/10.1021/jz100753x]

[78] Ma, J.; Tang, J.; Zhang, H.; Shinya, N.; Qin, L.C. Ultrathin carbon nanotube fibrils of high electrochemical capacitance. *ACS Nano,* **2009**, *3*(11), 3679-3683.
[http://dx.doi.org/10.1021/nn900787h] [PMID: 19877714]

[79] Bethune, D.S.; Kiang, C.H.; de Vries, M.S.; Gorman, G.; Savoy, R.; Vazquez, J.; Beyers, R. Cobalt-catalysed growth of carbon nanotubes with single-atomic-layer walls. *Nature,* **1993**, *363*(6430), 605-607.
[http://dx.doi.org/10.1038/363605a0]

[80] Xiong, H.; Slater, M.D.; Balasubramanian, M.; Johnson, C.S.; Rajh, T. Amorphous TiO_2 nanotube anode for rechargeable sodium ion batteries. *J. Phys. Chem. Lett.,* **2011**, *2*(20), 2560-2565. [http://dx.doi.org/10.1021/jz2012066]

[81] Perebeinos, V.; Tersoff, J. Carbon nanotube deformation and collapse under metal contacts. *Nano Lett.,* **2014**, *14*(8), 4376-4380.
[http://dx.doi.org/10.1021/nl5012646] [PMID: 25014612]

[82] Dillon, A.C.; Heben, M.J. Hydrogen storage using carbon adsorbents: past, present and future. *Appl. Phys., A Mater. Sci. Process.,* **2001**, *72*(2), 133-142.
[http://dx.doi.org/10.1007/s003390100788]

[83] Chen, Y.; Shaw, D.T.; Bai, X.D.; Wang, E.G.; Lund, C.; Lu, W.M.; Chung, D.D.L. Hydrogen storage in aligned carbon nanotubes. *Appl. Phys. Lett.,* **2001**, *78*(15), 2128-2130.
[http://dx.doi.org/10.1063/1.1341224]

[84] Svedružić, D.; Blackburn, J.L.; Tenent, R.C.; Rocha, J.D.R.; Vinzant, T.B.; Heben, M.J.; King, P.W. High-performance hydrogen production and oxidation electrodes with hydrogenase supported on metallic single-wall carbon nanotube networks. *J. Am. Chem. Soc.,* **2011**, *133*(12), 4299-4306. [http://dx.doi.org/10.1021/ja104785e] [PMID: 21384925]

[85] Novoselov, K.S.; Geim, A.K.; Morozov, S.V.; Jiang, D.; Zhang, Y.; Dubonos, S.V.; Grigorieva, I.V.; Firsov, A.A. Electric field effect in atomically thin carbon films. *Science,* **2004**, *306*(5696), 666-669. [http://dx.doi.org/10.1126/science.1102896] [PMID: 15499015]

[86] Li, X.; Cai, W.; An, J.; Kim, S.; Nah, J.; Yang, D.; Piner, R.; Velamakanni, A.; Jung, I.; Tutuc, E.; Banerjee, S.K.; Colombo, L.; Ruoff, R.S. Large-area synthesis of high-quality and uniform graphene films on copper foils. *Science,* **2009**, *324*(5932), 1312-1314.
[http://dx.doi.org/10.1126/science.1171245] [PMID: 19423775]

[87] Dreyer, D.R.; Ruoff, R.S.; Bielawski, C.W. From conception to realization: an historial account of graphene and some perspectives for its future. *Angew. Chem. Int. Ed.,* **2010**, *49*(49), 9336-9344. [http://dx.doi.org/10.1002/anie.201003024] [PMID: 21110353]

[88] Geim, A.K.; Novoselov, K.S. The rise of graphene. *Nat. Mater.*, **2007**, *6*(3), 183-191. [http://dx.doi.org/10.1038/nmat1849] [PMID: 17330084]

[89] Novoselov, K.S.; Geim, A.K.; Morozov, S.V.; Jiang, D.; Katsnelson, M.I.; Grigorieva, I.V.; Dubonos, S.V.; Firsov, A.A. Two-dimensional gas of massless Dirac fermions in graphene. *Nature,* **2005**, *438*(7065), 197-200. [http://dx.doi.org/10.1038/nature04233] [PMID: 16281030]

[90] Demming, A. Demming. *King of the elements.Nanotechnology,* **2010**. [http://dx.doi.org/10.1088/0957-4484/21/30/300201]

[91] Boyes, M.; Alieva, A.; Tong, J.; Nagyte, V.; Melle-Franco, M.; Vetter, T.; Casiraghi, C. Exploiting the surface properties of graphene for polymorph selectivity. *ACS Nano,* **2020**, *14*(8), 10394-10401. [http://dx.doi.org/10.1021/acsnano.0c04183] [PMID: 32692539]

[92] Xu, H.; Xie, L.; Zhang, H.; Zhang, J. Effect of graphene Fermi level on the Raman scattering intensity of molecules on graphene. *ACS Nano,* **2011**, *5*(7), 5338-5344. [http://dx.doi.org/10.1021/nn103237x] [PMID: 21678950]

[93] Neto, A.H.C.; Guinea, F.; Peres, N.M.R.; Novoselov, K.S.; Geim, A.K. *The electronic properties of graphene.*, **2009**, *109*(54), 81.

[94] Rao, C.N.R.; Matte, H.S.S.R.; Subrahmanyam, K.S. Synthesis and selected properties of graphene and graphene mimics. *Acc. Chem. Res.,* **2013**, *46*(1), 149-159. [http://dx.doi.org/10.1021/ar300033m] [PMID: 22738406]

[95] Kim, Y.A.; Fujisawa, K.; Muramatsu, H.; Hayashi, T.; Endo, M.; Fujimori, T.; Kaneko, K.; Terrones, M.; Behrends, J.; Eckmann, A.; Casiraghi, C.; Novoselov, K.S.; Saito, R.; Dresselhaus, M.S. Raman spectroscopy of boron-doped single-layer graphene. *ACS Nano,* **2012**, *6*(7), 6293-6300. [http://dx.doi.org/10.1021/nn301728j] [PMID: 22695033]

[96] Prathish, K.P.; Barsan, M.M.; Geng, D.; Sun, X.; Brett, C.M.A. Chemically modified graphene and nitrogen-doped graphene: Electrochemical characterisation and sensing applications. *Electrochim. Acta,* **2013**, *114*, 533-542. [http://dx.doi.org/10.1016/j.electacta.2013.10.080]

[97] Wang, H.; Maiyalagan, T.; Wang, X. Review on recent progress in nitrogen-doped graphene: Synthesis, characterization, and its potential applications. *ACS Catal.,* **2012**, *2*(5), 781-794. [http://dx.doi.org/10.1021/cs200652y]

[98] Wang, Y.; Shao, Y.; Matson, D.W.; Li, J.; Lin, Y. Nitrogen-doped graphene and its application in electrochemical biosensing. *ACS Nano,* **2010**, *4*(4), 1790-1798. [http://dx.doi.org/10.1021/nn100315s] [PMID: 20373745]

[99] Deng, D.; Pan, X.; Yu, L.; Cui, Y.; Jiang, Y.; Qi, J.; Li, W.X.; Fu, Q.; Ma, X.; Xue, Q.; Sun, G.; Bao, X. Toward N-doped graphene *via* solvothermal synthesis. *Chem. Mater.,* **2011**, *23*(5), 1188-1193. [http://dx.doi.org/10.1021/cm102666r]

[100] Usachov, D.; Vilkov, O.; Grüneis, A.; Haberer, D.; Fedorov, A.; Adamchuk, V.K.; Preobrajenski, A.B.; Dudin, P.; Barinov, A.; Oehzelt, M.; Laubschat, C.; Vyalikh, D.V. Nitrogen-doped graphene: efficient growth, structure, and electronic properties. *Nano Lett.,* **2011**, *11*(12), 5401-5407. [http://dx.doi.org/10.1021/nl2031037] [PMID: 22077830]

[101] Georgakilas, V.; Otyepka, M.; Bourlinos, A.B.; Chandra, V.; Kim, N.; Kemp, K.C.; Hobza, P.; Zboril, R.; Kim, K.S. Functionalization of graphene: covalent and non-covalent approaches, derivatives and applications. *Chem. Rev.,* **2012**, *112*(11), 6156-6214. [http://dx.doi.org/10.1021/cr3000412] [PMID: 23009634]

[102] Koehler, F.M.; Stark, W.J. Organic synthesis on graphene. *Acc. Chem. Res.,* **2013**, *46*(10), 2297-2306. [http://dx.doi.org/10.1021/ar300125w] [PMID: 23210633]

[103] Su, C.; Loh, K.P. Carbocatalysts: graphene oxide and its derivatives. *Acc. Chem. Res.,* **2013**, *46*(10),

2275-2285.
[http://dx.doi.org/10.1021/ar300118v] [PMID: 23270430]

[104] Quintana, M.; Vazquez, E.; Prato, M. Organic functionalization of graphene in dispersions. *Acc. Chem. Res.,* **2013**, *46*(1), 138-148.
[http://dx.doi.org/10.1021/ar300138e] [PMID: 22872046]

[105] Mirza-Aghayan, M.; Kashef-Azar, E.; Boukherroub, R. Graphite oxide: an efficient reagent for oxidation of alcohols under sonication. *Tetrahedron Lett.,* **2012**, *53*(37), 4962-4965.
[http://dx.doi.org/10.1016/j.tetlet.2012.07.016]

[106] Jia, H.P.; Dreyer, D.R.; Bielawski, C.W. C–H oxidation using graphite oxide. *Tetrahedron,* **2011**, *67*(24), 4431-4434.
[http://dx.doi.org/10.1016/j.tet.2011.02.065]

[107] Lücking, F.; Köser, H.; Jank, M.; Ritter, A. Iron powder, graphite and activated carbon as catalysts for the oxidation of 4-chlorophenol with hydrogen peroxide in aqueous solution. *Water Res.,* **1998**, *32*(9), 2607-2614.
[http://dx.doi.org/10.1016/S0043-1354(98)00016-5]

[108] Byung, H.H.; Dae, H.S.; Sung, Y.C. Graphite catalyzed reduction of aromatic and aliphatic nitro compounds with hydrazine hydrate. *Tetrahedron Lett.,* **1985**, *26*(50), 6233-6234.
[http://dx.doi.org/10.1016/S0040-4039(00)95060-3]

[109] Hu, F.; Patel, M.; Luo, F.; Flach, C.; Mendelsohn, R.; Garfunkel, E.; He, H.; Szostak, M. Graphene-catalyzed direct friedel–crafts alkylation reactions: mechanism, selectivity, and synthetic utility. *J. Am. Chem. Soc.,* **2015**, *137*(45), 14473-14480.
[http://dx.doi.org/10.1021/jacs.5b09636] [PMID: 26496423]

[110] Sachdeva, H. Recent advances in the catalytic applications of GO/rGO for green organic synthesis. *Green Processing and Synthesis,* **2020**, *9*(1), 515-537.
[http://dx.doi.org/10.1515/gps-2020-0055]

[111] Lombardi, L.; Bandini, M. Graphene oxide as a mediator in organic synthesis: a mechanistic focus. *Angew. Chem. Int. Ed.,* **2020**, *59*(47), 20767-20778.
[http://dx.doi.org/10.1002/anie.202006932] [PMID: 32516475]

[112] Zhao, J.; Wang, H.; Gao, B.; Wang, X.; Cai, Q.; Wang, X. Chemical functionalization of graphene *via* aryne cycloaddition: a theoretical study. *J. Mol. Model.,* **2012**, *18*(6), 2861-2868.
[http://dx.doi.org/10.1007/s00894-011-1309-8] [PMID: 22127614]

[113] Park, J.; Yan, M. Covalent functionalization of graphene with reactive intermediates. *Acc. Chem. Res.,* **2013**, *46*(1), 181-189.
[http://dx.doi.org/10.1021/ar300172h] [PMID: 23116448]

[114] Maktedar, S.S.; Mehetre, S.S.; Singh, M.; Kale, R.K. Ultrasound irradiation: A robust approach for direct functionalization of graphene oxide with thermal and antimicrobial aspects. *Ultrason. Sonochem.,* **2014**, *21*(4), 1407-1416.
[http://dx.doi.org/10.1016/j.ultsonch.2014.02.022] [PMID: 24613466]

[115] Mallakpour, S.; Abdolmaleki, A.; Borandeh, S. Covalently functionalized graphene sheets with biocompatible natural amino acids. *Appl. Surf. Sci.,* **2014**, *307*, 533-542.
[http://dx.doi.org/10.1016/j.apsusc.2014.04.070]

[116] Salavagione, H.J.; Gómez, M.A.; Martínez, G. Polymeric Modification of graphene through esterification of graphite oxide and poly(vinyl alcohol). *Macromolecules,* **2009**, *42*(17), 6331-6334.
[http://dx.doi.org/10.1021/ma900845w]

[117] Sanchez, V.C.; Jachak, A.; Hurt, R.H.; Kane, A.B. Biological interactions of graphene-family nanomaterials: an interdisciplinary review. *Chem. Res. Toxicol.,* **2012**, *25*(1), 15-34.
[http://dx.doi.org/10.1021/tx200339h] [PMID: 21954945]

[118] Chung, C.; Kim, Y.K.; Shin, D.; Ryoo, S.R.; Hong, B.H.; Min, D.H. Biomedical applications of

graphene and graphene oxide. *Acc. Chem. Res.*, **2013**, *46*(10), 2211-2224. [http://dx.doi.org/10.1021/ar300159f] [PMID: 23480658]

[119] Mao, H.Y.; Laurent, S.; Chen, W.; Akhavan, O.; Imani, M.; Ashkarran, A.A.; Mahmoudi, M. Graphene: promises, facts, opportunities, and challenges in nanomedicine. *Chem. Rev.*, **2013**, *113*(5), 3407-3424. [http://dx.doi.org/10.1021/cr300335p] [PMID: 23452512]

[120] Dai, L. Functionalization of graphene for efficient energy conversion and storage. *Acc. Chem. Res.*, **2013**, *46*(1), 31-42. [http://dx.doi.org/10.1021/ar300122m] [PMID: 23030244]

[121] Liu, Y.; Dong, X.; Chen, P. Biological and chemical sensors based on graphene materials. *Chem. Soc. Rev.*, **2012**, *41*(6), 2283-2307. [http://dx.doi.org/10.1039/C1CS15270J] [PMID: 22143223]

[122] Kim, F.; Luo, J.; Cruz-Silva, R.; Cote, L.J.; Sohn, K.; Huang, J. Self-propagating domino-like reactions in oxidized graphite. *Adv. Funct. Mater.*, **2010**, *20*(17), 2867-2873. [http://dx.doi.org/10.1002/adfm.201000736]

[123] Liu, C.; Hao, F.; Zhao, X.; Zhao, Q.; Luo, S.; Lin, H. Low temperature reduction of free-standing graphene oxide papers with metal iodides for ultrahigh bulk conductivity. *Scientific Reports*, **2014**, *3996*(6), 4.

[124] Pei, S.; Cheng, H.M. The reduction of graphene oxide. *Carbon*, **2012**, *50*(9), 3210-3228. [http://dx.doi.org/10.1016/j.carbon.2011.11.010]

[125] Chua, C.K.; Pumera, M. Chemical reduction of graphene oxide: a synthetic chemistry viewpoint. *Chem. Soc. Rev.*, **2014**, *43*(1), 291-312. [http://dx.doi.org/10.1039/C3CS60303B] [PMID: 24121318]

[126] Gao, X.; Jang, J.; Nagase, S. Hydrazine and thermal reduction of graphene oxide: Reaction mechanisms, product structures, and reaction design. *J. Phys. Chem. C*, **2010**, *114*(2), 832-842. [http://dx.doi.org/10.1021/jp909284g]

[127] Kim, M.; Hwang, Y.; Min, K.; Kim, J. Concentration dependence of graphene oxide–nanoneedle manganese oxide composites reduced by hydrazine hydrate for an electrochemical supercapacitor. *Phys. Chem. Chem. Phys.*, **2013**, *15*(37), 15602-15611. [http://dx.doi.org/10.1039/c3cp52126e] [PMID: 23942656]

[128] Dreyer, D.R.; Park, S.; Bielawski, C.W.; Ruoff, R.S. The chemistry of graphene oxide. *Chem. Soc. Rev.*, **2010**, *39*(1), 228-240. [http://dx.doi.org/10.1039/B917103G] [PMID: 20023850]

[129] Lee, D.W.; De Los Santos V, L.; Seo, J.W.; Felix, L.L.; Bustamante D, A.; Cole, J.M.; Barnes, C.H.W. The structure of graphite oxide: investigation of its surface chemical groups. *J. Phys. Chem. B*, **2010**, *114*(17), 5723-5728. [http://dx.doi.org/10.1021/jp1002275] [PMID: 20380401]

[130] He, H.; Riedl, T.; Lerf, A.; Klinowski, J. Solid-state nmr studies of the structure of graphite oxide. *J. Phys. Chem.*, **1996**, *100*(51), 19954-19958. [http://dx.doi.org/10.1021/jp961563t]

[131] Lerf, A.; He, H.; Forster, M.; Klinowski, J. Structure of graphite oxide revisited. *J. Phys. Chem. B*, **1998**, *102*(23), 4477-4482. [http://dx.doi.org/10.1021/jp9731821]

[132] Casabianca, L.B.; Shaibat, M.A.; Cai, W.W.; Park, S.; Piner, R.; Ruoff, R.S.; Ishii, Y. NMR-based structural modeling of graphite oxide using multidimensional ^{13}C solid-state NMR and ab initio chemical shift calculations. *J. Am. Chem. Soc.*, **2010**, *132*(16), 5672-5676. [http://dx.doi.org/10.1021/ja9030243] [PMID: 20359218]

[133] Cai, W.; Piner, R.D.; Stadermann, F.J.; Park, S.; Shaibat, M.A.; Ishii, Y.; Yang, D.; Velamakanni, A.;

An, S.J.; Stoller, M.; An, J.; Chen, D.; Ruoff, R.S. Synthesis and solid-state NMR structural characterization of ^{13}C-labeled graphite oxide. *Science,* **2008**, *321*(5897), 1815-1817. [http://dx.doi.org/10.1126/science.1162369] [PMID: 18818353]

[134] He, H.; Klinowski, J.; Forster, M.; Lerf, A. A new structural model for graphite oxide. *Chem. Phys. Lett.,* **1998**, *287*(1-2), 53-56. [http://dx.doi.org/10.1016/S0009-2614(98)00144-4]

[135] Ang, P.K.; Wang, S.; Bao, Q.; Thong, J.T.L.; Loh, K.P. High-throughput synthesis of graphene by intercalation-exfoliation of graphite oxide and study of ionic screening in graphene transistor. *ACS Nano,* **2009**, *3*(11), 3587-3594. [http://dx.doi.org/10.1021/nn901111s] [PMID: 19788171]

[136] Hernandez, Y.; Nicolosi, V.; Lotya, M.; Blighe, F.M.; Sun, Z.; De, S.; McGovern, I.T.; Holland, B.; Byrne, M.; Gun'Ko, Y.K.; Boland, J.J.; Niraj, P.; Duesberg, G.; Krishnamurthy, S.; Goodhue, R.; Hutchison, J.; Scardaci, V.; Ferrari, A.C.; Coleman, J.N. High-yield production of graphene by liquid-phase exfoliation of graphite. *Nat. Nanotechnol.,* **2008**, *3*(9), 563-568. [http://dx.doi.org/10.1038/nnano.2008.215] [PMID: 18772919]

[137] Parvez, K.; Wu, Z.S.; Li, R.; Liu, X.; Graf, R.; Feng, X.; Müllen, K. Exfoliation of graphite into graphene in aqueous solutions of inorganic salts. *J. Am. Chem. Soc.,* **2014**, *136*(16), 6083-6091. [http://dx.doi.org/10.1021/ja5017156] [PMID: 24684678]

[138] Paredes, J.I.; Villar-Rodil, S.; Martínez-Alonso, A.; Tascón, J.M.D. Graphene oxide dispersions in organic solvents. *Langmuir,* **2008**, *24*(19), 10560-10564. [http://dx.doi.org/10.1021/la801744a] [PMID: 18759411]

[139] Hirsch, A.; Englert, J.M.; Hauke, F. Wet chemical functionalization of graphene. *Acc. Chem. Res.,* **2013**, *46*(1), 87-96. [http://dx.doi.org/10.1021/ar300116q] [PMID: 22946482]

[140] Szabó, T.; Berkesi, O.; Forgó, P.; Josepovits, K.; Sanakis, Y.; Petridis, D.; Dékány, I. Evolution of surface functional groups in a series of progressively oxidized graphite oxides. *Chem. Mater.,* **2006**, *18*(11), 2740-2749. [http://dx.doi.org/10.1021/cm060258+]

[141] Niyogi, S.; Bekyarova, E.; Itkis, M.E.; Zhang, H.; Shepperd, K.; Hicks, J.; Sprinkle, M.; Berger, C.; Lau, C.N.; deHeer, W.A.; Conrad, E.H.; Haddon, R.C. Spectroscopy of covalently functionalized graphene. *Nano Lett.,* **2010**, *10*(10), 4061-4066. [http://dx.doi.org/10.1021/nl1021128] [PMID: 20738114]

[142] Pattabiraman, V.R.; Bode, J.W. Rethinking amide bond synthesis. *Nature,* **2011**, *480*(7378), 471-479. [http://dx.doi.org/10.1038/nature10702] [PMID: 22193101]

[143] Liu, L.H.; Lerner, M.M.; Yan, M. Derivitization of pristine graphene with well-defined chemical functionalities. *Nano Lett.,* **2010**, *10*(9), 3754-3756. [http://dx.doi.org/10.1021/nl1024744] [PMID: 20690657]

[144] Chen, D.; Feng, H.; Li, J. Graphene oxide: preparation, functionalization, and electrochemical applications. *Chem. Rev.,* **2012**, *112*(11), 6027-6053. [http://dx.doi.org/10.1021/cr300115g] [PMID: 22889102]

[145] Hsiao, M.C.; Liao, S.H.; Yen, M.Y.; Liu, P.I.; Pu, N.W.; Wang, C.A.; Ma, C.C.M. Preparation of covalently functionalized graphene using residual oxygen-containing functional groups. *ACS Appl. Mater. Interfaces,* **2010**, *2*(11), 3092-3099. [http://dx.doi.org/10.1021/am100597d] [PMID: 20949901]

[146] Vovusha, H.; Sanyal, S.; Sanyal, B. Interaction of nucleobases and aromatic amino acids with graphene oxide and graphene flakes. *J. Phys. Chem. Lett.,* **2013**, *4*(21), 3710-3718. [http://dx.doi.org/10.1021/jz401929h]

[147] Bourlinos, A.B.; Gournis, D.; Petridis, D.; Szabó, T.; Szeri, A.; Dékány, I. Graphite oxide: Chemical

reduction to graphite and surface modification with primary aliphatic amines and amino acids. *Langmuir,* **2003**, *19*(15), 6050-6055. [http://dx.doi.org/10.1021/la026525h]

[148] Zhang, F.; Cao, H.; Yue, D.; Zhou, Z. Graphene covalently modified by DNA G-base. *J. Phys. Chem. C,* **2013**, *117*(7), 3513-3519. [http://dx.doi.org/10.1021/jp307298p]

[149] Yang, H.; Shan, C.; Li, F.; Han, D.; Zhang, Q.; Niu, L. Covalent functionalization of polydisperse chemically-converted graphene sheets with amine-terminated ionic liquid. *Chem. Commun. (Camb.),* **2009**, *26*(26), 3880-3882. [http://dx.doi.org/10.1039/b905085j] [PMID: 19662239]

[150] Georgakilas, V.; Bourlinos, A.B.; Zboril, R.; Steriotis, T.A.; Dallas, P.; Stubos, A.K.; Trapalis, C. Organic functionalisation of graphenes. *Chem. Commun. (Camb.),* **2010**, *46*(10), 1766-1768. [http://dx.doi.org/10.1039/b922081j] [PMID: 20177643]

[151] McGrail, B.T.; Rodier, B.J.; Pentzer, E. Rapid functionalization of graphene oxide in water. *Chem. Mater.,* **2014**, *26*(19), 5806-5811. [http://dx.doi.org/10.1021/cm5031409]

[152] Shim, S.H.; Kim, K.T.; Lee, J.U.; Jo, W.H. Facile method to functionalize graphene oxide and its application to poly(ethylene terephthalate)/graphene composite. *ACS Appl. Mater. Interfaces,* **2012**, *4*(8), 4184-4191. [http://dx.doi.org/10.1021/am300906z] [PMID: 22764836]

[153] Park, S.; Dikin, D.A.; Nguyen, S.T.; Ruoff, R.S. Graphene oxide sheets chemically cross-linked by polyallylamine. *J. Phys. Chem. C,* **2009**, *113*(36), 15801-15804. [http://dx.doi.org/10.1021/jp907613s]

[154] Liao, W.H.; Yang, S.Y.; Wang, J.Y.; Tien, H.W.; Hsiao, S.T.; Wang, Y.S.; Li, S.M.; Ma, C.C.M.; Wu, Y.F. Effect of molecular chain length on the mechanical and thermal properties of amine-functionalized graphene oxide/polyimide composite films prepared by *in situ* polymerization. *ACS Appl. Mater. Interfaces,* **2013**, *5*(3), 869-877. [http://dx.doi.org/10.1021/am302494c] [PMID: 23286332]

[155] Xu, Z.; Wang, S.; Li, Y.; Wang, M.; Shi, P.; Huang, X. Covalent functionalization of graphene oxide with biocompatible poly(ethylene glycol) for delivery of paclitaxel. *ACS Appl. Mater. Interfaces,* **2014**, *6*(19), 17268-17276. [http://dx.doi.org/10.1021/am505308f] [PMID: 25216036]

[156] Roy, S.; Tang, X.; Das, T.; Zhang, L.; Li, Y.; Ting, S.; Hu, X.; Yue, C.Y. Enhanced molecular level dispersion and interface bonding at low loading of modified graphene oxide to fabricate super nylon 12 composites. *ACS Appl. Mater. Interfaces,* **2015**, *7*(5), 3142-3151. [http://dx.doi.org/10.1021/am5074408] [PMID: 25545112]

[157] Yang, Q.; Pan, X.; Clarke, K.; Li, K. Covalent functionalization of graphene with polysaccharides. *Ind. Eng. Chem. Res.,* **2012**, *51*(1), 310-317. [http://dx.doi.org/10.1021/ie201391e]

[158] Bekyarova, E.; Itkis, M.E.; Ramesh, P.; Berger, C.; Sprinkle, M.; de Heer, W.A.; Haddon, R.C. Chemical modification of epitaxial graphene: spontaneous grafting of aryl groups. *J. Am. Chem. Soc.,* **2009**, *131*(4), 1336-1337. [http://dx.doi.org/10.1021/ja8057327] [PMID: 19173656]

[159] Lomeda, J.R.; Doyle, C.D.; Kosynkin, D.V.; Hwang, W.F.; Tour, J.M. Diazonium functionalization of surfactant-wrapped chemically converted graphene sheets. *J. Am. Chem. Soc.,* **2008**, *130*(48), 16201-16206. [http://dx.doi.org/10.1021/ja806499w] [PMID: 18998637]

[160] Zhu, Y.; Higginbotham, A.L.; Tour, J.M. Covalent functionalization of surfactant-wrapped graphene nanoribbons. *Chem. Mater.,* **2009**, *21*(21), 5284-5291.

[http://dx.doi.org/10.1021/cm902939n]

[161] Sun, Z.; Kohama, S.; Zhang, Z.; Lomeda, J.R.; Tour, J.M. Soluble graphene through edge-selective functionalization. *Nano Res.,* **2010**, *3*(2), 117-125. [http://dx.doi.org/10.1007/s12274-010-1016-2]

[162] Avinash, M.B.; Subrahmanyam, K.S.; Sundarayya, Y.; Govindaraju, T. Covalent modification and exfoliation of graphene oxide using ferrocene. *Nanoscale,* **2010**, *2*(9), 1762-1766. [http://dx.doi.org/10.1039/c0nr00024h] [PMID: 20820707]

[163] Fang, M.; Wang, K.; Lu, H.; Yang, Y.; Nutt, S. Covalent polymer functionalization of graphene nanosheets and mechanical properties of composites. *J. Mater. Chem.,* **2009**, *19*(38), 7098-7105. [http://dx.doi.org/10.1039/b908220d]

[164] Chua, C.K.; Pumera, M. Covalent chemistry on graphene. *Chem. Soc. Rev.,* **2013**, *42*(8), 3222-3233. [http://dx.doi.org/10.1039/c2cs35474h] [PMID: 23403465]

[165] Yuan, C.; Chen, W.; Yan, L. Amino-grafted graphene as a stable and metal-free solid basic catalyst. *J. Mater. Chem.,* **2012**, *22*(15), 7456-7460. [http://dx.doi.org/10.1039/c2jm30442b]

[166] Pham, V.H.; Cuong, T.V.; Hur, S.H.; Oh, E.; Kim, E.J.; Shin, E.W.; Chung, J.S. Chemical functionalization of graphene sheets by solvothermal reduction of a graphene oxide suspension in N-methyl-2-pyrrolidone. *J. Mater. Chem.,* **2011**, *21*(10), 3371-3377. [http://dx.doi.org/10.1039/C0JM02790A]

[167] Xu, Y.; Liu, Z.; Zhang, X.; Wang, Y.; Tian, J.; Huang, Y.; Ma, Y.; Zhang, X.; Chen, Y. A Graphene hybrid material covalently functionalized with porphyrin: Synthesis and optical limiting property. *Adv. Mater.,* **2009**, *21*(12), 1275-1279. [http://dx.doi.org/10.1002/adma.200801617]

[168] Liu, Z.; Robinson, J.T.; Sun, X.; Dai, H. PEGylated nanographene oxide for delivery of water-insoluble cancer drugs. *J. Am. Chem. Soc.,* **2008**, *130*(33), 10876-10877. [http://dx.doi.org/10.1021/ja803688x] [PMID: 18661992]

[169] Shen, J.; Shi, M.; Yan, B.; Ma, H.; Li, N.; Hu, Y.; Ye, M. Covalent attaching protein to graphene oxide *via* diimide-activated amidation. *Colloids Surf. B Biointerfaces,* **2010**, *81*(2), 434-438. [http://dx.doi.org/10.1016/j.colsurfb.2010.07.035] [PMID: 20728319]

[170] Xue, Y.; Liu, Y.; Lu, F.; Qu, J.; Chen, H.; Dai, L. Functionalization of graphene oxide with polyhedral oligomeric silsesquioxane (POSS) for multifunctional applications. *J. Phys. Chem. Lett.,* **2012**, *3*(12), 1607-1612. [http://dx.doi.org/10.1021/jz3005877] [PMID: 26285716]

[171] Stankovich, S.; Piner, R.D.; Nguyen, S.T.; Ruoff, R.S. Synthesis and exfoliation of isocyanate-treated graphene oxide nanoplatelets. *Carbon,* **2006**, *44*(15), 3342-3347. [http://dx.doi.org/10.1016/j.carbon.2006.06.004]

[172] Yoonessi, M.; Shi, Y.; Scheiman, D.A.; Lebron-Colon, M.; Tigelaar, D.M.; Weiss, R.A.; Meador, M.A. Graphene polyimide nanocomposites; thermal, mechanical, and high-temperature shape memory effects. *ACS Nano,* **2012**, *6*(9), 7644-7655. [http://dx.doi.org/10.1021/nn302871y] [PMID: 22931435]

[173] Quintana, M.; Spyrou, K.; Grzelczak, M.; Browne, W.R.; Rudolf, P.; Prato, M. Functionalization of graphene *via* 1,3-dipolar cycloaddition. *ACS Nano,* **2010**, *4*(6), 3527-3533. [http://dx.doi.org/10.1021/nn100883p] [PMID: 20503982]

[174] Wang, A.; Yu, W.; Xiao, Z.; Song, Y.; Long, L.; Cifuentes, M.P.; Humphrey, M.G.; Zhang, C. A 1,3-dipolar cycloaddition protocol to porphyrin-functionalized reduced graphene oxide with a push-pull motif. *Nano Res.,* **2015**, *8*(3), 870-886. [http://dx.doi.org/10.1007/s12274-014-0569-x]

[175] Zhong, X.; Jin, J.; Li, S.; Niu, Z.; Hu, W.; Li, R.; Ma, J. Aryne cycloaddition: highly efficient chemical

modification of graphene. *Chem. Commun. (Camb.),* **2010**, *46*(39), 7340-7342. [http://dx.doi.org/10.1039/c0cc02389b] [PMID: 20820532]

[176] Seo, J.M.; Baek, J.B. A solvent-free Diels–Alder reaction of graphite into functionalized graphene nanosheets. *Chem. Commun. (Camb.),* **2014**, *50*(93), 14651-14653. [http://dx.doi.org/10.1039/C4CC07173E] [PMID: 25314623]

[177] Sarkar, S.; Bekyarova, E.; Niyogi, S.; Haddon, R.C. Diels-Alder chemistry of graphite and graphene: graphene as diene and dienophile. *J. Am. Chem. Soc.,* **2011**, *133*(10), 3324-3327. [http://dx.doi.org/10.1021/ja200118b] [PMID: 21341649]

[178] He, H.; Gao, C. General approach to individually dispersed, highly soluble, and conductive graphene nanosheets functionalized by nitrene chemistry. *Chem. Mater.,* **2010**, *22*(17), 5054-5064. [http://dx.doi.org/10.1021/cm101634k]

[179] Yang, X.; Li, J.; Wen, T.; Ren, X.; Huang, Y.; Wang, X. Adsorption of naphthalene and its derivatives on magnetic graphene composites and the mechanism investigation. *Colloids Surf. A Physicochem. Eng. Asp.,* **2013**, *422*, 118-125. [http://dx.doi.org/10.1016/j.colsurfa.2012.11.063]

[180] Kamada, S.; Nomoto, H.; Fukuda, K.; Fukawa, T.; Shirai, H.; Kimura, M. Noncovalent wrapping of chemically modified graphene with π-conjugated disk-like molecules. *Colloid Polym. Sci.,* **2011**, *289*(8), 925-932. [http://dx.doi.org/10.1007/s00396-011-2414-z]

[181] Qi, X.; Pu, K.Y.; Li, H.; Zhou, X.; Wu, S.; Fan, Q.L.; Liu, B.; Boey, F.; Huang, W.; Zhang, H. Amphiphilic graphene composites. *Angew. Chem. Int. Ed.,* **2010**, *49*(49), 9426-9429. [http://dx.doi.org/10.1002/anie.201004497] [PMID: 21031384]

[182] Choi, E.Y.; Han, T.H.; Hong, J.; Kim, J.E.; Lee, S.H.; Kim, H.W.; Kim, S.O. Noncovalent functionalization of graphene with end-functional polymers. *J. Mater. Chem.,* **2010**, *20*(10), 1907-1912. [http://dx.doi.org/10.1039/b919074k]

[183] Zhang, J.; Lei, J.; Pan, R.; Xue, Y.; Ju, H. Highly sensitive electrocatalytic biosensing of hypoxanthine based on functionalization of graphene sheets with water-soluble conducting graft copolymer. *Biosens. Bioelectron.,* **2010**, *26*(2), 371-376. [http://dx.doi.org/10.1016/j.bios.2010.07.127] [PMID: 20729055]

[184] Bai, H.; Xu, Y.; Zhao, L.; Li, C.; Shi, G. Non-covalent functionalization of graphene sheets by sulfonated polyaniline. *Chem. Commun. (Camb.),* **2009**, (13), 1667-1669. [http://dx.doi.org/10.1039/b821805f] [PMID: 19294256]

[185] Liu, H.; Gao, J.; Xue, M.; Zhu, N.; Zhang, M.; Cao, T. Processing of graphene for electrochemical application: noncovalently functionalize graphene sheets with water-soluble electroactive methylene green. *Langmuir,* **2009**, *25*(20), 12006-12010. [http://dx.doi.org/10.1021/la9029613] [PMID: 19764798]

[186] Geng, J.; Jung, H.T. Porphyrin functionalized graphene sheets in aqueous suspensions: From the preparation of graphene sheets to highly conductive graphene films. *J. Phys. Chem. C,* **2010**, *114*(18), 8227-8234. [http://dx.doi.org/10.1021/jp1008779]

[187] Ghosh, A.; Rao, K.V.; George, S.J.; Rao, C.N.R. Noncovalent functionalization, exfoliation, and solubilization of graphene in water by employing a fluorescent coronene carboxylate. *Chemistry,* **2010**, *16*(9), 2700-2704. [http://dx.doi.org/10.1002/chem.200902828] [PMID: 20108284]

[188] Jana, M.; Saha, S.; Khanra, P.; Samanta, P.; Koo, H.; Chandra Murmu, N.; Kuila, T. Non-covalent functionalization of reduced graphene oxide using sulfanilic acid azocromotrop and its application as a supercapacitor electrode material. *J. Mater. Chem. A Mater. Energy Sustain.,* **2015**, *3*(14), 7323-7331. [http://dx.doi.org/10.1039/C4TA07009G]

[189] Komeilizadeh, H. Does Nature prefer heterocycles? *Iran. J. Pharm. Sci.,* **2006**, *4*, 229-230.

[190] Busto, E.; Gotor-Fernández, V.; Gotor, V. Hydrolases in the stereoselective synthesis of N-heterocyclic amines and amino acid derivatives. *Chem. Rev.,* **2011**, *111*(7), 3998-4035. [http://dx.doi.org/10.1021/cr100287w] [PMID: 21526748]

[191] Lawrence, S.A. Amines: Synthesis, properties and applications. *J. Am. Chem. Soc.,* **2005**, *127*, 9655-9658.

[192] Cheng, K.W.; Chen, F.; Wang, M. Heterocyclic amines: Chemistry and health. *Mol. Nutr. Food Res.,* **2006**, *50*(12), 1150-1170. [http://dx.doi.org/10.1002/mnfr.200600086] [PMID: 17131456]

[193] Liu, P.; Zhang, X.; Feng, L.; Xiong, H.; Wang, S. Direct electrochemistry of hemoglobin on graphene nanosheet-based modified electrode and its electrocatalysis to nitrite. *Am. J. Biomed. Sci.,* **2011**, *3*, 70-77. [http://dx.doi.org/10.5099/aj110100070]

[194] Chen, S.Y.; Lu, Y.Y.; Shih, F.Y.; Ho, P.H.; Chen, Y.F.; Chen, C.W.; Chen, Y.T.; Wang, W.H. Biologically inspired graphene-chlorophyll phototransistors with high gain. *Carbon,* **2013**, *63*, 23-29. [http://dx.doi.org/10.1016/j.carbon.2013.06.031]

[195] Wu, T.Y.; Guo, N.; Teh, C.Y.; Hay, J.X.W. *Advances in Ultrasound Technology for Environmental Remediation*; Springer: Dordrecht, Heidelberg, New York, London, **2013**, pp. 5-12. [http://dx.doi.org/10.1007/978-94-007-5533-8_2]

[196] Bang, J.H.; Suslick, K.S. Applications of ultrasound to the synthesis of nanostructured materials. *Adv. Mater.,* **2010**, *22*(10), 1039-1059. [http://dx.doi.org/10.1002/adma.200904093] [PMID: 20401929]

[197] Griffing, V. Ultrasonic chemical effects. *Ind. Eng. Chem.,* **1955**, *47*(6), 1180. [http://dx.doi.org/10.1021/ie50546a025]

[198] Yang, H.; Li, H.; Zhai, J.; Sun, L.; Yu, H. Simple synthesis of graphene oxide using ultrasonic cleaner from expanded graphite. *Ind. Eng. Chem. Res.,* **2014**, *53*(46), 17878-17883. [http://dx.doi.org/10.1021/ie503586v]

[199] de la Hoz, A. *Microwaves in Organic Synthesis, Wiley-VCH Verlag GmbH & Co. KGaA.,* **2012**.

[200] Gawande, M.B.; Shelke, S.N.; Zboril, R.; Varma, R.S. Microwave-assisted chemistry: synthetic applications for rapid assembly of nanomaterials and organics. *Acc. Chem. Res.,* **2014**, *47*(4), 1338-1348. [http://dx.doi.org/10.1021/ar400309b] [PMID: 24666323]

[201] Zuo, P.P.; Feng, H.F.; Xu, Z.Z.; Zhang, L.F.; Zhang, Y.L.; Xia, W.; Zhang, W.Q. Fabrication of biocompatible and mechanically reinforced graphene oxide-chitosan nanocomposite films. *Chem. Cent. J.,* **2013**, *7*(1), 39. [http://dx.doi.org/10.1186/1752-153X-7-39] [PMID: 23442350]

[202] Yang, Z.; Zheng, Q.; Qiu, H.; Li, J.; Yang, J. A simple method for the reduction of graphene oxide by sodium borohydride with CaCl2 as a catalyst. *N. Carbon Mater.,* **2015**, *30*(1), 41-47. [http://dx.doi.org/10.1016/S1872-5805(15)60174-3]

[203] Chua, C.K.; Pumera, M. Reduction of graphene oxide with substituted borohydrides. *J. Mater. Chem. A Mater. Energy Sustain.,* **2013**, *1*(5), 1892-1898. [http://dx.doi.org/10.1039/C2TA00665K]

[204] Ambrosi, A.; Chua, C.K.; Bonanni, A.; Pumera, M. Lithium aluminum hydride as reducing agent for chemically reduced graphene oxides. *Chem. Mater.,* **2012**, *24*(12), 2292-2298. [http://dx.doi.org/10.1021/cm300382b]

[205] Fan, X.; Peng, W.; Li, Y.; Li, X.; Wang, S.; Zhang, G.; Zhang, F. Deoxygenation of exfoliated graphite oxide under alkaline conditions: A green route to graphene preparation. *Adv. Mater.,* **2008**,

20(23), 4490-4493.
[http://dx.doi.org/10.1002/adma.200801306]

[206] Cataldo, F.; Ursini, O.; Angelini, G. Graphite oxide and graphene nanoribbons reduction with hydrogen iodide. *Fuller. Nanotub. Carbon Nanostruct.*, **2011**, *19*(5), 461-468.
[http://dx.doi.org/10.1080/1536383X.2010.481064]

[207] Domingues, S.H.; Kholmanov, I.N.; Kim, T.; Kim, J.; Tan, C.; Chou, H.; Alieva, Z.A.; Piner, R.; Zarbin, A.J.G.; Ruoff, R.S. Reduction of graphene oxide films on Al foil for hybrid transparent conductive film applications. *Carbon,* **2013**, *63*, 454-459.
[http://dx.doi.org/10.1016/j.carbon.2013.07.007]

[208] Dreyer, D.R.; Murali, S.; Zhu, Y.; Ruoff, R.S.; Bielawski, C.W. Reduction of graphite oxide using alcohols. *J. Mater. Chem.*, **2011**, *21*(10), 3443-3447.
[http://dx.doi.org/10.1039/C0JM02704A]

[209] Chua, C.K.; Ambrosi, A.; Pumera, M. Graphene oxide reduction by standard industrial reducing agent: thiourea dioxide. *J. Mater. Chem.*, **2012**, *22*(22), 11054-11061.
[http://dx.doi.org/10.1039/c2jm16054d]

[210] Ma, H.L.; Zhang, H.B.; Hu, Q.H.; Li, W.J.; Jiang, Z.G.; Yu, Z.Z.; Dasari, A. Functionalization and reduction of graphene oxide with p-phenylene diamine for electrically conductive and thermally stable polystyrene composites. *ACS Appl. Mater. Interfaces,* **2012**, *4*(4), 1948-1953.
[http://dx.doi.org/10.1021/am201654b] [PMID: 22422027]

[211] Zhang, J.; Yang, H.; Shen, G.; Cheng, P.; Zhang, J.; Guo, S. Reduction of graphene oxide *via* l-ascorbic acid. *Chem. Commun. (Camb.),* **2010**, *46*(7), 1112-1114.
[http://dx.doi.org/10.1039/B917705A] [PMID: 20126730]

[212] Pham, V.H.; Pham, H.D.; Dang, T.T.; Hur, S.H.; Kim, E.J.; Kong, B.S.; Kim, S.; Chung, J.S. Chemical reduction of an aqueous suspension of graphene oxide by nascent hydrogen. *J. Mater. Chem.*, **2012**, *22*(21), 10530-10536.
[http://dx.doi.org/10.1039/c2jm30562c]

[213] Fan, Z.J.; Kai, W.; Yan, J.; Wei, T.; Zhi, L.J.; Feng, J.; Ren, Y.; Song, L.P.; Wei, F. Facile synthesis of graphene nanosheets *via* Fe reduction of exfoliated graphite oxide. *ACS Nano,* **2011**, *5*(1), 191-198.
[http://dx.doi.org/10.1021/nn102339t] [PMID: 21230006]

[214] Salas, E.C.; Sun, Z.; Lüttge, A.; Tour, J.M. Reduction of graphene oxide *via* bacterial respiration. *ACS Nano,* **2010**, *4*(8), 4852-4856.
[http://dx.doi.org/10.1021/nn101081t] [PMID: 20731460]

[215] Amarnath, C.A.; Hong, C.E.; Kim, N.H.; Ku, B.C.; Kuila, T.; Lee, J.H. Efficient synthesis of graphene sheets using pyrrole as a reducing agent. *Carbon,* **2011**, *49*(11), 3497-3502.
[http://dx.doi.org/10.1016/j.carbon.2011.04.048]

[216] Georgakilas, V.; Perman, J.A.; Tucek, J.; Zboril, R. Broad family of carbon nanoallotropes: classification, chemistry, and applications of fullerenes, carbon dots, nanotubes, graphene, nanodiamonds, and combined superstructures. *Chem. Rev.*, **2015**, *115*(11), 4744-4822.
[http://dx.doi.org/10.1021/cr500304f] [PMID: 26012488]

[217] Yin, P.T.; Shah, S.; Chhowalla, M.; Lee, K.B. Design, synthesis, and characterization of graphene-nanoparticle hybrid materials for bioapplications. *Chem. Rev.*, **2015**, *115*(7), 2483-2531.
[http://dx.doi.org/10.1021/cr500537t] [PMID: 25692385]

[218] Nguyen, P.; Berry, V. Graphene interfaced with biological cells: Opportunities and challenges. *J. Phys. Chem. Lett.*, **2012**, *3*(8), 1024-1029.
[http://dx.doi.org/10.1021/jz300033g] [PMID: 26286566]

[219] Maji, S.K.; Mandal, A.K.; Nguyen, K.T.; Borah, P.; Zhao, Y. Cancer cell detection and therapeutics using peroxidase-active nanohybrid of gold nanoparticle-loaded mesoporous silica-coated graphene. *ACS Appl. Mater. Interfaces,* **2015**, *7*(18), 9807-9816.

[http://dx.doi.org/10.1021/acsami.5b01758] [PMID: 25909624]

[220] Liu, X.; Ma, D.; Tang, H.; Tan, L.; Xie, Q.; Zhang, Y.; Ma, M.; Yao, S. Polyamidoamine dendrimer and oleic acid-functionalized graphene as biocompatible and efficient gene delivery vectors. *ACS Appl. Mater. Interfaces,* **2014**, *6*(11), 8173-8183.
[http://dx.doi.org/10.1021/am500812h] [PMID: 24836601]

[221] Paul, A.; Hasan, A.; Kindi, H.A.; Gaharwar, A.K.; Rao, V.T.S.; Nikkhah, M.; Shin, S.R.; Krafft, D.; Dokmeci, M.R.; Shum-Tim, D.; Khademhosseini, A. Injectable graphene oxide/hydrogel-based angiogenic gene delivery system for vasculogenesis and cardiac repair. *ACS Nano,* **2014**, *8*(8), 8050-8062.
[http://dx.doi.org/10.1021/nn5020787] [PMID: 24988275]

[222] de Faria, A.F.; Perreault, F.; Shaulsky, E.; Arias Chavez, L.H.; Elimelech, M. Antimicrobial electrospun biopolymer nanofiber mats functionalized with graphene oxide-silver nanocomposites. *ACS Appl. Mater. Interfaces,* **2015**, *7*(23), 12751-12759.
[http://dx.doi.org/10.1021/acsami.5b01639] [PMID: 25980639]

[223] Perreault, F.; de Faria, A.F.; Nejati, S.; Elimelech, M. Antimicrobial properties of graphene oxide nanosheets: Why size matters. *ACS Nano,* **2015**, *9*(7), 7226-7236.
[http://dx.doi.org/10.1021/acsnano.5b02067] [PMID: 26091689]

[224] Zhan, S.; Zhu, D.; Ma, S.; Yu, W.; Jia, Y.; Li, Y.; Yu, H.; Shen, Z. Highly efficient removal of pathogenic bacteria with magnetic graphene composite. *ACS Appl. Mater. Interfaces,* **2015**, *7*(7), 4290-4298.
[http://dx.doi.org/10.1021/am508682s] [PMID: 25634911]

[225] Musico, Y.L.F.; Santos, C.M.; Dalida, M.L.P.; Rodrigues, D.F. Surface modification of membrane filters using graphene and graphene oxide-based nanomaterials for bacterial inactivation and removal. *ACS Sustain. Chem.& Eng.,* **2014**, *2*(7), 1559-1565.
[http://dx.doi.org/10.1021/sc500044p]

[226] Chung, D.D.L. Electrical applications of carbon materials. *J. Mater. Sci.,* **2004**, *39*(8), 2645-2661.
[http://dx.doi.org/10.1023/B:JMSC.0000021439.18202.ea]

[227] Xu, Y.; Shi, G.; Duan, X. Self-assembled three-dimensional graphene macrostructures: Synthesis and applications in supercapacitors. *Acc. Chem. Res.,* **2015**, *48*(6), 1666-1675.
[http://dx.doi.org/10.1021/acs.accounts.5b00117] [PMID: 26042764]

[228] Lee, J.H.; Park, N.; Kim, B.G.; Jung, D.S.; Im, K.; Hur, J.; Choi, J.W. Restacking-inhibited 3D reduced graphene oxide for high performance supercapacitor electrodes. *ACS Nano,* **2013**, *7*(10), 9366-9374.
[http://dx.doi.org/10.1021/nn4040734] [PMID: 24001136]

[229] Guo, Y.; Guo, S.; Ren, J.; Zhai, Y.; Dong, S.; Wang, E. Cyclodextrin functionalized graphene nanosheets with high supramolecular recognition capability: synthesis and host-guest inclusion for enhanced electrochemical performance. *ACS Nano,* **2010**, *4*(7), 4001-4010.
[http://dx.doi.org/10.1021/nn100939n] [PMID: 20583782]

[230] Wu, W.; Li, Y.; Yang, L.; Ma, Y.; Pan, D.; Li, Y. A facile one-pot preparation of dialdehyde starch reduced graphene oxide/polyaniline composite for supercapacitors. *Electrochim. Acta,* **2014**, *139*, 117-126.
[http://dx.doi.org/10.1016/j.electacta.2014.06.166]

[231] Bhattacharjya, D.; Jeon, I.Y.; Park, H.Y.; Panja, T.; Baek, J.B.; Yu, J.S. Graphene nanoplatelets with selectively functionalized edges as electrode material for electrochemical energy storage. *Langmuir,* **2015**, *31*(20), 5676-5683.
[http://dx.doi.org/10.1021/acs.langmuir.5b00195] [PMID: 25942431]

[232] Goh, B.M.; Wang, Y.; Reddy, M.V.; Ding, Y.L.; Lu, L.; Bunker, C.; Loh, K.P. Filling the voids of graphene foam with graphene "eggshell" for improved lithium-ion storage. *ACS Appl. Mater. Interfaces,* **2014**, *6*(12), 9835-9841.

[http://dx.doi.org/10.1021/am5022655] [PMID: 24857682]

[233] Guldi, D.M.; Sgobba, V. Carbon nanostructures for solar energy conversion schemes. *Chem. Commun. (Camb.),* **2011**, *47*(2), 606-610.
[http://dx.doi.org/10.1039/C0CC02411B] [PMID: 20871887]

[234] Qiu, S.; Hu, W.; Yu, B.; Yuan, B.; Zhu, Y.; Jiang, S.; Wang, B.; Song, L.; Hu, Y. Effect of functionalized graphene oxide with organophosphorus oligomer on the thermal and mechanical properties and fire safety of polystyrene. *Ind. Eng. Chem. Res.,* **2015**, *54*(13), 3309-3319.
[http://dx.doi.org/10.1021/ie504511f]

[235] Lim, J.; Yeo, H.; Goh, M.; Ku, B.C.; Kim, S.G.; Lee, H.S.; Park, B.; You, N.H. Grafting of polyimide onto chemically-functionalized graphene nanosheets for mechanically-strong barrier membranes. *Chem. Mater.,* **2015**, *27*(6), 2040-2047.
[http://dx.doi.org/10.1021/cm5044254]

[236] Sun, Z.; Chang, H. Graphene and graphene-like two-dimensional materials in photodetection: mechanisms and methodology. *ACS Nano,* **2014**, *8*(5), 4133-4156.
[http://dx.doi.org/10.1021/nn500508c] [PMID: 24716438]

[237] Gupta, V.; Chaudhary, N.; Srivastava, R.; Sharma, G.D.; Bhardwaj, R.; Chand, S. Luminscent graphene quantum dots for organic photovoltaic devices. *J. Am. Chem. Soc.,* **2011**, *133*(26), 9960-9963.
[http://dx.doi.org/10.1021/ja2036749] [PMID: 21650464]

[238] Balcioglu, M.; Rana, M.; Robertson, N.; Yigit, M.V. DNA-length-dependent quenching of fluorescently labeled iron oxide nanoparticles with gold, graphene oxide and MoS_2 nanostructures. *ACS Appl. Mater. Interfaces,* **2014**, *6*(15), 12100-12110.
[http://dx.doi.org/10.1021/am503553h] [PMID: 25014711]

[239] Dinda, D.; Shaw, B.K.; Saha, S.K. Thymine functionalized graphene oxide for fluorescence "turn-of--on" sensing of Hg^{2+} and I^- in aqueous medium. *ACS Appl. Mater. Interfaces,* **2015**, *7*(27), 14743-14749.
[http://dx.doi.org/10.1021/acsami.5b02603] [PMID: 26094997]

[240] Cai, B.; Wang, S.; Huang, L.; Ning, Y.; Zhang, Z.; Zhang, G.J. Ultrasensitive label-free detection of PNA-DNA hybridization by reduced graphene oxide field-effect transistor biosensor. *ACS Nano,* **2014**, *8*(3), 2632-2638.
[http://dx.doi.org/10.1021/nn4063424] [PMID: 24528470]

[241] Alonso-Cristobal, P.; Vilela, P.; El-Sagheer, A.; Lopez-Cabarcos, E.; Brown, T.; Muskens, O.L.; Rubio-Retama, J.; Kanaras, A.G. Highly sensitive DNA sensor based on upconversion nanoparticles and graphene oxide. *ACS Appl. Mater. Interfaces,* **2015**, *7*(23), 12422-12429.
[http://dx.doi.org/10.1021/am507591u] [PMID: 25622622]

[242] Huang, K.J.; Niu, D.J.; Sun, J.Y.; Han, C.H.; Wu, Z.W.; Li, Y.L.; Xiong, X.Q. Novel electrochemical sensor based on functionalized graphene for simultaneous determination of adenine and guanine in DNA. *Colloids Surf. B Biointerfaces,* **2011**, *82*(2), 543-549.
[http://dx.doi.org/10.1016/j.colsurfb.2010.10.014] [PMID: 21050729]

[243] Zhan, L.; Zhang, Y.; Zeng, Q.L.; Liu, Z.D.; Huang, C.Z. Facile one-pot synthesis of folic acid-modified graphene to improve the performance of graphene-based sensing strategy. *J. Colloid Interface Sci.,* **2014**, *426*, 293-299.
[http://dx.doi.org/10.1016/j.jcis.2014.03.056] [PMID: 24863796]

[244] Kurkina, T.; Sundaram, S.; Sundaram, R.S.; Re, F.; Masserini, M.; Kern, K.; Balasubramanian, K. Self-assembled electrical biodetector based on reduced graphene oxide. *ACS Nano,* **2012**, *6*(6), 5514-5520.
[http://dx.doi.org/10.1021/nn301429k] [PMID: 22545858]

[245] Borini, S.; White, R.; Wei, D.; Astley, M.; Haque, S.; Spigone, E.; Harris, N.; Kivioja, J.; Ryhänen, T. Ultrafast graphene oxide humidity sensors. *ACS Nano,* **2013**, *7*(12), 11166-11173.

[http://dx.doi.org/10.1021/nn404889b] [PMID: 24206232]

[246] Kundu, A.; Nandi, S.; Das, P.; Nandi, A.K. Fluorescent graphene oxide *via* polymer grafting: an efficient nanocarrier for both hydrophilic and hydrophobic drugs. *ACS Appl. Mater. Interfaces,* **2015**, *7*(6), 3512-3523.
[http://dx.doi.org/10.1021/am507110r] [PMID: 25612470]

[247] Yao, K.; Tan, P.; Luo, Y.; Feng, L.; Xu, L.; Liu, Z.; Li, Y.; Peng, R. Graphene oxide selectively enhances thermostability of trypsin. *ACS Appl. Mater. Interfaces,* **2015**, *7*(22), 12270-12277.
[http://dx.doi.org/10.1021/acsami.5b03118] [PMID: 25985836]

[248] Wang, J.; Gao, X.; Wang, J.; Wei, Y.; Li, Z.; Gao, C. O-(carboxymethyl)-chitosan nanofiltration membrane surface functionalized with graphene oxide nanosheets for enhanced desalting properties. *ACS Appl. Mater. Interfaces,* **2015**, *7*(7), 4381-4389.
[http://dx.doi.org/10.1021/am508903g] [PMID: 25635511]

[249] Cote, L.J.; Kim, J.; Tung, V.C.; Luo, J.; Kim, F.; Huang, J. Graphene oxide as surfactant sheets. *Pure Appl. Chem.,* **2010**, *83*(1), 95-110.
[http://dx.doi.org/10.1351/PAC-CON-10-10-25]

[250] Xu, J.; Wang, L.; Zhu, Y. Decontamination of bisphenol A from aqueous solution by graphene adsorption. *Langmuir,* **2012**, *28*(22), 8418-8425.
[http://dx.doi.org/10.1021/la301476p] [PMID: 22571829]

[251] Dinda, D.; Gupta, A.; Saha, S.K. Removal of toxic Cr(vi) by UV-active functionalized graphene oxide for water purification. *J. Mater. Chem. A Mater. Energy Sustain.,* **2013**, *1*(37), 11221-11228.
[http://dx.doi.org/10.1039/c3ta12504a]

[252] Tan, M.; Liu, X.; Li, W.; Li, H. Enhancing sorption capacities for copper(II) and lead(II) under weakly acidic conditions by l-tryptophan-functionalized graphene oxide. *J. Chem. Eng. Data,* **2015**, *60*(5), 1469-1475.
[http://dx.doi.org/10.1021/acs.jced.5b00015]

[253] Li, P.; Xing, C.; Qu, S.; Li, B.; Shen, W. Carbon dioxide capturing by nitrogen-doping microporous carbon. *ACS Sustain. Chem.& Eng.,* **2015**, *3*(7), 1434-1442.
[http://dx.doi.org/10.1021/acssuschemeng.5b00291]

[254] Xu, H.; Qu, Z.; Zong, C.; Huang, W.; Quan, F.; Yan, N. MnO_x/graphene for the catalytic oxidation and adsorption of elemental mercury. *Environ. Sci. Technol.,* **2015**, *49*(11), 6823-6830.
[http://dx.doi.org/10.1021/es505978n] [PMID: 25922870]

[255] Whitener, K.E., Jr; Stine, R.; Robinson, J.T.; Sheehan, P.E. Jr.; Stine, R.; Robinson, J.T.; Sheehan, P.E. Graphene as electrophile: reactions of graphene fluoride. *J. Phys. Chem. C,* **2015**, *119*(19), 10507-10512.
[http://dx.doi.org/10.1021/acs.jpcc.5b02730]

[256] Mehetre, S.S.; Maktedar, S.S.; Singh, M. Understanding the mechanism of surface modification through enhanced thermal and electrochemical stabilities of N-doped graphene oxide. *Appl. Surf. Sci.,* **2016**, *366*, 514-522.
[http://dx.doi.org/10.1016/j.apsusc.2016.01.108]

[257] Maktedar, S.S.; Mehetre, S.S.; Avashthi, G.; Singh, M.; Jayaraman, T.; Jagannathan, M. Functionalized graphene oxide: an advanced material for the photocatalytic application. *Int. J. Phtochem.,* **2016**, *2*, 1-4.

[258] Maktedar, S.S.; Mehetre, S.S.; Avashthi, G.; Singh, M. In situ sonochemical reduction and direct functionalization of graphene oxide: A robust approach with thermal and biomedical applications. *Ultrason. Sonochem.,* **2017**, *34*, 67-77.
[http://dx.doi.org/10.1016/j.ultsonch.2016.05.015] [PMID: 27773294]

[259] Maktedar, S.S.; Avashthi, G.; Singh, M. Ultrasound assisted simultaneous reduction and direct functionalization of graphene oxide with thermal and cytotoxicity profile. *Ultrason. Sonochem.,* **2017**,

34, 856-864.
[http://dx.doi.org/10.1016/j.ultsonch.2016.07.016] [PMID: 27773313]

[260] Maktedar, S.S.; Avashthi, G.; Singh, M. Understanding the significance of O-doped graphene towards biomedical applications. *RSC Advances,* **2016**, *6*(115), 114264-114275. [http://dx.doi.org/10.1039/C6RA23416J]

[261] Marrella, A.; Giannoni, P.; Pulsoni, I.; Quarto, R.; Raiteri, R.; Scaglione, S. Topographical features of graphene-oxide-functionalized substrates modulate cancer and healthy cell adhesion based on the cell tissue of origin. *ACS Appl. Mater. Interfaces,* **2018**, *10*(49), 41978-41985. [http://dx.doi.org/10.1021/acsami.8b15036] [PMID: 30479135]

[262] Zhang, J.; Chen, L.; Chen, J.; Zhang, Q.; Feng, J. Stability, cellular uptake, and in vivo tracking of zwitterion modified graphene oxide as a drug carrier. *Langmuir,* **2019**, *35*(5), 1495-1502. [http://dx.doi.org/10.1021/acs.langmuir.8b01995] [PMID: 30089359]

[263] Zhao, R.; Lv, M.; Li, Y.; Sun, M.; Kong, W.; Wang, L.; Song, S.; Fan, C.; Jia, L.; Qiu, S.; Sun, Y.; Song, H.; Hao, R. Stable nanocomposite based on pegylated and silver nanoparticles loaded graphene oxide for long-term antibacterial activity. *ACS Appl. Mater. Interfaces,* **2017**, *9*(18), 15328-15341. [http://dx.doi.org/10.1021/acsami.7b03987] [PMID: 28422486]

[264] Szunerits, S.; Boukherroub, R. Antibacterial activity of graphene-based materials. *J. Mater. Chem. B Mater. Biol. Med.,* **2016**, *4*(43), 6892-6912. [http://dx.doi.org/10.1039/C6TB01647B] [PMID: 32263558]

[265] Zhang, L.; Zhou, Q.; Song, W.; Wu, K.; Zhang, Y.; Zhao, Y. Dual-functionalized graphene oxide based sirna delivery system for implant surface biomodification with enhanced osteogenesis. *ACS Appl. Mater. Interfaces,* **2017**, *9*(40), 34722-34735. [http://dx.doi.org/10.1021/acsami.7b12079] [PMID: 28925678]

[266] Sun, J.; Deng, Y.; Li, J.; Wang, G.; He, P.; Tian, S.; Bu, X.; Di, Z.; Yang, S.; Ding, G.; Xie, X. A new graphene derivative: hydroxylated graphene with excellent biocompatibility. *ACS Appl. Mater. Interfaces,* **2016**, *8*(16), 10226-10233. [http://dx.doi.org/10.1021/acsami.6b02032] [PMID: 27052945]

[267] Wu, X.; Ding, S.J.; Lin, K.; Su, J. A review on the biocompatibility and potential applications of graphene in inducing cell differentiation and tissue regeneration. *J. Mater. Chem. B Mater. Biol. Med.,* **2017**, *5*(17), 3084-3102. [http://dx.doi.org/10.1039/C6TB03067J] [PMID: 32263706]

[268] Upadhyay, R.; Naskar, S.; Bhaskar, N.; Bose, S.; Basu, B. Modulation of protein adsorption and cell proliferation on polyethylene immobilized graphene oxide reinforced HDPE bionanocomposites. *ACS Appl. Mater. Interfaces,* **2016**, *8*(19), 11954-11968. [http://dx.doi.org/10.1021/acsami.6b00946] [PMID: 27108739]

[269] Akhavan, O. Graphene scaffolds in progressive nanotechnology/stem cell-based tissue engineering of the nervous system. *J. Mater. Chem. B Mater. Biol. Med.,* **2016**, *4*(19), 3169-3190. [http://dx.doi.org/10.1039/C6TB00152A] [PMID: 32263253]

[270] Dai, C.; Zhang, S.; Liu, Z.; Wu, R.; Chen, Y. Two-dimensional graphene augments nanosonosensitized sonocatalytic tumor eradication. *ACS Nano,* **2017**, *11*(9), 9467-9480. [http://dx.doi.org/10.1021/acsnano.7b05215] [PMID: 28829584]

[271] Liang, H.F.; Smith, C.T.G.; Mills, C.A.; Silva, S.R.P. The band structure of graphene oxide examined using photoluminescence spectroscopy. *J. Mater. Chem. C Mater. Opt. Electron. Devices,* **2015**, *3*(48), 12484-12491. [http://dx.doi.org/10.1039/C5TC00307E]

[272] Du, D.; Song, H.; Nie, Y.; Sun, X.; Chen, L.; Ouyang, J. Photoluminescence of graphene oxide in visible range arising from excimer formation. *J. Phys. Chem. C,* **2015**, *119*(34), 20085-20090. [http://dx.doi.org/10.1021/acs.jpcc.5b04529]

[273] Chiu, N.F.; Yang, C.D. Real-time and stepwise deoxidization processes to tune the photoluminescence properties of graphene oxide using EC-SPR spectroscopy. *RSC Advances,* **2018**, *8*(21), 11557-11565. [http://dx.doi.org/10.1039/C7RA13594G] [PMID: 35542802]

[274] Li, S.S.; Tu, K.H.; Lin, C.C.; Chen, C.W.; Chhowalla, M. Solution-processable graphene oxide as an efficient hole transport layer in polymer solar cells. *ACS Nano,* **2010**, *4*(6), 3169-3174. [http://dx.doi.org/10.1021/nn100551j] [PMID: 20481512]

[275] Bonaccorso, F.; Sun, Z.; Hasan, T.; Ferrari, A.C. Graphene photonics and optoelectronics. *Nat. Photonics,* **2010**, *4*(9), 611-622. [http://dx.doi.org/10.1038/nphoton.2010.186]

[276] Loh, K.P.; Bao, Q.; Eda, G.; Chhowalla, M. Graphene oxide as a chemically tunable platform for optical applications. *Nat. Chem.,* **2010**, *2*(12), 1015-1024. [http://dx.doi.org/10.1038/nchem.907] [PMID: 21107364]

[277] Eda, G.; Lin, Y.Y.; Mattevi, C.; Yamaguchi, H.; Chen, H.A.; Chen, I.S.; Chen, C.W.; Chhowalla, M. Blue photoluminescence from chemically derived graphene oxide. *Adv. Mater.,* **2010**, *22*(4), 505-509. [http://dx.doi.org/10.1002/adma.200901996] [PMID: 20217743]

[278] Chien, C.T.; Li, S.S.; Lai, W.J.; Yeh, Y.C.; Chen, H.A.; Chen, I.S.; Chen, L.C.; Chen, K.H.; Nemoto, T.; Isoda, S.; Chen, M.; Fujita, T.; Eda, G.; Yamaguchi, H.; Chhowalla, M.; Chen, C.W. Tunable photoluminescence from graphene oxide. *Angew. Chem. Int. Ed.,* **2012**, *51*(27), 6662-6666. [http://dx.doi.org/10.1002/anie.201200474] [PMID: 22623281]

[279] Kochmann, S.; Hirsch, T.; Wolfbeis, O.S. The pH dependence of the total fluorescence of graphite oxide. *J. Fluoresc.,* **2012**, *22*(3), 849-855. [http://dx.doi.org/10.1007/s10895-011-1019-8] [PMID: 22143963]

[280] Li, M.; Cushing, S.K.; Zhou, X.; Guo, S.; Wu, N. Fingerprinting photoluminescence of functional groups in graphene oxide. *J. Mater. Chem.,* **2012**, *22*(44), 23374-23379. [http://dx.doi.org/10.1039/c2jm35417a]

[281] Cushing, S.K.; Li, M.; Huang, F.; Wu, N. Origin of strong excitation wavelength dependent fluorescence of graphene oxide. *ACS Nano,* **2014**, *8*(1), 1002-1013. [http://dx.doi.org/10.1021/nn405843d] [PMID: 24359152]

[282] Demchenko, A.P. The red-edge effects: 30 years of exploration. *Luminescence,* **2002**, *17*(1), 19-42. [http://dx.doi.org/10.1002/bio.671] [PMID: 11816059]

[283] Enoki, T.; Takai, K. The edge state of nanographene and the magnetism of the edge-state spins. *Solid State Commun.,* **2009**, *149*(27-28), 1144-1150. [http://dx.doi.org/10.1016/j.ssc.2009.02.054]

[284] Diamantopoulou, A.; Glenis, S.; Zolnierkiwicz, G.; Guskos, N., Likodimos, V. *Magnetism in pristine and chemically reduced graphene oxide.,* **2017**, 1-11.

[285] Tang, T.; Liu, F.; Liu, Y.; Li, X.; Xu, Q.; Feng, Q.; Tang, N.; Du, Y. Identifying the magnetic properties of graphene oxide. *Identifying the magnetic properties of graphene oxide.,* **2014**, *104*, 1-5. [http://dx.doi.org/10.1063/1.4869827]

[286] Miao, Q.; Wang, L.; Liu, Z.; Wei, B.; Xu, F.; Fei, W. Magnetic properties of N-doped graphene with high Curie temperature.*21832*, 1-10.**2016**,

[287] Sahoo, P.K.; Panigrahy, B.; Li, D.; Bahadur, D. *Magnetic properties of N-doped graphene with high Curie temperature.,* **2013**, *113*, 1-3.

[288] Zhang, H.; Zhao, M.; Yang, X.; Xia, H.; Liu, X.; Xia, Y. Diffusion and coalescence of vacancies and interstitials in graphite: A first-principles study. *Diamond Related Materials,* **2010**, *19*(10), 1240-1244. [http://dx.doi.org/10.1016/j.diamond.2010.06.010]

[289] Sahoo, R.K.; Jeyapandiarajan, P.; Devi Chandrasekhar, K.; Daniel, B.S.S.; Venimadhav, A.; Sant, S.B.; Jacob, C. Single-step synthesis of graphene-carbon nanofiber hybrid material and its synergistic

magnetic behaviour. *J. Alloys Compd.*, **2014**, *615*, 348-354.
[http://dx.doi.org/10.1016/j.jallcom.2014.06.171]

[290] Mansour, M.; Bechelany, M.; Habchi, R.; Eid, C. Influence of graphene oxide doping on the morphology and the magnetic properties of $Ni_{0.8}$ $Gd_{0.2}$ Fe_2 O_4 nanofibers prepared by electrospinning. *Phys. Lett. A,* **2017**, *381*(6), 658-662.
[http://dx.doi.org/10.1016/j.physleta.2016.12.014]

[291] Bahadur, J.; Pal, K. Structural and magnetic properties of reduced graphene oxide-TiO_2 nanoflower composite. *Physica E,* **2017**, *90*, 98-103.
[http://dx.doi.org/10.1016/j.physe.2017.02.022]

[292] Frumkin, A.; Polianovskaya, N.; Bagotskaya, I.; Grigoryev, N. Electrocatalysis and electrode surface properties. *J. Electroanal. Chem. Interfacial Electrochem.*, **1971**, *33*(2), 319-328.
[http://dx.doi.org/10.1016/S0022-0728(71)80120-1]

[293] Azzam, A.M.; Bockris, J.O.M. O'M. Cathodic evolution of hydrogen at very high current densities. *Nature,* **1950**, *165*(4193), 403-404.
[http://dx.doi.org/10.1038/165403b0]

[294] Xia, B.; Yan, Y.; Wang, X.; Lou, X.W.D. Recent progress on graphene-based hybrid electrocatalysts. *Mater. Horiz.*, **2014**, *1*(4), 379-399.
[http://dx.doi.org/10.1039/C4MH00040D]

[295] Duan, J.; Chen, S.; Jaroniec, M.; Qiao, S.Z. Heteroatom-doped graphene-based materials for energy-relevant electrocatalytic processes. *ACS Catal.*, **2015**, *5*(9), 5207-5234.
[http://dx.doi.org/10.1021/acscatal.5b00991]

[296] Debe, M.K. Electrocatalyst approaches and challenges for automotive fuel cells. *Nature,* **2012**, *486*(7401), 43-51.
[http://dx.doi.org/10.1038/nature11115] [PMID: 22678278]

[297] Armand, M.; Tarascon, J.M. Building better batteries. *Nature,* **2008**, *451*(7179), 652-657.
[http://dx.doi.org/10.1038/451652a] [PMID: 18256660]

[298] Gray, H.B. Powering the planet with solar fuel. *Nat. Chem.,* **2009**, *1*(1), 7.
[http://dx.doi.org/10.1038/nchem.141] [PMID: 21378780]

[299] Li, Y.; Dai, H. Recent advances in zinc–air batteries. *Chem. Soc. Rev.*, **2014**, *43*(15), 5257-5275.
[http://dx.doi.org/10.1039/C4CS00015C] [PMID: 24926965]

[300] Qiao, J.; Liu, Y.; Hong, F.; Zhang, J. A review of catalysts for the electroreduction of carbon dioxide to produce low-carbon fuels. *Chem. Soc. Rev.*, **2014**, *43*(2), 631-675.
[http://dx.doi.org/10.1039/C3CS60323G] [PMID: 24186433]

[301] Liang, Y.; Li, Y.; Wang, H.; Zhou, J.; Wang, J.; Regier, T.; Dai, H. Co_3O_4 nanocrystals on graphene as a synergistic catalyst for oxygen reduction reaction. *Nat. Mater.*, **2011**, *10*(10), 780-786.
[http://dx.doi.org/10.1038/nmat3087] [PMID: 21822263]

[302] Mirzakulova, E.; Khatmullin, R.; Walpita, J.; Corrigan, T.; Vargas-Barbosa, N.M.; Vyas, S.; Oottikkal, S.; Manzer, S.F.; Hadad, C.M.; Glusac, K.D. Electrode-assisted catalytic water oxidation by a flavin derivative. *Nat. Chem.,* **2012**, *4*(10), 794-801.
[http://dx.doi.org/10.1038/nchem.1439] [PMID: 23000992]

[303] Suntivich, J.; May, K.J.; Gasteiger, H.A.; Goodenough, J.B.; Shao-Horn, Y. A perovskite oxide optimized for oxygen evolution catalysis from molecular orbital principles. *Science,* **2011**, *334*(6061), 1383-1385.
[http://dx.doi.org/10.1126/science.1212858] [PMID: 22033519]

[304] Yin, Q.; Tan, J.M.; Besson, C.; Geletii, Y.V.; Musaev, D.G.; Kuznetsov, A.E.; Luo, Z.; Hardcastle, K.I.; Hill, C.L. A fast soluble carbon-free molecular water oxidation catalyst based on abundant metals. *Science,* **2010**, *328*(5976), 342-345.
[http://dx.doi.org/10.1126/science.1185372] [PMID: 20223949]

[305] Abdelhafiz, A.; Vitale, A.; Joiner, C.; Vogel, E.; Alamgir, F.M. Layer-by-layer evolution of structure, strain, and activity for the oxygen evolution reaction in graphene-templated Pt monolayers. *ACS Appl. Mater. Interfaces,* **2015**, *7*(11), 6180-6188.
[http://dx.doi.org/10.1021/acsami.5b00182] [PMID: 25730297]

[306] Ardizzone, S.; Fregonara, G.; Trasatti, S. "Inner" and "outer" active surface of RuO_2 electrodes. *Electrochim. Acta,* **1990**, *35*(1), 263-267.
[http://dx.doi.org/10.1016/0013-4686(90)85068-X]

[307] Cherevko, S.; Zeradjanin, A.R.; Topalov, A.A.; Kulyk, N.; Katsounaros, I.; Mayrhofer, K.J.J. Dissolution of noble metals during oxygen evolution in acidic media. *ChemCatChem,* **2014**, *6*(8), 2219-2223.
[http://dx.doi.org/10.1002/cctc.201402194]

[308] Jiao, Y.; Zheng, Y.; Jaroniec, M.; Qiao, S.Z. Design of electrocatalysts for oxygen- and hydrogen-involving energy conversion reactions. *Chem. Soc. Rev.,* **2015**, *44*(8), 2060-2086.
[http://dx.doi.org/10.1039/C4CS00470A] [PMID: 25672249]

[309] Bergmann, A.; Zaharieva, I.; Dau, H.; Strasser, P. Electrochemical water splitting by layered and 3D cross-linked manganese oxides: correlating structural motifs and catalytic activity. *Energy Environ. Sci.,* **2013**, *6*(9), 2745-2755.
[http://dx.doi.org/10.1039/c3ee41194j]

[310] Gong, M.; Li, Y.; Wang, H.; Liang, Y.; Wu, J.Z.; Zhou, J.; Wang, J.; Regier, T.; Wei, F.; Dai, H. An advanced Ni-Fe layered double hydroxide electrocatalyst for water oxidation. *J. Am. Chem. Soc.,* **2013**, *135*(23), 8452-8455.
[http://dx.doi.org/10.1021/ja4027715] [PMID: 23701670]

[311] Zhao, Y.; Nakamura, R.; Kamiya, K.; Nakanishi, S.; Hashimoto, K. Nitrogen-doped carbon nanomaterials as non-metal electrocatalysts for water oxidation. *Nat. Commun.,* **2013**, *4*, 1-7.
[http://dx.doi.org/10.1038/ncomms3390]

[312] Barnett, S.M.; Goldberg, K.I.; Mayer, J.M. A soluble copper–bipyridine water-oxidation electrocatalyst. *Nat. Chem.,* **2012**, *4*(6), 498-502.
[http://dx.doi.org/10.1038/nchem.1350] [PMID: 22614386]

[313] Fillol, J.L.; Codolà, Z.; Garcia-Bosch, I.; Gómez, L.; Pla, J.J.; Costas, M. Efficient water oxidation catalysts based on readily available iron coordination complexes. *Nat. Chem.,* **2011**, *3*(10), 807-813.
[http://dx.doi.org/10.1038/nchem.1140] [PMID: 21941254]

[314] Geletii, Y.V.; Botar, B.; Kögerler, P.; Hillesheim, D.A.; Musaev, D.G.; Hill, C.L. An all-inorganic, stable, and highly active tetraruthenium homogeneous catalyst for water oxidation. *Angew. Chem. Int. Ed. Engl.,* **2008**, *47*(21), 3896-3899.
[http://dx.doi.org/10.1002/anie.200705652] [PMID: 18351608]

[315] Thomas, J.G.N. Kinetics of electrolytic hydrogen evolution and the adsorption of hydrogen by metals. *Trans. Faraday Soc.,* **1961**, *57*, 1603-1611.
[http://dx.doi.org/10.1039/tf9615701603]

[316] Greeley, J.; Jaramillo, T.F.; Bonde, J.; Chorkendorff, I.; Nørskov, J.K. Computational high-throughput screening of electrocatalytic materials for hydrogen evolution. *Nat. Mater.,* **2006**, *5*(11), 909-913.
[http://dx.doi.org/10.1038/nmat1752] [PMID: 17041585]

[317] Sheng, W.; Zhuang, Z.; Gao, M.; Zheng, J.; Chen, J.G.; Yan, Y. *Correlating hydrogen oxidation and evolution activity on platinum at different pH with measured hydrogen binding energy.,* **2015**, *6*(5848), 1-6.
[http://dx.doi.org/10.1038/ncomms6848]

[318] McKone, J.R.; Sadtler, B.F.; Werlang, C.A.; Lewis, N.S.; Gray, H.B. Ni–Mo nanopowders for efficient electrochemical hydrogen evolution. *ACS Catal.,* **2013**, *3*(2), 166-169.
[http://dx.doi.org/10.1021/cs300691m]

[319] Jaramillo, T.F.; Jørgensen, K.P.; Bonde, J.; Nielsen, J.H.; Horch, S.; Chorkendorff, I. Identification of active edge sites for electrochemical H_2 evolution from MoS_2 nanocatalysts. *Science,* **2007**, *317*(5834), 100-102.
[http://dx.doi.org/10.1126/science.1141483] [PMID: 17615351]

[320] Huang, Z.; Chen, Z.; Chen, Z.; Lv, C.; Meng, H.; Zhang, C. $Ni_{12}P_5$ nanoparticles as an efficient catalyst for hydrogen generation *via* electrolysis and photoelectrolysis. *ACS Nano,* **2014**, *8*(8), 8121-8129.
[http://dx.doi.org/10.1021/nn5022204] [PMID: 25065948]

[321] Vrubel, H.; Hu, X. Molybdenum boride and carbide catalyze hydrogen evolution in both acidic and basic solutions. *Angew. Chem. Int. Ed.,* **2012**, *51*(51), 12703-12706.
[http://dx.doi.org/10.1002/anie.201207111] [PMID: 23143996]

[322] Xin, N.; Huang, H.; Zhang, J.; Dai, Z.; Gan, L. Fullerene doping: preparation of azafullerene C59NH and oxafulleroids $C_{59}O_3$ and $C_{60}O_4$. *Angew. Chem. Int. Ed.,* **2012**, *51*(25), 6163-6166.
[http://dx.doi.org/10.1002/anie.201202777] [PMID: 22573566]

[323] Tian, J.; Liu, Q.; Asiri, A.M.; Sun, X. Self-supported nanoporous cobalt phosphide nanowire arrays: an efficient 3D hydrogen-evolving cathode over the wide range of pH 0-14. *J. Am. Chem. Soc.,* **2014**, *136*(21), 7587-7590.
[http://dx.doi.org/10.1021/ja503372r] [PMID: 24830333]

[324] Duan, J.; Chen, S.; Jaroniec, M.; Qiao, S.Z. Porous C_3N_4 nanolayers@N-graphene films as catalyst electrodes for highly efficient hydrogen evolution. *ACS Nano,* **2015**, *9*(1), 931-940.
[http://dx.doi.org/10.1021/nn506701x] [PMID: 25559360]

[325] Putri, L.K.; Ong, W.J.; Chang, W.S.; Chai, S.P. Heteroatom doped graphene in photocatalysis: A review. *Appl. Surf. Sci.,* **2015**, *358*, 2-14.
[http://dx.doi.org/10.1016/j.apsusc.2015.08.177]

[326] Turchi, C.; Ollis, D.F. Photocatalytic degradation of organic water contaminants: Mechanisms involving hydroxyl radical attack. *J. Catal.,* **1990**, *122*(1), 178-192.
[http://dx.doi.org/10.1016/0021-9517(90)90269-P]

[327] Xiong, Z.; Zhang, L.L.; Ma, J.; Zhao, X.S. Photocatalytic degradation of dyes over graphene–gold nanocomposites under visible light irradiation. *Chem. Commun. (Camb.),* **2010**, *46*(33), 6099-6101.
[http://dx.doi.org/10.1039/c0cc01259a] [PMID: 20661492]

[328] Singh, P.; Shandilya, P.; Raizada, P.; Sudhaik, A.; Rahmani-Sani, A.; Hosseini-Bandegharaei, A. Review on various strategies for enhancing photocatalytic activity of graphene based nanocomposites for water purification. *Arab. J. Chem.,* **2020**, *13*(1), 3498-3520.
[http://dx.doi.org/10.1016/j.arabjc.2018.12.001]

[329] Posa, V.R.; Annavaram, V.; Koduru, J.R.; Bobbala, P.; V, M.; Somala, A.R. Preparation of graphene–TiO $_2$ nanocomposite and photocatalytic degradation of Rhodamine-B under solar light irradiation. *J. Exp. Nanosci.,* **2016**, *11*(9), 722-736.
[http://dx.doi.org/10.1080/17458080.2016.1144937]

[330] Posa, V.R.; Annavaram, V.; Somala, A.R. Fabrication of graphene–TiO_2 nanocomposite with improved photocatalytic degradation for acid orange 7 dye under solar light irradiation. *Bull. Mater. Sci.,* **2016**, *39*(3), 759-767.
[http://dx.doi.org/10.1007/s12034-016-1215-x]

[331] Khare, P.; Singh, A.; Verma, S.; Bhati, A.; Sonker, A.K.; Tripathi, K.M.; Sonkar, S.K. Sunlight-induced selective photocatalytic degradation of methylene blue in bacterial culture by pollutant soot derived nontoxic graphene nanosheets. *ACS Sustain. Chem.& Eng.,* **2018**, *6*(1), 579-589.
[http://dx.doi.org/10.1021/acssuschemeng.7b02929]

[332] Liu, G.; Sun, C.; Yang, H.G.; Smith, S.C.; Wang, L.; Lu, G.Q.M.; Cheng, H.M. Nanosized anatase TiO_2 single crystals for enhanced photocatalytic activity. *Chem. Commun. (Camb.),* **2010**, *46*(5), 755-

757.
[http://dx.doi.org/10.1039/B919895D] [PMID: 20087510]

[333] Wu, Z.; Dong, F.; Zhao, W.; Wang, H.; Liu, Y.; Guan, B. The fabrication and characterization of novel carbon doped TiO_2 nanotubes, nanowires and nanorods with high visible light photocatalytic activity. **2009**, *20*(235701), 1-9.

[334] Wang, J.; Tafen, D.N.; Lewis, J.P.; Hong, Z.; Manivannan, A.; Zhi, M.; Li, M.; Wu, N. Origin of photocatalytic activity of nitrogen-doped TiO_2 nanobelts. *J. Am. Chem. Soc.,* **2009**, *131*(34), 12290-12297.
[http://dx.doi.org/10.1021/ja903781h] [PMID: 19705915]

[335] Ullah, R.; Dutta, J. Photocatalytic degradation of organic dyes with manganese-doped ZnO nanoparticles. *J. Hazard. Mater.,* **2008**, *156*(1-3), 194-200.
[http://dx.doi.org/10.1016/j.jhazmat.2007.12.033] [PMID: 18221834]

[336] Sakthivel, S.; Shankar, M.V.; Palanichamy, M.; Arabindoo, B.; Bahnemann, D.W.; Murugesan, V. Enhancement of photocatalytic activity by metal deposition: characterisation and photonic efficiency of Pt, Au and Pd deposited on TiO_2 catalyst. *Water Res.,* **2004**, *38*(13), 3001-3008.
[http://dx.doi.org/10.1016/j.watres.2004.04.046] [PMID: 15261537]

[337] Arai, T.; Yanagida, M.; Konishi, Y.; Iwasaki, Y.; Sugihara, H.; Sayama, K. Efficient complete oxidation of acetaldehyde into CO_2 over $CuBi_2O_4/WO_3$ composite photocatalyst under visible and uv light irradiation. *J. Phys. Chem. C,* **2007**, *111*(21), 7574-7577.
[http://dx.doi.org/10.1021/jp0725533]

CHAPTER 6

Graphene-based Nanocomposite Catalysts: Synthesis, Properties and Applications

Pradeep Kumar Jaiswal[1,*], Munsaf Ali[2], Siddharth Sharma[3], Dinesh Kumar Yadav[3] and **Nitinkumar Satyadev Upadhyay[2]**

[1] *Department of Biochemistry and Biophysics, Texas A&M University, College Station, Texas 77843, USA*

[2] *Colleges of Nanoscale Science and Engineering, State University of New York, Polytechnic Institute, 257 Fuller Rd. Albany, NY 12203, New York, USA*

[3] *Department of Chemistry, Mohanlal Sukhadia University, Udaipur, 313001, India*

Abstract: In recent years, the progress of doped carbon catalysts (such as graphene-based nanocomposites), has attracted the tremendous attention of the scientific community because of their broad area of applications involving unique mechanical, electrical and industrial chemical production processes. The catalytic nature of metal-doped graphene-based nanocomposites has significantly improved, and provides alternatives to traditional transition metal-based catalysts. In addition, the progress of simple and easy synthetic approaches for graphene-based nanocomposites provides a wide range of opportunities for the synthesis of graphene-based nanocomposites *via* incorporating various metal or polymer-based nanomaterials for diverse applications. In this context, the recent development in the synthesis of graphene-based nanocomposites, their properties and applications are summarized in this chapter. The future insights and challenges towards the design and utilization of graphene-based nanocomposites with decent stability and enhanced catalytic performance are also outlined in this chapter.

Keywords: Electrochemical Biosensor, Graphene, Graphene Nanocomposite Catalysts, Graphene Oxide, Reduced Graphene Oxide.

INTRODUCTION

Since the discovery of carbon-related catalysts, the hydrocarbon-based compounds such as activated carbon, carbon fibers, graphite, and graphite-intercalation compounds, have gained significant attraction from chemists, in the last decades. Nowadays, carbon-based materials like graphene-based nanomaterial synthesis have become an attractive field in chemistry, physics, biology, and

* **Corresponding author Pradeep Kumar Jaiswal:** Department of Biochemistry and Biophysics, Texas A&M University, College Station, Texas 77843, USA; E-mail:pradeepjais@gmail.com

Pinki Bala Punjabi, Rakshit Ameta & Sharoni Gupta (Eds.)

materials science [1]. Graphene is a two-dimensional carbon-based material having a single layer of sp^2 hybridized bonded carbon atoms that are closely packed in a honeycomb lattice [2]. Since the remarkable work of Prof. Novoselov's group in carbon-based chemistry research, the fast development in graphene-based research has envisioned the great potential of graphene in optoelectronics, nanoelectronics, electrochemical drug delivery, and biosensor applications due to its unique 2D structure, excellent electrical capacity, high adsorption susceptibility, high mechanical strength, and high thermal conductivity [3 - 5]. Although graphene sheets were well explored in fundamentals by a physicist at the discovery time, graphene metal-nanocomposites are drawing more attention to real applications for industry [6 - 9]. Until now, various synthetic procedures, such as covalent as well as non-covalent methods, electrochemical, micro-mechanical, solvo-thermal, chemical oxidation–reduction, chemical vapour deposition, photochemical reaction, physical deposition, and mixing [10 - 14] have been fabricated for the synthesis of well-defined functional graphene-based nanocomposites [15 - 20]. It was investigated that the graphene-based sheets can be doped with several components such as metal, metal oxides, or polymers [15, 21, 22] and developed as graphene-based-hybrids which can be employed in multiple applications [23 - 29]. In this view, extensive efforts have been devoted towards the study of graphene-based hybrids and many reviews on their design, development, synthesis, properties, characterization, and applications in recent decades are documented [29 - 32]. The significant potential growth of graphene-based nanocomposites toward their application in a variety of fields motivated us to present its synthesis, properties, and potential applications in a single chapter.

It is well established that the design of a heterogeneous catalyst plays an important role in its catalytic activity [33, 34]. As a new carbon family member, the extraordinary properties of graphene-based nanocomposites provide an ideal catalytic support [32, 35]. Therefore, we hope that this chapter will draw significant attention to the recent developments and advancements in the synthetic protocols and catalytic utilization of graphene-based nanocomposites. It is noteworthy that almost all graphene supports in the nanocomposites are developed *via* chemical reduction–oxidation methodology. The catalytic reactions, which are demonstrated, are an extension of the model reactions towards industrial-based reactions and the utilization of non-selective reactions towards more challenging and selective hydrogenation, coupling, oxidation, or reduction reactions. In the present chapter, a brief overview of the characteristic synthetic approaches and catalytic potential of graphene-based nanocomposites is presented, along with the enhanced catalytic behavior. A summary of some of the scientific opportunities and challenges is also presented at the end of the chapter.

APPROACHES TOWARDS GRAPHENE-BASED NANOCOMPOSITES SYNTHESES

Many versatile synthetic approaches have been documented to synthesize the graphene-based nanocomposites, such as covalent/non-covalent interaction, photochemical reactions, chemical/physical deposition or mixing, hydrothermal/solvothermal growth, electrophoresis, and electrochemical deposition. These developed approaches have several advantages as well as limitations, along with the improved recent development of specific graphene nanocomposites by adoption of appropriate preparation methods.

Non-covalent Interactive Approach

Non-covalent interaction is a leading applied methodology to design and synthesize graphene-based nanocomposites. Non-covalent association of organic and inorganic species with graphene leads to the attachment of organic and inorganic species on graphene incorporated surfaces. *in-situ* Small molecules, such as surfactants and polymers, are bonded to graphene surfaces *via* - stacking and/or hydrophobic interactions, providing an important way to modify graphene surfaces for graphene nanocomposites [36 - 40]. Because of the strong -stacking interaction with the graphene sheet of the pyrene moiety in the molecule, Xu and colleagues used PB- to synthesize enhanced hydrophilic graphene [40]. This was provided by *in-situ* chemical reduction of GO by hydrazine in the presence of 1-pyrenebutyrate (PB-).The resultant water-soluble graphene can be utilized to develop large scale conductive graphene films with higher conductivity in comparison to graphene oxide.

Metal ions can also be doped in graphene sheets through non-covalent congregation of graphene to form 3-D composites or hydrogel [41, 42]. For example, graphene oxide and metal precursors are hydrothermally decomposed to furnish three dimensional graphene-supported metal nanocomposites. Novel metal nanocomposites (Ag, Au, Pd, Rh, Ir, or Pt) on graphene material exhibit active sites, which enhance the porosity in graphene metal composites. The catalytic activity of 3-D composite with respect to Heck reaction has shown excellent and selective physical behavior [41]. GO sheets also undergo metal ions promoted self-assembly based reaction *via* the reduction of GO and produce magnetic Fe_3O_4 nanoparticles with macroscopic graphene based nanocomposites [42].

Covalent Interactive Approach

Another method that can be used to design and synthesis graphene nanocomposites is covalent reaction. Ionic liquids, polymers, porphyrin, azidotrimethylsilane, *p*-phenyl-SO_3H, perfluorophenylazide (PFPA), and organic nitrite can all be covalently bonded to graphene *via* atom-transfer radical polymerization, amide-bonding, or click chemistry reaction. There are limited examples reported in the literature for the synthesis of graphene metal composites *via* covalent assembly [43-45].

Click chemistry is one of the best methods for the formation of covalent bond with metal composites and graphene surfaces. Various photo-induced moieties such as ruthenium-phenanthroline and zinc-porphyrin chromophores can be attached to graphene sheet using click chemistry. Firstly, the graphene surface was modified by unsaturated moieties *via* exposing alkyne groups followed by click chemistry reaction. In the result, these hybrids showed significantly improved photocurrent responses [46].

In-situ Chemical Electroless Deposition-based Approach

In the last few years, various efforts have been made to deposit nano-objects on graphene sheets to fabricate nanocomposites *via* chemical electroless deposition. Specific nanorod, nanoparticle and nanosheet can be fabricated *via* doping of metal on graphene surface, through *in situ* chemical deposition method, of the selective precursors. These process are straight forward and convenient for designing of a large number of graphene-based nanocomposites using metal oxides, metals and dichalcogenide precursors, such as, Au, Pt, Ag, Cu, TiO_2, Fe_2O_3, Mn_3O_4, SnO_2, and CdS [47 - 55]. Using the electroless chemical deposition method, Li *et al.* created graphene-based bimetallic nanocomposites with improved electrochemical surface area [48]. High-quality 3D-bimetallic nano-dendrites Pt-on-Pd were doped on graphene surfaces in two steps (Fig. **1**). Pt precursors (K_2PtCl_4) were reduced by earlier doped Pd-NPs on graphene-based material and fabricated into branched Pt-nano-dendrites with the number of branches being tunable with a synthetic framework [48]. Similarly, Ag-deposited graphene nanocomposites were created by heating a silver nitrate-graphene oxide aqueous solution at 75°C or by in situ reduction of silver precursor in the presence of GO by glucose or other reducing agents [50, 51, 56]. Copper-based graphene composites were also created by heating a copper precursor in the presence of graphene oxide and potassium borohydride [57].

The resultant Cu-nanoparticles are well dispersed on a graphene sheet having a diameter of about 20.8 nm. In another report, graphene-based macroscopic hydrogel-iron oxide nanocomposite with robust interconnected three-dimensional frame was constructed by the synergistic effect *via in-situ* reduction of graphene oxide from ferrous ions [42]. The amount of reducing agent and pH value play a very significant role in the synthesis of graphene-based nanocomposites. In another investigation, TiO_2 deposited graphene nanocomposites have been designed and synthesized by *in-situ* hydrolysis of absorbed Ti (IV)-butoxide on rGO surfaces with oxygenated functional groups [74]. To achieve both better deposition and high conductivity of TiO_2 NPs, the oxidization time in the chemical exfoliation process was essential [53]. *In-situ* Metal chalcogenides like cadmium sulfide-NPs can be produced by *in-situ* chemical deposition on graphene sheets [58]. CdS nanoparticles were also synthesized from cadmium nitrate and CH_4N_2S at 150 °C heating in microwave irradiation [59]. This method was also used to synthesize graphene-zinc sulphide nanocomposites from zinc ions and H_2S gas reactions [54].

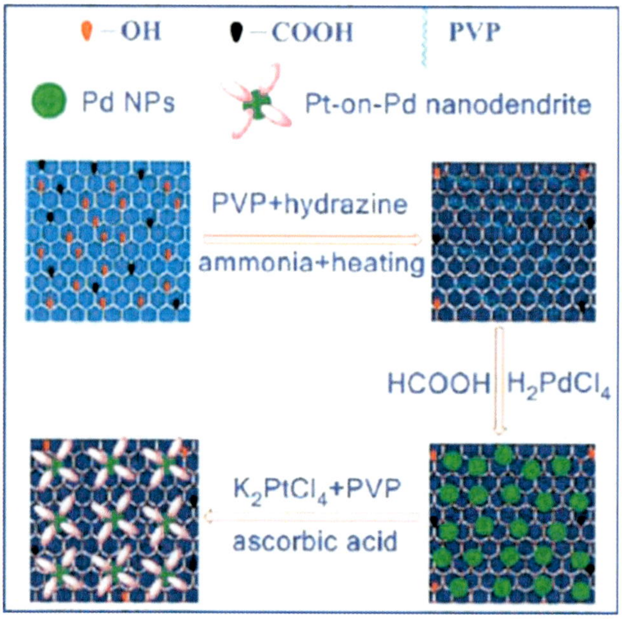

Fig. (1). Schematic representation for designing graphene nanosheet/Pt on Pd bimetallic nano-dendrite hybrids. (Reproduced with permission from Ref [48]. Copyright 2010, American Chemical Society).

Solvothermal and Hydrothermal-growth Based Approach

Solvothermal and hydrothermal methods are also extensively employed to construct graphene nanocomposites. Due to high pressure in solvothermal and hydrothermal methodologies, the acquired materials have close covalent network between deposited composites and graphene, which shows various useful applications.

To fabricate graphene metal composites by solvothermal or hydrothermal methodology, the metal/metal oxide precursor is mixed with graphene or GO in a solution followed by solvothermal/hydrothermal deposition of the precursor on graphene sheet at high pressure in an autoclave. Solvothermal and hydrothermal processes are very flexible and can be used to design various kinds of nanomaterials (Fig. **2**) [60 - 65].

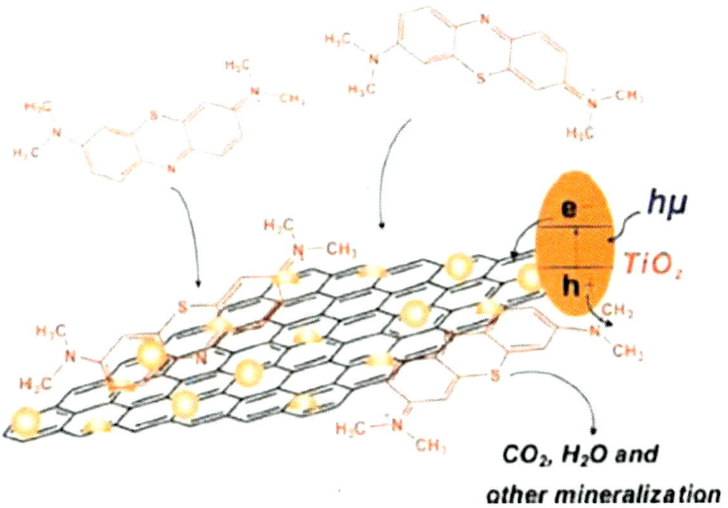

Fig. (2). Schematic representation of P25 nanoparticles and photodegradation of methylene blue (MB) over P25-GR. (Reproduced with permission from Ref [64]. Copyright 2010, American Chemical Society).

Metals like Ag and Pt-Ru, metal oxide (TiO_2, SnO_2, VO_2, Mn_3O_4), metal chalcogenides (*e.g*: ZnS, SnS) and complex compounds ($LiFePO_4$) based nanocomposites have been successfully decorated on the 2D carbon layer of graphene with well controlled solvothermal methods. Ye and his co-workers have reported a hydrothermal synthesis of reduced graphene oxide based Ag-nanocomposites by using ascorbic acid as a reducing agent and silver nitrate and graphene oxide as precursors [60]. Yoon and co-workers fabricated platinum–ruthenium (Pt-Ru) nanoparticles on graphene sheet *via* a hydrothermal process and the composites showed improved tolerance towards poisoning issues of oxidation of methanol with enhanced electrocatalytic activity [61].

Metal oxides are also used to synthesize graphene-based metal oxide composites on 2D carbon sheets by hydrothermal or solvothermal assembly. Zou *et al.* utilized seed-directed hydrothermal process to synthesize a large number of graphene-based metal oxide (such as TiO_2, ZnO, MnO_2, CuO, and ZrO_2) nanocomposites *via* designing a novel series of metal oxide-graphene-metal oxide hetero-structures. The hetero-structures exhibit four times increased photo-catalytic properties compared to pure metal oxide nanoparticles [62]. TiO_2 attached graphene nanocomposites were synthesized from GO and a reducing agent using a hydrothermal process (Fig. **3**) [63]. The crystal facets of TiO_2 nano-objects in graphene based-nanocomposites, can also be controlled in a hydrothermal fashion [66]. Recently, graphene-based TiO_2 nanotubes and TiO_2 nanosheets with exposed facets were developed under a hydrothermal process in alkaline solution/ aqueous ethanol (Fig. **3**) [67, 68].

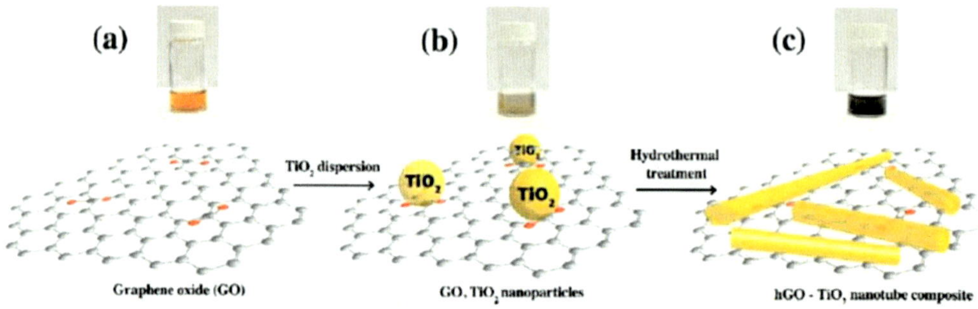

Fig. (3). Graphical presentation of the synthesis of TiO_2 nanotubes on rGO sheets. The GO sheets (a) coordinated with TiO_2 nanoparticles (b) converted to TNTs on hGO (c) The digital images show the color variations of the aqueous dispersions in each stage. (Reproduced with permission from Ref [68]. Copyright 2012, American Chemical Society).

After 2010, several research groups broadly investigated the synthesis of TiO_2 and VO_2 nanotubes, MoO_3 nano-objects, Mn_3O_4 and Fe_3O_4 nanoparticles having spherical geometry and hollow interiors with graphene *via* hydrothermal approach (Fig. **4**) [13, 65, 68 - 71]. Graphitic layers decorated with metal chalcogenides were also synthesized by solvothermal method. Cao and his co-workers have reported synthesis of CdS quantum dots attached to graphene sheets [72]. The nanocomposites formed have demonstrated tremendous performance in electron transfer from quantum dots (QDts) to graphene surfaces and have significant importance in optoelectronic materials. Later on, Gong *et al.* adopted the same methodology for the fabrication of graphene CdS nanocluster for hydrogen generation from water [72]. ZnSe mixed graphene nanocomposites were designed and synthesized by one-pot hydrothermal based methodology by using graphene-oxide (GO) and zinc-selenide (ZnSe) as precursors. The nanobelts were deposited from GO followed by heating at 180°C temperature for 12 h under a solvothermal approach [73]. In addition, the complex compounds, such as $Li_4Ti_5O_{12}$, $LiFePO_4$ and $Bi_2WO_{6;}$ were also synthesized by hydrothermal and solvothermal processes [74 - 76].

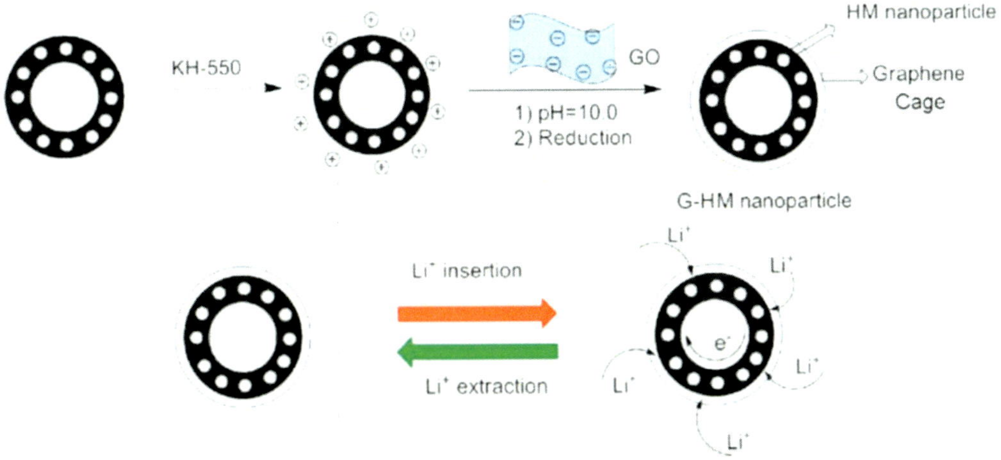

Fig. (4). Schematic illustration of the fabrication process along with the structure of the resulting graphene nanocomposite. (Reproduced with permission from Ref [69]. Copyright 2011, American Chemical Society).

Microwave- assisted Approach

Under this approach, microwave irradiation (MWI) is widely utilized towards the design and synthesis of graphene-based nanocomposites [8, 77, 78]. Particularly, the construction of graphene-based metal nanocomposites by *in-situ* reduction of various metal salts and graphitic materials, exhibited significant catalytic activity. In 2009, Hassan *et al.* reported the scalable synthesis of RG/NP composites by microwave-induced reduction [79]. This strategy represents a fast reduction of graphene oxide with several reducing agents in water and/or organic media, and the instantaneous reduction of GO with metal precursors furnishes the construction of metallic/bimetallic NPs doped graphene sheets. Under MWI conditions, metal- attached graphene nanocomposites of Pd, Cu, Ag, and Au have been made by *in-situ* reduction of appropriate metallic salts (Pd nitrate, $HAuCl_4$) and graphitic material with several reducing agents like ethylenediamine, ammonium hydroxide, and hydrazine hydrate [78]. *In-situ* reduction leads to homogeneous dispersions of nanostructures on graphene sheets *via* specific interaction and physical mixing between the graphene sheets and nanocomposites. In addition, the *in-situ* MWI approach has also been applied to fabricate graphene-based metal nanocomposites such as Graphene–SnO_2 nanocomposites. (Fig. **5**) [80]. Liu *et al.* investigated improved conditions for the synthesis of ZnS nanospheres on graphene, and the reaction was optimized in an aqueous medium induced by MWI [81]. Additionally, in 2014, a MWI-based approach for an ionic liquid supported Pd-Pt nanostructures on HRG sheets was investigated [82].

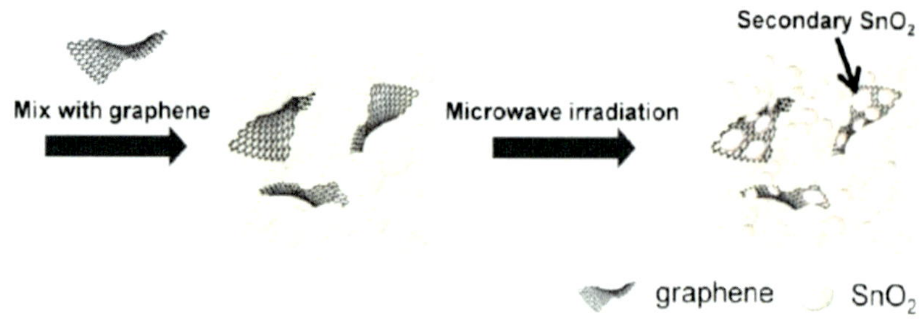

Fig. (5). Schematic illustration of the morphologies of the graphene–SnO_2 nanocomposites developed by the MWI. (Reproduced with permission from Ref [80]. Copyright 2011, American Chemical Society).

Electrochemical and Electrophoresis Deposition-based Approach

Electrochemical and electrophoresis depositions are another widely used methodologies for the synthesis of graphene-based metal nanostructures. Metal salts and oxides easily mix with 2D carbon layers by electrochemical assemblies [83 - 89].

Using electric current, electrochemical reduction of metal nanomaterial was also discovered. Some semiconductors such as Cd-Se was electrochemically doped on graphene surfaces and resulted the graphene nanocomposites. ZnO nanoparticles were also deposited on conductive reduced graphene oxide (rGO) under electrochemical deposition assembly [83]. In brief for the first time, a constant potential was used with ZnO on reduce graphene oxide for the synthesis of nanocomposite.

The conductivity of rGO has a strong influence on zinc oxide-based graphene nanocomposite. Particle-shaped nanocomposites were synthesized with rGO having low conductivity, whereas the hexagonal nanorods were constructed with rGO having greater conductivity. A graphene-based MnO_2 nanocomposite was also fabricated by electrochemical method (Fig. **6**) [85]. In 2011, Graphene composite having Cu_2O/ZrO_2 was also prepared by using electrochemical methodology [86, 90]. Metals like Ni, Cu, and Pt were doped on graphene sheet by an electrochemical approach [87, 91 - 93]. Yang *et al.* developed a mixed metal Au-Pd nanocomposite on graphene surface by constant-potential reduction using electrochemical approach with $HAuCl_4$, $PdCl_2$, and 0.1 M KCl [93]. In 2013, Ma and co-workers fabricated graphene-based NiOH nanocomposite under the electrophoresis method with Ni^{2+} ions and GO colloidal solutions. The positively charged graphene oxide, by nickel ions, was deposited on substrates *via* electrochemical reduction of GO to rGO under negative current [94].

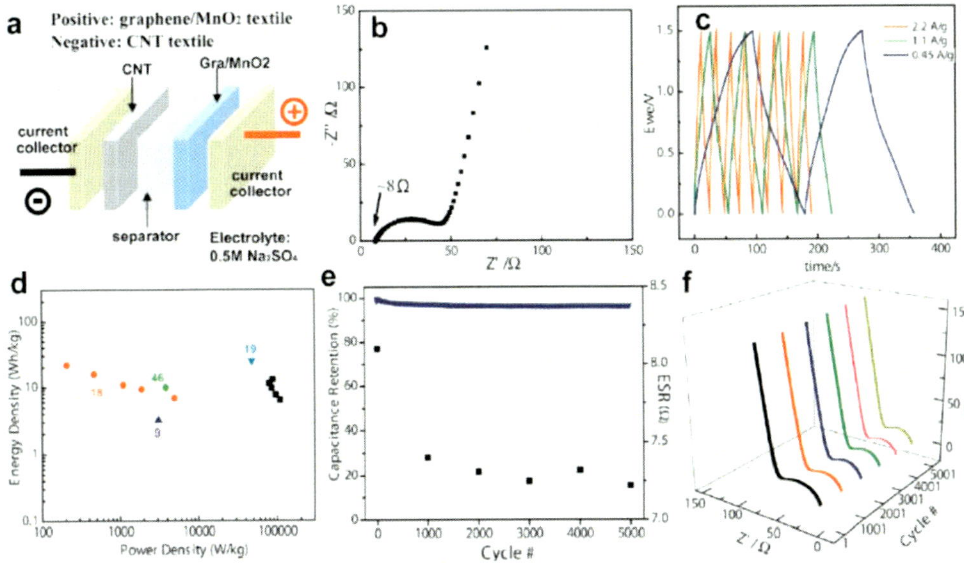

Fig. (6). An Asymmetric ECs based CNT-textile as negative electrode and graphene/MnO_2-textile as positive electrode. (a) Representation of the structure of hybrid EC-cells in an aqueous electrolyte (0.5 M Na_2SO_4). (b) Nyquist plot for hybrid EC (the frequency range was 100 kHz-0.1 Hz). (c) Galvanostatic discharge/charge curves with different current densities. (d) Ragone plot of energy-density vs power-density. (e) Cycling performance of hybrid ECs (capacitance retention having □95% after 5000 cycles). (f) Nyquist plots demonstrating the corresponding impedance curves for assembled ECs (observed after every 1000 cycles). (Reproduced with permission from Ref [85], Copyright 2011, American Chemical Society).

Chemical-vapor Deposition (CVD) Based Approach

Chemical-vapor deposition is another valuable approach for constructing graphene-based nanocomposites. Normally, graphene is utilized as a base material in the normal chemical-vapor deposition (CVD) based approach for the synthesis of graphene based hetero-nanostructures having other materials [95 - 102]. In 2009, Kim and his co-workers fabricated graphene-based ZnO nanoneedles employing vapor-phase epitaxy [95]. ZnO deposited graphene heterogeneous nano-objects were also synthesized by tuning the temperature in CVD assembly [96]. Recently, Fan and coworkers published a report on the CVD induced 3D CNTs/graphene sandwich compound prepared by using graphene oxide with uniform catalysts for CNT growth (which improved surface area) [97]. In the year 2010, pillared graphene mixed CNTs were also fabricated [98]. The same CVD processing was used to make metal-doped graphene-based CNTs composites and three-dimensional graphene mixed CNTs carpet micro supercapacitor electrodes [99, 100, 102]. Tavakkoli and colleagues recently developed a one-step catalytic CVD assembly to make a highly graphitized/CNT hybrid-material that was doped with single atoms (N-Co-GF/CNT). The enhanced surface area composites have mesoporous structures, which facilitates oxygen mass transfer within the catalyst film and also act as an electrocatalyst for oxygen reduction in basic medium [101].

Photochemical- based Approach

Photochemical reaction is another approach for designing and synthesizing graphene-based metal-nanostructures [103 - 111]. Under UV light irradiation, graphene surfaces can be modified with reducing agents or metal precursors, which may provide the graphene-based nanoarchitectures. For example, the graphene-based TiO_2 nanocomposite was fabricated by mixing GO and TiO_2 NPs in ethanol under UV irradiation conditions. The photoactive TiO_2 transfers the electrons to graphene oxide and affords reduced graphene/TiO_2 nanocomposites. (Fig. **7**) [103].

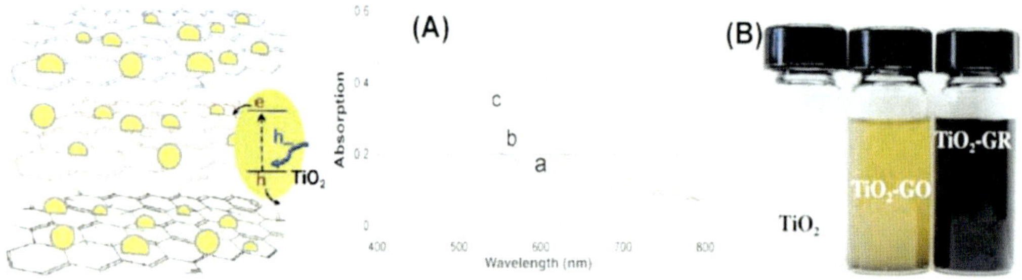

Fig. (7). (**A**) TiO_2-graphene nanocomposite having response in UV excitation/absorption spectra of GO and TiO_2 in ethanol suspension (a) before UV irradiation, (b) after 5 min (c) after 15 min under UV irradiation; (**B**) The color change of TiO_2 nanoparticles of 0.5 mg/mL GO of 10 mM solution before/ after UV irradiation. (Reproduced with permission from Ref [103], Copyright 2008, American Chemical Society).

In 2011, Amal and his co-workers adopted the same strategy to fabricate WO^{3-} and $BiVO^{4-}$ based graphene nanocomposites by irradiation with light amongst their precursors and graphene oxide [104]. Additionally, in metal nanoparticle based graphene nanocomposites, the precursors were doped on 2D carbon sheet under photochemical irradiation [104, 106 - 109, 111]. A photochemical strategy was also applied for the photolysis of phosphor-tungstic acid (PTA) to reduce GO for graphene based Au, Ag, and Pt NPs synthesis [105].

At the same time, two- step photochemical process was also developed for the dispersed noble metals (Pd, Ag, and Au) NPs on graphene sheet in dimethylformamide solvent. In the first step, the photolysis of phosphor-tungstate provides reduced graphene from graphene oxide. The noble metal ion precursors were doped with above metals to fabricate graphene-nano composites [106]. In

2012, Chen and his colleagues used an Hg lamp (450 W) to initiate the photochemical reaction for the synthesis of Ag/rG nano-structures using a mixture of $Ag(NH_3)_2OH$ complex and GO [107]. In the same year, the Ag NPs were also deposited on graphene oxide using similar methodology [108]. Ag NPs deposited on graphene oxide and Pt-graphene based-nanocomposites were synthesized *via* same method without using any reducing agent (Fig. **8**) [109 - 111].

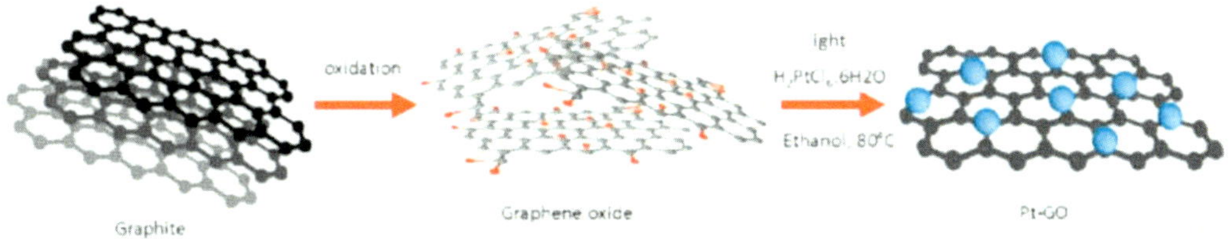

Fig. (8). Schematic depiction of light mediated Pt-Salt reduction on GO-sheet in ethanolic solution. (Reproduced with permission from Ref [110]. Copyright 2012, American Chemical Society).

Physical Deposition/Mixing Based Approach

The straight forward and simple process to construct graphene based-nanostructures, directly attached on 2D carbon sheet, *via* physical loading (physical vapor deposition/atomic layer deposition) is also discovered [112 - 116]. Metal oxides like Al_2O_3 was physically loaded on graphene sheet with atomic layer deposition (ALD) assembly (Fig. **9**) [112]. Metal based nanostructures were also doped physically on modified graphene surfaces under chemical environment by physical-vapor deposition based methodology [113]. The NPs growth on 2D (two dimensional) carbon layer was illustrated below, which exhibits significantly difference of activity for diverse metals (Ag, Fe, Au, Pt, Pd, Ti). Physical mixing is also one of the simplest approaches to synthesize the graphene-based nanocomposites. In the general procedure, the three-dimensional graphene material was obtained, when $LiFePO_4$ nanoparticles were loaded on graphene sheet by using annealing and spray-drying methodology [114].

Alternatively, the commercially available $LiFePO_4$ material was deposited on graphene sheet through a dropwise loading of $LiFePO_4$ particles at 180 °C in DMF solution. This process was stimulated by a non-covalent interaction between the graphene surface and the amorphous carbon shell of $LiFePO_4$ particles. These modifications resulted improved stability and more efficient capacity of lithium battery of $LiFePO_4$ [115, 116].

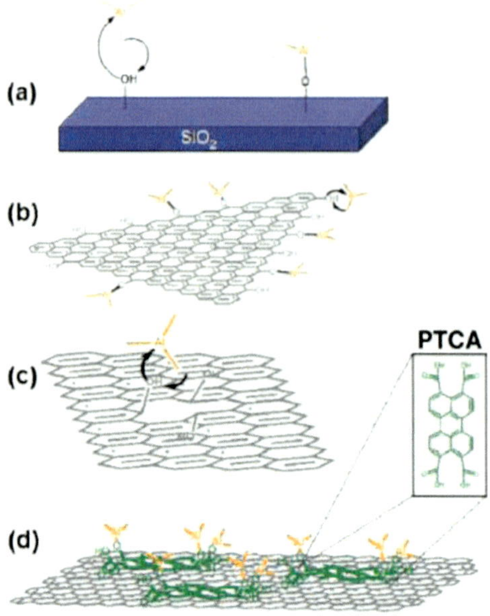

Fig. (9). Schematics representation using trimethylaluminum precursor for atomic layer deposition of aluminum oxide on (**a**) plain SiO_2 substrate, (**b**) a single-layered graphene sheet with their edges, (**c**) graphene having a defect site, (**d**) PTCA (perylene tetracarboxylic acid)-capped graphene. (Reproduced with permission from Ref [112], Copyright 2008, American Chemical Society).

PROPERTIES AND APPLICATIONS OF GRAPHENE-BASED NANOCOMPOSITE CATALYSTS

The diverse nature of graphene-based nanocomposites exhibits a huge range of properties towards various aspects. The novel properties and applications of graphene-based nanocomposites are the main criteria for catalyst design and synthesis. Therefore, only those potential graphene-based nanocomposites that exhibit significant catalytic potential are shortlisted and discussed in the present chapter.

Based on theoretical and experimental calculations for graphene and graphene-based nanocomposites (using carbon nanotubes (CNT)), the properties and applications were found to be outstanding with respect to several aspects. In graphene, the carbon atoms are arranged hexagonally, which results in a 2D honeycomb scaffold. While the strong bonds in graphene sheets lead to the stability and rigidity in hexagonal structures, the other out of plane π bonds provide interaction between different 2D carbon layers [117]. In addition, another significant property of graphene is the adsorption on the surfaces. It was found that the adsorption and interaction between 2D carbon surfaces and adsorbed objects are unique and specific parameters towards the design and synthesis of the graphene-based nanocomposites. The theoretical calculations are of paramount importance in order to emphasize this property of graphene sheets. The stable configuration of embedded metal atoms in graphene displays the charge transfer between graphene and the adsorbed metal atoms. Generally, the binding of the adsorbed metal atom on graphene takes place *via* three sites of symmetry: a) at the center of a hollow hexagon (H); b) at the center of a C-C bond bridge (B); c) and directly on the carbon atom top (T) [118]. In 2010, Hu and his coworkers developed a good correlation between the adsorption of a huge number of doped metal atoms on graphene surface using DFT energy calculations [119]. They outlined that hollow hexagon sites are well suited for Ti, Sc, V, Co, Ni, and Fe deposition on graphene sheet, whereas B or T are the most striking sites for metals that have a completely filled or near- to- filled d-shell (Pd, Pt and Cu). Having half-filled d-shell transition metals (such as Au, Ag and Zn) has lesser adsorption energies. The preferred adsorption site is based on the intensity of the adsorption energy, which represents the remarkable effect on the chemical bond between the graphene surface and deposited materials. So, the adsorption of these metals (such as Ag, Au, and Cu) was reflected as physisorption, whereas the adsorption of Ni, Pt, Co, and Pd showed chemisorption on the graphene sheet surface. Similar results were reported for Ag, Au and Pt metals by Tang and his coworkers [120]. In 2011, the Nakada research group discovered the migration and adsorption energies for almost all atoms in the periodic table from H to Bi (except Ln, noble gases) on 2D carbon sheets using DFT calculations. The computational data demonstrated that the chemisorption of metals includes hybridization of d-orbitals of deposited atoms with the graphene surface. For the doped atom, there is a decrease in the magnetic moment for adsorbed metal atoms compared to the isolated metal atom, which depends on charge transfer between graphene and the adsorbed atom [119]. Several other research groups have also demonstrated the relationship between chemical/biologically privileged compounds with graphene sheet or metal deposited graphene sheet [121 - 123]. On the basis of theoretical as well as experimental observations, enormous works have been documented for the synthesis of graphene- based inorganic nanocomposites, typically using transition metal/metal oxide nanoparticles [8, 32, 78, 118, 124]. Moreover, massive interest has been reported in synthesizing graphene-based nanocomposites and utilizing them for diverse applications [8, 82, 91, 100, 125].

For Energy Storage and Conversion

The discovery of novel methodologies for the manufacturing of clean energy is an interesting and more challenging area for the scientific research community. Due to the increased air pollution and global warming conditions, the development of high energy density devices for energy storage and its conversion has increased tremendously [126, 127].

As a Basic Component in Lithium-ion Batteries (LIBs)

Graphene-based nanocomposites have attracted significant popularity in the electrochemical energy storage research area because of their unique physico-chemical properties, such as enhanced chemical and thermal stability, better electrical conducting behavior and mechanical properties. Graphene-based nanocomposites have been utilized as electrode materials and investigated for energy storage and conversion devices like lithium ion batteries (LIBs), supercapacitors, and fuel/solar cells [128 - 133]. TiO_2 has gained tremendous interest in high-performance LIBs. Several TiO_2 doped nanomaterials with increased lithium storage capabilities compared to non-doped pure TiO_2 have been developed [134, 135].

In 2011, Li *et al.* investigated a simple one-pot route *via* template free self-assembly methodology to fabricate mesoporous TiO_2 nanocomposite on a graphene surface. These TiO_2 nanocomposites are mono-crystalline in nature compared to the conventionally based poly-crystalline mesoporous anatase particles. The nanocomposite exhibits a significant improvement in the specific capacity, which is 6-fold higher than the pure TiO_2 [136]. Due to the layered structure and abundant availability in nature along with the high energy density, the vanadium pentoxide (V_2O_5) has been discovered as a cathode material in rechargeable LIBs. Therefore, a huge number of V_2O_5-based graphene nanocomposites have been investigated [137 - 139]. A V_2O_5 doped HRG nanocomposite having significant electrochemical properties was synthesized *via* hydrothermal reactions [140]. Yang *et al.* demonstrated in 2011 that graphene-based nanocomposites decorated with MoO_3 act like n-type semiconductors with energy storage in electrochromic and photochromic devices [141]. It was later investigated that MoO_3-based graphene nanocomposites showed enhanced discharge capacity and an outstanding high specific capacitance activity [142]. Mn_3O_4 has been investigated as an anode material for LIBs, having higher specific capacitance than graphite [143]. In order to achieve higher capacity, Wang *et al.* produced Mn_3O_4-based graphene materials in two steps.

The hybrid nanocomposites exhibited significant high capacity of ~ 900 mA h g^{-1} as well as good cycling stability and rate capability [144]. Fe_3O_4- based graphene nano-composite was developed by Su *et al.* with excellent reversible capacity, high cyclic performance, and enhanced rate capability compared to the Fe_3O_4 nanoparticles (Fig. **10**) [145]. Zou *et al.* reported Fe_2O_3 doped graphite nanocomposites as an anode material in LIBs. This metal nanocomposite revealed a good reversible capacity (1184 mA h g^{-1}) upon cycling along with an outstanding rate capability at increased currents [146]. In addition, graphene-based nanocomposites of Fe_2O_3 and Fe_3O_4 and other iron-based precursors (like $LiFePO_4$) have also been utilized in LIBs [146 - 148]. Moreover, Qu *et al.* made sandwich-like Fe_3O_4 doped graphene nanocomposites *via* FeOOH nanostructures with graphene [149].

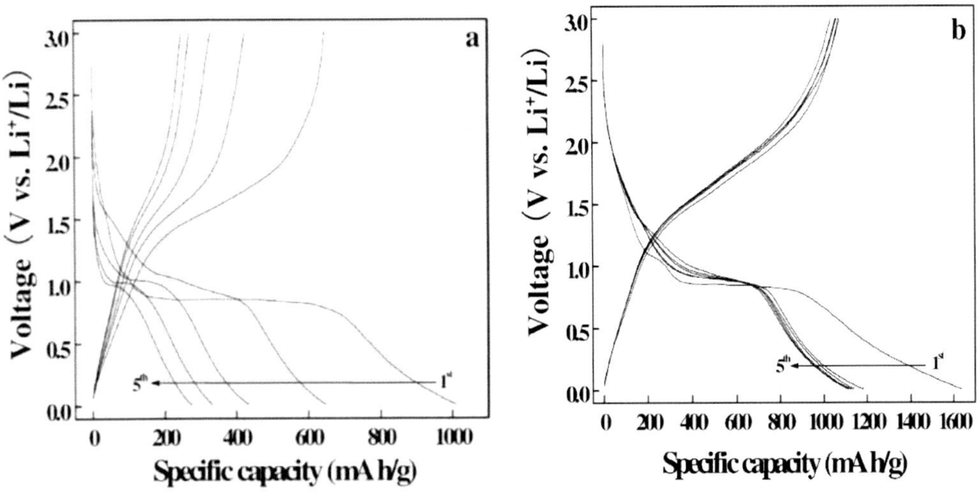

Fig. (10). Charge and discharge profiles of graphene oxide nanocomposite with Fe_3O_4 (**a**) GN-Fe_3O_4 nanocomposites (**b**) electrodes from one to fifth cycles. (Reproduced with permission from Ref [145], Copyright 2011, American Chemical Society).

As a Basic Component in Supercapacitors

Supercapacitors are utilized as brilliant energy storage materials and display enhanced power density and increased life cycle along with improved performance in comparison to common batteries. The energy storage capacity in supercapacitors contains a carbon-based EDL (electric double layer). In EDLCs, the charge storage depends on the material of electrode, because the stored energy depends on the adsorption of ionic charges. Therefore, the property of electrode material is highly responsible to increase the efficacy of capacitance of EDLCs. In this point of view, graphene as well as graphene-based nanocomposite was found to be a good alternative for making improved electrodes [150]. For example, the electrochemical capacity of RuO_2 hybrids with activated carbon, exfoliated graphite and CNTs, exhibited significant potential towards efficacy [151]. Wu *et al.* synthesized RuO_2 doped HRG nanocomposites by solvo-thermal process by varying the Ru loading amount [152]. Later on, using high theoretical capacitance, the expensive amorphous RuO_2 was replaced by cost effective and easily available $Ni(OH)_2$ and NiO [153, 154].

As an Electrocatalyst in Fuel Cells

Several graphene-based nanocomposites have been well explored as an electrocatalysts in ORR. Pd incorporated rGO nanocomposites were used as electrocatalysts in ORR along with the oxidation of alcohols in various fuel cells, such as DEFCs, DMFCs and DFAFCs [25, 155, 156]. Bimetallic Pt–Au graphene nanocomposites are an important electro-catalyst in DFAFCs fuel cells, which are also utilized for formic acid oxidation. Au loading on Pt nanoparticles has an specific effect on the direct oxidation to CO_2 from formic acid without CO intermediates [157]. Later on, Pt decorated graphene nanocomposites with perfluorosulfonic acid (PFSA) were utilized as electro-catalysts for ORR. PFSA accelerates the Pt–graphene interactions and opens more channels for H^+ on Pt surfaces, and thus generates more active catalytic property (Fig. **11**) [158]. In addition, PFSA activates the absorption of H_2O and CO on graphene composite surfaces and leads to an active Pt–OH compound's synthesis, which enables the oxidation to CO_2 from CO. Thus, PFSA-anchored HRG/Pt nanocomposites showed an enhanced catalytic oxidation of CO along with minimum electrochemical active area reduction. Rao and his coworkers discovered graphene-based Pt_3Cr and Pt_3Co nanocomposites, having 3 to 4 times enhanced catalytic activity than HRG/Pt, as electrocatalysts for ORR [159].

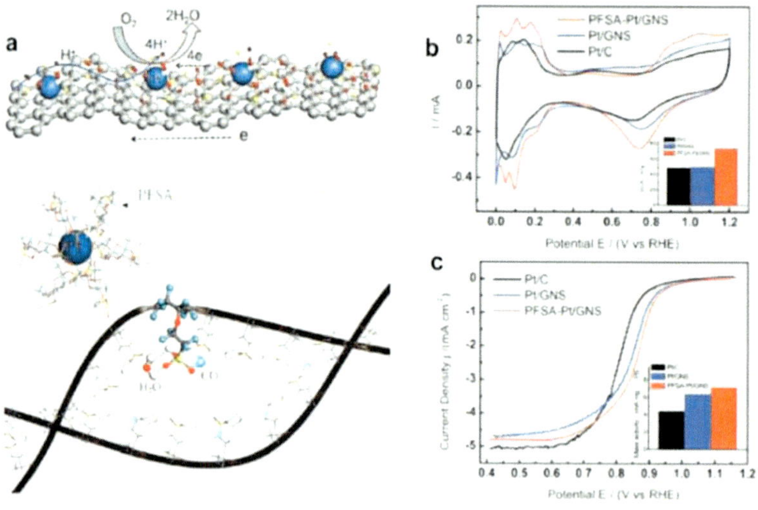

Fig. (11). (**a**) Schematic demonstration of the electrocatalytic mechanism with Graphene/Pt nanocomposites. (**b**) CV curves of nano-catalysts (PFSA–Pt/GNS, Pt/GNS and Pt/C). (**c**) Current potential curves for ORR and having comparisons between ORR mass-activities and specific-activity of catalysts. (Reproduced with permission from Ref [158], Copyright 2012, American Chemical Society).

As an Electrochemical Sensor

Various graphene-based nanocomposites have also been used as electrochemical sensors for detecting numerous chemicals. In 2011, Du and his co-workers synthesized ZrO_2 decorated graphene composites, where ZrO_2 nanoparticles were utilized as a selective sorbent for SPE (solid-phase extraction) and detection of pesticides (such as OP pesticides and nitroaromatic OP pesticides) [160]. Graphene based ZrO_2 electrochemical sensor was also utilized for rapid extraction for analyte, due to its strong affinity to phosphorus based species. Recently, Gong and his co-workers investigated HRG/ZrO_2 nanomaterial as SPE for enzyme less methyl parathion sensing [161]. Subramanian *et al.* discovered a non-enzymatic $Ni(OH)_2$ decorated graphene nanocomposite as a sensor for glucose detection. The nanocomposite is composed of leaf-shaped $Ni(OH)_2$ (diameter =~150 nm) dispersed between the graphene layers. The sensing limit (up to 0.6 mM) having a good linear range (2 mM-3.1 mM) was attained in this study [162]. Moreover, the excellent sensing capability of graphene and the detection capability of Pd nanoparticles have been also utilized simultaneously for numerous sensing applications [163]. The enhanced catalytic property of Pd-decorated graphene nanocomposites was achieved by charge transfer from molecules wrapped on graphene surface *via* change in the electrical conductive behavior of graphene. In 2010, Johnson and his co-workers investigated graphene-based Pd nanocomposites as hydrogen sensors in fuel cells [164]. This research group prepared multi-layered graphene as a nanoribbon *via* dispersing graphite material along with functionalized Pd-nanostructure in a water-surfactant solution. The nanoribbon, having porous structure, provides improved surface area, which permits better detection of the adsorbed gas. HRG based Au-nanocomposite was investigated as an electrochemical sensor, for sensing organic species, chemicals and metallic ions (such as lead, organophosphate pesticides, hydroquinone *etc.*) [165, 166]. The quenching potential of Au-nanocomposite was utilized for the detection of lead ions in organophosphate pesticides as a chemo-sensor. The composite electrode having acetylcholinesterase, was stabilized from aqueous soluble poly-diallyldimethylammonium chloride (PDDA electrolyte) [167]. The graphene and ZnO, together exhibit electronic and semiconducting nature so, a ZnO-graphene composite was used to fabricate a highly sensitive photo-sensor [168]. Graphene nanocomposites having ZnO NPs work as visible-blind UV-photosensors, where graphene behaves as a conductive matrix and the ZnO NPs were used as a charge carrier generating a UV absorbing material also [169]. The HRG nano-material having ZnO with photoluminescence (PL) nature was explored for gas detection [170]. HRG having ZnO nanocomposites showed optoelectronic and semiconducting features [171]. The ZnO nanomaterial with luminescent nature present between the graphene layers showed sensing property for various gases (such as NH_3, CO, and NO). In addition, SnO_2 based graphene nanocomposites with excellent selectivity and sensitivity were also developed [172].

Lu and his co-workers discovered graphene-based Pd decorated electrode, for the detection of glucose, as an electrochemical biosensor [173]. This research group modified graphene with chitosan covalently which improved the solubility and biocompatibility [174]. Dey and his co-workers reported the synthesis of graphene-Pt based nanocomposite as an amperometry biosensor for detecting cholesterol and H_2O_2 *via* catalyzing H_2O_2 oxidation and electron transfer [175]. In 2012, Chen and his coworkers fabricated Cu decorated graphene nanocomposites for carbohydrate sensing *via* reduction of graphene oxide with KBH_4. (Fig. **12**) [176].

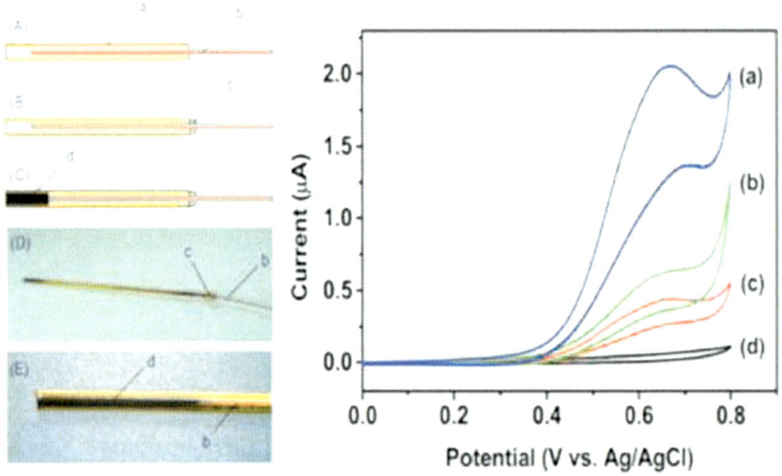

Fig. (12). (**A-E**): Schematic illustration of fabrication of graphene-copper nanoparticle composite electrode, and cyclic voltammetry diagram of graphene-copper nanocomposites. (Reproduced with permission from Ref [176], Copyright 2012, American Chemical Society).

Ag- based HRG nanocomposites were also prepared for sensing of H_2O_2, glucose and NO_2-based aromatic compounds [177, 178]. Graphene decorated with Ag nanocomposite was also explored for the detection of glucose, H_2O_2, and SERS substrate [179]. Zhang *et al.* investigated Au decorated graphene nanocomposites as an electrochemical bio-sensor, for sensing H_2O_2 (detection range 0.1mM to 2.3mM; and detection limit 0.01mM at 3s) in food *via* the reduction of H_2O_2 at pH 5.8 in buffer solution [180]. In 2012, another study was performed related to epinephrine (EP), an important neurotransmitter, using an Au- based GCE electrode with improved sensitivity [181].

Graphene-based nanomaterials also show significant biomedical applications [182]. Various studies have been reported towards the biomedical potential of graphene such as drug and gene delivery systems, anti-tumor and antibacterial properties with scanning and imaging nature *etc* [183 - 186]. Paramagnetic Fe_2O_3 decorated graphene nanocomposites were applied for magnetic resonance imaging (MRI) and in hyperthermia therapy having improved physiological stability and biocompatibility [187 - 189]. In 2011, Zhang and his co-workers investigated aminodextran-coated Fe_3O_4 graphene nanocomposites for T2-weighted cellular MRI detection having improved T2 relativity *via* the aggregation of Fe_3O_4 nanoparticles on graphene surface (Fig. **13**) [190]. Oleylamine with Ag nanoparticles decorated on graphene oxide surface displayed improved antibacterial property compared to pristine based Ag NPs against Gram-negative bacteria [191]. In 2011, Ma and his co-workers reported the enhanced anti-bacterial activity of Ag-based graphene oxide due to H-bonding between GO and the cell membrane of *E. coli.*, which prevents the nutrients uptake from the cell membrane and causes cell death [191].

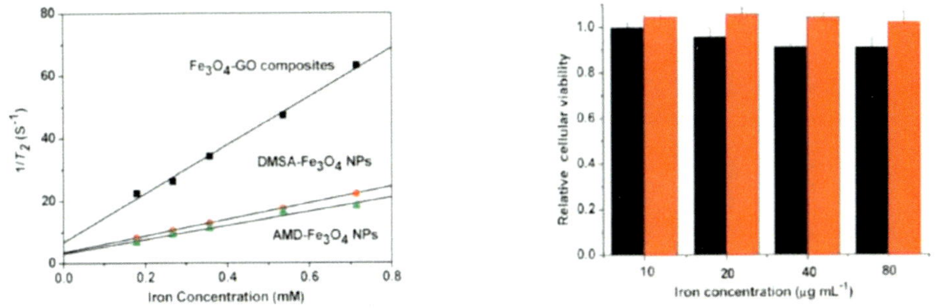

Fig. (13). (**a**) T-2 relaxation rates of the GO-Fe_3O_4, DMSA-Fe_3O_4NPs, AMD-Fe_3O_4NPs nanocomposites having iron concentration (mM) at 11.7 T. (**b**) Relative cellular viability of HeLa cells of AMD-Fe_3O_4 (red) and GO-Fe_3O_4 (black) incubated for 24 h, respectively. (Reproduced with permission from Ref [190], Copyright 2011, American Chemical Society).

Catalytic Potential of Graphene-based Nanocomposites

CO Oxidation

CO oxidation is one of the most useful reactions due to its significant potential towards the removal of CO and emission of pollutants [192].

In 2012, Liu and his co-workers synthesized a graphene-based Pd nanocomposite, loaded with 10 wt% Pd on graphitic material which showed the complete conversion of CO at 393 K, and improved stability without significant Pd size change [193]. El-Shall's research group developed bimetallic Pd–CoO based-graphene nanocomposites without using a chemical reductant [194]. They showed that the deoxygenation property of GO and light-induced metallic reduction property can be enhanced during the laser irradiation due to the formation of alloy structure, which also improved the catalytic property of graphene surface in comparison to the separately deposited graphene surface with Pd or CoO. Similar catalytic performance was also reported for bimetallic Pt–Ni based graphene nanocomposite *via* high dispersion of Pt–Ni on graphene surface [195]. Liu and his co-workers showed that the positively charged Pd-Gr can display a good relation of temperature with CO oxidation (Fig. **14**) [196].

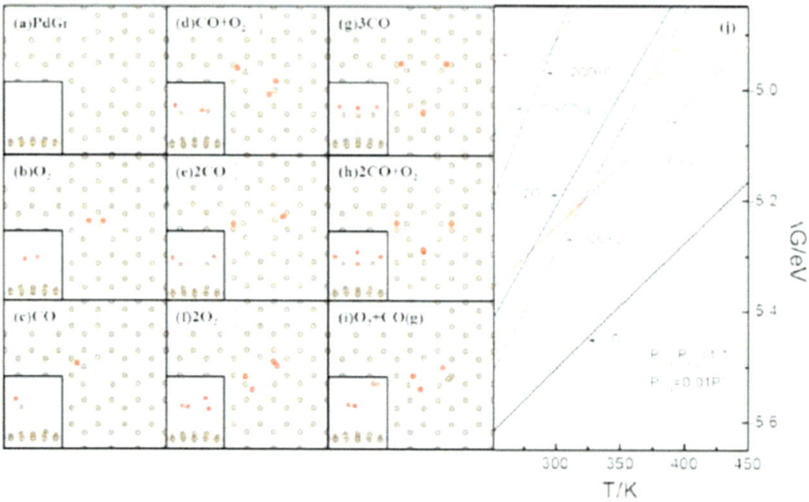

Fig. (14). Left and right-side panels show relation of atomic structures and temperature dependence with Gibbs free energies (ΔG) change over Pd-Gr in CO oxidation. (Reproduced with permission from Ref [196], Copyright 2020, American Chemical Society).

Oxidation of Alcohols

The carbonyl compounds (aldehydes and ketones) are an important group of compounds for the synthesis of several fine chemicals and biologically privileged molecules. The conventional oxidation of alcohol to carbonyl compounds is done by the use of homogeneous transition metal-based catalysts or several organometallic complexes (Fig. **15**). However, these homogeneous catalysts have several potential drawbacks such as moderate to high amount of catalyst loading, difficult removal of catalyst from the reaction medium (due to its toxic nature) and high cost [197]. To overcome these limitations, several research groups demonstrated that the Pd decorated on various carbon based material (like activated carbon [198], CNTs [199], and CMK-3) performed excellent catalytic property for oxidation of benzyl alcohol to benzaldehyde. In this context, graphene was utilized as a support for these transition metals for oxidation of alcohols to carbonyl compounds under a heterogeneous catalytic system.

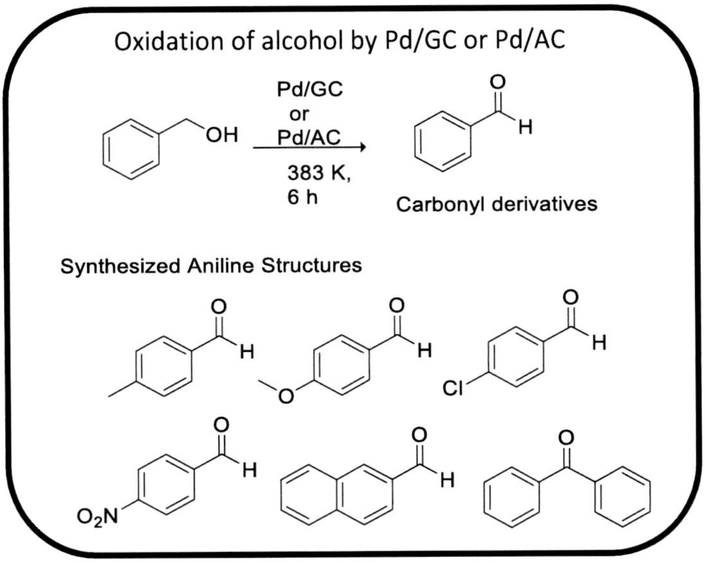

Fig. (15). Benzyl alcohol oxidation by palladium-based nanocomposite on different carbon-based material.

Li and his co-workers synthesized a Pd-based nanocomposite on carbon materials (such as carbon nanotubes, graphene (Pd/G), and activated carbon) *via* impregnation, which showed significant catalytic potential for benzyl alcohol oxidation [200]. The synthesized Pd-based nanocomposites having different carbon materials were approximately of the same size (≈ 4 nm) having the catalytic property in the following decreasing order: Pd/G > Pd/AC > Pd/CNT > commercial Pd/AC.

The observed catalytic activities of nanocomposites were significantly dependent on the use of carbon materials, and graphene was found superior in comparison to the other carbon materials. Huang *et al.* got similar results by decorating Au nanoparticles on several carbon materials, in which the rGO supported nanocomposite displayed the best catalytic performance (selectivity of 93% and conversion rate of 65%) for the oxidation of benzyl alcohol to benzaldehyde [201]. Furthermore, graphene-based metal oxides (eg: NiOx, CoOx, and FeOx,) [202], bimetallic Au–Pd based nanomaterial [203], and magnetically retrievable Fe_3O_4–Pt decorated rGO nanocomposites [204] were also synthesized and utilized for the selective oxidation to carbonyl compounds from alcohols. The bimetallic Au–Pd based graphene nanocomposites were found to be more active in comparison to the others towards selective oxidation to methyl formate from methanol [203].

Graphene-based Nanocomposite for Hydrogenation Reactions

The hydrogenation of various unsaturated organic compounds (such as alkenes, alkynes, nitriles, carbonyl compounds, and nitroarenes) to other privileged compounds is very challenging and has significant industrial applications [205]. In this context, the carbon- based catalytic hydrogenation is one of the most striking pathways due to its several advantages [206, 207] and in recent decades, graphene-based nanocomposites have displayed significant potential in this area.

Selective Reduction of Nitroarenes Using Graphene-based Nanocomposite

Aromatic amines are important precursors utilized for the synthesis of a variety of biologically important materials and industrial chemicals. Selective reduction of nitroarenes *via* catalytic hydrogenation is found to be one of the most effective routes for the synthesis of aromatic amines.

Pd-based nanocomposites display excellent catalytic potential for selective hydrogenation to amines from nitroarenes. In 2014, Wang and his co-workers synthesized graphene-based Pd nanocomposites *via* dispersion on graphene support by a hydrothermal process [206]. Pd-based rGO displayed improved catalytic potential in comparison to the commercial Pd/C towards the synthesis of 4-aminophenol (4-AP) from 4-nitrophenol (4-NP) with a reduction from $NaBH_4$.

Xu and his co-workers discovered similar findings *via* a simple redox reaction of H_2PdCl_4 and GO [208]. In addition, graphene-based Ni-Pd [209] or bimetallic Pd-Cu NPs [210] also displayed efficient catalytic properties towards the hydrogenation of nitroarenes (Fig. **16**).

Ni-Pd based graphene nanocomposites catalyze the tandem reduction of nitroarenes and dehydrogenation of ammonia borane (AB) [209]. Hou and his colleagues synthesized Pt-doped rGO *via* ethylene glycol reduction and used this composite to hydrogenate nitrobenzene *via* H_2 gas [211]. Additionally, it was also found that the hydrogenation efficacy was improved *via* the activation of H_2 gas by using a mixture of Pt and graphene [212]. Fan *et al.* synthesized Ru-decorated rGO nanocomposite *via* a hydrothermal process and utilized this nanocomposite for the selective hydrogenation of halo-nitroarenes [213].

Later on, the same research group also demonstrated higher selectivity of p-CAN over Ru anchored rGO nano-catalyst [214]. Other metal-based nanoparticles such as nickel [215, 216], gold [217, 218], silver [219] and Ni-Co alloy [220], metal oxide nanocomposites like Fe_3O_4 [221] were also anchored on the graphene surface and utilized for selective hydrogenation of aromatic nitro-compounds (Fig. **17**).

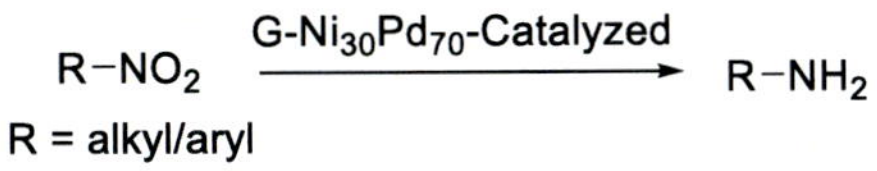

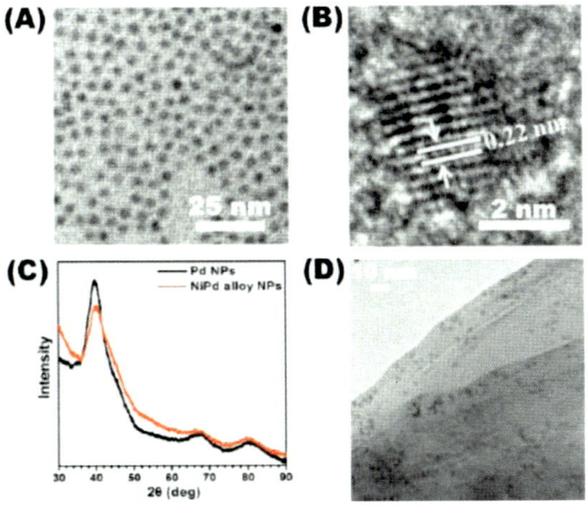

Fig. (16). (**A**) TEM (**B**) HRTEM diagram of $Ni_{30}Pd_{70}$ NPs, (**C**) XRD of Pd and Ni-Pd NPs, and (**D**) TEM image of Ni-Pd NPs decorated on graphene. (Reproduced with permission from Ref [209], Copyright 2014, American Chemical Society).

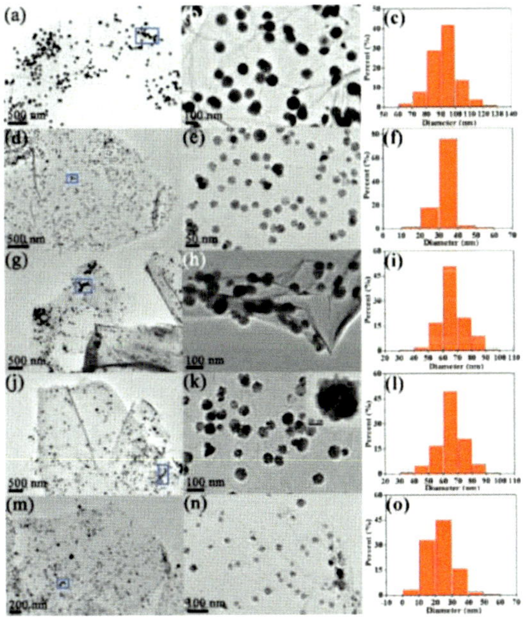

Fig. (17). Typical TEM images and Ni_xCo_{100-x} particle size distribution of RGO-Ni_xCo_{100-x} nanocomposites under different molar ratios of Ni/Co: (**a–c**) Co (**d–f**) $Ni_{25}Co_{75}$ (**g–i**) $Ni_{50}Co_{50}$ (**j–l**) $Ni_{75}Co_{25}$; and (**m–o**) Ni. (Reproduced with permission from Ref [220], Copyright 2012, American Chemical Society).

Magnetic retrievability and better recyclability are some of the advantageous features for catalysts. In this context, the magnetically separable catalysts were also synthesized by the use of metal NPs such as Pt, Pd, or Pt-Pd on Fe_3O_4/graphene surface, which was functionalized with L-lysine *via* selective doping. The NH_2 and COOH functional groups of L-lysine displayed unique roles as linkers between Fe_3O_4/graphene nanocomposites and the noble metals [222].

Hydrogenation of the C=C Bond using Graphene-based Nanocomposite

Conventionally, the hydrogenation of carbon-carbon unsaturated bonds is performed by metals such as Ru, Rh, Pt, Pd, and Au. In 2013, Maser and his co-workers developed Pd NPs based graphene nanocomposites having improved catalytic properties for the hydrogenation of unsaturated compounds [223].

In another study, graphene-based Pt nanocomposites were also prepared from H_2PtCl_6 having excellent catalytic reductive potential towards reduction of alkenes and alkynes to corresponding alkanes [224]. Han and his co-workers synthesized Ru NPs decorated with rGO for the hydrogenation of toluene [225]. Later on, several other Ru-decorated graphene nano-catalysts were also reported for the reduction of benzene/cyclohexene to cyclohexane [226, 227]. Begin and his co-workers fabricated graphene-based Pd nanocomposites for hydrogenation of cinnamaldehyde *via* H_2 reduction [228]. Furthermore, under microwave irradiated reduction conditions [229] and in solvothermal process [230], graphene anchored-Pd nanocomposites catalyst for the selective hydrogenation of the C=C bond of vinyl acetate and isophorone were reported.

Hydrogenation of the C=O Bond using Graphene-based Nanocomposite

The selective hydrogenation of the C=O bond of α,β-unsaturated carbonyl compounds over C=C bonds is more challenging due to comparably more favorable thermodynamic and kinetic stability of the C=O moiety over C=C bond [231, 232]. In 2014, Fu and his coworkers synthesized graphene-supported Pt nanocomposites for selective hydrogenation of cinnamic aldehyde (CAL) to cinnamyl alcohol (COL) [233]. The authors proposed that the electronic network of graphene based nanocomposite may form the inter-molecular π–π interactions with CAL and the aromatic moiety. Notably, the electron transfer behavior was also found for Pt decorated rGO nanocomposite [234]. The uniform dispersion of Pt on rGO in a specific solvent also displayed enhanced catalytic activity [235]. In addition, Fe [236] and Ni [237] based graphene nanocomposites for C=O over C=C hydrogenation were also reported. Furthermore, chiral-functionalized Ni [238] or Pd [239] nanocomposites based on rGO demonstrated exceptional catalytic potential for asymmetric hydrogenation, yielding optically pure enantiomers from prochiral substrates. Furthermore, graphene-based Pd [240], Pt [241], and RuO_2 [242] nanocomposites were also reported for selective hydrogenation of cellulose to sorbitol [241], catalytic hydrodesulfurization [240], and selective hydrogenation of carbonyl compounds [242].

Graphene Based Nanocomposite in Fischer–Tropsch Synthesis

CO hydrogenation in Fischer-Tropsch synthesis (FTS) has received a lot of attention [243, 244], and carbon-based materials have been discovered to be good supports for FTS catalysts due to their low interaction with active metals. In this context, a variety of carbon supports (such as carbon spheres, activated carbon, CNTs, mesoporous carbon, and carbon nanofibers) have been utilized for the FTS catalyst synthesis (Fig. **18**) [245 - 247].

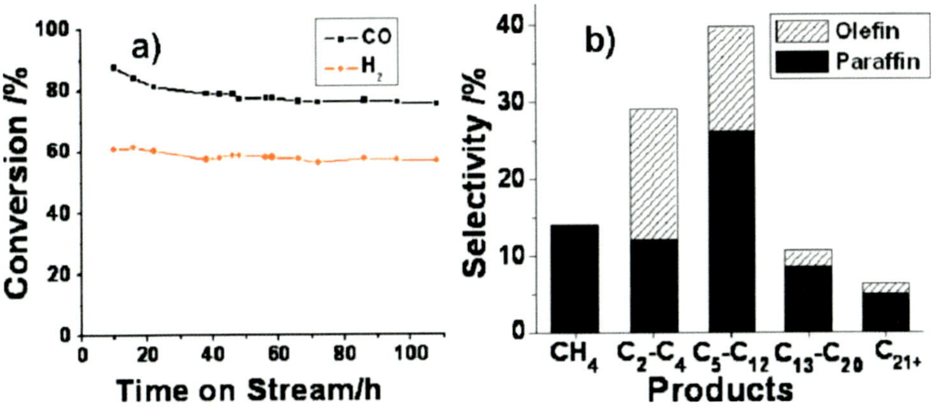

Fig. (18). Fischer–Tropsch synthesis (FTS) on reduced Fe*x*O*y*@C spheres. (**a**) conversion of CO and H_2 with time. (**b**) hydrocarbon selectivity. (Reproduced with permission from Ref [245], Copyright 2010, American Chemical Society).

Qiao and his co-workers synthesized Fe-based rGO nanocomposites *via* hydrothermal hydrolysis and reduction process by the use of GO and $Fe(acac)_3$ [248]. The nanocomposites synthesized demonstrated exceptional catalytic potential as well as thermodynamic stability. From p-rGO, a Fe-anchored pre-reduced GO (p-rGO) was synthesized, which demonstrated improved selectivity and stability to C_{5+} products. This was the first documented report which showed the application of graphene-based nanocomposite having mixed metals at high-temperatures. The role of the oxygen-containing functionalities on graphite support is authenticated over GO decorated with pyrolytic iron in the FTS process [249]. El-Shall and his colleagues created microwave-irradiated graphene-based K/Mn-functionalized iron nano-catalysts [250]. The graphene support improved the C_{8+} products as well as reduced the production of CO_2 and CH_4 as compared to carbon nanotubes. The synthesis of CO_2 from CO was reduced from 14% to 2.6%. The catalytic selectivity (without CO_2) towards C_{8+} products increased (61.5% to 86.7%), and for CH_4, it was reduced (11.5% to 2.6%). In 2013, Yang and his co-workers fabricated cobalt-based graphene nanocomposites for the hydrogenation of CO_2 *via* a solvothermal process [251]. In the solvothermal process, as the temperature changed from 453 to 493 K, the cobalt nano-catalysts transformed from cubic shape to hexagonal shape, followed by an irregular arrangement to a sea-urchin like structure, and ultimately to a flower like final structure. In CO_2 reduction, the nanocomposites formed at 493 K displayed excellent catalytic properties due to the active hexagonal Co phase formation (Fig. **19**) [252].

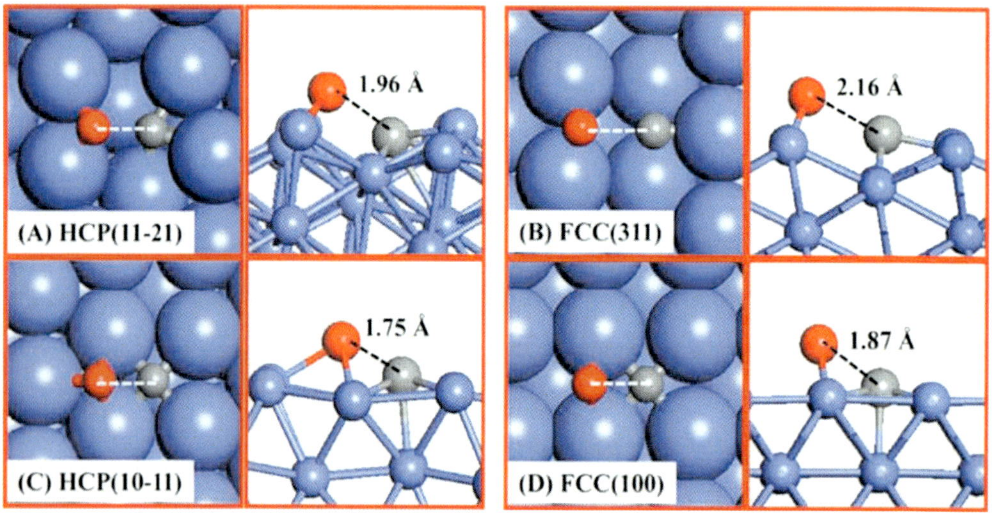

Fig. (19). Top and side views of CO dissociation from optimized TSs. (Blue balls represent Co, red balls represent O, and gray balls represent C atoms). (Reproduced with permission from Ref [252], Copyright 2013, American Chemical Society).

Graphene-supported Nanocomposites for Hydrolysis of AB (Ammonia Borane)

Hydrogen is a well-documented clean fuel due to its safe, efficient delivery and storage point of view [253]. The efficient chemical synthesis of H_2 was performed by using B–N based compounds, in which AB (ammonia borane) was found to be the best one due to its good stability, less hazardous nature, and high hydrogen content (up to 19.6 wt%) [254]. Thus, several research groups have worked towards the synthesis of modified graphene-based nanocomposites having good catalytic potential towards the hydrolysis of AB.

In 2012, Metin *et al.* fabricated Pd-anchored graphene nanocomposites for AB hydrolysis [255]. In the same year, Zeng and his co-workers synthesized Pd-anchored rGO nanocomposite *via* a solvothermal process [256]. Graphene-based

Co [257] and Cu [258] NPs were also developed using AB as a reductant, which displayed improved catalytic properties in comparison to the rGO-free Co or Cu NPs. Similarly, co-anchored rGO [259] and Ru- decorated rGO [260] nanocomposites were also synthesized from methylamine borane (Me-AB) using $NaBH_4$ [260]. Moreover, the bimetallic nanocomposites with graphene support were also fabricated and utilized for the hydrolysis of ammonia borane. Several research groups have fabricated bimetallic graphene-based nanocomposites (like Co-Fe [261], Ni-Co [262], Co-Cu [263], and Ru-Cu [264]. Ciftci and Metin reported Ni-Pd nanocomposites with rGO *via* reduction from $Pd(acac)_2$ and $Ni(OAc)_2$ [265]. Furthermore, other bimetallic graphene-supported nanocomposites such as Cu-Co [266], Ag-Co [267, 268], Pd-Co [269], Ru-Ni [270], and Ru-Co [271], were also synthesized *via* one-step co-reduction employing ammoniaborane as a reductant depending on the change of the two metal substrates in their reduction potentials (Fig. **20**).

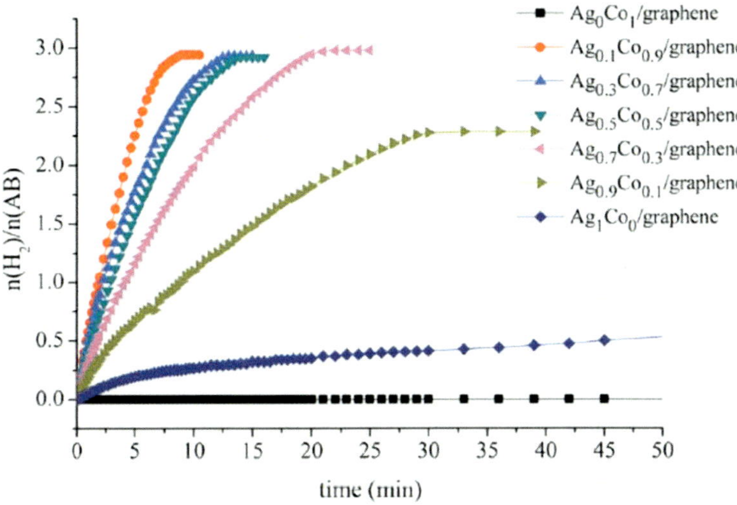

Fig. (20). Representation of H_2 gas generation after hydrolysis from graphene based nano-catalysts. (Reproduced with permission from Ref [268], Copyright 2013, American Chemical Society).

It is also observed that the graphene-based nanocomposites show improved catalytic nature compared to their graphene-free counterparts. Recent studies have synthesized multi-metallic graphene-based nanocomposite such as Ni-Cu-Co [272], Ni-Ag-Co [273], and Fe-Ag-Co [274] *via* a co-reduction process using NaBH4, Me-AB, or AB as a reductant. In a similar study, it was reported that the catalytic property can be tuned according to the composite size by changing the graphite-based surface of the Ag-Co-Ni nanocomposite for AB hydrolysis [275]. Graphene-based nanocomposites were also utilized for H_2 production from $NaBH_4$ [276], Me-AB [268], and propane [277]. It is noteworthy that the design of graphene-based nanocomposites was utilized as a key feature for the efficient hydrolysis of AB [278].

Graphene-supported Nanocomposite for Coupling Reactions

C–C bond construction in organic scaffolds *via* coupling reactions is a very demandable approach for both academic research and industrial production. In 2010, the Nobel Prize in Chemistry was awarded to Heck, Negishi, and Suzuki for their potential discovery of Pd-catalyzed cross-coupling reactions. In general, coupling reactions are carried out under homogeneous catalytic conditions; however, due to the severe toxicity profile, poor recovery and recyclability, as well as higher water and air-sensitive properties of homogeneous catalysts, their applications to large scale industrial processes are limited [279]. In this context, reusable heterogeneous catalysts provide potential benefits to overcome this problem.

Typically Pd NPs anchored on some materials like zeolites [280], activated carbon [281], and carbon nanotubes [282] have been utilized for coupling reactions. In recent years, a huge number of graphene-based nanocomposites have been synthesized and utilized for the coupling reactions. In 2009, Mülhaupt and his co-workers developed a Pd-functionalized graphene support by immobilizing the Pd^{2+} on GO *via* cation exchange followed by the reduction from $N_2H_4{\cdot}H_2O$, H_2 or thermal reduction (Fig. **21**) [283].

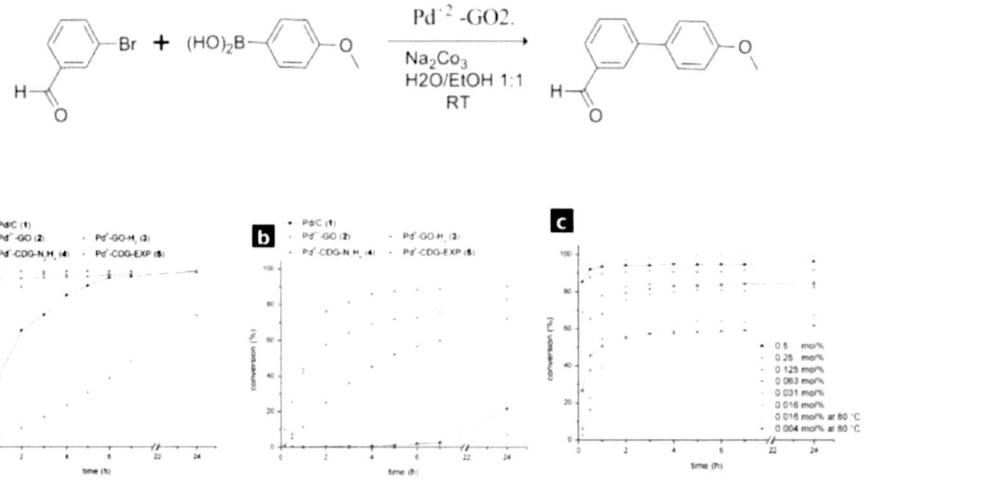

Fig. (21). Screening of Pd-GO/CDG catalysts for Suzuki-Miyaura reaction. Coupling reaction of 4-methoxyphenylboronic acid with (**a**) 4-bromo-1,2-difluorobenzene (**b**) 4-bromo-1,2-(methylenedioxy) benzene (**c**) 3-bromobenzaldehyde. (Reproduced with permission from Ref [283], Copyright 2009, American Chemical Society).

Furthermore, the graphene-based nanocomposites displayed enhanced catalytic activity in comparison to the typical Pd/C catalyst, and very little Pd leaching (<1 ppm) was found. Later on, graphene-based Pd nanocomposites having improved catalytic property were also synthesized with various reductants like H_2 [284], $NaBH_4$ [285], $N_2H_4{\cdot}H_2O$ [286, 287], and ascorbic acid [288].

Other approaches were investigated, including CO-reduction with GO and H_2PtCl_6 from Ti^{3+} [289], electrochemical-based CO-deposition [290], thermal exfoliation process [291], and pulsed laser based irradiation [292]. The graphene support based mixed hybrid also demonstrated excellent catalytic properties for Sonogashira and Heck cross-coupling reactions. Notably, it was also documented that the Pd-based graphene nanocomposite displayed improved stability to polymeric materials like polyvinyl pyrrolidone, polystyrene-polyethylene oxide, and poly(amido-amine) dendrimers compared to the Pd nanoparticles (Fig. **22**) [292]. In 2013, Metin and his co-workers developed a synthetic route for graphene-based Ni/Pd nanocomposites *via* a two-step methodology and utilized this synthesized nanocomposite for Suzuki–Miyaura cross-coupling of phenyl iodide and phenylboronic acid [293]. Xu *et al.* developed graphene-supported Pd–Co nanocomposites, which were utilized in the Suzuki coupling reaction of phenylboronic acid with iodobenzene and conversion up to 96% and almost 100% selectivity with respect to biphenyl was achieved [294]. Fan *et al.* developed uniformly deposited Au NPs on graphene support having enhanced catalytic properties compared to unsupported Au NPs by reducing $HAuCl_4$ using sodium dodecyl sulfate [295, 296].

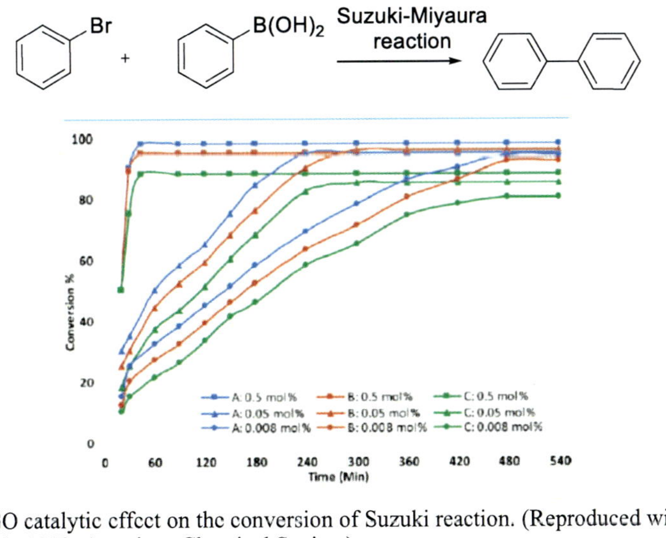

Fig. (22). Pd/PRGO catalytic effect on the conversion of Suzuki reaction. (Reproduced with permission from Ref [292], Copyright 2012, American Chemical Society).

In 2013, Wang and his co-workers investigated a facile synthetic route for Fe_3O_4 doped graphene nanocomposite synthesis *via* thermal decomposition of $Fe(CO)_5$ on GO surface [297]. In 2014, another report was published for graphene-based Ni nanocomposite synthesis and utilization of its catalytic potential in the Kumada-Corriu cross-coupling reaction of 4-iodoanisole and phenyl magnesium chloride. Therefore, the progress of graphene-based nanocomposites with improved catalytic activity and stability towards coupling reactions warrants further research and development.

CONCLUSION AND SUMMARY

Graphene is utilized as a two-dimensional solid support for the growth and nucleation of metal or metal oxide nano-particles, and the synergistic effects of these graphene-supported nanocomposites dramatically enhance the catalytic performance of the composite materials. The better knowledge of graphene surface-based chemistry leads to the development of potential synthetic routes for graphene-based inorganic nanocomposites synthesis by introducing several metal-based NPs. Thus, graphene-based metal nanocomposites with decorated, anchored, doped, encapsulated, layered, wrapped, and mixed structures have demonstrated huge potential for numerous industrial applications and are commercially achievable compared to less active nanocomposites using other carbon family members. The synergistic effects of graphene support and metal NPs in these nanocomposites result in improved catalytic potential in a variety of applications such as medicine, energy, biosensors, and industrial applications. The synthesis of these graphene-based nanocomposites, with well-defined shape and size along with the crystallinity of nanocomposites, not only prevents the restacking of graphene-based nanomaterial but also offers better support for the synthesis of 2D porous materials. This exceptional and remarkable catalytic potential of graphene-based nanocomposite has captured the scientific community's attention enormously, and as a result, a large number of publications and patents related to graphene as well as graphene-based nanocomposite synthesis and utilization have been reported in recent decades. In this context, the recent developments towards the synthesis of graphene-based metal nanocomposites and their catalytic utilization are broadly discussed in the present chapter.

Despite the significant advancement in the preparation of graphene-based nanocomposites, challenges still exist in their catalytic utilization on a large scale. For instance, advanced utilization of graphene-based metal nanocomposites needs further extensive exploration to understand the interactions between metal nanomaterials and the graphene/graphene oxide surface, which have a direct correlation with the catalytic nature of these nanocomposites. A thorough investigation of these surface interactions will undoubtedly improve the catalytic potential of graphene-based nanocomposites in a variety of fields, including biosensing, drug delivery, imaging, and catalysis. Moreover, the advancement in graphene surface modification will ultimately allow the fabrication of mono/multi-layered nanocomposites and also improve the protocols for the production of graphene-based nanocomposites on a large scale to get an advanced form of nanotechnology to build devices for several potential applications.

LIST OF ABBREVIATIONS

2D structure	Two-dimensional structure
3D composites	Three-dimensional structure
CDG	Chemically derived graphene
CNTs	Carbon nanotubes
DEFCs	Direct ethanol fuel cells
DFAFCs	Direct formic acid fuel cells
DFT	Density functional theory
DMFCs	Direct methanol fuel cell
EDLC	Electric double layer carbon
ERGO	Electrochemically reduced graphene oxide
FLG	Few-layer graphene
GC	Graphene composites
GNP	Graphene nanoparticle
GNS	Graphene nano-sheet
GO	Graphene oxide
HRG	Highly reduced graphene oxide
MWI	Microwave irradiation
OP pesticides	Organophosphorus pesticides
ORR	Oxygen reduction reaction
Pd/AC	Palladium/activated carbon
Pd/PRGO	Pd-partially reduced graphene oxide catalysts
RG/NP	Reduced graphene nanoparticles
r-GO	Reduced graphene oxide

CONSENT FOR PUBLICATION

Not applicable.

CONFLICT OF INTEREST

The authors declare no conflict of interest, financial or otherwise.

ACKNOWLEDGEMENTS

The corresponding author P. K. J. gratefully acknowledges Prof. (Dr.) James Sacchettini, Department of Biochemistry & Biophysics, Texas A&M University for support and encouragement and Dr. Aashish Srivastava, Senior Research Scientist, Department of Biochemistry and Biophysics for technical help and discussion.

REFERENCES

[1] Wallace, P.R. The Band Theory of Graphite. *Phys. Rev.*, **1947**, *71*(9), 622-634. [http://dx.doi.org/10.1103/PhysRev.71.622]

[2] Bodenmann, A.K.; MacDonald, A.H. Graphene: Exploring carbon flatland. *Phys. Today,* **2007**, *60*(8), 35-41. [http://dx.doi.org/10.1063/1.2774096]

[3] Rao, C.N.R.; Sood, A.K.; Subrahmanyam, K.S.; Govindaraj, A. Graphene: the new two-dimensional nanomaterial. *Angew. Chem. Int. Ed.,* **2009**, *48*(42), 7752-7777. [http://dx.doi.org/10.1002/anie.200901678] [PMID: 19784976]

[4] Allen, M.J.; Tung, V.C.; Kaner, R.B. Honeycomb carbon: a review of graphene. *Chem. Rev.,* **2010**, *110*(1), 132-145. [http://dx.doi.org/10.1021/cr900070d] [PMID: 19610631]

[5] Yuan, W.; Huang, L.; Zhou, Q.; Shi, G. Ultrasensitive and selective nitrogen dioxide sensor based on self-assembled graphene/polymer composite nanofibers. *ACS Appl. Mater. Interfaces,* **2014**, *6*(19), 17003-17008. [http://dx.doi.org/10.1021/am504616c] [PMID: 25208097]

[6] Stankovich, S.; Dikin, D.A.; Dommett, G.H.B.; Kohlhaas, K.M.; Zimney, E.J.; Stach, E.A.; Piner, R.D.; Nguyen, S.T.; Ruoff, R.S. Graphene-based composite materials. *Nature,* **2006**, *442*(7100), 282-286. [http://dx.doi.org/10.1038/nature04969] [PMID: 16855586]

[7] Ding, S.; Chen, J.S.; Luan, D.; Boey, F.Y.C.; Madhavi, S.; Lou, X.W.D. Graphene-supported anatase TiO_2 nanosheets for fast lithium storage. *Chem. Commun. (Camb.),* **2011**, *47*(20), 5780-5782. [http://dx.doi.org/10.1039/c1cc10687b] [PMID: 21494738]

[8] Mohan, V.B.; Lau, K.; Hui, D.; Bhattacharyya, D. Graphene-based materials and their composites: A review on production, applications and product limitations. *Compos., Part B Eng.,* **2018**, *142*, 200-220. [http://dx.doi.org/10.1016/j.compositesb.2018.01.013]

[9] Torres, T. Graphene chemistry. *Chem. Soc. Rev.,* **2017**, *46*(15), 4385-4386. [http://dx.doi.org/10.1039/C7CS90061A] [PMID: 28661520]

[10] Shang, N.G.; Papakonstantinou, P.; McMullan, M.; Chu, M.; Stamboulis, A.; Potenza, A.; Dhesi, S.S.; Marchetto, H. Catalyst-Free Efficient Growth, Orientation and Biosensing Properties of Multilayer Graphene Nanoflake Films with Sharp Edge Planes. *Adv. Funct. Mater.,* **2008**, *18*(21), 3506-3514.

[http://dx.doi.org/10.1002/adfm.200800951]

[11] Su, C.Y.; Lu, A.Y.; Xu, Y.; Chen, F.R.; Khlobystov, A.N.; Li, L.J. High-quality thin graphene films from fast electrochemical exfoliation. *ACS Nano,* **2011**, *5*(3), 2332-2339. [http://dx.doi.org/10.1021/nn200025p] [PMID: 21309565]

[12] Novoselov, K.S.; Geim, A.K.; Morozov, S.V.; Jiang, D.; Zhang, Y.; Dubonos, S.V.; Grigorieva, I.V.; Firsov, A.A. Electric field effect in atomically thin carbon films. *Science,* **2004**, *306*(5696), 666-669. [http://dx.doi.org/10.1126/science.1102896] [PMID: 15499015]

[13] Wang, H.; Cui, L.F.; Yang, Y.; Sanchez Casalongue, H.; Robinson, J.T.; Liang, Y.; Cui, Y.; Dai, H. Mn3O4-graphene hybrid as a high-capacity anode material for lithium ion batteries. *J. Am. Chem. Soc.,* **2010**, *132*(40), 13978-13980. [http://dx.doi.org/10.1021/ja105296a] [PMID: 20853844]

[14] Li, H.; Chi, Z.; Li, J. Covalent bonding synthesis of magnetic graphene oxide nanocomposites for Cr(III) removal. *Desalination Water Treat.,* **2014**, *52*(10-12), 1937-1946. [http://dx.doi.org/10.1080/19443994.2013.806224]

[15] Singh, V.; Joung, D.; Zhai, L.; Das, S.; Khondaker, S.I.; Seal, S. Graphene based materials: Past, present and future. *Prog. Mater. Sci.,* **2011**, *56*(8), 1178-1271. [http://dx.doi.org/10.1016/j.pmatsci.2011.03.003]

[16] Tabarsa, M. ZareNezhad, B. An Eco-Friendly Air–Water Plasma Surface Treatment Technique for Improving the Stability of Graphene Oxide Nanosheets in Aqueous Solutions. *Mater. Today Commun.,* **2021**, *101940*, 26.

[17] Spilarewicz-Stanek, K.; Jakimińska, A.; Kisielewska, A.; Dudek, M.; Piwoński, I. Graphene Oxide Photochemical Transformations Induced by UV Irradiation during Photocatalytic Processes. *Mater. Sci. Semicond. Process.,* **2020**.

[18] Kuila, T.; Bose, S.; Mishra, A.K.; Khanra, P.; Kim, N.H.; Lee, J.H. Chemical functionalization of graphene and its applications. *Prog. Mater. Sci.,* **2012**, *57*(7), 1061-1105. [http://dx.doi.org/10.1016/j.pmatsci.2012.03.002]

[19] Velásquez-Rojas, M.M.; Contreras-Torres, F.F.; Meza-Laguna, V.; Álvarez-Zauco, E.; Farías, M.H.; Basiuk, V.A.; Basiuk, E.V. Solvent-free functionalization of graphene oxide powder and paper with aminobenzo-crown ethers and complexation with alkali metal cations. *Mater. Chem. Phys.,* **2021**, *260*, 124127. [http://dx.doi.org/10.1016/j.matchemphys.2020.124127]

[20] Garg, B.; Bisht, T., Ling, Y.C. Graphene-based nanomaterials as heterogeneous acid catalysts: a comprehensive perspective. *Molecules,* **2014**, *19*(9), 14582-14614. [http://dx.doi.org/10.3390/molecules190914582] [PMID: 25225721]

[21] Huang, X.; Yin, Z.; Wu, S.; Qi, X.; He, Q.; Zhang, Q.; Yan, Q.; Boey, F.; Zhang, H. Graphene-based materials: synthesis, characterization, properties, and applications. *Small,* **2011**, *7*(14), 1876-1902. [http://dx.doi.org/10.1002/smll.201002009] [PMID: 21630440]

[22] Srivastava, S.; Jain, S.K.; Gupta, G.; Senguttuvan, T.D.; Gupta, B.K. Boron-doped few-layer graphene nanosheet gas sensor for enhanced ammonia sensing at room temperature. *RSC Advances,* **2020**, *10*(2), 1007-1014. [http://dx.doi.org/10.1039/C9RA08707A] [PMID: 35494469]

[23] Hong, W.; Xu, Y.; Lu, G.; Li, C.; Shi, G. Transparent graphene/PEDOT–PSS composite films as counter electrodes of dye-sensitized solar cells. *Electrochem. Commun.,* **2008**, *10*(10), 1555-1558. [http://dx.doi.org/10.1016/j.elecom.2008.08.007]

[24] La Notte, L.; Giordano, L.; Calabrò, E.; Bedini, R.; Colla, G.; Puglisi, G.; Reale, A. Hybrid and organic photovoltaics for greenhouse applications. *Appl. Energy,* **2020**, *278*, 115582. [http://dx.doi.org/10.1016/j.apenergy.2020.115582]

[25] Liu, M.; Zhang, R.; Chen, W. Graphene-supported nanoelectrocatalysts for fuel cells: synthesis,

properties, and applications. *Chem. Rev.*, **2014**, *114*(10), 5117-5160. [http://dx.doi.org/10.1021/cr400523y] [PMID: 24666160]

[26] Yang, C.; Jiang, Q.; Huang, H.; He, H.; Yang, L.; Li, W.; Li, W. Polyelectrolyte-Induced Stereoassembly of Grain Boundary-Enriched Platinum Nanoworms on $Ti_3C_2T_x$ MXene Nanosheets for Efficient Methanol Oxidation. *ACS Appl. Mater. Interfaces*, **2020**, *12*(21), 23822-23830. [http://dx.doi.org/10.1021/acsami.0c02806] [PMID: 32356656]

[27] Yang, H.; Li, S.; Feng, F.; Ou, S.; Li, F.; Yang, M.; Qian, K.; Jin, J.; Ma, J. Palladium Nanoparticles with Surface Enrichment of Palladium Oxide Species Immobilized on the Aniline-Functionalized Graphene As an Advanced Electrocatalyst of Ethanol Oxidation. *ACS Sustain. Chem.& Eng.*, **2019**, *7*(17), 14621-14628. [http://dx.doi.org/10.1021/acssuschemeng.9b02397]

[28] Li, X.; Huang, X.; Liu, D.; Wang, X.; Song, S.; Zhou, L.; Zhang, H. Synthesis of 3D Hierarchical Fe_3O_4/Graphene Composites with High Lithium Storage Capacity and for Controlled Drug Delivery. *J. Phys. Chem. C*, **2011**, *115*(44), 21567-21573. [http://dx.doi.org/10.1021/jp204502n]

[29] Adepu, A.K.; Goskula, S.; Chirra, S.; Siliveri, S.; Gujjula, S.R.; Venkatathri, N. Magnetically separable porous titanosilicate/Fe3O4 hybrid nanocomposites with enhanced photocatalytic performance under UV light irradiation. *J. Porous Mater.*, **2019**, *26*(5), 1259-1267. [http://dx.doi.org/10.1007/s10934-019-00719-9]

[30] Wu, Z.S.; Zhou, G.; Yin, L.C.; Ren, W.; Li, F.; Cheng, H.M. Graphene/metal oxide composite electrode materials for energy storage. *Nano Energy*, **2012**, *1*(1), 107-131. [http://dx.doi.org/10.1016/j.nanoen.2011.11.001]

[31] Chen, D.; Zhang, H.; Liu, Y.; Li, J. Graphene and its derivatives for the development of solar cells, photoelectrochemical, and photocatalytic applications. *Energy Environ. Sci.*, **2013**, *6*(5), 1362-1387. [http://dx.doi.org/10.1039/c3ee23586f]

[32] Hu, M.; Yao, Z.; Wang, X. Graphene-Based Nanomaterials for Catalysis. *Ind. Eng. Chem. Res.*, **2017**, *56*(13), 3477-3502. [http://dx.doi.org/10.1021/acs.iecr.6b05048]

[33] Khodakov, A.; Olthof, B.; Bell, A.T.; Iglesia, E. Structure and Catalytic Properties of Supported Vanadium Oxides: Support Effects on Oxidative Dehydrogenation Reactions. *J. Catal.*, **1999**, *181*(2), 205-216. [http://dx.doi.org/10.1006/jcat.1998.2295]

[34] Santander, P.; Bravo, L.; Pecchi, G.; Karelovic, A. The consequences of support identity on the oxidative conversion of furfural to maleic anhydride on vanadia catalysts. *Appl. Catal. A Gen.*, **2020**, *595*, 117513. [http://dx.doi.org/10.1016/j.apcata.2020.117513]

[35] Luo, D.; Zhang, G.; Liu, J.; Sun, X. Evaluation Criteria for Reduced Graphene Oxide. *J. Phys. Chem. C*, **2011**, *115*(23), 11327-11335. [http://dx.doi.org/10.1021/jp110001y]

[36] Xu, Y.; Bai, H.; Lu, G.; Li, C.; Shi, G. Flexible graphene films *via* the filtration of water-soluble noncovalent functionalized graphene sheets. *J. Am. Chem. Soc.*, **2008**, *130*(18), 5856-5857. [http://dx.doi.org/10.1021/ja800745y] [PMID: 18399634]

[37] Su, Q.; Pang, S.; Alijani, V.; Li, C.; Feng, X.; Müllen, K. Composites of Graphene with Large Aromatic Molecules. *Adv. Mater.*, **2009**, *21*(31), 3191-3195. [http://dx.doi.org/10.1002/adma.200803808]

[38] Kim, J.; Cha, J.; Chung, B.; Ryu, S.; Hong, S.H. Fabrication and mechanical properties of carbon fiber/epoxy nanocomposites containing high loadings of noncovalently functionalized graphene nanoplatelets. *Compos. Sci. Technol.*, **2020**, *192*, 108101. [http://dx.doi.org/10.1016/j.compscitech.2020.108101]

[39] An, X.; Simmons, T.; Shah, R.; Wolfe, C.; Lewis, K.M.; Washington, M.; Nayak, S.K.; Talapatra, S.; Kar, S. Stable aqueous dispersions of noncovalently functionalized graphene from graphite and their multifunctional high-performance applications. *Nano Lett.*, **2010**, *10*(11), 4295-4301. [http://dx.doi.org/10.1021/nl903557p] [PMID: 20557029]

[40] Shang, X.; Park, C.H.; Jung, G.Y.; Kwak, S.K.; Oh, J.H. Highly Enantioselective Graphene-Based Chemical Sensors Prepared by Chiral Noncovalent Functionalization. *ACS Appl. Mater. Interfaces*, **2018**, *10*(42), 36194-36201. [http://dx.doi.org/10.1021/acsami.8b13517] [PMID: 30270614]

[41] Tang, Z.; Shen, S.; Zhuang, J.; Wang, X. Noble-metal-promoted three-dimensional macroassembly of single-layered graphene oxide. *Angew. Chem. Int. Ed.*, **2010**, *49*(27), 4603-4607. [http://dx.doi.org/10.1002/anie.201000270] [PMID: 20491102]

[42] Cong, H.P.; Ren, X.C.; Wang, P.; Yu, S.H. Macroscopic multifunctional graphene-based hydrogels and aerogels by a metal ion induced self-assembly process. *ACS Nano*, **2012**, *6*(3), 2693-2703. [http://dx.doi.org/10.1021/nn300082k] [PMID: 22303866]

[43] Stankovich, S.; Dikin, D.A.; Piner, R.D.; Kohlhaas, K.A.; Kleinhammes, A.; Jia, Y.; Wu, Y.; Nguyen, S.T.; Ruoff, R.S. Synthesis of graphene-based nanosheets *via* chemical reduction of exfoliated graphite oxide. *Carbon*, **2007**, *45*(7), 1558-1565. [http://dx.doi.org/10.1016/j.carbon.2007.02.034]

[44] Xu, Y.; Liu, Z.; Zhang, X.; Wang, Y.; Tian, J.; Huang, Y.; Ma, Y.; Zhang, X.; Chen, Y. A Graphene Hybrid Material Covalently Functionalized with Porphyrin: Synthesis and Optical Limiting Property. *Adv. Mater.*, **2009**, *21*(12), 1275-1279. [http://dx.doi.org/10.1002/adma.200801617]

[45] Liu, J.; Liang, Q.; Zhao, R.; Lei, S.; Hu, W. Application of organic–graphene hybrids in high performance photodetectors. *Mater. Chem. Front.*, **2020**, *4*(2), 354-368. [http://dx.doi.org/10.1039/C9QM00517J]

[46] Wang, H.X.; Zhou, K.G.; Xie, Y.L.; Zeng, J.; Chai, N.N.; Li, J.; Zhang, H.L. Photoactive graphene sheets prepared by "click" chemistry. *Chem. Commun. (Camb.)*, **2011**, *47*(20), 5747-5749. [http://dx.doi.org/10.1039/c1cc11121c] [PMID: 21503310]

[47] Li, Y.; Tang, L.; Li, J. Preparation and electrochemical performance for methanol oxidation of pt/graphene nanocomposites. *Electrochem. Commun.*, **2009**, *11*(4), 846-849. [http://dx.doi.org/10.1016/j.elecom.2009.02.009]

[48] Guo, S.; Dong, S., Wang, E. Three-dimensional Pt-on-Pd bimetallic nanodendrites supported on graphene nanosheet: facile synthesis and used as an advanced nanoelectrocatalyst for methanol oxidation. *ACS Nano*, **2010**, *4*(1), 547-555. [http://dx.doi.org/10.1021/nn9014483] [PMID: 20000845]

[49] Peng, X.; Lu, D.; Qin, Y.; Li, M.; Guo, Y.; Guo, S. Pt-on-Pd Dendritic Nanosheets with Enhanced Bifunctional Fuel Cell Catalytic Performance. *ACS Appl. Mater. Interfaces*, **2020**, *12*(27), 30336-30342. [http://dx.doi.org/10.1021/acsami.0c05868] [PMID: 32525299]

[50] Zhou, X.; Huang, X.; Qi, X.; Wu, S.; Xue, C.; Boey, F.Y.C.; Yan, Q.; Chen, P.; Zhang, H. *In Situ* Synthesis of Metal Nanoparticles on Single-Layer Graphene Oxide and Reduced Graphene Oxide Surfaces. *J. Phys. Chem. C*, **2009**, *113*(25), 10842-10846. [http://dx.doi.org/10.1021/jp903821n]

[51] Xu, C.; Wang, X. Fabrication of flexible metal-nanoparticle films using graphene oxide sheets as substrates. *Small*, **2009**, *5*(19), 2212-2217. [http://dx.doi.org/10.1002/smll.200900548] [PMID: 19662647]

[52] Gnanaprakasam, P.; Jeena, S.E.; Selvaraju, T. Hierarchical electroless Pt deposition at Au decorated reduced graphene oxide *via* a galvanic exchanged process: an electrocatalytic nanocomposite with

enhanced mass activity for methanol and ethanol oxidation. *J. Mater. Chem. A Mater. Energy Sustain.*, **2015**, *3*(35), 18010-18018.
[http://dx.doi.org/10.1039/C5TA04293C]

[53] Tang, Y.B.; Lee, C.S.; Xu, J.; Liu, Z.T.; Chen, Z.H.; He, Z.; Cao, Y.L.; Yuan, G.; Song, H.; Chen, L.; Luo, L.; Cheng, H.M.; Zhang, W.J.; Bello, I.; Lee, S.T. Incorporation of graphenes in nanostructured $TiO(_2)$ films *via* molecular grafting for dye-sensitized solar cell application. *ACS Nano,* **2010**, *4*(6), 3482-3488.
[http://dx.doi.org/10.1021/nn100449w] [PMID: 20455548]

[54] Nethravathi, C.; Nisha, T.; Ravishankar, N.; Shivakumara, C.; Rajamathi, M. Graphene - nanocrystalline metal sulphide composites produced by a one-pot reaction starting from graphite oxide. *Carbon,* **2009**, *47*(8), 2054-2059.
[http://dx.doi.org/10.1016/j.carbon.2009.03.055]

[55] Mao, S.; Wen, Z.; Kim, H.; Lu, G.; Hurley, P.; Chen, J. A general approach to one-pot fabrication of crumpled graphene-based nanohybrids for energy applications. *ACS Nano,* **2012**, *6*(8), 7505-7513.
[http://dx.doi.org/10.1021/nn302818j] [PMID: 22838735]

[56] Soysal, F.; Çıplak, Z.; Gökalp, C.; Getiren, B.; Yıldız, N. One-step hydrothermal synthesis of nitrogen doped reduced graphene oxide-silver nanocomposites: Catalytic performance. *Appl. Organomet. Chem.,* **2020**, *34*(5), 34.
[http://dx.doi.org/10.1002/aoc.5621]

[57] Gowthaman, N.S.K.; Raj, M.A.; John, S.A. Nitrogen-Doped Graphene as a Robust Scaffold for the Homogeneous Deposition of Copper Nanostructures: A Nonenzymatic Disposable Glucose Sensor. ACS Sustainable Chemistry &amp. *Engineering,* **2017**, *5*(2), 1648-1658.

[58] Chang, H.; Lv, X.; Zhang, H.; Li, J. Quantum dots sensitized graphene: *In Situ* growth and application in photoelectrochemical cells. *Electrochem. Commun.,* **2010**, *12*(3), 483-487.
[http://dx.doi.org/10.1016/j.elecom.2010.01.025]

[59] Liu, X.; Pan, L.; Lv, T.; Zhu, G.; Sun, Z.; Sun, C. Microwave-assisted synthesis of CdS–reduced graphene oxide composites for photocatalytic reduction of Cr(vi). *Chem. Commun. (Camb.),* **2011**, *47*(43), 11984-11986.
[http://dx.doi.org/10.1039/c1cc14875c] [PMID: 21960206]

[60] Shen, J.; Shi, M.; Yan, B.; Ma, H.; Li, N.; Ye, M. One-pot hydrothermal synthesis of Ag-reduced graphene oxide composite with ionic liquid. *J. Mater. Chem.,* **2011**, *21*(21), 7795-7801.
[http://dx.doi.org/10.1039/c1jm10671f]

[61] Lee, S.H.; Kakati, N.; Jee, S.H.; Maiti, J.; Yoon, Y.S. Hydrothermal synthesis of PtRu nanoparticles supported on graphene sheets for methanol oxidation in direct methanol fuel cell. *Mater. Lett.,* **2011**, *65*(21-22), 3281-3284.
[http://dx.doi.org/10.1016/j.matlet.2011.07.025]

[62] Zou, R.; Zhang, Z.; Yu, L.; Tian, Q.; Chen, Z.; Hu, J. A general approach for the growth of metal oxide nanorod arrays on graphene sheets and their applications. *Chemistry,* **2011**, *17*(49), 13912-13917.
[http://dx.doi.org/10.1002/chem.201101981] [PMID: 22038954]

[63] Shen, J.; Yan, B.; Shi, M.; Ma, H.; Li, N.; Ye, M. One step hydrothermal synthesis of TiO_2-reduced graphene oxide sheets. *J. Mater. Chem.,* **2011**, *21*(10), 3415-3421.
[http://dx.doi.org/10.1039/c0jm03542d]

[64] Zhang, H.; Lv, X.; Li, Y.; Wang, Y.; Li, J. P25-graphene composite as a high performance photocatalyst. *ACS Nano,* **2010**, *4*(1), 380-386.
[http://dx.doi.org/10.1021/nn901221k] [PMID: 20041631]

[65] Zou, Y.; Wang, Y. NiO nanosheets grown on graphene nanosheets as superior anode materials for Li-ion batteries. *Nanoscale,* **2011**, *3*(6), 2615-2620.
[http://dx.doi.org/10.1039/c1nr10070j] [PMID: 21523266]

[66] Wang, Z.; Huang, B.; Dai, Y.; Liu, Y.; Zhang, X.; Qin, X.; Wang, J.; Zheng, Z.; Cheng, H. Crystal facets controlled synthesis of graphene@TiO_2 nanocomposites by a one-pot hydrothermal process. *CrystEngComm*, **2012**, *14*(5), 1687-1692. [http://dx.doi.org/10.1039/C1CE06193C]

[67] Xiang, Q.; Yu, J.; Jaroniec, M. Enhanced photocatalytic H_2-production activity of graphene-modified titania nanosheets. *Nanoscale*, **2011**, *3*(9), 3670-3678. [http://dx.doi.org/10.1039/c1nr10610d] [PMID: 21826308]

[68] Perera, S.D.; Mariano, R.G.; Vu, K.; Nour, N.; Seitz, O.; Chabal, Y.; Balkus, K.J., Jr Hydrothermal Synthesis of Graphene-TiO_2 Nanotube Composites with Enhanced Photocatalytic Activity. *ACS Catal.*, **2012**, *2*(6), 949-956. [http://dx.doi.org/10.1021/cs200621c]

[69] Chen, D.; Ji, G.; Ma, Y.; Lee, J.Y.; Lu, J. Graphene-encapsulated hollow Fe_3O_4 nanoparticle aggregates as a high-performance anode material for lithium ion batteries. *ACS Appl. Mater. Interfaces*, **2011**, *3*(8), 3078-3083. [http://dx.doi.org/10.1021/am200592r] [PMID: 21749101]

[70] Nethravathi, C.; Viswanath, B.; Michael, J.; Rajamath, M. Hydrothermal synthesis of a monoclinic VO_2 nanotube–graphene hybrid for use as cathode material in lithium ion batteries. *Carbon*, **2012**, *50*(13), 4839-4846. [http://dx.doi.org/10.1016/j.carbon.2012.06.010]

[71] Tang, Q.; Wang, L.; Zhu, K.; Shan, Z.; Qin, X. Synthesis and electrochemical properties of H-MoO3/graphene composite. *Mater. Lett.*, **2013**, *100*, 127-129. [http://dx.doi.org/10.1016/j.matlet.2013.03.005]

[72] Cao, A.; Liu, Z.; Chu, S.; Wu, M.; Ye, Z.; Cai, Z.; Chang, Y.; Wang, S.; Gong, Q.; Liu, Y. A facile one-step method to produce graphene-CdS quantum dot nanocomposites as promising optoelectronic materials. *Adv. Mater.*, **2010**, *22*(1), 103-106. [http://dx.doi.org/10.1002/adma.200901920] [PMID: 20217706]

[73] Chen, P.; Xiao, T.Y.; Li, H.H.; Yang, J.J.; Wang, Z.; Yao, H.B.; Yu, S.H. Nitrogen-doped graphene/ZnSe nanocomposites: hydrothermal synthesis and their enhanced electrochemical and photocatalytic activities. *ACS Nano*, **2012**, *6*(1), 712-719. [http://dx.doi.org/10.1021/nn204191x] [PMID: 22136425]

[74] Shen, L.; Yuan, C.; Luo, H.; Zhang, X.; Yang, S.; Lu, X. *In Situ* synthesis of high-loading $Li_4Ti_5O_{12}$ –graphene hybrid nanostructures for high rate lithium ion batteries. *Nanoscale*, **2011**, *3*(2), 572-574. [http://dx.doi.org/10.1039/C0NR00639D] [PMID: 21076732]

[75] Shi, Y.; Chou, S.L.; Wang, J.Z.; Wexler, D.; Li, H.J.; Liu, H.K.; Wu, Y. Graphene wrapped $LiFePO_4$/C composites as cathode materials for Li-ion batteries with enhanced rate capability. *J. Mater. Chem.*, **2012**, *22*(32), 16465-16470. [http://dx.doi.org/10.1039/c2jm32649c]

[76] Gao, E.; Wang, W.; Shang, M.; Xu, J. Synthesis and enhanced photocatalytic performance of graphene-Bi_2WO_6 composite. *Phys. Chem. Chem. Phys.*, **2011**, *13*(7), 2887-2893. [http://dx.doi.org/10.1039/C0CP01749C] [PMID: 21161101]

[77] Li, J.; Yang, Z.; Qiu, H.; Dai, Y.; Zheng, Q.; Zheng, G.P.; Yang, J. Microwave-assisted simultaneous reduction and titanate treatment of graphene oxide. *J. Mater. Chem. A Mater. Energy Sustain.*, **2013**, *1*(37), 11451-11456. [http://dx.doi.org/10.1039/c3ta12228j]

[78] Huang, X.; Qi, X.; Boey, F.; Zhang, H. Graphene-based composites. *Chem. Soc. Rev.*, **2012**, *41*(2), 666-686. [http://dx.doi.org/10.1039/C1CS15078B] [PMID: 21796314]

[79] Hassan, H.M.A.; Abdelsayed, V.; Khder, A.E.R.S.; AbouZeid, K.M.; Terner, J.; El-Shall, M.S.; Al-

Resayes, S.I.; El-Azhary, A.A. Microwave synthesis of graphene sheets supporting metal nanocrystals in aqueous and organic media. *J. Mater. Chem.*, **2009**, *19*(23), 3832-3837. [http://dx.doi.org/10.1039/b906253j]

[80] Kim, H.W.; Na, H.G.; Kwon, Y.J.; Kang, S.Y.; Choi, M.S.; Bang, J.H.; Wu, P.; Kim, S.S. Microwave-Assisted Synthesis of Graphene–SnO_2 Nanocomposites and Their Applications in Gas Sensors. *ACS Appl. Mater. Interfaces*, **2017**, *9*(37), 31667-31682. [http://dx.doi.org/10.1021/acsami.7b02533] [PMID: 28846844]

[81] Hu, H.; Wang, X.; Liu, F.; Wang, J.; Xu, C. Rapid microwave-assisted synthesis of graphene nanosheets–zinc sulfide nanocomposites: Optical and photocatalytic properties. *Synth. Met.*, **2011**, *161*(5-6), 404-410. [http://dx.doi.org/10.1016/j.synthmet.2010.12.018]

[82] Li, M.; Bo, X.; Zhang, Y.; Han, C.; Guo, L. One-pot ionic liquid-assisted synthesis of highly dispersed PtPd nanoparticles/reduced graphene oxide composites for nonenzymatic glucose detection. *Biosens. Bioelectron.*, **2014**, *56*, 223-230. [http://dx.doi.org/10.1016/j.bios.2014.01.030] [PMID: 24508545]

[83] Yin, Z.; Wu, S.; Zhou, X.; Huang, X.; Zhang, Q.; Boey, F.; Zhang, H. Electrochemical deposition of ZnO nanorods on transparent reduced graphene oxide electrodes for hybrid solar cells. *Small*, **2010**, *6*(2), 307-312. [http://dx.doi.org/10.1002/smll.200901968] [PMID: 20039255]

[84] Nandi, D.; Mohan, V.B.; Bhowmick, A.K.; Bhattacharyya, D. Metal/metal oxide decorated graphene synthesis and application as supercapacitor: a review. *J. Mater. Sci.*, **2020**, *55*(15), 6375-6400. [http://dx.doi.org/10.1007/s10853-020-04475-z]

[85] Yu, G.; Hu, L.; Vosgueritchian, M.; Wang, H.; Xie, X.; McDonough, J.R.; Cui, X.; Cui, Y.; Bao, Z. Solution-processed graphene/MnO_2 nanostructured textiles for high-performance electrochemical capacitors. *Nano Lett.*, **2011**, *11*(7), 2905-2911. [http://dx.doi.org/10.1021/nl2013828] [PMID: 21667923]

[86] Wu, S.; Yin, Z.; He, Q.; Lu, G.; Zhou, X.; Zhang, H. Electrochemical deposition of Cl-doped n-type Cu_2O on reduced graphene oxide electrodes. *J. Mater. Chem.*, **2011**, *21*(10), 3467-3470. [http://dx.doi.org/10.1039/C0JM02267E]

[87] Wu, S.; Yin, Z.; He, Q.; Lu, G.; Yan, Q.; Zhang, H. Nucleation Mechanism of Electrochemical Deposition of Cu on Reduced Graphene Oxide Electrodes. *J. Phys. Chem. C*, **2011**, *115*(32), 15973-15979. [http://dx.doi.org/10.1021/jp201667p]

[88] Liu, A.; Li, C.; Bai, H.; Shi, G. Electrochemical Deposition of Polypyrrole/Sulfonated Graphene Composite Films. *J. Phys. Chem. C*, **2010**, *114*(51), 22783-22789. [http://dx.doi.org/10.1021/jp108826e]

[89] Zhu, G.; Pan, L.; Lu, T.; Xu, T.; Sun, Z. Electrophoretic deposition of reduced graphene-carbon nanotubes composite films as counter electrodes of dye-sensitized solar cells. *J. Mater. Chem.*, **2011**, *21*(38), 14869-14875. [http://dx.doi.org/10.1039/c1jm12433a]

[90] ReddyPrasad, P.; Naidoo, E.B.; Sreedhar, N.Y. Electrochemical preparation of a novel type of C-dots/ZrO2 nanocomposite onto glassy carbon electrode for detection of organophosphorus pesticide. *Arab. J. Chem.*, **2019**, *12*(8), 2300-2309. [http://dx.doi.org/10.1016/j.arabjc.2015.02.012]

[91] Liu, S.; Wang, J.; Zeng, J.; Ou, J.; Li, Z.; Liu, X.; Yang, S. "Green" electrochemical synthesis of Pt/graphene sheet nanocomposite film and its electrocatalytic property. *J. Power Sources*, **2010**, *195*(15), 4628-4633. [http://dx.doi.org/10.1016/j.jpowsour.2010.02.024]

[92] Zhang, Y.; Xiao, X.; Sun, Y.; Shi, Y.; Dai, H.; Ni, P.; Hu, J.; Li, Z.; Song, Y.; Wang, L.

Electrochemical Deposition of Nickel Nanoparticles on Reduced Graphene Oxide Film for Nonenzymatic Glucose Sensing. *Electroanalysis,* **2013**, *25*(4), 959-966. [http://dx.doi.org/10.1002/elan.201200479]

[93] Yang, J.; Deng, S.; Lei, J.; Ju, H.; Gunasekaran, S. Electrochemical synthesis of reduced graphene sheet–AuPd alloy nanoparticle composites for enzymatic biosensing. *Biosens. Bioelectron.,* **2011**, *29*(1), 159-166. [http://dx.doi.org/10.1016/j.bios.2011.08.011] [PMID: 21903376]

[94] Zhang, H.; Zhang, X.; Zhang, D.; Sun, X.; Lin, H.; Wang, C.; Ma, Y. One-step electrophoretic deposition of reduced graphene oxide and $Ni(OH)_2$ composite films for controlled syntheses supercapacitor electrodes. *J. Phys. Chem. B,* **2013**, *117*(6), 1616-1627. [http://dx.doi.org/10.1021/jp305198j] [PMID: 22994913]

[95] Kim, Y.J.; Lee, J.H.; Yi, G.C. Vertically aligned ZnO nanostructures grown on graphene layers. *Appl. Phys. Lett.,* **2009**, *95*(21), 213101. [http://dx.doi.org/10.1063/1.3266836]

[96] Lin, J.; Penchev, M.; Wang, G.; Paul, R.K.; Zhong, J.; Jing, X.; Ozkan, M.; Ozkan, C.S. Heterogeneous graphene nanostructures: ZnO nanostructures grown on large-area graphene layers. *Small,* **2010**, *6*(21), 2448-2452. [http://dx.doi.org/10.1002/smll.201000250] [PMID: 20878792]

[97] Fan, Z.; Yan, J.; Zhi, L.; Zhang, Q.; Wei, T.; Feng, J.; Zhang, M.; Qian, W.; Wei, F. A three-dimensional carbon nanotube/graphene sandwich and its application as electrode in supercapacitors. *Adv. Mater.,* **2010**, *22*(33), 3723-3728. [http://dx.doi.org/10.1002/adma.201001029] [PMID: 20652901]

[98] Paul, R.K.; Ghazinejad, M.; Penchev, M.; Lin, J.; Ozkan, M.; Ozkan, C.S. Synthesis of a pillared graphene nanostructure: a counterpart of three-dimensional carbon architectures. *Small,* **2010**, *6*(20), 2309-2313. [http://dx.doi.org/10.1002/smll.201000525] [PMID: 20862676]

[99] Zhu, Y.; Li, L.; Zhang, C.; Casillas, G.; Sun, Z.; Yan, Z.; Ruan, G.; Peng, Z.; Raji, A.R.O.; Kittrell, C.; Hauge, R.H.; Tour, J.M. A seamless three-dimensional carbon nanotube graphene hybrid material. *Nat. Commun.,* **2012**, *3*(1), 1225. [http://dx.doi.org/10.1038/ncomms2234] [PMID: 23187625]

[100] Yan, Z.; Ma, L.; Zhu, Y.; Lahiri, I.; Hahm, M.G.; Liu, Z.; Yang, S.; Xiang, C.; Lu, W.; Peng, Z.; Sun, Z.; Kittrell, C.; Lou, J.; Choi, W.; Ajayan, P.M.; Tour, J.M. Three-dimensional metal-graphen--nanotube multifunctional hybrid materials. *ACS Nano,* **2013**, *7*(1), 58-64. [http://dx.doi.org/10.1021/nn3015882] [PMID: 23194106]

[101] Tavakkoli, M.; Flahaut, E.; Peljo, P.; Sainio, J.; Davodi, F.; Lobiak, E.V.; Mustonen, K.; Kauppinen, E.I. Mesoporous Single-Atom-Doped Graphene–Carbon Nanotube Hybrid: Synthesis and Tunable Electrocatalytic Activity for Oxygen Evolution and Reduction Reactions. *ACS Catal.,* **2020**, *10*(8), 4647-4658. [http://dx.doi.org/10.1021/acscatal.0c00352]

[102] Lin, J.; Zhang, C.; Yan, Z.; Zhu, Y.; Peng, Z.; Hauge, R.H.; Natelson, D.; Tour, J.M. 3-Dimensional graphene carbon nanotube carpet-based microsupercapacitors with high electrochemical performance. *Nano Lett.,* **2013**, *13*(1), 72-78. [http://dx.doi.org/10.1021/nl3034976] [PMID: 23237453]

[103] Williams, G.; Seger, B.; Kamat, P.V. TiO_2-graphene nanocomposites. UV-assisted photocatalytic reduction of graphene oxide. *ACS Nano,* **2008**, *2*(7), 1487-1491. [http://dx.doi.org/10.1021/nn800251f] [PMID: 19206319]

[104] Ng, Y.H.; Iwase, A.; Bell, N.J.; Kudo, A.; Amal, R. Semiconductor/Reduced Graphene Oxide Nanocomposites Derived from Photocatalytic Reactions In: *Proceedings of the Catalysis Today*; Elsevier, **2011**; pp. 353-357.

[http://dx.doi.org/10.1016/j.cattod.2010.10.090]

[105] Pasricha, R.; Gupta, S.; Joshi, A.G.; Bahadur, N.; Haranath, D.; Sood, K.N.; Singh, S.; Singh, S. Directed nanoparticle reduction on graphene. *Mater. Today,* **2012**, *15*(3), 118-125. [http://dx.doi.org/10.1016/S1369-7021(12)70047-0]

[106] Moon, G.; Park, Y.; Kim, W.; Choi, W. Photochemical loading of metal nanoparticles on reduced graphene oxide sheets using phosphotungstate. *Carbon,* **2011**, *49*(11), 3454-3462. [http://dx.doi.org/10.1016/j.carbon.2011.04.042]

[107] Zhang, Y.; Yuan, X.; Wang, Y.; Chen, Y. One-pot photochemical synthesis of graphene composites uniformly deposited with silver nanoparticles and their high catalytic activity towards the reduction of 2-nitroaniline. *J. Mater. Chem.,* **2012**, *22*(15), 7245-7251. [http://dx.doi.org/10.1039/c2jm16455h]

[108] Moon, G.; Kim, H.; Shin, Y.; Choi, W. Chemical-free growth of metal nanoparticles on graphene oxide sheets under visible light irradiation. *RSC Advances,* **2012**, *2*(6), 2205-2207. [http://dx.doi.org/10.1039/c2ra00875k]

[109] Chen, J.; Zheng, X.; Wang, H.; Zheng, W. Graphene oxide-Ag nanocomposite: *In Situ* photochemical synthesis and application as a surface-enhanced Raman scattering substrate. *Thin Solid Films,* **2011**, *520*(1), 179-185. [http://dx.doi.org/10.1016/j.tsf.2011.07.012]

[110] Tjoa, V.; Chua, J.; Pramana, S.S.; Wei, J.; Mhaisalkar, S.G.; Mathews, N. Facile photochemical synthesis of graphene-pt nanoparticle composite for counter electrode in dye sensitized solar cell. *ACS Appl. Mater. Interfaces,* **2012**, *4*(7), 3447-3452. [http://dx.doi.org/10.1021/am300437g] [PMID: 22646232]

[111] Xu, F.; Sun, Y.; Zhang, Y.; Shi, Y.; Wen, Z.; Li, Z. Graphene–Pt nanocomposite for nonenzymatic detection of hydrogen peroxide with enhanced sensitivity. *Electrochem. Commun.,* **2011**, *13*(10), 1131-1134. [http://dx.doi.org/10.1016/j.elecom.2011.07.017]

[112] Wang, X.; Tabakman, S.M.; Dai, H. Atomic layer deposition of metal oxides on pristine and functionalized graphene. *J. Am. Chem. Soc.,* **2008**, *130*(26), 8152-8153. [http://dx.doi.org/10.1021/ja8023059] [PMID: 18529002]

[113] Pandey, P.A.; Bell, G.R.; Rourke, J.P.; Sanchez, A.M.; Elkin, M.D.; Hickey, B.J.; Wilson, N.R. Physical vapor deposition of metal nanoparticles on chemically modified graphene: observations on metal-graphene interactions. *Small,* **2011**, *7*(22), 3202-3210. [http://dx.doi.org/10.1002/smll.201101430] [PMID: 21953833]

[114] Zhou, X.; Wang, F.; Zhu, Y.; Liu, Z. Graphene modified LiFePO4 cathode materials for high power lithium ion batteries. *J. Mater. Chem.,* **2011**, *21*(10), 3353-3358. [http://dx.doi.org/10.1039/c0jm03287e]

[115] Lung-Hao Hu, B.; Wu, F.Y.; Lin, C.T.; Khlobystov, A.N.; Li, L.J. Graphene-modified LiFePO4 cathode for lithium ion battery beyond theoretical capacity. *Nat. Commun.,* **2013**, *4*(1), 1687. [http://dx.doi.org/10.1038/ncomms2705] [PMID: 23575691]

[116] Chang, H.; Wang, G.; Yang, A.; Tao, X.; Liu, X.; Shen, Y.; Zheng, Z. A Transparent, Flexible, Low-Temperature, and Solution-Processible Graphene Composite Electrode. *Adv. Funct. Mater.,* **2010**, *20*(17), 2893-2902. [http://dx.doi.org/10.1002/adfm.201000900]

[117] Geim, A.K.; Novoselov, K.S. The Rise of Graphene. In: *Nanoscience and Technology: A Collection of Reviews from Nature Journals*; World Scientific Publishing Co., **2009**; pp. 11-19. [http://dx.doi.org/10.1142/9789814287005_0002]

[118] Machado, B.F.; Serp, P. Graphene-based materials for catalysis. *Catal. Sci. Technol.,* **2012**, *2*(1), 54-75.

[http://dx.doi.org/10.1039/C1CY00361E]

[119] Hu, L.; Hu, X.; Wu, X.; Du, C.; Dai, Y.; Deng, J. Density functional calculation of transition metal adatom adsorption on graphene. *Physica B,* **2010**, *405*(16), 3337-3341. [http://dx.doi.org/10.1016/j.physb.2010.05.001]

[120] Tang, Y.; Yang, Z.; Dai, X. Noble metals induced magnetic properties of graphene. *J. Magn. Magn. Mater.,* **2011**, *323*(20), 2441-2447. [http://dx.doi.org/10.1016/j.jmmm.2011.05.004]

[121] Leenaerts, O.; Partoens, B.; Peeters, F.M. Adsorption of H_2O, NH_3, CO, NO_2, and NO on graphene: A first-principles study. *Phys. Rev. B Condens. Matter Mater. Phys.,* **2008**, *77*(12), 125416. [http://dx.doi.org/10.1103/PhysRevB.77.125416]

[122] Ma, F.; Zhang, Z.; Jia, H.; Liu, X.; Hao, Y.; Xu, B. Adsorption of cysteine molecule on intrinsic and Pt-doped graphene: A first-principle study. *J. Mol. Struct. THEOCHEM,* **2010**, *955*(1-3), 134-139. [http://dx.doi.org/10.1016/j.theochem.2010.06.007]

[123] Voloshina, E.N.; Mollenhauer, D.; Chiappisi, L.; Paulus, B. Theoretical study on the adsorption of pyridine derivatives on graphene. *Chem. Phys. Lett.,* **2011**, *510*(4-6), 220-223. [http://dx.doi.org/10.1016/j.cplett.2011.05.025]

[124] Bai, S.; Shen, X. Graphene–inorganic nanocomposites. *RSC Advances,* **2012**, *2*(1), 64-98. [http://dx.doi.org/10.1039/C1RA00260K]

[125] Chang, H.; Wu, H. Graphene-based nanocomposites: preparation, functionalization, and energy and environmental applications. *Energy Environ. Sci.,* **2013**, *6*(12), 3483-3507. [http://dx.doi.org/10.1039/c3ee42518e]

[126] Mahmood, N.; Zhang, C.; Yin, H.; Hou, Y. Graphene-based nanocomposites for energy storage and conversion in lithium batteries, supercapacitors and fuel cells. *J. Mater. Chem. A Mater. Energy Sustain.,* **2014**, *2*(1), 15-32. [http://dx.doi.org/10.1039/C3TA13033A]

[127] Chen, P.; Guo, L.; Wang, Y. Graphene wrapped SnCo nanoparticles for high-capacity lithium ion storage. *J. Power Sources,* **2013**, *222*, 526-532. [http://dx.doi.org/10.1016/j.jpowsour.2012.09.026]

[128] Li, Q.; Mahmood, N.; Zhu, J.; Hou, Y.; Sun, S. Graphene and its composites with nanoparticles for electrochemical energy applications. *Nano Today,* **2014**, *9*(5), 668-683. [http://dx.doi.org/10.1016/j.nantod.2014.09.002]

[129] Lee, J.; Kim, S.M.; Lee, I.S. Functionalization of hollow nanoparticles for nanoreactor applications. *Nano Today,* **2014**, *9*(5), 631-667. [http://dx.doi.org/10.1016/j.nantod.2014.09.003]

[130] Bonaccorso, F.; Colombo, L.; Yu, G.; Stoller, M.; Tozzini, V.; Ferrari, A.C.; Ruoff, R.S.; Pellegrini, V. *Graphene, Related Two-Dimensional Crystals, and Hybrid Systems for Energy Conversion and Storage. Science (80).,* **2015**, 347.

[131] Dubal, D.P.; Ayyad, O.; Ruiz, V.; Gómez-Romero, P. Hybrid energy storage: the merging of battery and supercapacitor chemistries. *Chem. Soc. Rev.,* **2015**, *44*(7), 1777-1790. [http://dx.doi.org/10.1039/C4CS00266K] [PMID: 25623995]

[132] Hu, C.; Song, L.; Zhang, Z.; Chen, N.; Feng, Z.; Qu, L. Tailored graphene systems for unconventional applications in energy conversion and storage devices. *Energy Environ. Sci.,* **2015**, *8*(1), 31-54. [http://dx.doi.org/10.1039/C4EE02594F]

[133] Peng, X.; Peng, L.; Wu, C.; Xie, Y. Two dimensional nanomaterials for flexible supercapacitors. *Chem. Soc. Rev.,* **2014**, *43*(10), 3303-3323. [http://dx.doi.org/10.1039/c3cs60407a] [PMID: 24614864]

[134] Qiu, B.; Xing, M.; Zhang, J. Mesoporous TiO_2 nanocrystals grown *in situ* on graphene aerogels for

high photocatalysis and lithium-ion batteries. *J. Am. Chem. Soc.*, **2014**, *136*(16), 5852-5855. [http://dx.doi.org/10.1021/ja500873u] [PMID: 24712676]

[135] Etacheri, V.; Yourey, J.E.; Bartlett, B.M. Chemically bonded TiO_2-bronze nanosheet/reduced graphene oxide hybrid for high-power lithium ion batteries. *ACS Nano,* **2014**, *8*(2), 1491-1499. [http://dx.doi.org/10.1021/nn405534r] [PMID: 24446910]

[136] Li, N.; Liu, G.; Zhen, C.; Li, F.; Zhang, L.; Cheng, H.M. Battery Performance and Photocatalytic Activity of Mesoporous Anatase TiO_2 Nanospheres/Graphene Composites by Template-Free Self-Assembly. *Adv. Funct. Mater.*, **2011**, *21*(9), 1717-1722. [http://dx.doi.org/10.1002/adfm.201002295]

[137] Lee, M.; Balasingam, S.K.; Jeong, H.Y.; Hong, W.G.; Lee, H.B.R.; Kim, B.H.; Jun, Y. One-step hydrothermal synthesis of graphene decorated V_2O_5 nanobelts for enhanced electrochemical energy storage. *Sci. Rep.*, **2015**, *5*(1), 8151. [http://dx.doi.org/10.1038/srep08151] [PMID: 25633147]

[138] Li, M.; Sun, G.; Yin, P.; Ruan, C.; Ai, K. Controlling the formation of rodlike V_2O_5 nanocrystals on reduced graphene oxide for high-performance supercapacitors. *ACS Appl. Mater. Interfaces,* **2013**, *5*(21), 11462-11470. [http://dx.doi.org/10.1021/am403739g] [PMID: 24138545]

[139] Cheng, J.; Wang, B.; Xin, H.L.; Yang, G.; Cai, H.; Nie, F.; Huang, H. Self-assembled V_2O_5 nanosheets/reduced graphene oxide hierarchical nanocomposite as a high-performance cathode material for lithium ion batteries. *J. Mater. Chem. A Mater. Energy Sustain.*, **2013**, *1*(36), 10814-10820. [http://dx.doi.org/10.1039/c3ta12066j]

[140] Liu, Q.; Li, Z.F.; Liu, Y.; Zhang, H.; Ren, Y.; Sun, C.J.; Lu, W.; Zhou, Y.; Stanciu, L.; Stach, E.A.; Xie, J. Graphene-modified nanostructured vanadium pentoxide hybrids with extraordinary electrochemical performance for Li-ion batteries. *Nat. Commun.*, **2015**, *6*(1), 6127. [http://dx.doi.org/10.1038/ncomms7127] [PMID: 25600907]

[141] Yang, X.; Ding, H.; Zhang, D.; Yan, X.; Lu, C.; Qin, J.; Zhang, R.; Tang, H.; Song, H. Hydrothermal synthesis of MoO_3 nanobelt-graphene composites. *Cryst. Res. Technol.*, **2011**, *46*(11), 1195-1201. [http://dx.doi.org/10.1002/crat.201100302]

[142] Lu, P.J.; Lei, M.; Liu, J. Graphene nanosheets encapsulated α-MoO_3 nanoribbons with ultrahigh lithium ion storage properties. *CrystEngComm,* **2014**, *16*(29), 6745-6755. [http://dx.doi.org/10.1039/C4CE00252K]

[143] Chen, C.; Jian, H.; Fu, X.; Ren, Z.; Yan, M.; Qian, G.; Wang, Z. Facile synthesis of graphene-supported mesoporous Mn_3O_4 nanosheets with a high-performance in Li-ion batteries. *RSC Advances,* **2014**, *4*(11), 5367-5370. [http://dx.doi.org/10.1039/c3ra45976d]

[144] Wang, B.; Park, J.; Wang, C.; Ahn, H.; Wang, G. Mn_3O_4 nanoparticles embedded into graphene nanosheets: Preparation, characterization, and electrochemical properties for supercapacitors. *Electrochim. Acta,* **2010**, *55*(22), 6812-6817. [http://dx.doi.org/10.1016/j.electacta.2010.05.086]

[145] Su, J.; Cao, M.; Ren, L.; Hu, C. Fe_3O_4–Graphene Nanocomposites with Improved Lithium Storage and Magnetism Properties. *J. Phys. Chem. C,* **2011**, *115*(30), 14469-14477. [http://dx.doi.org/10.1021/jp201666s]

[146] Zou, Y.; Kan, J.; Wang, Y. Fe_2O_3 -Graphene Rice-on-Sheet Nanocomposite for High and Fast Lithium Ion Storage. *J. Phys. Chem. C,* **2011**, *115*(42), 20747-20753. [http://dx.doi.org/10.1021/jp206876t]

[147] Kim, H.; Choi, W.; Yoon, J.; Um, J.H.; Lee, W.; Kim, J.; Cabana, J.; Yoon, W.S. Exploring Anomalous Charge Storage in Anode Materials for Next-Generation Li Rechargeable Batteries. *Chem. Rev.*, **2020**, *120*(14), 6934-6976.

[http://dx.doi.org/10.1021/acs.chemrev.9b00618] [PMID: 32101429]

[148] Yang, J.; Wang, J.; Tang, Y.; Wang, D.; Li, X.; Hu, Y.; Li, R.; Liang, G.; Sham, T.K.; Sun, X. $LiFePO_4$–graphene as a superior cathode material for rechargeable lithium batteries: impact of stacked graphene and unfolded graphene. *Energy Environ. Sci.,* **2013**, *6*(5), 1521-1528. [http://dx.doi.org/10.1039/c3ee24163g]

[149] Qu, Q.; Yang, S.; Feng, X. 2D sandwich-like sheets of iron oxide grown on graphene as high energy anode material for supercapacitors. *Adv. Mater.,* **2011**, *23*(46), 5574-5580. [http://dx.doi.org/10.1002/adma.201103042] [PMID: 22052661]

[150] Bharali, P.; Kuratani, K.; Takeuchi, T.; Kiyobayashi, T.; Kuriyama, N. Capacitive behavior of amorphous and crystalline RuO_2 composite electrode fabricated by spark plasma sintering technique. *J. Power Sources,* **2011**, *196*(18), 7878-7881. [http://dx.doi.org/10.1016/j.jpowsour.2011.03.097]

[151] Vellacheri, R.; Pillai, V.K.; Kurungot, S. Hydrous RuO_2–carbon nanofiber electrodes with high mass and electrode-specific capacitance for efficient energy storage. *Nanoscale,* **2012**, *4*(3), 890-896. [http://dx.doi.org/10.1039/C2NR11479H] [PMID: 22159715]

[152] Wu, Z.S.; Wang, D.W.; Ren, W.; Zhao, J.; Zhou, G.; Li, F.; Cheng, H.M. Anchoring Hydrous RuO_2 on Graphene Sheets for High-Performance Electrochemical Capacitors. *Adv. Funct. Mater.,* **2010**, *20*(20), 3595-3602. [http://dx.doi.org/10.1002/adfm.201001054]

[153] Wu, Z.S.; Sun, Y.; Tan, Y.Z.; Yang, S.; Feng, X.; Müllen, K. Three-dimensional graphene-based macro- and mesoporous frameworks for high-performance electrochemical capacitive energy storage. *J. Am. Chem. Soc.,* **2012**, *134*(48), 19532-19535. [http://dx.doi.org/10.1021/ja308676h] [PMID: 23148416]

[154] Wang, Y.M.; Zhao, D.D.; Zhao, Y.Q.; Xu, C.L.; Li, H.L. Effect of electrodeposition temperature on the electrochemical performance of a $Ni(OH)_2$electrode. *RSC Advances,* **2012**, *2*(3), 1074-1082. [http://dx.doi.org/10.1039/C1RA00613D]

[155] Liang, H.W.; Wei, W.; Wu, Z.S.; Feng, X.; Müllen, K. Mesoporous metal-nitrogen-doped carbon electrocatalysts for highly efficient oxygen reduction reaction. *J. Am. Chem. Soc.,* **2013**, *135*(43), 16002-16005. [http://dx.doi.org/10.1021/ja407552k] [PMID: 24128393]

[156] Singh, R.N.; Awasthi, R. Graphene support for enhanced electrocatalytic activity of Pd for alcohol oxidation. *Catal. Sci. Technol.,* **2011**, *1*(5), 778-783. [http://dx.doi.org/10.1039/c1cy00021g]

[157] Zhang, S.; Shao, Y.; Yin, G.; Lin, Y. Electrostatic self-assembly of a Pt-around-Au nanocomposite with high activity towards formic acid oxidation. *Angew. Chem. Int. Ed.,* **2010**, *49*(12), 2211-2214. [http://dx.doi.org/10.1002/anie.200906987] [PMID: 20186901]

[158] He, D.; Cheng, K.; Li, H.; Peng, T.; Xu, F.; Mu, S.; Pan, M. Highly active platinum nanoparticles on graphene nanosheets with a significant improvement in stability and CO tolerance. *Langmuir,* **2012**, *28*(8), 3979-3986. [http://dx.doi.org/10.1021/la2045493] [PMID: 22276697]

[159] Rao, C.V.; Reddy, A.L.M.; Ishikawa, Y.; Ajayan, P.M. Synthesis and electrocatalytic oxygen reduction activity of graphene-supported Pt_3Co and Pt_3Cr alloy nanoparticles. *Carbon,* **2011**, *49*(3), 931-936. [http://dx.doi.org/10.1016/j.carbon.2010.10.056]

[160] Du, D.; Liu, J.; Zhang, X.; Cui, X.; Lin, Y. One-step electrochemical deposition of a graphene-ZrO_2 nanocomposite: Preparation, characterization and application for detection of organophosphorus agents. *J. Mater. Chem.,* **2011**, *21*(22), 8032-8037. [http://dx.doi.org/10.1039/c1jm10696a]

[161] Zhu, M.; Chen, P.; Liu, M. Graphene oxide enwrapped Ag/AgX (X = Br, Cl) nanocomposite as a highly efficient visible-light plasmonic photocatalyst. *ACS Nano,* **2011**, *5*(6), 4529-4536. [http://dx.doi.org/10.1021/nn200088x] [PMID: 21524132]

[162] Subramanian, P.; Niedziolka-Jonsson, J.; Lesniewski, A.; Wang, Q.; Li, M.; Boukherroub, R.; Szunerits, S. Preparation of reduced graphene oxide–Ni(OH) $_2$ composites by electrophoretic deposition: application for non-enzymatic glucose sensing. *J. Mater. Chem. A Mater. Energy Sustain.,* **2014**, *2*(15), 5525-5533. [http://dx.doi.org/10.1039/C4TA00123K]

[163] Fu, L.; Yu, S.; Thompson, L.; Yu, A. Development of a novel nitrite electrochemical sensor by stepwise *in situ* formation of palladium and reduced graphene oxide nanocomposites. *RSC Advances,* **2015**, *5*(50), 40111-40116. [http://dx.doi.org/10.1039/C5RA02661J]

[164] Johnson, J.L.; Behnam, A.; Pearton, S.J.; Ural, A. Hydrogen sensing using pd-functionalized multi-layer graphene nanoribbon networks. *Adv. Mater.,* **2010**, *22*(43), 4877-4880. [http://dx.doi.org/10.1002/adma.201001798] [PMID: 20803539]

[165] Fu, X.; Lou, T.; Chen, Z.; Lin, M.; Feng, W.; Chen, L. "Turn-on" fluorescence detection of lead ions based on accelerated leaching of gold nanoparticles on the surface of graphene. *ACS Appl. Mater. Interfaces,* **2012**, *4*(2), 1080-1086. [http://dx.doi.org/10.1021/am201711j] [PMID: 22264012]

[166] Wang, F.; Zhu, L.; Zhang, J. Electrochemical sensor for levofloxacin based on molecularly imprinted polypyrrole–graphene–gold nanoparticles modified electrode. *Sens. Actuators B Chem.,* **2014**, *192*, 642-647. [http://dx.doi.org/10.1016/j.snb.2013.11.037]

[167] Wang, Y.; Zhang, S.; Du, D.; Shao, Y.; Li, Z.; Wang, J.; Engelhard, M.H.; Li, J.; Lin, Y. Self assembly of acetylcholinesterase on a gold nanoparticles–graphene nanosheet hybrid for organophosphate pesticide detection using polyelectrolyte as a linker. *J. Mater. Chem.,* **2011**, *21*(14), 5319-5325. [http://dx.doi.org/10.1039/c0jm03441j]

[168] Liu, X.; Du, H.; Sun, X.W. High-performance photoresponse of carbon-doped ZnO/reduced graphene oxide hybrid nanocomposites under UV and visible illumination. *RSC Advances,* **2014**, *4*(10), 5136-5140. [http://dx.doi.org/10.1039/c3ra46864j]

[169] Chang, H.; Sun, Z.; Ho, K.Y.F.; Tao, X.; Yan, F.; Kwok, W.M.; Zheng, Z. A highly sensitive ultraviolet sensor based on a facile *in situ* solution-grown ZnO nanorod/graphene heterostructure. *Nanoscale,* **2011**, *3*(1), 258-264. [http://dx.doi.org/10.1039/C0NR00588F] [PMID: 20976323]

[170] Anand, K.; Singh, O.; Singh, M.P.; Kaur, J.; Singh, R.C. Hydrogen sensor based on graphene/ZnO nanocomposite. *Sens. Actuators B Chem.,* **2014**, *195*, 409-415. [http://dx.doi.org/10.1016/j.snb.2014.01.029]

[171] Singh, G.; Choudhary, A.; Haranath, D.; Joshi, A.G.; Singh, N.; Singh, S.; Pasricha, R. ZnO decorated luminescent graphene as a potential gas sensor at room temperature. *Carbon,* **2012**, *50*(2), 385-394. [http://dx.doi.org/10.1016/j.carbon.2011.08.050]

[172] Zhu, C.; Fang, Y.; Wen, D.; Dong, S. One-pot synthesis of functional two-dimensional graphene/SnO_2 composite nanosheets as a building block for self-assembly and an enhancing nanomaterial for biosensing. *J. Mater. Chem.,* **2011**, *21*(42), 16911-16917. [http://dx.doi.org/10.1039/c1jm11612f]

[173] Lu, L.M.; Li, H.B.; Qu, F.; Zhang, X.B.; Shen, G.L.; Yu, R.Q. *In situ* synthesis of palladium nanoparticle–graphene nanohybrids and their application in nonenzymatic glucose biosensors. *Biosens. Bioelectron.,* **2011**, *26*(8), 3500-3504.

[http://dx.doi.org/10.1016/j.bios.2011.01.033] [PMID: 21342759]

[174] Zeng, Q.; Cheng, J.S.; Liu, X.F.; Bai, H.T.; Jiang, J.H. Palladium nanoparticle/chitosan-grafted graphene nanocomposites for construction of a glucose biosensor. *Biosens. Bioelectron.*, **2011**, *26*(8), 3456-3463.
[http://dx.doi.org/10.1016/j.bios.2011.01.024] [PMID: 21324668]

[175] Dey, R.S.; Raj, C.R. Development of an Amperometric Cholesterol Biosensor Based on Graphene−Pt Nanoparticle Hybrid Material. *J. Phys. Chem. C*, **2010**, *114*(49), 21427-21433.
[http://dx.doi.org/10.1021/jp105895a]

[176] Chen, Q.; Zhang, L.; Chen, G. Facile preparation of graphene-copper nanoparticle composite by *in situ* chemical reduction for electrochemical sensing of carbohydrates. *Anal. Chem.*, **2012**, *84*(1), 171-178.
[http://dx.doi.org/10.1021/ac2022772] [PMID: 22098222]

[177] Lu, X.; Qi, H.; Zhang, X.; Xue, Z.; Jin, J.; Zhou, X.; Liu, X. Highly dispersive Ag nanoparticles on functionalized graphene for an excellent electrochemical sensor of nitroaromatic compounds. *Chem. Commun. (Camb.)*, **2011**, *47*(46), 12494-12496.
[http://dx.doi.org/10.1039/c1cc15697g] [PMID: 22024647]

[178] Huang, K.J.; Liu, Y.J.; Wang, H.B.; Wang, Y.Y. A sensitive electrochemical DNA biosensor based on silver nanoparticles-polydopamine@graphene composite. *Electrochim. Acta*, **2014**, *118*, 130-137.
[http://dx.doi.org/10.1016/j.electacta.2013.12.019]

[179] Zhang, Y.; Liu, S.; Wang, L.; Qin, X.; Tian, J.; Lu, W.; Chang, G.; Sun, X. One-pot green synthesis of Ag nanoparticles-graphene nanocomposites and their applications in SERS, H_2O_2, and glucose sensing. *RSC Advances*, **2012**, *2*(2), 538-545.
[http://dx.doi.org/10.1039/C1RA00641J]

[180] Zhang, B.; Cui, Y.; Chen, H.; Liu, B.; Chen, G.; Tang, D. A New Electrochemical Biosensor for Determination of Hydrogen Peroxide in Food Based on Well-Dispersive Gold Nanoparticles on Graphene Oxide. *Electroanalysis*, **2011**, *23*(8), 1821-1829.
[http://dx.doi.org/10.1002/elan.201100171]

[181] Cui, F.; Zhang, X. Electrochemical sensor for epinephrine based on a glassy carbon electrode modified with graphene/gold nanocomposites. *J. Electroanal. Chem. (Lausanne)*, **2012**, *669*, 35-41.
[http://dx.doi.org/10.1016/j.jelechem.2012.01.021]

[182] Chen, Y.; Tan, C.; Zhang, H.; Wang, L. Two-dimensional graphene analogues for biomedical applications. *Chem. Soc. Rev.*, **2015**, *44*(9), 2681-2701.
[http://dx.doi.org/10.1039/C4CS00300D] [PMID: 25519856]

[183] Jakus, A.E.; Secor, E.B.; Rutz, A.L.; Jordan, S.W.; Hersam, M.C.; Shah, R.N. Three-dimensional printing of high-content graphene scaffolds for electronic and biomedical applications. *ACS Nano*, **2015**, *9*(4), 4636-4648.
[http://dx.doi.org/10.1021/acsnano.5b01179] [PMID: 25858670]

[184] Ahadian, S.; Estili, M.; Surya, V.J.; Ramón-Azcón, J.; Liang, X.; Shiku, H.; Ramalingam, M.; Matsue, T.; Sakka, Y.; Bae, H.; Nakajima, K.; Kawazoe, Y.; Khademhosseini, A. Facile and green production of aqueous graphene dispersions for biomedical applications. *Nanoscale*, **2015**, *7*(15), 6436-6443.
[http://dx.doi.org/10.1039/C4NR07569B] [PMID: 25779762]

[185] Dan, L.; Cheng, Q.; Narain, R.; Krause, B.; Pötschke, P.; Elias, A. Three-Dimensional Printed and Biocompatible Conductive Composites Comprised of Polyhydroxybutyrate and Multiwalled Carbon Nanotubes. *Indus & Eng. Chem. Res*, **2021**, *60*(2), 885-897.
[http://dx.doi.org/10.1021/acs.iecr.0c04753]

[186] Ajiteru, O.; Sultan, M.T.; Lee, Y.J.; Seo, Y.B.; Hong, H.; Lee, J.S.; Lee, H.; Suh, Y.J.; Ju, H.W.; Lee, O.J.; Park, H.S.; Jang, M.; Kim, S.H.; Park, C.H. A 3D Printable Electroconductive Biocomposite Bioink Based on Silk Fibroin-Conjugated Graphene Oxide. *Nano Lett.*, **2020**, *20*(9), 6873-6883.
[http://dx.doi.org/10.1021/acs.nanolett.0c02986] [PMID: 32794720]

[187] Shen, J.M.; Huang, G.; Zhou, X.; Zou, J.; Yang, Y.; Chen, Y.F.; Men, S.K. Safety evaluation of graphene oxide-based magnetic nanocomposites as MRI contrast agents and drug delivery vehicles. *RSC Advances,* **2014**, *4*(92), 50464-50477. [http://dx.doi.org/10.1039/C4RA09657F]

[188] Ren, L.; Huang, S.; Fan, W.; Liu, T. One-step preparation of hierarchical superparamagnetic iron oxide/graphene composites *via* hydrothermal method. *Appl. Surf. Sci.,* **2011**, *258*(3), 1132-1138. [http://dx.doi.org/10.1016/j.apsusc.2011.09.049]

[189] Fan, X.; Jiao, G.; Zhao, W.; Jin, P.; Li, X. Magnetic Fe_3O_4–graphene composites as targeted drug nanocarriers for pH-activated release. *Nanoscale,* **2013**, *5*(3), 1143-1152. [http://dx.doi.org/10.1039/c2nr33158f] [PMID: 23288110]

[190] Chen, W.; Yi, P.; Zhang, Y.; Zhang, L.; Deng, Z.; Zhang, Z. Composites of aminodextran-coated Fe_3O_4 nanoparticles and graphene oxide for cellular magnetic resonance imaging. *ACS Appl. Mater. Interfaces,* **2011**, *3*(10), 4085-4091. [http://dx.doi.org/10.1021/am2009647] [PMID: 21882840]

[191] Liu, L.; Liu, J.; Wang, Y.; Yan, X.; Sun, D.D. Facile synthesis of monodispersed silver nanoparticles on graphene oxide sheets with enhanced antibacterial activity. *New J. Chem.,* **2011**, *35*(7), 1418-1423. [http://dx.doi.org/10.1039/c1nj20076c]

[192] Freund, H.J.; Meijer, G.; Scheffler, M.; Schlögl, R.; Wolf, M. CO oxidation as a prototypical reaction for heterogeneous processes. *Angew. Chem. Int. Ed.,* **2011**, *50*(43), 10064-10094. [http://dx.doi.org/10.1002/anie.201101378] [PMID: 21960461]

[193] Li, Y.; Yu, Y.; Wang, J.G.; Song, J.; Li, Q.; Dong, M.; Liu, C.J. CO oxidation over graphene supported palladium catalyst. *Appl. Catal. B,* **2012**, *125*, 189-196. [http://dx.doi.org/10.1016/j.apcatb.2012.05.023]

[194] Moussa, S.; Abdelsayed, V.; Samy El-Shall, M. Laser synthesis of Pt, Pd, CoO and Pd–CoO nanoparticle catalysts supported on graphene. *Chem. Phys. Lett.,* **2011**, *510*(4-6), 179-184. [http://dx.doi.org/10.1016/j.cplett.2011.05.026]

[195] Zhang, C.; Lv, W.; Yang, Q.; Liu, Y. Graphene supported nano particles of Pt–Ni for CO oxidation. *Appl. Surf. Sci.,* **2012**, *258*(20), 7795-7800. [http://dx.doi.org/10.1016/j.apsusc.2012.03.131]

[196] Liu, X.; Xu, M.; Wan, L.; Zhu, H.; Yao, K.; Linguerri, R.; Chambaud, G.; Han, Y.; Meng, C. Superior Catalytic Performance of Atomically Dispersed Palladium on Graphene in CO Oxidation. *ACS Catal.,* **2020**, *10*(5), 3084-3093. [http://dx.doi.org/10.1021/acscatal.9b04840]

[197] Sheldon, R.A.; Arends, I.W.C.E.; Dijksman, A. New developments in catalytic alcohol oxidations for fine chemicals synthesis. *Catal. Today,* **2000**, *57*(1-2), 157-166. [http://dx.doi.org/10.1016/S0920-5861(99)00317-X]

[198] Villa, A.; Wang, D.; Dimitratos, N.; Su, D.; Trevisan, V.; Prati, L. Pd on carbon nanotubes for liquid phase alcohol oxidation. *Catal. Today,* **2010**, *150*(1-2), 8-15. [http://dx.doi.org/10.1016/j.cattod.2009.06.009]

[199] Harada, T.; Ikeda, S.; Miyazaki, M.; Sakata, T.; Mori, H.; Matsumura, M. A simple method for preparing highly active palladium catalysts loaded on various carbon supports for liquid-phase oxidation and hydrogenation reactions. *J. Mol. Catal. Chem.,* **2007**, *268*(1-2), 59-64. [http://dx.doi.org/10.1016/j.molcata.2006.12.010]

[200] Wu, G.; Wang, X.; Guan, N.; Li, L. Palladium on graphene as efficient catalyst for solvent-free aerobic oxidation of aromatic alcohols: Role of graphene support. *Appl. Catal. B,* **2013**, *136-137*, 177-185. [http://dx.doi.org/10.1016/j.apcatb.2013.01.067]

[201] Yu, X.; Huo, Y.; Yang, J.; Chang, S.; Ma, Y.; Huang, W. Reduced graphene oxide supported Au

nanoparticles as an efficient catalyst for aerobic oxidation of benzyl alcohol. *Appl. Surf. Sci.*, **2013**, *280*, 450-455.
[http://dx.doi.org/10.1016/j.apsusc.2013.05.008]

[202] Wu, Y.; Yu, H.; Wang, H.; Peng, F. Controllable synthesis and catalytic performance of graphene-supported metal oxide nanoparticles. *Chin. J. Catal.*, **2014**, *35*(6), 952-959.
[http://dx.doi.org/10.1016/S1872-2067(14)60114-5]

[203] Wang, R.; Wu, Z.; Chen, C.; Qin, Z.; Zhu, H.; Wang, G.; Wang, H.; Wu, C.; Dong, W.; Fan, W.; Wang, J. Graphene-supported Au–Pd bimetallic nanoparticles with excellent catalytic performance in selective oxidation of methanol to methyl formate. *Chem. Commun. (Camb.)*, **2013**, *49*(74), 8250-8252.
[http://dx.doi.org/10.1039/c3cc43948h] [PMID: 23925488]

[204] Wu, S.; He, Q.; Zhou, C.; Qi, X.; Huang, X.; Yin, Z.; Yang, Y.; Zhang, H. Synthesis of Fe_3O_4 and Pt nanoparticles on reduced graphene oxide and their use as a recyclable catalyst. *Nanoscale*, **2012**, *4*(7), 2478-2483.
[http://dx.doi.org/10.1039/c2nr11992g] [PMID: 22388949]

[205] Pei, Y.; Zhou, G.; Luan, N.; Zong, B.; Qiao, M.; Tao, F.F. Synthesis and catalysis of chemically reduced metal–metalloid amorphous alloys. *Chem. Soc. Rev.*, **2012**, *41*(24), 8140-8162.
[http://dx.doi.org/10.1039/c2cs35182j] [PMID: 22907172]

[206] Sun, J.; Fu, Y.; He, G.; Sun, X.; Wang, X. Catalytic hydrogenation of nitrophenols and nitrotoluenes over a palladium/graphene nanocomposite. *Catal. Sci. Technol.*, **2014**, *4*(6), 1742-1748.
[http://dx.doi.org/10.1039/C4CY00048J]

[207] Zhang, H.; Zhao, Y.; Liu, W.; Gao, S.; Shang, N.; Wang, C.; Wang, Z. Preparation of magnetically separable Cu6/7Co1/7Fe_2O_4–graphene catalyst and its application in selective reduction of nitroarenes. *Catal. Commun.*, **2015**, *59*, 161-165.
[http://dx.doi.org/10.1016/j.catcom.2014.10.016]

[208] Yang, M.Q.; Pan, X.; Zhang, N.; Xu, Y.J. A facile one-step way to anchor noble metal (Au, Ag, Pd) nanoparticles on a reduced graphene oxide mat with catalytic activity for selective reduction of nitroaromatic compounds. *CrystEngComm*, **2013**, *15*(34), 6819-6828.
[http://dx.doi.org/10.1039/c3ce40694f]

[209] Göksu, H.; Ho, S.F.; Metin, Ö.; Korkmaz, K.; Mendoza Garcia, A.; Gültekin, M.S.; Sun, S. Tandem Dehydrogenation of Ammonia Borane and Hydrogenation of Nitro/Nitrile Compounds Catalyzed by Graphene-Supported NiPd Alloy Nanoparticles. *ACS Catal.*, **2014**, *4*(6), 1777-1782.
[http://dx.doi.org/10.1021/cs500167k]

[210] Feng, Y.S.; Ma, J.J.; Kang, Y.M.; Xu, H.J. PdCu nanoparticles supported on graphene: an efficient and recyclable catalyst for reduction of nitroarenes. *Tetrahedron*, **2014**, *70*(36), 6100-6105.
[http://dx.doi.org/10.1016/j.tet.2014.04.034]

[211] Nie, R.; Wang, J.; Wang, L.; Qin, Y.; Chen, P.; Hou, Z. Platinum supported on reduced graphene oxide as a catalyst for hydrogenation of nitroarenes. *Carbon*, **2012**, *50*(2), 586-596.
[http://dx.doi.org/10.1016/j.carbon.2011.09.017]

[212] Wu, J.; Ong, S.W.; Kang, H.C.; Tok, E.S. Hydrogen Adsorption on Mixed Platinum and Nickel Nanoclusters: The Influence of Cluster Composition and Graphene Support. *J. Phys. Chem. C*, **2010**, *114*(49), 21252-21261.
[http://dx.doi.org/10.1021/jp108517k]

[213] Fan, G.; Huang, W.; Wang, C. Insitu synthesis of Ru/RGO nanocomposites as a highly efficient catalyst for selective hydrogenation of halonitroaromatics. *Nanoscale*, **2013**, *5*(15), 6819-6825.
[http://dx.doi.org/10.1039/c3nr01532g] [PMID: 23771438]

[214] Fan, G.Y.; Huang, W.J. Synthesis of ruthenium/reduced graphene oxide composites and application for the selective hydrogenation of halonitroaromatics. *Chin. Chem. Lett.*, **2014**, *25*(2), 359-363.
[http://dx.doi.org/10.1016/j.cclet.2013.11.044]

[215] Ji, Z.; Shen, X.; Zhu, G.; Zhou, H.; Yuan, A. Reduced graphene oxide/nickel nanocomposites: facile synthesis, magnetic and catalytic properties. *J. Mater. Chem.*, **2012**, *22*(8), 3471-3477. [http://dx.doi.org/10.1039/c2jm14680k]

[216] Wu, Y.; Wen, M.; Wu, Q.; Fang, H. Ni/graphene Nanostructure and Its Electron-Enhanced Catalytic Action for Hydrogenation Reaction of Nitrophenol. *J. Phys. Chem. C,* **2014**, *118*(12), 6307-6313. [http://dx.doi.org/10.1021/jp412711b]

[217] Wang, W.; Gu, J.; Hua, W.; Jia, X.; Xi, K. A novel high efficiency composite catalyst: single crystal triangular Au nanoplates supported by functional reduced graphene oxide. *Chem. Commun. (Camb.),* **2014**, *50*(64), 8889-8891. [http://dx.doi.org/10.1039/C4CC03306J] [PMID: 24968811]

[218] Huang, J.; Zhang, L.; Chen, B.; Ji, N.; Chen, F.; Zhang, Y.; Zhang, Z. Nanocomposites of size-controlled gold nanoparticles and graphene oxide: Formation and applications in SERS and catalysis. *Nanoscale,* **2010**, *2*(12), 2733-2738. [http://dx.doi.org/10.1039/c0nr00473a] [PMID: 20936236]

[219] Ji, Z.; Shen, X.; Yang, J.; Zhu, G.; Chen, K. A novel reduced graphene oxide/Ag/CeO_2 ternary nanocomposite: Green synthesis and catalytic properties. *Appl. Catal. B,* **2014**, *144*, 454-461. [http://dx.doi.org/10.1016/j.apcatb.2013.07.052]

[220] Bai, S.; Shen, X.; Zhu, G.; Li, M.; Xi, H.; Chen, K. *In situ* growth of $Ni_{(x)}Co_{(100-x)}$ nanoparticles on reduced graphene oxide nanosheets and their magnetic and catalytic properties. *ACS Appl. Mater. Interfaces,* **2012**, *4*(5), 2378-2386. [http://dx.doi.org/10.1021/am300310d] [PMID: 22486337]

[221] Feng, C.; Zhang, H.Y.; Shang, N.Z.; Gao, S.T.; Wang, C. Magnetic graphene nanocomposite as an efficient catalyst for hydrogenation of nitroarenes. *Chin. Chem. Lett.*, **2013**, *24*(6), 539-541. [http://dx.doi.org/10.1016/j.cclet.2013.03.036]

[222] Li, X.; Wang, X.; Song, S.; Liu, D.; Zhang, H. Selectively deposited noble metal nanoparticles on Fe_3O_4/graphene composites: stable, recyclable, and magnetically separable catalysts. *Chemistry,* **2012**, *18*(24), 7601-7607. [http://dx.doi.org/10.1002/chem.201103726] [PMID: 22508188]

[223] Cano, M.; Benito, A.M.; Urriolabeitia, E.P.; Arenal, R.; Maser, W.K. Reduced graphene oxide: firm support for catalytically active palladium nanoparticles and game changer in selective hydrogenation reactions. *Nanoscale,* **2013**, *5*(21), 10189-10193. [http://dx.doi.org/10.1039/c3nr02822d] [PMID: 24056941]

[224] Sheng, B.; Hu, L.; Yu, T.; Cao, X.; Gu, H. Highly-dispersed ultrafine Pt nanoparticles on graphene as effective hydrogenation catalysts. *RSC Advances,* **2012**, *2*(13), 5520-5523. [http://dx.doi.org/10.1039/c2ra20400b]

[225] Yao, K.X.; Liu, X.; Li, Z.; Li, C.C.; Zeng, H.C.; Han, Y. Preparation of a Ru-Nanoparticles/Defectiv--Graphene Composite as a Highly Efficient Arene-Hydrogenation Catalyst. *ChemCatChem,* **2012**, *4*(12), 1938-1942. [http://dx.doi.org/10.1002/cctc.201200354]

[226] Zhao, J.; Hu, W.; Li, H.; Ji, M.; Zhao, C.; Wang, Z.; Hu, H. One-step green synthesis of a ruthenium/graphene composite as a highly efficient catalyst. *RSC Advances,* **2015**, *5*(10), 7679-7686. [http://dx.doi.org/10.1039/C4RA11397G]

[227] Marquardt, D.; Vollmer, C.; Thomann, R.; Steurer, P.; Mülhaupt, R.; Redel, E.; Janiak, C. The use of microwave irradiation for the easy synthesis of graphene-supported transition metal nanoparticles in ionic liquids. *Carbon,* **2011**, *49*(4), 1326-1332. [http://dx.doi.org/10.1016/j.carbon.2010.09.066]

[228] Truong-Huu, T.; Chizari, K.; Janowska, I.; Moldovan, M.S.; Ersen, O.; Nguyen, L.D.; Ledoux, M.J.; Pham-Huu, C.; Begin, D. Few-Layer Graphene Supporting Palladium Nanoparticles with a Fully

Accessible Effective Surface for Liquid-Phase Hydrogenation Reaction In: *Proceedings of the Catalysis Today*; Elsevier, **2012**; Vol. 189, pp. 77-82. [http://dx.doi.org/10.1016/j.cattod.2012.04.005]

[229] Yang, J.H.; Ma, D. Graphene-supported Pd nanoparticles: microwave-assisted synthesis and as microwave-active selective hydrogenation catalysts. *RSC Advances,* **2013**, *3*(26), 10131-10134. [http://dx.doi.org/10.1039/c3ra41340c]

[230] Chandra, S.; Bag, S.; Das, P.; Bhattacharya, D.; Pramanik, P. Fabrication of magnetically separable palladium–graphene nanocomposite with unique catalytic property of hydrogenation. *Chem. Phys. Lett.*, **2012**, *519-520*, 59-63. [http://dx.doi.org/10.1016/j.cplett.2011.11.017]

[231] Pei, Y.; Guo, P.; Qiao, M.; Li, H.; Wei, S.; He, H.; Fan, K. The modification effect of Fe on amorphous CoB alloy catalyst for chemoselective hydrogenation of crotonaldehyde. *J. Catal.*, **2007**, *248*(2), 303-310. [http://dx.doi.org/10.1016/j.jcat.2007.03.024]

[232] Zhu, Y.; Tian, L.; Jiang, Z.; Pei, Y.; Xie, S.; Qiao, M.; Fan, K. Heteroepitaxial growth of gold on flowerlike magnetite: An efficacious and magnetically recyclable catalyst for chemoselective hydrogenation of crotonaldehyde to crotyl alcohol. *J. Catal.,* **2011**, *281*(1), 106-118. [http://dx.doi.org/10.1016/j.jcat.2011.04.007]

[233] Ji, X.; Niu, X.; Li, B.; Han, Q.; Yuan, F.; Zaera, F.; Zhu, Y.; Fu, H. Selective Hydrogenation of Cinnamaldehyde to Cinnamal Alcohol over Platinum/Graphene Catalysts. *ChemCatChem,* **2014**, *6*(11), 3246-3253. [http://dx.doi.org/10.1002/cctc.201402573]

[234] Sun, Z.; Rong, Z.; Wang, Y.; Xia, Y.; Du, W.; Wang, Y. Selective hydrogenation of cinnamaldehyde over Pt nanoparticles deposited on reduced graphene oxide. *RSC Advances,* **2014**, *4*(4), 1874-1878. [http://dx.doi.org/10.1039/C3RA44962A]

[235] Shi, J.; Nie, R.; Chen, P.; Hou, Z. Selective hydrogenation of cinnamaldehyde over reduced graphene oxide supported Pt catalyst. *Catal. Commun.*, **2013**, *41*, 101-105. [http://dx.doi.org/10.1016/j.catcom.2013.07.012]

[236] Stein, M.; Wieland, J.; Steurer, P.; Tölle, F.; Mülhaupt, R.; Breit, B. Iron Nanoparticles Supported on Chemically-Derived Graphene: Catalytic Hydrogenation with Magnetic Catalyst Separation. *Adv. Synth. Catal.*, **2011**, *353*(4), 523-527. [http://dx.doi.org/10.1002/adsc.201000877]

[237] Xie, R.; Fan, G.; Ma, Q.; Yang, L.; Li, F. Facile synthesis and enhanced catalytic performance of graphene-supported Ni nanocatalyst from a layered double hydroxide-based composite precursor. *J. Mater. Chem. A Mater. Energy Sustain.*, **2014**, *2*(21), 7880-7889. [http://dx.doi.org/10.1039/c4ta00395k]

[238] Ding, C.; Wei, W.; Sun, H.; Ding, J.; Ren, J.; Qu, X. Reduced graphene oxide supported chiral Ni particles as magnetically reusable and enantioselective catalyst for asymmetric hydrogenation. *Carbon,* **2014**, *79*, 615-622. [http://dx.doi.org/10.1016/j.carbon.2014.08.022]

[239] Szőri, K.; Puskás, R.; Szőllősi, G.; Bertóti, I.; Szépvölgyi, J.; Bartók, M. Palladium Nanoparticle–Graphene Catalysts for Asymmetric Hydrogenation. *Catal. Lett.,* **2013**, *143*(6), 539-546. [http://dx.doi.org/10.1007/s10562-013-1006-6]

[240] Xu, W.; Wang, X.; Zhou, Q.; Meng, B.; Zhao, J.; Qiu, J.; Gogotsi, Y. Low-temperature plasma-assisted preparation of graphene supported palladium nanoparticles with high hydrodesulfurization activity. *J. Mater. Chem.,* **2012**, *22*(29), 14363-14368. [http://dx.doi.org/10.1039/c2jm16479e]

[241] Wang, D.; Niu, W.; Tan, M.; Wu, M.; Zheng, X.; Li, Y.; Tsubaki, N. Pt nanocatalysts supported on reduced graphene oxide for selective conversion of cellulose or cellobiose to sorbitol. *ChemSusChem,*

2014, *7*(5), 1398-1406.
[http://dx.doi.org/10.1002/cssc.201301123] [PMID: 24648252]

[242] Gopiraman, M.; Babu, S.G.; Khatri, Z.; Wei, K.; Endo, M.; Karvembu, R.; Kim, I.S. Facile and homogeneous decoration of RuO2 nanorods on graphene nanoplatelets for transfer hydrogenation of carbonyl compounds. *Catal. Sci. Technol.*, **2013**, *3*(6), 1485-1489.
[http://dx.doi.org/10.1039/c3cy20735h]

[243] Sun, B.; Qiao, M.; Fan, K.; Ulrich, J.; Tao, F.F. Fischer-Tropsch Synthesis over Molecular Sieve Supported Catalysts. *ChemCatChem,* **2011**, *3*(3), 542-550.
[http://dx.doi.org/10.1002/cctc.201000352]

[244] Xu, K.; Sun, B.; Lin, J.; Wen, W.; Pei, Y.; Yan, S.; Qiao, M.; Zhang, X.; Zong, B. ε-Iron carbide as a low-temperature Fischer–Tropsch synthesis catalyst. *Nat. Commun.*, **2014**, *5*(1), 5783.
[http://dx.doi.org/10.1038/ncomms6783] [PMID: 25503569]

[245] Yu, G.; Sun, B.; Pei, Y.; Xie, S.; Yan, S.; Qiao, M.; Fan, K.; Zhang, X.; Zong, B. Fe(x)O(y)@C spheres as an excellent catalyst for Fischer-Tropsch synthesis. *J. Am. Chem. Soc.*, **2010**, *132*(3), 935-937.
[http://dx.doi.org/10.1021/ja906370b] [PMID: 20028126]

[246] Sun, B.; Xu, K.; Nguyen, L.; Qiao, M.; Tao, F.F. Preparation and Catalysis of Carbon-Supported Iron Catalysts for Fischer-Tropsch Synthesis. *ChemCatChem,* **2012**, *4*(10), 1498-1511.
[http://dx.doi.org/10.1002/cctc.201200241]

[247] Sun, Z.; Sun, B.; Qiao, M.; Wei, J.; Yue, Q.; Wang, C.; Deng, Y.; Kaliaguine, S.; Zhao, D. A general chelate-assisted co-assembly to metallic nanoparticles-incorporated ordered mesoporous carbon catalysts for Fischer-Tropsch synthesis. *J. Am. Chem. Soc.*, **2012**, *134*(42), 17653-17660.
[http://dx.doi.org/10.1021/ja306913x] [PMID: 23020275]

[248] Sun, B.; Jiang, Z.; Fang, D.; Xu, K.; Pei, Y.; Yan, S.; Qiao, M.; Fan, K.; Zong, B. One-Pot Approach to a Highly Robust Iron Oxide/Reduced Graphene Oxide Nanocatalyst for Fischer-Tropsch Synthesis. *ChemCatChem,* **2013**, *5*(3), 714-719.
[http://dx.doi.org/10.1002/cctc.201200653]

[249] Zhao, H.; Zhu, Q.; Gao, Y.; Zhai, P.; Ma, D. Iron oxide nanoparticles supported on pyrolytic graphene oxide as model catalysts for Fischer Tropsch synthesis. *Appl. Catal. A Gen.*, **2013**, *456*, 233-239.
[http://dx.doi.org/10.1016/j.apcata.2013.03.006]

[250] Moussa, S.O.; Panchakarla, L.S.; Ho, M.Q.; El-Shall, M.S. Graphene-Supported, Iron-Based Nanoparticles for Catalytic Production of Liquid Hydrocarbons from Synthesis Gas: The Role of the Graphene Support in Comparison with Carbon Nanotubes. *ACS Catal.*, **2014**, *4*(2), 535-545.
[http://dx.doi.org/10.1021/cs4010198]

[251] He, F.; Niu, N.; Qu, F.; Wei, S.; Chen, Y.; Gai, S.; Gao, P.; Wang, Y.; Yang, P. Synthesis of three-dimensional reduced graphene oxide layer supported cobalt nanocrystals and their high catalytic activity in F–T CO2 hydrogenation. *Nanoscale,* **2013**, *5*(18), 8507-8516.
[http://dx.doi.org/10.1039/c3nr03038e] [PMID: 23892431]

[252] Liu, J.X.; Su, H.Y.; Sun, D.P.; Zhang, B.Y.; Li, W.X. Crystallographic dependence of CO activation on cobalt catalysts: HCP versus FCC. *J. Am. Chem. Soc.*, **2013**, *135*(44), 16284-16287.
[http://dx.doi.org/10.1021/ja408521w] [PMID: 24147726]

[253] Graetz, J. New approaches to hydrogen storage. *Chem. Soc. Rev.*, **2009**, *38*(1), 73-82.
[http://dx.doi.org/10.1039/B718842K] [PMID: 19088966]

[254] Hamilton, C.W.; Baker, R.T.; Staubitz, A.; Manners, I. B–N compounds for chemical hydrogenstorage. *Chem. Soc. Rev.*, **2009**, *38*(1), 279-293.
[http://dx.doi.org/10.1039/B800312M] [PMID: 19088978]

[255] Metin, Ö.; Kayhan, E.; Özkar, S.; Schneider, J.J. Palladium nanoparticles supported on chemically derived graphene: An efficient and reusable catalyst for the dehydrogenation of ammonia borane. *Int.*

J. Hydrogen Energy, **2012**, *37*(10), 8161-8169.
[http://dx.doi.org/10.1016/j.ijhydene.2012.02.128]

[256] Xi, P.; Chen, F.; Xie, G.; Ma, C.; Liu, H.; Shao, C.; Wang, J.; Xu, Z.; Xu, X.; Zeng, Z. Surfactant free RGO/Pd nanocomposites as highly active heterogeneous catalysts for the hydrolytic dehydrogenation of ammonia borane for chemical hydrogen storage. *Nanoscale,* **2012**, *4*(18), 5597-5601.
[http://dx.doi.org/10.1039/c2nr31010d] [PMID: 22732933]

[257] Yu-Wen, Y.A.N.G.; Gang, F.E.N.G.; Zhang-Hui, L.U.; Na, H.U.; Fei, Z.H.A.N.G.; Xiang-Shu, C.H.E.N. *In situ* Synthesis of Reduced Graphene Oxide Supported Co Nanoparticles as Efficient Catalysts for Hydrogen Generation from NH_3BH_3. *Wuli Huaxue Xuebao,* **2014**, *30*(6), 1180-1186.
[http://dx.doi.org/10.3866/PKU.WHXB201404141]

[258] Yang, Y.; Lu, Z.H.; Hu, Y.; Zhang, Z.; Shi, W.; Chen, X.; Wang, T. Facile *in situ* synthesis of copper nanoparticles supported on reduced graphene oxide for hydrolytic dehydrogenation of ammonia borane. *RSC Advances,* **2014**, *4*(27), 13749-13752.
[http://dx.doi.org/10.1039/C3RA47023G]

[259] Yang, L.; Cao, N.; Du, C.; Dai, H.; Hu, K.; Luo, W.; Cheng, G. Graphene supported cobalt(0) nanoparticles for hydrolysis of ammonia borane. *Mater. Lett.,* **2014**, *115*, 113-116.
[http://dx.doi.org/10.1016/j.matlet.2013.10.039]

[260] Cao, N.; Luo, W.; Cheng, G. One-step synthesis of graphene supported Ru nanoparticles as efficient catalysts for hydrolytic dehydrogenation of ammonia borane. *Int. J. Hydrogen Energy,* **2013**, *38*(27), 11964-11972.
[http://dx.doi.org/10.1016/j.ijhydene.2013.06.125]

[261] Qiu, F.; Li, L.; Liu, G.; Wang, Y.; An, C.; Xu, C.; Xu, Y.; Wang, Y.; Jiao, L.; Yuan, H. Synthesis of $Fe_{0.3}Co_{0.7}$/rGO nanoparticles as a high performance catalyst for the hydrolytic dehydrogenation of ammonia borane. *Int. J. Hydrogen Energy,* **2013**, *38*(18), 7291-7297.
[http://dx.doi.org/10.1016/j.ijhydene.2013.03.168]

[262] Feng, W.; Yang, L.; Cao, N.; Du, C.; Dai, H.; Luo, W.; Cheng, G. *In situ* facile synthesis of bimetallic CoNi catalyst supported on graphene for hydrolytic dehydrogenation of amine borane. *Int. J. Hydrogen Energy,* **2014**, *39*(7), 3371-3380.
[http://dx.doi.org/10.1016/j.ijhydene.2013.12.113]

[263] Yan, J.M.; Wang, Z.L.; Wang, H.L.; Jiang, Q. Rapid and energy-efficient synthesis of a graphene–CuCo hybrid as a high performance catalyst. *J. Mater. Chem.,* **2012**, *22*(22), 10990-10993.
[http://dx.doi.org/10.1039/c2jm31042b]

[264] Cao, N.; Hu, K.; Luo, W.; Cheng, G. RuCu nanoparticles supported on graphene: A highly efficient catalyst for hydrolysis of ammonia borane. *J. Alloys Compd.,* **2014**, *590*, 241-246.
[http://dx.doi.org/10.1016/j.jallcom.2013.12.134]

[265] Çiftci, N.S.; Metin, Ö. Monodisperse nickel–palladium alloy nanoparticles supported on reduced graphene oxide as highly efficient catalysts for the hydrolytic dehydrogenation of ammonia borane. *Int. J. Hydrogen Energy,* **2014**, *39*(33), 18863-18870.
[http://dx.doi.org/10.1016/j.ijhydene.2014.09.060]

[266] Du, Y.; Cao, N.; Yang, L.; Luo, W.; Cheng, G. One-step synthesis of magnetically recyclable rGO supported Cu@Co core–shell nanoparticles: highly efficient catalysts for hydrolytic dehydrogenation of ammonia borane and methylamine borane. *New J. Chem.,* **2013**, *37*(10), 3035-3042.
[http://dx.doi.org/10.1039/c3nj00552f]

[267] Du, C.; Su, J.; Luo, W.; Cheng, G. Graphene supported Ag@Co core–shell nanoparticles as efficient catalysts for hydrolytic dehydrogenation of amine boranes. *J. Mol. Catal. Chem.,* **2014**, *383-384*, 38-45.
[http://dx.doi.org/10.1016/j.molcata.2013.11.018]

[268] Yang, L.; Luo, W.; Cheng, G. Graphene-supported Ag-based core-shell nanoparticles for hydrogen generation in hydrolysis of ammonia borane and methylamine borane. *ACS Appl. Mater. Interfaces,*

2013, *5*(16), 8231-8240.
[http://dx.doi.org/10.1021/am402373p] [PMID: 23927435]

[269] Wang, J.; Qin, Y.L.; Liu, X.; Zhang, X.B. *In situ* synthesis of magnetically recyclable graphene-supported Pd@Co core–shell nanoparticles as efficient catalysts for hydrolytic dehydrogenation of ammonia borane. *J. Mater. Chem.*, **2012**, *22*(25), 12468-12470.
[http://dx.doi.org/10.1039/c2jm32229c]

[270] Cao, N.; Su, J.; Luo, W.; Cheng, G. Hydrolytic dehydrogenation of ammonia borane and methylamine borane catalyzed by graphene supported Ru@Ni core–shell nanoparticles. *Int. J. Hydrogen Energy,* **2014**, *39*(1), 426-435.
[http://dx.doi.org/10.1016/j.ijhydene.2013.10.059]

[271] Cao, N.; Su, J.; Luo, W.; Cheng, G. Graphene supported Ru@Co core–shell nanoparticles as efficient catalysts for hydrogen generation from hydrolysis of ammonia borane and methylamine borane. *Catal. Commun.*, **2014**, *43*, 47-51.
[http://dx.doi.org/10.1016/j.catcom.2013.09.003]

[272] Meng, X.; Yang, L.; Cao, N.; Du, C.; Hu, K.; Su, J.; Luo, W.; Cheng, G. Graphene-Supported Trimetallic Core-Shell Cu@CoNi Nanoparticles for Catalytic Hydrolysis of Amine Borane. *ChemPlusChem,* **2014**, *79*(2), 325-332.
[http://dx.doi.org/10.1002/cplu.201300336] [PMID: 31986590]

[273] Yang, L.; Su, J.; Meng, X.; Luo, W.; Cheng, G. *In situ* synthesis of graphene supported Ag@CoNi core–shell nanoparticles as highly efficient catalysts for hydrogen generation from hydrolysis of ammonia borane and methylamine borane. *J. Mater. Chem. A Mater. Energy Sustain.,* **2013**, *1*(34), 10016-10023.
[http://dx.doi.org/10.1039/c3ta11835e]

[274] Yang, L.; Su, J.; Luo, W.; Cheng, G. Strategic synthesis of graphene supported trimetallic Ag-based core–shell nanoparticles toward hydrolytic dehydrogenation of amine boranes. *Int. J. Hydrogen Energy,* **2014**, *39*(7), 3360-3370.
[http://dx.doi.org/10.1016/j.ijhydene.2013.12.089]

[275] Qiu, F.; Li, L.; Liu, G.; Xu, C.; An, C.; Xu, Y.; Wang, Y.; Huang, Y.; Chen, C.; Wang, Y.; Jiao, L.; Yuan, H. Synthesis of size-controlled Ag@Co@Ni/graphene core-shell nanoparticles for the catalytic hydrolysis of ammonia borane. *Chem. Asian J.,* **2014**, *9*(2), 487-493.
[http://dx.doi.org/10.1002/asia.201301034] [PMID: 24273129]

[276] Zhang, F.; Hou, C.; Zhang, Q.; Wang, H.; Li, Y. Graphene sheets/cobalt nanocomposites as low-cost/high-performance catalysts for hydrogen generation. *Mater. Chem. Phys.,* **2012**, *135*(2-3), 826-831.
[http://dx.doi.org/10.1016/j.matchemphys.2012.05.065]

[277] López Guerra, E.; Shanmugharaj, A.M.; Choi, W.S.; Ryu, S.H. Thermally reduced graphene oxide-supported nickel catalyst for hydrogen production by propane steam reforming. *Appl. Catal. A Gen.,* **2013**, *468*, 467-474.
[http://dx.doi.org/10.1016/j.apcata.2013.09.025]

[278] Sun, D.; Mazumder, V.; Metin, Ö.; Sun, S. Catalytic hydrolysis of ammonia borane *via* cobalt palladium nanoparticles. *ACS Nano,* **2011**, *5*(8), 6458-6464.
[http://dx.doi.org/10.1021/nn2016666] [PMID: 21766875]

[279] Bhowmik, K.; Sengupta, D.; Basu, B.; De, G. Reduced graphene oxide supported Ni nanoparticles: a high performance reusable heterogeneous catalyst for Kumada–Corriu cross-coupling reactions. *RSC Advances,* **2014**, *4*(67), 35442-35448.
[http://dx.doi.org/10.1039/C4RA04834B]

[280] Djakovitch, L.; Koehler, K. Heck reaction catalyzed by PD-modified zeolites. *J. Am. Chem. Soc.,* **2001**, *123*(25), 5990-5999.
[http://dx.doi.org/10.1021/ja001087r] [PMID: 11414833]

[281] Zhao, F.; Bhanage, B.M.; Shirai, M.; Arai, M. Heck reactions of iodobenzene and methyl acrylate with conventional supported palladium catalysts in the presence of organic and/or inorganic bases without ligands. *Chemistry,* **2000**, *6*(5), 843-848. [http://dx.doi.org/10.1002/(SICI)1521-3765(20000303)6:5<843::AID-CHEM843>3.0.CO;2-G] [PMID: 10826606]

[282] Chen, X.; Hou, Y.; Wang, H.; Cao, Y.; He, J. Facile Deposition of Pd Nanoparticles on Carbon Nanotube Microparticles and Their Catalytic Activity for Suzuki Coupling Reactions. *J. Phys. Chem. C,* **2008**, *112*(22), 8172-8176. [http://dx.doi.org/10.1021/jp800610q]

[283] Scheuermann, G.M.; Rumi, L.; Steurer, P.; Bannwarth, W.; Mülhaupt, R. Palladium nanoparticles on graphite oxide and its functionalized graphene derivatives as highly active catalysts for the Suzuki-Miyaura coupling reaction. *J. Am. Chem. Soc.*, **2009**, *131*(23), 8262-8270. [http://dx.doi.org/10.1021/ja901105a] [PMID: 19469566]

[284] Li, N.; Wang, Z.; Zhao, K.; Shi, Z.; Xu, S.; Gu, Z. Graphene-Pd composite as highly active catalyst for the Suzuki-Miyaura coupling reaction. *J. Nanosci. Nanotechnol.*, **2010**, *10*(10), 6748-6751. [http://dx.doi.org/10.1166/jnn.2010.2638] [PMID: 21137791]

[285] Jafar Hoseini, S.; Dehghani, M.; Nasrabadi, H. Thin film formation of Pd/reduced-graphene oxide and Pd nanoparticles at oil–water interface, suitable as effective catalyst for Suzuki–Miyaura reaction in water. *Catal. Sci. Technol.*, **2014**, *4*(4), 1078-1083. [http://dx.doi.org/10.1039/c3cy01068f]

[286] Wang, P.; Zhang, G.; Jiao, H.; Liu, ; Deng, X.; Chen, Y.; Zheng, X. Pd/graphene nanocomposite as highly active catalyst for the Heck reactions. *Appl. Catal. A Gen.*, **2015**, *489*, 188-192. [http://dx.doi.org/10.1016/j.apcata.2014.10.044]

[287] Siamaki, A.R.; Khder, A.E.R.S.; Abdelsayed, V.; El-Shall, M.S.; Gupton, B.F. Microwave-assisted synthesis of palladium nanoparticles supported on graphene: A highly active and recyclable catalyst for carbon–carbon cross-coupling reactions. *J. Catal.*, **2011**, *279*(1), 1-11. [http://dx.doi.org/10.1016/j.jcat.2010.12.003]

[288] Kim, S.H.; Jeong, G.H.; Choi, D.; Yoon, S.; Jeon, H.B.; Lee, S.M.; Kim, S.W. Synthesis of noble metal/graphene nanocomposites without surfactants by one-step reduction of metal salt and graphene oxide. *J. Colloid Interface Sci.*, **2013**, *389*(1), 85-90. [http://dx.doi.org/10.1016/j.jcis.2012.08.064] [PMID: 23026300]

[289] Xiang, G.; He, J.; Li, T.; Zhuang, J.; Wang, X. Rapid preparation of noble metal nanocrystals *via* facile coreduction with graphene oxide and their enhanced catalytic properties. *Nanoscale,* **2011**, *3*(9), 3737-3742. [http://dx.doi.org/10.1039/c1nr10439j] [PMID: 21804982]

[290] Shendage, S.S.; Singh, A.S.; Nagarkar, J.M. Facile approach to the electrochemical synthesis of palladium-reduced graphene oxide and its application for Suzuki coupling reaction. *Tetrahedron Lett.*, **2014**, *55*(4), 857-860. [http://dx.doi.org/10.1016/j.tetlet.2013.12.022]

[291] Nishina, Y.; Miyata, J.; Kawai, R.; Gotoh, K. Recyclable Pd–graphene catalyst: mechanistic insights into heterogeneous and homogeneous catalysis. *RSC Advances,* **2012**, *2*(25), 9380-9382. [http://dx.doi.org/10.1039/c2ra21185h]

[292] Moussa, S.; Siamaki, A.R.; Gupton, B.F.; El-Shall, M.S. Pd-Partially Reduced Graphene Oxide Catalysts (Pd/PRGO): Laser Synthesis of Pd Nanoparticles Supported on PRGO Nanosheets for Carbon–Carbon Cross Coupling Reactions. *ACS Catal.*, **2012**, *2*(1), 145-154. [http://dx.doi.org/10.1021/cs200497e]

[293] Metin, Ö.; Ho, S.F.; Alp, C.; Can, H.; Mankin, M.N.; Gültekin, M.S.; Chi, M.; Sun, S. Ni/Pd core/shell nanoparticles supported on graphene as a highly active and reusable catalyst for Suzuki-Miyaura cross-coupling reaction. *Nano Res.*, **2013**, *6*(1), 10-18.

[http://dx.doi.org/10.1007/s12274-012-0276-4]

[294] Feng, Y.S.; Lin, X.Y.; Hao, J.; Xu, H.J. Pd–Co bimetallic nanoparticles supported on graphene as a highly active catalyst for Suzuki–Miyaura and Sonogashira cross-coupling reactions. *Tetrahedron,* **2014**, *70*(34), 5249-5253.
[http://dx.doi.org/10.1016/j.tet.2014.05.083]

[295] Li, Y.; Fan, X.; Qi, J.; Ji, J.; Wang, S.; Zhang, G.; Zhang, F. Gold nanoparticles–graphene hybrids as active catalysts for Suzuki reaction. *Mater. Res. Bull.,* **2010**, *45*(10), 1413-1418.
[http://dx.doi.org/10.1016/j.materresbull.2010.06.041]

[296] Movahed, S.K.; Fakharian, M.; Dabiri, M.; Bazgir, A. Gold nanoparticle decorated reduced graphene oxide sheets with high catalytic activity for Ullmann homocoupling. *RSC Advances,* **2014**, *4*(10), 5243-5247.
[http://dx.doi.org/10.1039/c3ra45518a]

[297] Huo, X.; Liu, J.; Wang, B.; Zhang, H.; Yang, Z.; She, X.; Xi, P. A one-step method to produce graphene–Fe_3O_4 composites and their excellent catalytic activities for three-component coupling of aldehyde, alkyne and amine. *J. Mater. Chem. A Mater. Energy Sustain.,* **2013**, *1*(3), 651-656.
[http://dx.doi.org/10.1039/C2TA00485B]

CHAPTER 7

Graphene Supported Palladium Nanocatalyst for Cross-Coupling Reactions

Mukesh Kumar[1], **Meenakshi Singh Solanki**[2] and **Siddharth Sharma**[3,*]

[1] *Department of Chemistry. Sahu Jain College, Najibabad, Bijnor 246763, India*

[2] *Department of Chemistry, PAHER University, Udaipur, Rajasthan 313003, India*

[3] *Department of Chemistry. Mohanlal Sukhadia University Udaipur, Rajasthan 313001, India*

Abstract: The catalytic potential of graphene oxide (GO) and graphite oxide has been well explored in recent years. The reactivity of metal-supported graphite oxide as a solid-phase heterogeneous catalyst has started to be an exceptionally powerful tool for the construction of C-C bonds in synthetic organic chemistry. Among them, palladium has been widely used in different catalysts for a variety of cross-coupling reactions such as Mizoroki-Heck, Suzuki-Miyaura, Kumada, Stille, Negishi, Hiyama, and Buchwald-Hartwig, *etc.*, because of its high catalytic activity and the rapid installation of a complex molecular structure with selectivity in comparison to other transition metal catalysts. A description of recent advances in graphene-supported palladium nanocatalysts for cross-coupling reactions is presented in this chapter. Catalyst synthesis and mechanistic aspects are also given, followed by comparisons with traditional methods.

Keywords: C-C Bond Formation, Cross-coupling Reaction, Graphene Oxide, Graphite Oxide, Reduced Graphene Oxide (r-GO).

INTRODUCTION

Transition metal-catalyzed cross-coupling reactions have emerged as an exceptionally powerful tool for the construction of C-C and C-Hetero atoms bonds in synthetic organic chemistry and continue to have widespread application, especially in drug development programmes and agrochemicals [1, 2]. In the last decades, many transition metals have been employed to carry out cross-coupling reactions. Among them, palladium has been widely used in different catalysts for a variety of cross-coupling reactions such as Mizoroki-Heck, Suzuki-Miyaura, Kumada, Stille, Negishi, Hiyama, Buchwald-Hartwig, *etc.*, because of its high catalytic activity and the rapid installation of complex molecular structure with

* **Corresponding author Siddharth Sharma:** Department of Chemistry. Mohanlal Sukhadia University Udaipur, Rajasthan 313001, India; E-mail: siddharth@mlsu.ac.in

Pinki Bala Punjabi, Rakshit Ameta & Sharoni Gupta (Eds.)

selectivity in comparison to other transition metal catalysts [3]. The impact of palladium-catalyzed reactions in the scientific community has been recognized after getting the Nobel Prize in chemistry by three renowned scientists in this area: Professor Richard F. Heck, Professor Eichi Negishi, and Professor Akira Suzuki [4].

Despite the high impact of homogenous palladium-catalyzed cross-coupling reactions in all fields of science, there are many issues associated with homogenous catalysis, such as the use of a high-cost catalytic system, lack of reusability of catalyst, and potential contamination of the drug by residual heavy metals, which strongly limit their use in the pharmaceutical industry [5, 6]. To address these issues, synthetic chemists started channeling their efforts to innovate environmentally benign processes by developing new catalysts with high efficiency and reusability, applying efficient energy sources, and avoiding the use of ligands and auxiliaries. For these reasons, synthetic chemists moved on to heterogeneous catalysis, in which palladium catalysts immobilized on solid support (organic, inorganic, and hybrid materials) have become an important alternative to address all the issues associated with homogenous catalysis. Several heterogeneous palladium catalysts have been generated using silica [7, 8], zeolites [9, 10], ionic liquids [11 - 13], alumina [14], clay [15], zinc ferrite [16], organic, and inorganic polymers as supports [17 - 22]. There are many carbon-based nanomaterials such as one-dimensional (1D) carbon nanotubes (CNTs), two-dimensional (2D) graphene, amorphous activated carbon (AC), and mesoporous carbon (MC), which have also been applied as supports for the synthesis of palladium nanoparticles catalysts [23 - 26].

Among them, graphene and modified graphene has recently become more attractive support in the field of catalysis and have attracted much attention from the scientific community because of their extraordinary properties. Graphene, a 2D material, is considered a "wonder material" formed by carbon atoms that are arranged in a hexagonal monolayer pattern (a honeycomb crystal lattice). Each carbon atom of graphene is sp^2 hybridized and 200 times harder than steel, very feathery, and flexible [27 - 29]. In addition, graphene possesses a large surface area, and excellent chemical, thermal and electrical stability. Moreover, it showed a strong hydrophobic interaction with metal and was easily modified to other forms [30]. Graphene-based architectures are highly desirable in the synthesis of metal nanoparticles, especially as a support for palladium nanoparticles (PdNPs), and displayed exceptional catalytic efficiency for C-C cross-coupling reactions because graphene has high binding energy with palladium and good charge storage performance [31]. Moreover, due to the presence of transparency and conductive properties of graphene it also plays a vital role in the areas of electrochemical applications such as sensors [32], electrodes [33], and fuel cells

[34]. Furthermore, graphene derivatives, such as graphene oxide (GO) and reduced graphene oxide (rGO) have also become useful support in nano metal-catalysis due to their structural defects, which allow an easy covalent and non-covalent (hydrogen-binding or π-stacking) functionalization of the material that can be used either as supports for immobilizing active species or as metal-free catalysts and enhancing the rate of interaction with anchored metal nanoparticles [35, 36]. Graphene oxide was first synthesized by oxidation of graphite in acidic conditions using the Hummers-Offeman method [37]. This method involves the treatment of graphite with sodium nitrite ($NaNO_3$), potassium permanganate ($KMnO_4$), and followed by hydrogen peroxide (H_2O_2).

Recently, Palladium nanoparticles (PdNPs) have established as an intensively investigated and rapidly growing area in ligand-free catalysis because of their high surface-to-volume ratio and highly active surface atoms in comparison to other bulk catalysts [38 - 42]. Palladium nanoparticles (PdNPs), used as a heterogeneous catalyst for the production of pharmaceutical agents, herbicides, fine chemicals, and polymers have a lot of advantages over the palladium-catalyzed homogenous reactions because of their remarkable features such as stability, selectivity, reusability, cost-effective, simple workup procedure and easy separation of the final product [43]. Palladium nanoparticles (PdNPs) supported on graphene/graphene oxide (GO) /reduced graphene oxide (rGO) increase the surface area of nanohybrids by expanding the distance between the sheets. This process enhances the catalytic performance of this type of heterogeneous catalyst. These heterogeneous catalysts have not only shown their catalytic efficiency for catalyzing the cross-coupling reactions in which carbon-carbon bond is formed but they have also been applied in other reactions, such as hydrogenation of alkenes, alkynes, and hydrogenation of carbonyl (-CO) and nitro ($-NO_2$) groups [44 - 52].

This chapter covers the graphene/graphene oxide (GO)/reduced graphene oxide (rGO) supported metal nanoparticles catalyzed cross-coupling reactions such as Suzuki, Heck, and Sonogashira, leading to the formation of biological importance molecules.

GRAPHENE SUPPORTED PALLADIUM NANOPARTICLES (PDNPS/G) CATALYZED CROSS-COUPLING REACTIONS

Palladium nanoparticles supported on graphene (PdNPs-G) and reduced graphene oxide (PdNPs-rGO) have been used as catalysts for Suzuki cross-coupling reactions between aryl bromides and potassium aryltrifluoroborates by Gómez-Martínez and his co-workers in 2015 [53]. Potassium aryl- and heteroaryl-trifluoroborates efficiently coupled with various aryl halides in the presence of 0.1

mol% catalyst, potassium carbonate as a base in MeOH:H_2O as solvent using conventional and microwave heating to generate a library of functionalized biphenyl derivatives in good to excellent yields (Scheme **1**). Sterically hindered substrate such as 2-bromo-1,3-dimethylbenzene was unable to provide the cross-coupling product in good yield with potassium phenyltrifluoroborate. Heterocyclic bromides such as 2-bromopyridine and 2-bromothiophene were smoothly coupled with potassium phenyltrifluoroborate to give the coupling products with slightly lower yields. However, less reactive chlorobenzene gave the lowest yield of coupling products. Microwave irradiation provides a lower accumulation of the PdNPs. Identical results of coupling products in terms of yields have been observed using microwave and conventional heating. It is interesting to note that this catalyst was used in Gram-scale synthesis and provides an excellent yield of Suzuki products without loss of catalytic activity. The reusability of the catalyst was also studied and found that the catalyst could be reused up to eight times without loss of significant activity under microwave irradiation, whereas four times when the conventional heating condition was applied.

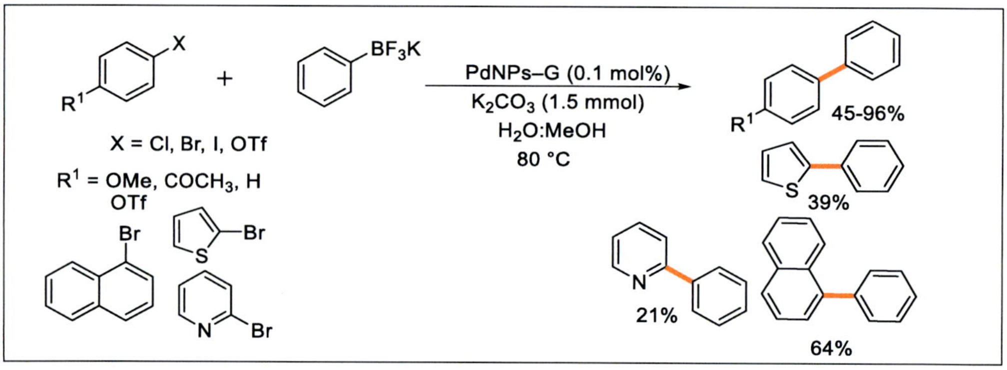

Scheme (1). Synthesis of biaryl derivatives using PdNPs-G catalyst.

Wang and co-workers disclosed the formation of graphene-supported Pd (Pd/graphene) catalyst derived from graphene oxide (GO) and $PdCl_2$ followed by reduction with hydrazine hydrate in an aqueous medium [54]. The performance of this catalyst was investigated in Heck cross-coupling reactions in which iodo- and bromobenzene efficiently coupled with *n*-butyl acrylate or styrene in the presence of 0.05 mol% Pd/G and K_2CO_3 as the base at 100 ^{0}C to provide the coupling products in good to excellent yields (Scheme **2**). However, chlorobenzene was unable to provide a satisfactory coupling product even after increasing the catalyst loading and reaction time. The reusability of the catalyst was also studied using the reaction between bromobenzene with *n*-butyl acrylate. The recycled catalyst could be used seven times without significant loss of activity. Pd NPs in

Pd/graphene were uniformly dispersed on the surface of graphene as displayed by the TEM image. The shape and size of Palladium nanoparticles were found to be spherical and ranged from 2-6, respectively.

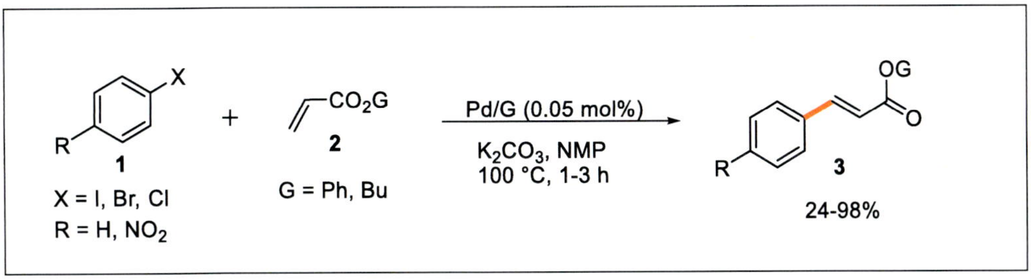

Scheme (2). Heck cross-coupling reaction catalyzed by Pd/G catalyst.

Siamaki *et al*. described a highly efficient protocol for the formation of graphene supported-palladium nanoparticles; in which an aqueous mixture of Pd $(NO_3)_2$ and GO sheets were reduced under microwave irradiation in the presence or absence of hydrazine hydrate respectively [55]. The catalytic exploration of these Pd/G and Pd/GO catalysts was tested using C-C cross-coupling reactions (Suzuki and Heck) without the use of any ligand in aqueous ethanol. The author observed that both catalysts (Pd/G and Pd/GO) exhibit excellent catalytic activity towards the cross-coupling reaction of phenylboronic acid with aryl halides (Scheme **3**), however, Pd/G displayed outstanding activity towards the Suzuki coupling even at low concentrations.

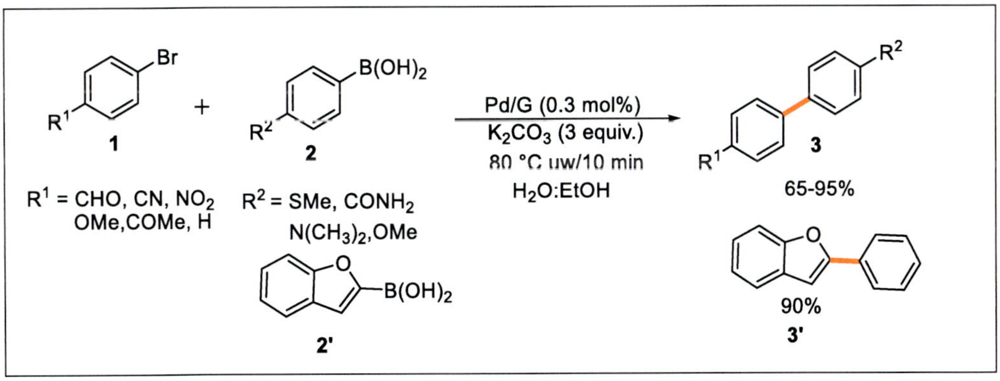

Scheme (3). Graphene-supported palladium catalyzed Suzuki reaction.

The Pd/G catalyst promotes the coupling reaction between bromobenzene and phenylboronic acid with a very low catalyst loading (0.007 mol% Pd/G) to form a biphenyl derivative with a TON of 9000, a TOF of 108 000 h^{-1}, and a 63%

conversion at 80 °C under microwave for 5 min radiation. It is interesting to note that this is the highest TOF for palladium-catalyzed Suzuki cross-coupling reaction under microwave irradiation. Pd/G catalyst also showed excellent catalytic activity towards the Heck cross-coupling. Many aryl bromides coupled with a variety of olefins in the presence of 0.3 mol% of Pd/G, K_2CO_3 as a base in H_2O:EtOH at 180 °C under microwave irradiation. Aryl bromides bearing electron-donating or electron-donating groups easily react with olefins even with inactivated olefins such as *tert*-butylstyrene to furnish excellent yield of the coupling products (Scheme **4**). The recyclability of the catalyst was also examined and found it could be reused eight times without loss of catalytic activity. The activity of the catalyst was dropped after eight times due to the agglomeration and accumulation of the Pd nanoparticles on the surface of graphene. The uniform dispersion of Pd nanoparticles on both graphene and graphene oxide was confirmed by TEM images. The size of NPs supported on graphene was found to be smaller than those supported on the GO sheets.

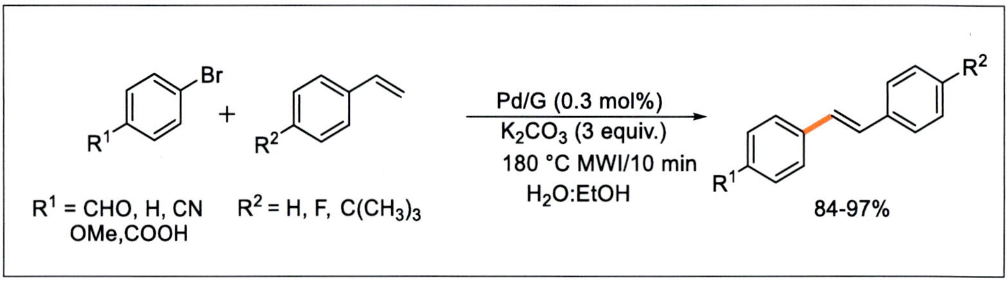

Scheme (4). Pd/G catalyzed Heck cross-coupling reactions.

Later, Gilliland III *et al.* research group developed the catalysts having several defect sites through strong electrostatic adsorption (SEA) with solvent-free microwave reduction method using four different types of commercially available graphene supports (graphene powder 1-5 layers, graphene oxide, reduced graphene oxide, and graphene Nanoplatelet aggregates) as starting materials [56]. The synthesized catalysts were characterized using different analytical techniques such as powder x-ray diffraction (XRD), transmission electron microscopy (TEM), and X-ray photoelectron spectroscopy (XPS). The prepared catalysts (graphene Nanoplatelet) were economical and showed excellent activity towards the Suzuki cross-coupling reactions than other graphene-based supports (Scheme **5**).

In 2013, Lee and co-workers demonstrated a multistep route for the synthesis of a high-performance graphene-based palladium nanoparticles catalyst and evaluated their catalytic power towards the copper and ligand-free Sonogashira reaction

[57]. The synthetic route involved the formation of multilayer Pd^{+2}/GO intermediate between the reaction of GO with $Pd(OAc)_2$ followed by reduction and exfoliation into Pd/G using H_2 gas as a reducing agent. TEM images displayed the uniform distribution of PdNPs on the surface of graphene sheets without aggregation. The mean particle size of PdNPs was found to be 1.4 ±0.6 nm with the help of ICP-MS. The synthesized catalyst exhibited excellent catalytic activity in the Sonogashira coupling reaction in which iodobenzene coupled with phenylacetate in the presence of the formed catalyst (Pd/G) without using any ligand to form biphenylacetylene derivative in a high (89%) yield. There are many advantages of this catalyst such as easy removal of catalyst from the reaction mixture and reusability; *i.e.* it can be reused five times without loss of significant activity.

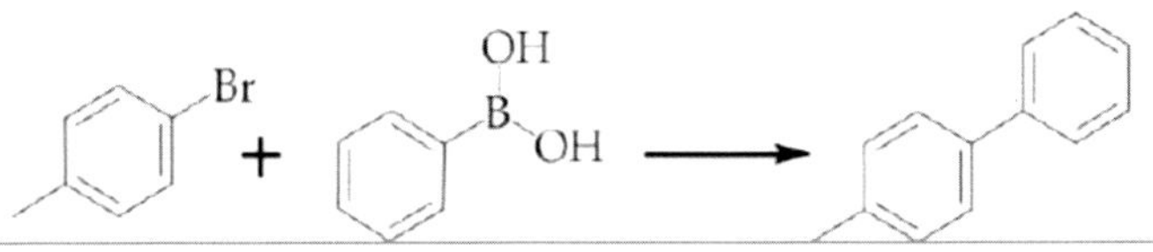

Scheme (5). Suzuki cross-coupling.

In 2015, Rana research groups fabricated a new type of nanocomposite catalyst by highly deposited PdNPs on organodiamine-functionalized graphene support. Organodiamine-functionalized graphene was synthesized by the reaction of [3-(-Aminoethylamino)propyl] trimethoxysilane (AAPTMS) with GO using toluene as solvent. Organodiamine-functionalized graphene-supported PdNPs were obtained by a reduction of $Pd(NO_3)_2$ using $NaBH_4$ as a reducing agent [58].

Organodiamine-functionalized graphene support worked as a stabilizer for Pd NPs and provides sufficient electron density to Pd(0) during oxidative addition of aryl halide. In addition, it also worked as excellent support that helped in eluding "Ostwald ripening". The newly fabricated catalyst exhibited outstanding activity toward the Suzuki cross-coupling reaction of aryl iodides with phenylboronic acids. The reaction was performed in the presence of 0.05 mol% of catalyst using $Ba(OH)_2$ or K_3PO_4 as a base in an aqueous medium providing 98% yield with 100% conversion of starting materials (aryl iodides with phenylboronic acids). 2-bromopyridine smoothly reacted with phenylboronic acid producing biaryl products in good yield, whereas chlorobenzene was unable to provide the desired products. The authors proposed the reaction mechanism in which oxidative addition of phenyl halide in the presence of a catalyst is the main step. The yield of coupling products was not increased when catalyst loading was increased from 0.05 mol% to 0.06 mol%. The catalyst was recycled up to six times without any change in catalytic activity. However, noticeable changes can be seen in the catalytic activity of the catalyst after the seventh and eighth repetitive uses in cross-coupling reaction, the leached amount of Pd was ~10% and ~15% respectively, which indicated that the redeposition of the Pd species was not fully completed during the release-redeposition cycle. TEM images displayed that Pd nanoparticles (3-7 nm) were uniformly distributed on the surface of amine functionalized graphene.

Mahdavi *et al.* employed a similar strategy for the development of polyacrylamide (PAM) functionalized Graphene oxide supported palladium nanoparticles (PAM-g-rGO/Pd) catalyst using ammonium cerium (IV) sulphate $[(NH_4)_4Ce(SO_4)_4 \cdot 2H_2O]$ induced by redox polymerization [59]. PAM helps to make effective dispersion of graphene oxide because of its high polarity and hydrophilic nature. Then PdNPs were synthesized on the surface of PAM-g-rGO using $NaBH_4$ as a reducing agent. The synthesized catalyst was used for Suzuki cross-coupling reactions to observe its catalytic activity. The reactions were performed between different aryl halides and boronic acid in the presence of PAM-g-rGO/Pd using potassium carbonate as a base in an aqueous medium. Iodobenzene-bearing electron-withdrawing substituents efficiently coupled with boronic acid to furnish the corresponding biphenyl products in 100% yields. However, bromobenzene and chlorobenzene needed a long reaction time to couple with the boronic acid and provide the coupling products in excellent yields. Boronic acid that incorporated electron-donating substituents smoothly reacted with aryl halides to provide the corresponding products in excellent yield except methyl boronic acid gave 100% yield of coupling product with iodobenzene. The most important thing about this type of heterogeneous catalyst is its reusability. The catalyst can be recycled successfully up to eight times with a minimum loss of activity (Scheme **6**).

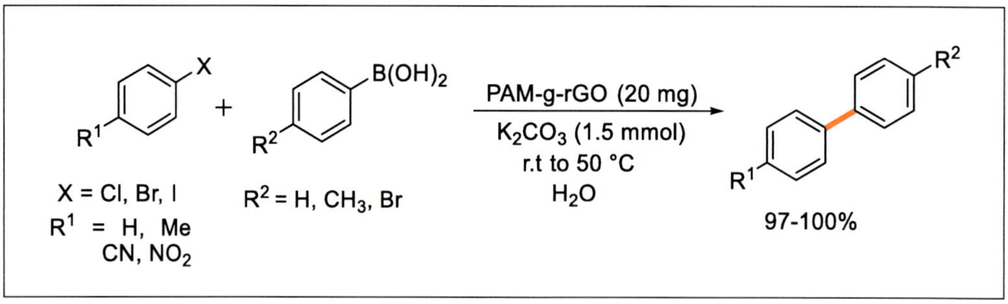

Scheme (6). Suzuki cross-coupling reaction catalyzed by PAM-g-rGO/Pd.

Another important approach has been described for the synthesis of reusable heterogeneous catalysts based on palladium nanoparticles supported on graphene-benzimidazole(G-BI-Pd) by Sarvestani and co-workers [60]. The catalyst was prepared in the following steps. In the first step, graphene oxide-benzimidazole was prepared by the reaction of graphene oxide with an excess amount of *ortho*-phenylenediamine in the presence of polyphosphoric acid which subsequently underwent reduction of graphene oxide using hydrazine as a reducing agent to generate benzimidazole-grafted graphene (G-BI). Finally, PdNPs were immobilized on the surface of benzimidazole-grafted graphene to form (G-BI-Pd) *via* the formation of the N-Pd complex. Several analytical techniques have been used for the confirmation of synthesized catalysts. The developed catalyst demonstrated superb catalytic activity toward the Suzuki cross-coupling reaction of aryl halides with phenylboronic acids (Scheme **7**). The best results were obtained in terms of yield when the reaction was performed in the presence of 0.45 mol% of G-BI-Pd catalyst using potassium carbonate as a base in an aqueous medium at 80 ^{0}C. Bromobenzene/iodobenzene having electron-withdrawing or electron-donating substituents smoothly transformed into the corresponding biphenyl products in short reaction times with excellent yields. Aryl bromide required a longer reaction time to complete the reaction in comparison to iodobenzene. 2-bromofuran and α-bromo-naphthalene were smoothly converted into the corresponding products in excellent yields under the same catalytic system, except for chlorobenzene which was unable to provide good results in terms of yields. This heterogeneous catalyst showed high stability and could be reused up to five times without loss of their catalytic activity. The uniform distribution of palladium nanoparticles on the surface of G-BI and no aggregation was observed in TEM images.

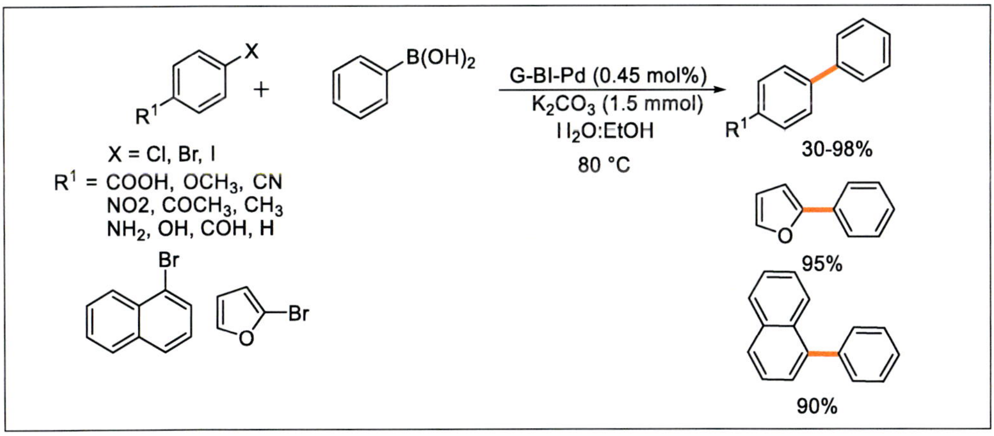

Scheme (7). Synthesis of biphenyl derivative using GBIPd catalyst.

In continuation, Fareghi-Alamdari and co-workers also applied a similar strategy for the construction of palladium nanoparticles catalyst decorated on the surface of phosphine functionalized graphene (PFG-Pd) and demonstrated its application as a highly efficient catalyst for the Suzuki cross-coupling reaction [61]. A phosphine functionalized graphene (PFG) was prepared by the treatment of graphene oxide (GO) with potassium hydroxide resulting in the formation of a free hydroxyl group on the surface of graphene (GOH). The generated hydroxy-functionalized graphene reacted with chlorodiphenylphosphine ($ClPPh_2$) to form phosphine grafted graphene oxide $GOPPh_2$ (PFG), which could be helpful in the preparation of reusable metal catalysts and for the stabilization of metal nanoparticles or the complexation of transition metals. The application of phosphine functionalized graphene (PFG-Pd) catalyst was tested for Suzuki cross-coupling reaction and found that different aryl halides bearing electron-donating or electron-withdrawing substituents smoothly reacted with arylboronic acids using 1.2 mol% of PFG-Pd catalyst and K_2CO_3 as a base in aqueous medium at 80 ^{0}C providing the corresponding biphenyl products in good to excellent yields. Heterocyclic substrates were also applied in place of aryl halides which smoothly coupled with arylboronic acid to furnish the coupling products in good to excellent yields but with more reaction time (Scheme **8**). It is interesting to note that this catalyst was also used for Heck cross-coupling and *N*-arylation. The catalyst can be consecutively recycled up to five times without significant leaching of the palladium catalyst.

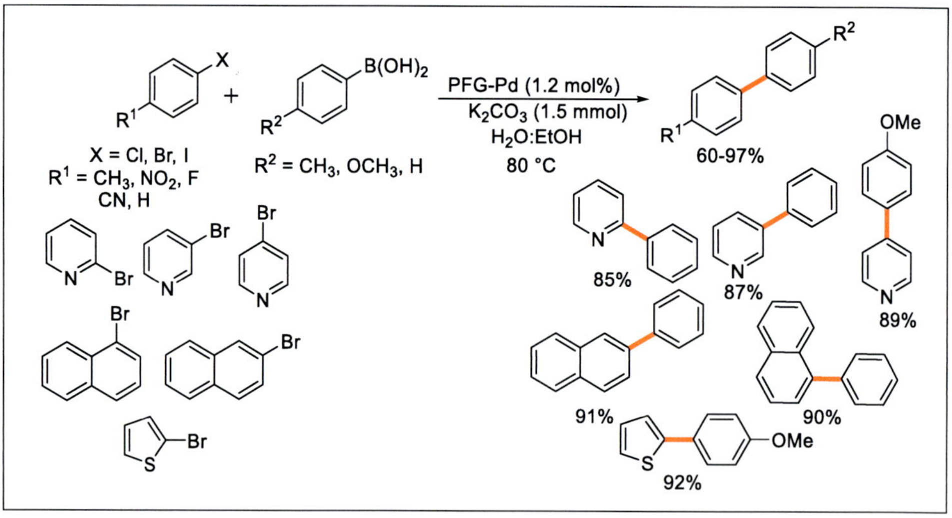

Scheme (8). Synthesis of biaryl derivatives using Suzuki cross-coupling reaction.

GRAPHENE OXIDE (GO) SUPPORTED PALLADIUM NANOPARTICLES (PDNPS/GO) CATALYZED CROSS-COUPLING REACTIONS

In 2014, Yamamoto and his research group described the synthesis of palladium nanoparticles supported on single layer oxide (Pd-slGO) using palladium (II) acetate ($Pd(OAc)_2$) and GO using ethanol as solvent [36]. Ethanol also acted as a reducing agent for palladium precursor at room temperature or 60 ^{0}C or 100 ^{0}C. TEM images displayed that Pd-slGO-60 (reaction at 60 ^{0}C) had a smaller particle size in comparison to PdslGO-100 (reaction at 100 ^{0}C), whereas Pd-slGO-r.t (reaction at room temperature) was not exit at room temperature. Pd-slGO-60 exhibited outstanding catalytic activity towards the Suzuki cross-coupling reaction between iodobenzenes/bromobenzenes with phenylboronic acids. Bromobenzene-bearing electron-withdrawing groups efficiently coupled with phenylboronic acids in the presence of 0.01 mol% of catalyst (Pd-slGO-60) using K_2CO_3 as a base in EtOH:H_2O at room temperature to furnish the corresponding biaryl products in moderate to excellent yields whereas bromobenzene with electron-donating substituents was less reactive towards the phenylboronic acids (Scheme **9**). However, heterocyclic bromides, such as thienyl bromide and indolyl bromide were easily transformed into the corresponding coupling products with phenylboronic acids using the same catalyst. It is interesting to note that iodobenzene showed much reactive towards the phenylboronic acid to provide the coupling product in good yield using 0.00016 mol% catalyst with a turnover number (TON) and TOF (h^{-1}) of 237 000. Moreover, the authors also proved that GO could be used as a scavenger to prevent Pd leaching and contamination of the product. The reusability of this catalyst was also investigated and found that the catalytic activity was dropped after three runs due to the aggregation of PdNPs on the surface of GO and reduction in particle size.

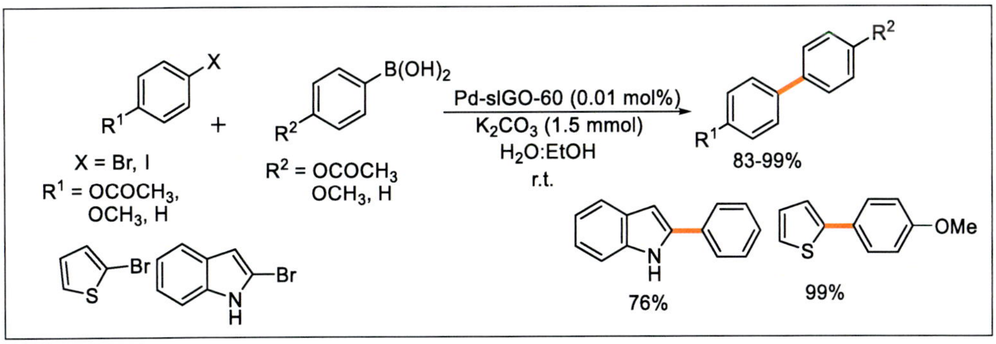

Scheme (9). Synthesis of the biaryl derivative using Pd-slGO-60 catalyst.

In recent years, three-dimensional graphene-based materials (3DGMs) have become an emerging support for the synthesis of heterogeneous nano-catalysts because of their high surface area, low mass density, mechanical strength, and flexible porosity [62]. The remarkable feature of this type of heterogeneous catalyst is its porous structure, which could be used in the stabilization of metal nanoparticles and control their growth [63, 64].

In 2017, Tran and co-workers have discovered a novel method for the preparation of palladium nanoparticles catalyst supported on graphene oxide framework (Pd@GOF) [65]. According to this approach, graphene oxide was first intercalated with benzene-1,4-diboronic through a covalent bond to achieve a three-dimensional (3D) structure. After that PdNPs were formed by the reduction of $PdCl_2$. The intercalation of graphene oxide (GO) with benzene-1,4-diboronic acid through a covalent link afforded a three-dimensional GOF structure having uniformly expanded interlayer spacing which provided thermal stability. The fabricated palladium nanoparticles were uniformly encapsulated and stabilized inside the rigid framework. The synthesized catalyst (Pd@GOF) was used for Suzuki's cross-coupling reaction of aryl halides with phenylboronic acids to evaluate its catalytic performance and found that it exhibits superb activity in both polar and non-polar solvents in terms of yields and could be recycled at least five times without significant loss of catalytic activity.

Functionalized graphene oxide-based materials have been considered as promising supports for nanoparticles and play a significant role in organic transformations such as C-C bond forming cross-coupling reactions [66]. The functional groups present in graphene oxide make the sorption and intercalation of ions and molecules possible and are very helpful in making promising materials for catalytic applications [67]. In 2014, a recyclable heterogeneous catalyst based on palladium nanoparticles supported on non-covalently functionalized graphene oxide (GO) using 1-pyrene carboxylic acid has been developed by Sharavath and Ghosh [68]. Graphene oxide was reacted with a 1-pyrene carboxylic acid in the presence of sodium hydroxide and hydrazine monohydrate to form functionalized graphene and was dispersed in ethanol for 30 min. After that, $PdCl_2$ was added to the above suspension which reduced to palladium nanoparticles (PdNPs). It is interesting to note that ethanol worked both as a solvent and as a reducing agent. The fabricated catalyst (PCA-GNS-Pd) was found to be effective in the promotion of the Suzuki cross-coupling reaction between aryl halides and phenylboronic acids in an aqueous medium. Aryl bromide-bearing electron-donating and electron-withdrawing substituents are easily coupled with phenylboronic acids to transform into the corresponding biphenyl products in good yields. The synthesized catalyst was also useful for unreactive substrates such as chlorobenzene, which reacted smoothly with arylboronic acids to provide coupling products in good yields (see Table **1**). The application of the PCA-GN--Pd catalyst was further demonstrated for the Heck cross-coupling reaction of aryl halides and olefins. A series of Heck coupling products were obtained in good yields using 0.2 mol% of nanocatalyst in water.

Table 1. Synthesis of biaryl analogous *via* Suzuki cross-coupling of aryl halides with phenylboronic acid using PCA-GNS-Pd catalyst[a].

Entry	R^1	X	Yield (%)[b]	Time (h)
1	H	Br	97	2
2	*p*-CHO	Br	96, 98[d]	2.5, 2.5[d]
3	*p*-OMe	Br	85[f]	6
4	*p*-OH	Br	90	4
5	*p*-Me	Br	85	7
6	2,4-OMe	Br	80	9
7	1-Naphthyl[d]	Br	85	6
8	2-Pyridyl	Br	80	5
9	3-Pyridyl	Br	82	4.5
10	5-Pyrimidine	Br	80	4.5
11	*m*-NO_2	Br	90[e], 75[c]	2[e], 12[c]
12	H	Cl	80	15
13	*p*-CHO	Cl	85	18
14	2-Pyridyl	Cl	60	15
15	3-Pyridyl	Cl	67	15
16	*m*-NO_2	Cl	75	15
17	*p*-OH	Cl	70	10
18	2,4-OMe	Cl	65	15

[a]Reaction conditions:aryl halide (1.435 mmol), phenylboronic acid (1.722 mmol) Na_2CO_3 (2.153 mmol), water (5 mL), PCA-GNS-Pd (3 mg, 0.2 mol %) (mol% Pd *vs.* halide), heated to 90 °C. [b]Isolated yield. [c]PCA-GNS-Pd (3 mg, 0.6 mol% Pd *vs.* halide) was used. [d]Isolated yield and time using Et_3N as base. [e]4-Cyanophenylboronic acid was used instead of $PhB(OH)_2$. [f]2-Napthylboronic acid was used instead of $PhB(OH)_2$.

The nanocatalyst was recovered and successfully reused up to five times without loss of catalytic performance. The catalytic efficiency of the catalyst was dropped after the fifth run due to the leaching of PdNPs from the support during the reaction and their agglomeration on the catalyst support, which was analyzed by ICP-AES.

Very recently, Mohazzab's research group has developed a novel and efficient method for the fabrication of separable palladium nanoparticles (PdNPs) supported on graphene oxide (GO)/stainless-steel mesh using a laser ablation in liquids (LAL) process and an electrophoretic deposition (EPD) technique [69]. In this method, no need of any chemical or capping agents was required during the preparation of a mesh-GO/Pd catalyst. The developed catalyst was characterized by using various analytical techniques. The catalytic performance of this catalyst was investigated for Suzuki cross-coupling reaction in ethanol-water without the use of any ligand. The results showed that aryl halides having either electron-attracting or electron-releasing groups smoothly coupled with phenylboronic acids and provided coupling products in excellent yields. In addition, less reactive chlorobenzene also reacted with phenylboronic acids affording the corresponding products in good yields. The reusability of this catalyst was also examined using the Suzuki cross-coupling reaction and found that it could be recycled up to five times with no remarkable loss of catalytic activity.

The research group of Premia has described another method for the functionalization of graphene oxide with heteroatoms such as sulfur and nitrogen on which Palladium nanoparticles (PdNPs) were uniformly dispersed [70]. The presence of functional groups in graphene oxide can increase the rate of dispersion of nanohybrids in organic solvents and can be used as a ligand to stabilize the active palladium species. The fabricated catalyst was applied for the Heck cross-coupling reaction to prove its catalytic performance. The coupling reactions between aryl iodides/bromides and substituted styrenes/acrylates were performed in the presence of 0.02 mol% of catalyst using potassium hydroxide as a base at 120 °C in DMF-H_2O (Scheme **10**).

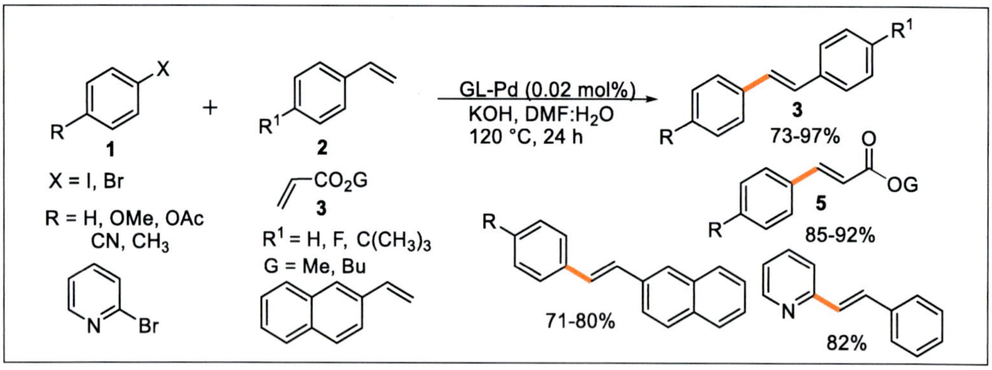

Scheme (10). GL-Pd catalyzed Heck coupling between aryl halides and olefins.

Under the catalytic system, aryl halides bearing electron-withdrawing groups reacted smoothly with olefins to give the corresponding stilbene derivatives, whereas lower yields of coupling products were obtained with aryl halide bearing electron-donating substituents. Heteroaryl halides such as 2-bromo-pyridine are also easily coupled with styrene to afford the corresponding coupling products in moderate yields. In addition, polycyclic aryl halide was also used to increase the scope and versatility of the reaction. As shown in Scheme (**11**), the proposed reaction mechanism for the Heck cross-coupling reaction was initiated by the formation of GL-Pd complex I in which Pd is stabilized through coordination with sulfur and nitrogen. Then, aryl halide was activated through oxidative addition with Pd(0) to form complex II followed by the coordination of olefin to give π-complex III and then migratory insertion of olefin takes place to generate IV. Subsequently, *β*-hydride elimination takes place to afford coupling product with the regeneration of the catalyst. The nano-catalyst was recovered, and reused up to three times without any significant loss of its catalytic efficiency, but catalytic activity was dropped after the third cycle due to the aggregation of PdNPs.

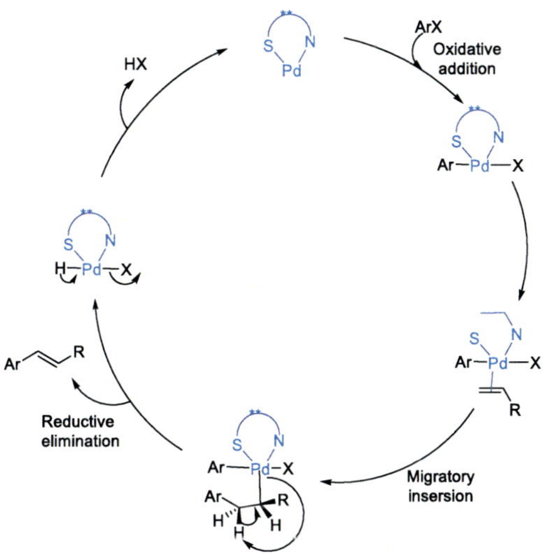

Scheme (11). Proposed mechanism for the Heck coupling reaction.

Recently, Veisi and co-workers reported the synthesis of palladium nanoparticles supported on the surface of graphene oxide, functionalized with metformin (MET) [71]. Metformin functionalized graphene oxide (Met-GO) was prepared between the condensation reaction of graphene oxide with metformin and then palladium nanoparticles (PdNPs) were decorated onto the surface of functionalized graphene oxide by the reduction of $PdCl_2$ using hydrazine hydrate as reducing agents. TEM and FESEM images showed that palladium nanoparticles were uniformly incorporated onto the surface of functionalized graphene oxide. The synthesized catalyst showed outstanding catalytic performance towards the Heck cross-coupling reaction of various aryl halides with olefins without the use of any phosphine or ligand coordination. The coupling reaction proceeded in good to excellent yields of trans-stilbene derivatives between the reactions of aryl halides with olefins using 0.1 mol% of catalyst, triethylamine as a base in DMF at 120 ^{0}C. The catalyst was found to be the most effective in the Gram-scale synthesis of trans-stilbene product. The most impressive feature of this catalyst was its reusability, and it could be recycled up to seven cycles without a noticeable loss of its catalytic performance. A hot filtration test was carried out to determine the heterogeneity of this synthesized catalyst (Met-GO-Pd) using Heck cross-coupling reaction between aryl bromide and styrene under optimized reaction conditions. Then, coupling products were obtained in 54% yield and the catalyst was separated out. After separating the catalyst, the reaction was continued for another hour. However, no significant improvement was observed in the yield of the coupling product. This observation proved that the catalytic process was heterogeneous.

A novel approach for the construction of palladium nanoparticles (PdNPs) supported on a carbon-based graphene oxide (GO) monolith has been described by Nakhate and his co-workers [72]. It was fabricated by the carbonization of glucose with a small amount of graphene oxide as a structure directing agent, which offers a large surface area for the dispersion of palladium nanoparticles. The carbon-based graphene oxide monolith reduced the leaching of PdNPs into the reaction mixture and increased the rate of interaction between the support and the nanoparticles.

The developed catalyst was found to be effective for Sonogashira cross-coupling reaction and hydrogenation of alkene and nitrobenzene. The catalytic efficiency of the Pd@CGO monolith was compared with 5% Pd/C and 5% Pd@GO for the Sonogashira coupling of iodobenzene with phenyl-acetylene. The Pd@CGO monolith exhibited superb reactivity towards the Sonogashira cross-coupling reaction of iodobenzene with phenylacetylene leading to 93% conversion of the coupling products without the use of copper and ligand at 90 ^{0}C (Scheme **12**). Iodobenzene, substituted with electron-withdrawing groups easily transformed to the corresponding diphenylethyne as coupling products in good to excellent yields. Good to excellent yields of desired products were also obtained with iodobenzene, substituted with electron-donating groups. It is interesting to note that less reactive bromobenzenes smoothly reacted with phenylacetylene to furnish desired products in good yields. The author also proposed the reaction mechanism and kinetic equation. The hot filtration test showed that the catalyst has high stability and could be reused at least four times without loss of activity.

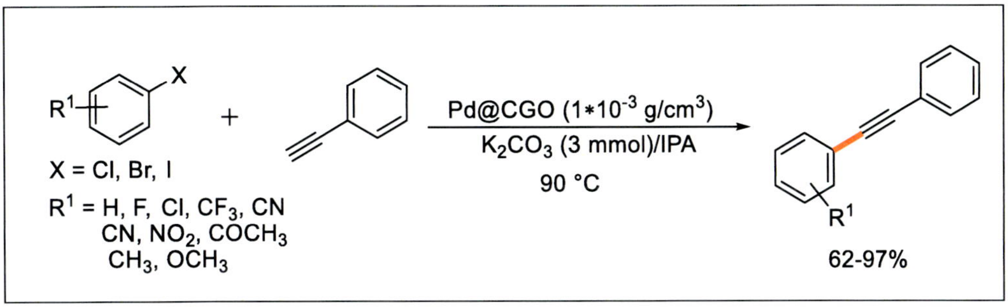

Scheme (12). Sonogashira cross-coupling catalyzed by Pd@CGO catalyst.

REDUCED GRAPHENE OXIDE (RGO) SUPPORTED PALLADIUM NANOPARTICLES (PDNPS-RGO) CATALYZED COUPLING REACTIONS

In 2014, Shendage and co-workers developed electrochemically co-deposited palladium nanoparticles (PdNPs) supported on a reduced graphene oxide (ERGO-Pd) catalyst and used for Suzuki cross-coupling reactions [73]. They observed that PdNPs with smaller sizes in ranges from 3 to 9 nm exhibited high catalytic activity. The prepared catalyst has major advantages such as a high degree of catalytic activity, chemical stability, selectivity, and reusability up to six times. This catalytic system generates a library of biphenyl derivatives using K_2CO_3 as a base in ethanol under reflux conditions, whereas lower yield was observed at room temperature (Scheme **13**). Several aryl halides having electron-withdrawing as well as electron-donating groups smoothly coupled with various phenylboronic acids to form coupling products in good to excellent. Aryl iodide provides better results in compassion to aryl bromides in terms of yield and time. There are many analytical techniques such as SEM (scanning electron microscopy), TEM (transmission electron microscopy), X-Ray analysis, and EDAX (energy dispersive X-ray analysis) have been used to characterize the catalyst.

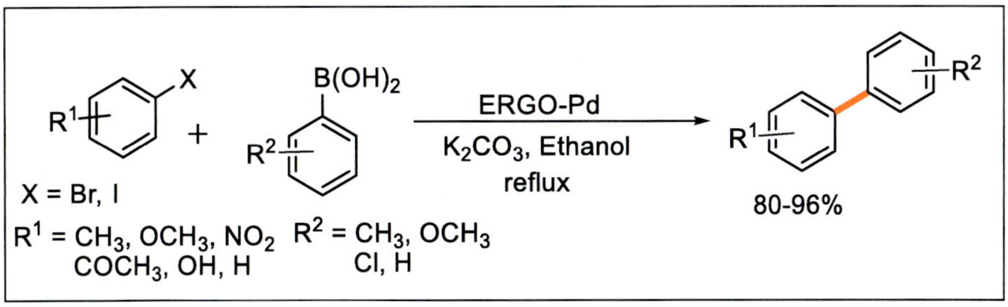

Scheme (13). Suzuki coupling reaction catalyzed by ERGO-Pd.

Later, the same research group reported co-deposited palladium nanoparticles (Pd NPs) supported on reduced graphene oxide (ERGO-Pd) catalyzed carbon-carbon cross-coupling Heck reactions [74]. Under this catalytic system, several aryl halide bearing either electron-withdrawing or electron-donating groups reacted with different styrene and acrylate to generate a series of substituted stilbenes and cinnamates in good to excellent yields (Scheme **14**). Aryl bromides take a longer time due to high bond dissociation energy in comparison to aryl iodide. The reusability of this catalyst is high and can be recycled up to four to five times without loss of activity.

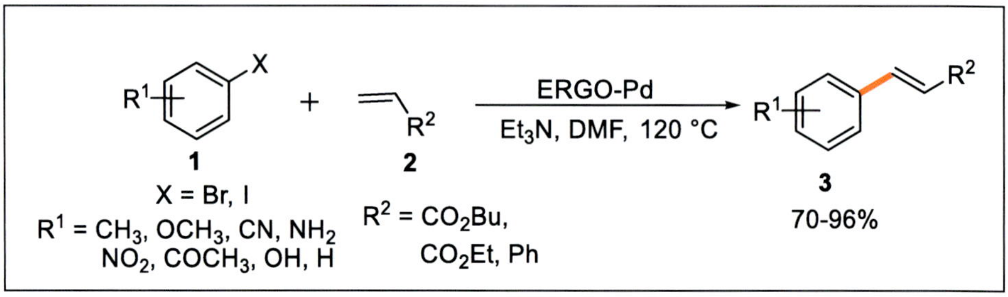

Scheme (14). Heck cross-coupling reaction catalyzed by ERGO-Pd.

In 2015, a new multi-step protocol for in-situ generation of palladium nanoparticles supported on reduced graphene oxide using 1-butyl-3-methylimidazolium hexafluorophosphate ([BMIM]PF_6) ionic liquid by phase-transfer method (Pd/rGO-IL) in aqueous medium was developed by Liu and co-workers [75]. In-situ generated catalyst having an average particle size of 2 nm and uniformly dispersed on the surface of reduced graphene oxide was confirmed by TEM images. The synthesized catalyst showed excellent activity for Heck cross-coupling reactions of iodobenzene with butyl acrylate. The reactions were performed using a catalytic amount of Pd/RGO-IL in the presence of Et_3N as a base at 130 ^{0}C; in which iodobenzene-bearing electron-withdrawing or electron-donating groups coupled with acrylate to provide a moderate yield of desired products with 76-87% conversion. However, neutral iodobenzene reacted with acrylate and styrene to give coupling products in excellent yields with 95% and 92% conversion respectively (Scheme **15**). The reusability of the catalyst was also studied and found that it can be recycled up to ten times without significant loss of catalytic activity, which showed that it has high catalytic stability. The high stability of the catalyst might be due to the double protection of the organic cation [BMIM]+ of IL and reduced graphene oxide (rGO) which could be helpful in the reduction of collision and agglomeration of Pd nanoparticles.

In 2015, the research group of Liu developed an effective and eco-friendly protocol for the synthesis of multimetallic nanoparticles catalyst, especially magnetically separable Pd NPs supported on graphene oxide [76]. The polymetallic graphene oxide supported PdNPs ($NiFe_2O_4$@GO-Pd) hybrids were obtained from $PdCl_2$, $FeCl_3.6H_2O$, and $NiCl_2.4H_2O$ in ethanol-ethylene glycol using hydrothermal treatment. The $NiFe_2O_4$ were uniformly dispersed on the surface of graphene oxide without help of any linker molecules, which help in providing ion transport channels to bridge the NPs and the matrix. The authors have also demonstrated the catalytic power of this synthesized catalyst for Heck cross-coupling reactions of aryl halides with substituted olefins. Iodobenzene-bearing electron-donating or electron-withdrawing groups efficiently reacted with olefins in the presence of (0.08 mol% Pd) $NiFe_2O_4$@GO-Pd catalyst and 2.0 equiv. of K_2CO_3 as a base in EtOH-H_2O at 80°C to provide coupling products in excellent yield (Scheme **16**). A slightly decreased yield of the coupling product was obtained with iodobenzene having a methyl group at the ortho position. Bromobenzene reacted with styrene to furnish a moderate yield of coupling product under the same catalytic system. However, chlorobenzene provides a lower yield of the desired product even after increasing the catalyst loading and reaction time. The most attractive feature of this catalyst was its recyclability and stability, it can be recycled up to six times without the loss of catalytic activity supported by ICP-MS analysis. The $NiFe_2O_4$@GO-Pd catalyst showed high catalytic power and stability towards the Heck cross-coupling reactions in comparison to $NiFe_2O_4$-Pd catalyst because of the supportive role of graphene oxide for $NiFe_2O_4$@GO-Pd, which reduces the agglomeration of Pd nanoparticles.

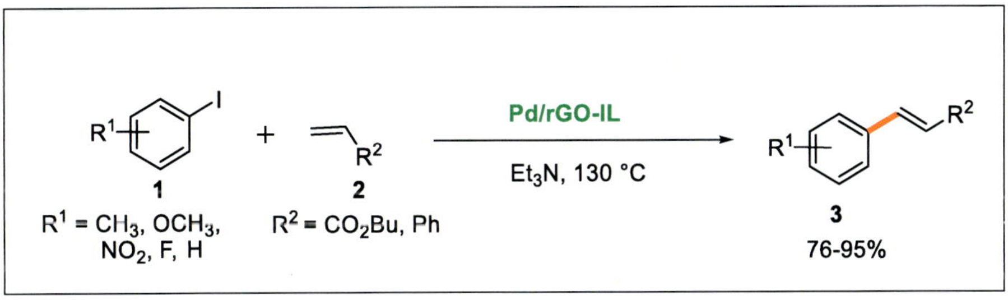

Scheme (15). Heck reaction catalyzed by Pd/RGO-IL.

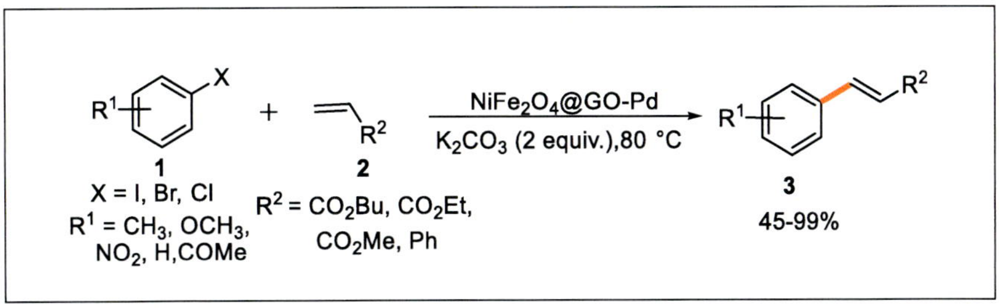

Scheme (16). $NiFe_2O_4$@GO-Pd catalyzed Heck cross-coupling reaction.

Another important approach for the construction of highly active Pd/Fe_3O_4 nanoparticles supported on graphene (Pd/Fe_3O_4/G) has been described by Elazab and co-workers in 2015 [77]. The catalyst (Pd/Fe_3O_4/G) was synthesized by microwave-assisted reduction of palladium nitrate, ferric nitrate, and graphene oxide using hydrazine hydrate as the reducing agent in an aqueous medium. The constructed magnetic catalyst (Pd/Fe_3O_4/G) showed exceptional catalytic activity for cross-coupling reactions (Heck and Suzuki cross-coupling reaction). The most striking feature of this magnetic catalyst is that it shows a high turnover number (TON) of 9250 and a turnover frequency (TOF) of 111,000 h^{-1} for Suzuki and Heck cross-coupling reactions. The catalytic performance was first studied for Suzuki cross-coupling reactions and found that a wide range of functionalized aryl halides efficiently coupled with aryl boronic acid in the presence of 0.3 mol% of catalyst using K_2CO_3 as a base in H_2O:EtOH as a solvent to provide the corresponding biphenyl products in moderate to excellent yield (Scheme **17**). Arylhalide-bearing electron-withdrawing groups such as aldehyde, ketone, nitro, and nitrile furnished the corresponding biphenyl derivatives in excellent yield, whereas chlorobenzene provided the Suzuki cross-coupling product in a relatively good yield. However, boronic acid containing electron-donating groups (dimethyl amino, thiomethyl, and *p*-methoxy) smoothly converted into biphenyl derivatives with aryl halides in excellent yield.

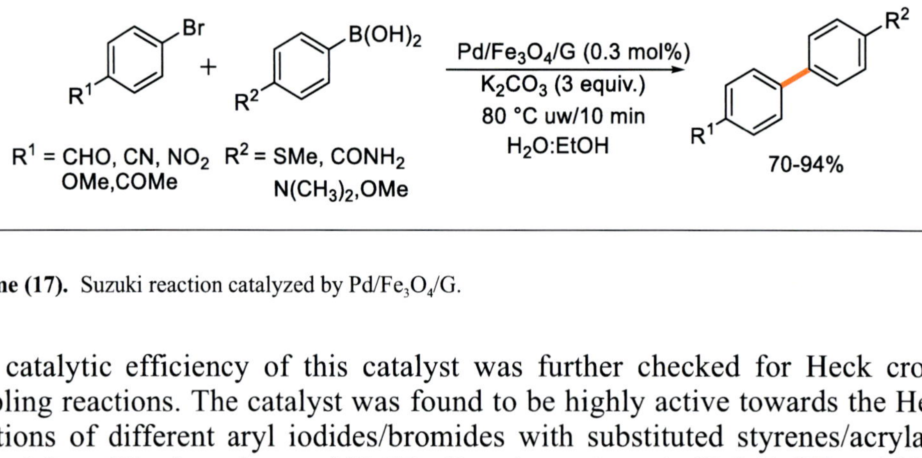

Scheme (17). Suzuki reaction catalyzed by Pd/Fe_3O_4/G.

The catalytic efficiency of this catalyst was further checked for Heck cross-coupling reactions. The catalyst was found to be highly active towards the Heck reactions of different aryl iodides/bromides with substituted styrenes/acrylates using 0.3 mol% of catalyst and K_2CO_3 (3 eq.) as a base in H_2O:EtOH at 150 ^{0}C under microwave irradiation. The reaction provided the corresponding products in excellent yields in cases of bromobenzene-bearing electron-donating and electron-withdrawing groups. These results prove the outstanding activity of this catalyst. The catalyst can be recovered and further used up to ten times without significant loss of catalytic activity (Scheme **18**).

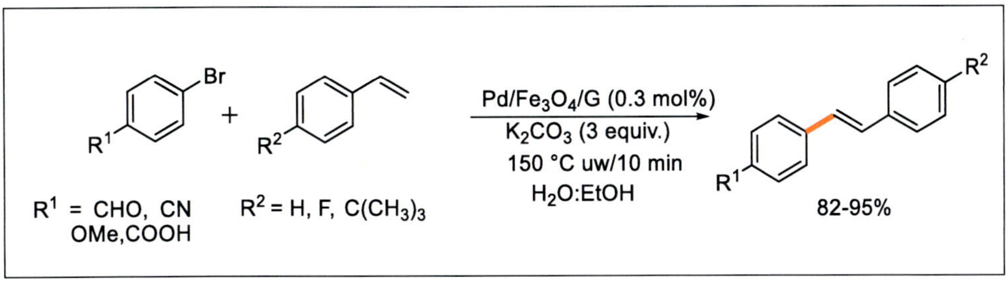

Scheme (18). Heck cross-coupling reaction catalyzed by Pd/Fe_3O_4/G.

Recently, a similar approach for the preparation of magnetically recoverable palladium nanoparticles supported on graphene oxide has been developed by Fu's research workers [78]. In the development of Pd-Fe_3O_4/rGO, $FeCl_2.4H_2O$ was used as a starting material reacting with ammonium hydroxide to form iron hydroxide followed by simultaneously oxidation and dehydration to form Fe_3O_4 nanoparticles at high temperature and pressure whereas Pd nanoparticles were dispersed onto the surface of graphene oxide using ethanol as the reducing agent under the ultrasonic and hydrothermal strategy. Concurrently, the Fe_3O_4 NPs were well dispersed on two-dimensional rGO nanosheets and formed Pd-Fe_3O_4/rGO nanohybrids materials through a self-assembly approach. The synthesized catalyst was characterized using various analytical techniques. TEM images displayed that both palladium nanoparticles and iron oxide nanoparticles were homogeneously distributed onto the surface of reduced graphene oxide nanosheets, whereas HRTEM images indicate the size of nanoparticles up to 2 nm and no agglomeration on the surface of graphene oxide.

The formed catalyst was used to evaluate its catalytic efficiency for Heck cross-coupling of 4-iodoanisole and styrene. A wide range of iodobenzene reacted with styrenes in the presence of 0.06 mol% catalyst using triethylamine as a base in DMF as solvent at 120 ^{0}C to provide the corresponding coupling products in moderate to excellent yields. Iodobenzene-bearing electron-withdrawing groups efficiently coupled with styrene to give the cross-coupling products in excellent yields, whereas iodobenzene with an amino group at the para position provides the corresponding product in lower yield (Scheme **19**). The reusability of this catalyst was also examined and found that the recovered catalyst can be consecutively recycled up to six times without significant loss of catalytic power. These results indicate that this catalyst has high catalytic efficiency and is beneficial to both the environmental and economic perspectives.

Another important approach for the synthesis of palladium nanoparticles supported on graphene oxide using hydrogen gas as the reducing agent has been developed by Hussain and his co-workers [79]. The catalyst was synthesized by the following methods, involving reduction of graphene oxide followed by dispersion of palladium nanoparticles on the surface of reduced graphene oxide nanosheets and simultaneously reduction of olefin. The synthesized catalyst was found to be more effective towards the cross-coupling reactions of aryl iodides with phenylboronic acids (Scheme **20**). Iodobenzene that incorporated electron-donating and electron-withdrawing substituents smoothly coupled with arylboronic acids under the catalytic system (Pd-rGO-H_2 (25 mg), Na_2CO_3 (2 mmol), H_2O:EtOH (10 mL, 1:1), r.t) providing the corresponding biphenyl analogous in moderate to excellent yields. It is interesting to note that phenylboronic acid neopentylglycol ester was used in place of boronic acid as a coupling partner for the Suzuki cross-coupling with bromobenzene to give the coupling product in an excellent yield under the same catalytic system. However, the use of chlorobenzene as a coupling partner with arylboronic acid gave a slightly lower yield of the coupling product even after increasing the reaction time and reaction temperature. The most important feature of this catalyst was its reusability. The recovered catalyst can be recycled up to four times without loss of catalytic activity for Suzuki cross-coupling reactions. However, the catalytic performance decreased up to 62% after the fourth run might be due to the decrease in the surface area.

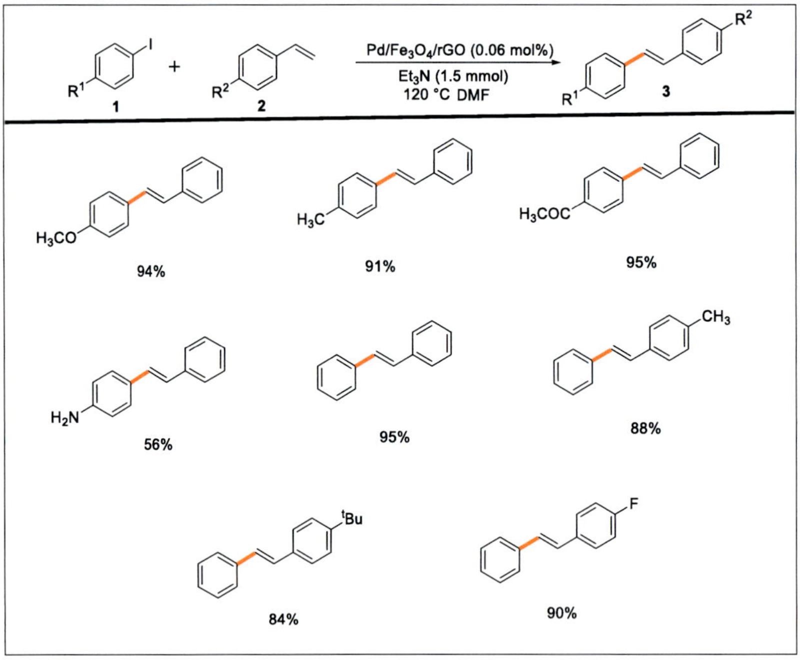

Scheme (19). Synthesis of biphenyl styrenes by Heck cross-coupling reaction using Pd-Fe_3O_4/rGO.

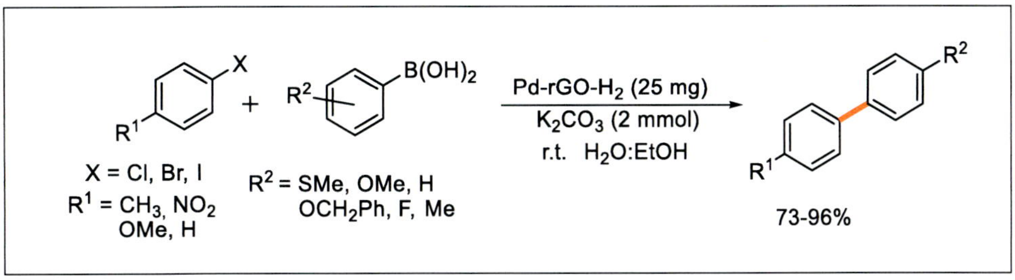

Scheme (20). Suzuki cross-coupling reaction catalyzed by Pd-rGO-H_2.

In 2018, the Raiza research group reported a biogenic one-pot strategy for the synthesis of graphene-supported palladium nanoparticle (Pd Nps/rGO) catalyst using reduced graphene oxide (rGO) and H_2PdCl_4 as starting material followed by reduction with *Ficus carica* fruit juice as the reducing agent and Poly(*N*-vinyl-2-pyrrolidone) (PVP) as stabilizing agent [80]. The Ultraviolet-Visible spectroscopy (UV-Vis) displaced the complete reduction of ions and the formation of PdNP and PdNP/rGO. TEM images showed the homogeneous distribution of PdNPs on the surface of rGO with a smaller particle size in a spherical shape. The HRTEM image confirmed the crystal lattice of PdNPs and the edges of rGO nanosheets to be ~16 nm. The crystalline nature of observed using the SAED image (Fig. **1**). The formed catalyst was examined in Suzuki cross-coupling reactions using aryl halides with phenylboronic acids. A variety of biphenyl derivatives have been synthesized in good to excellent yields using 0.96 mol%, K_2CO_3 as a base and water: ethanol as solvent at 180 °C. The reduced yield was observed after decreasing the catalyst loading from 0.96 mol% to 0.5 mol% whereas no yield of coupling products were observed after increasing catalyst loading. The reusability of the catalyst in Suzuki cross-coupling was examined and found that the catalyst can be used ten times without loss of activity.

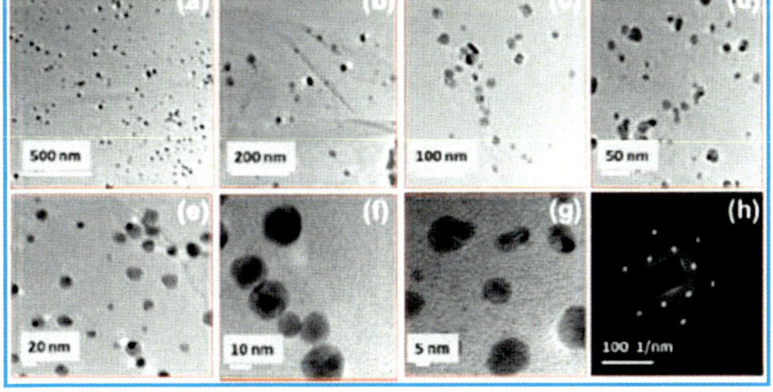

Fig. (1). Transmission electron microscopy images. (**a-c**) Images along with particle size distribution; (**g**) HETEM image along with fringe spacing; and (**f**) SAED image of the PdNP/rGO.

In 2018, Kulkarni and co-workers developed an eco-friendly protocol for the synthesis of graphene oxide supported palladium nanoparticle catalyst using graphene oxide, $PdCl_2$ as the precursor, followed by reduction with aloe vera extract as the reducing agent in the presence of sunlight as an energy source [81]. The prepared catalyst was characterized using different analytical techniques.

SEM image confirmed the formation of PdNPs on reduced graphene oxide with a uniform dispersion of PdNPs whereas the TEM image indicated the presence of PdNPs on the surface of rGO with an average particle size of 2-19 nm. The catalytic performance of the synthesized catalyst was examined in Suzuki cross-coupling reactions. The authors obtained several coupling products in good to excellent yields, when the reaction was performed using aryl halides with phenylboronic acid in the presence of 0.2 mol% of catalyst K_2CO_3 as a base and water: ethanol as solvent at 80 ^{0}C (Scheme **21**). Aryl iodide having electron-donating or electron-withdrawing groups was more reactive towards the phenylboronic acid providing exceptional results in terms of yield and reaction time, whereas aryl bromide/aryl chloride was less reactive with phenyl boronic acid under same the reaction conditions. The recovery of the catalyst was also studied and found that it could be reused seven times without significant loss of activity. A considerable loss of activity of the catalyst was seen after the 7th cycle due to agglomeration and an increase in particle size of palladium nanoparticles on reduced graphene oxide.

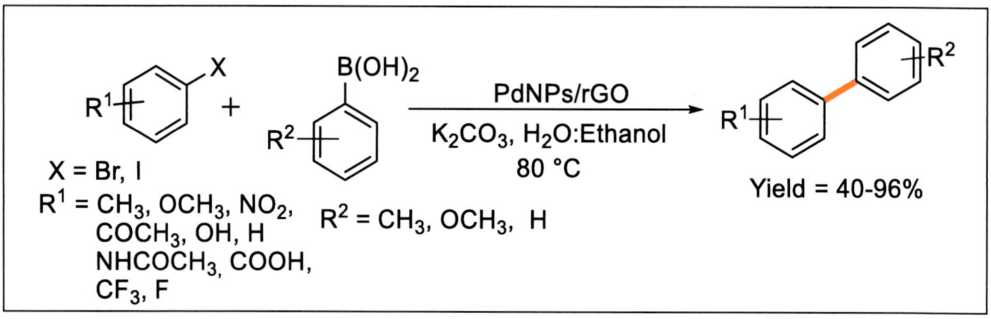

Scheme (21). Suzuki cross-coupling reaction catalyzed by PdNPs/rGO.

In 2016, Mahanta and co-workers described an eco-friendly copper-free Sonogashira cross-coupling reaction catalyzed by palladium nanoparticles supported on reduced graphene oxide at room temperature in ethanol [82]. PdNPs were synthesized using hydrogen gas as a reducing agent to generate PdNPs supported on reduced graphene oxide sheets. The catalytic performance of the catalyst was investigated for the Sonogashira cross-coupling reaction. The reactions proceeded smoothly using 5 wt% of the catalyst in the presence of

K_2CO_3 as a base at room temperature in ethanol to provide the coupling products with good to excellent yield (see Table 2). The alkyne bearing electron-donating or electron-withdrawing groups was coupled with aryl halides to furnish the biphenylacetylene derivatives in excellent to good yield. However, non-aromatic alkynes such as cyclohexylacetylene were coupled with different iodides to provide the coupling products in moderate yield under the same catalytic system. Moreover, bromobenzene did not provide a satisfactory yield of the coupling products even after increasing the reaction time (Table **2**, entries 11 and 12). The reusability of the catalyst was also studied and observed that it could be recycled up to six times without loss of minimal activity.

Recently, the research group of Tarahomi has employed the Pd catalyst supported on dendrimer-functionalized graphene oxide for Sonogashira cross-coupling reaction [83]. The dendrimer-based palladium catalyst was developed in multistep. The first step involves the reaction of GO with $FeCl_2.4H_2O$, and $FeCl_3.4H_2O$ followed by the addition of 1 mL of NH_3-H_2O to get magnetic graphene oxide (MGO), then MGO was covalently attached with modified PAMAM dendrimer (G. 2) in the presence of EDC/NHS [84].

Table 2. Scope of Sonogashira coupling reaction catalyzed by PdNPs@GO.

Entry	R^1	R^2	X	Time	Yield
1	4-NO_2	C_6H_5	I	7	94
2	3-Me	C_6H_5	I	7	96
3	4-Me	C_6H_5	I	8	95
4	3-NO_2	C_6H_5	I	8	95
5	4-OMe	C_6H_5	I	8	92
6	H	C_6H_5	I	7	96
7	4-Me	4-MeC_6H_5	I	7	98
8	4-Me	C_6H_{11}	I	10	85
9	H	C_6H_{11}	I	10	80
10	4-OMe	4-MeC_6H_5	I	7.5	98
11	4-NO_2	C_6H_5	Br	12	47
12	4-OMe	C_6H_5	Br	12	53

[a]Reaction conditions: acetylene (0.75 mmol), aryl halide (0.5 mmol) [b]Isolated yield: purified by column chromatography.

Subsequently, PdNPs were dispersed on the surface of dendrimer-functionalized graphene oxide through graft interactions between free chelating amino groups of dendrimer and Pd precursor. The most attractive feature of these nanocomposite materials was not only solubility in an aqueous medium but also stabilized metal NPs on their surfaces. The structural characterization of this catalyst was confirmed by using different analytical techniques such as TEM, BET, FTIR, and elemental mapping analysis. The catalytic power of this catalyst was checked for the Sonogashira cross-coupling reaction. The Sonogashira cross-coupling reaction works very well for alkynes and aryl halides in the presence of dendrimer-supported palladium catalyst to furnish corresponding products in excellent yields, except for chlorobenzene which gave a lower yield of the desired product (Scheme **22**). Regarding the mechanism, it was proposed that the reaction is initiated by the coordination of the Pd@MGO-D-NH_2 catalyst to the C-C triple bond of phenylacetylene followed by the deprotonation of terminal C-H of alkyne with K_2CO_3 to form potassium acetylide. Pd@MGO-D-NH_2 catalyst also activates the aryl halide (ArX) and forms aryl palladium halide complex (ArPdX) and then halide displacement takes place to form an aryl palladium acetylide. Subsequently, reductive elimination takes place to give a cross-coupling product with a regeneration catalyst. The reusability of this catalyst was also studied and found that it could be reused up to four times without significant loss of catalytic power.

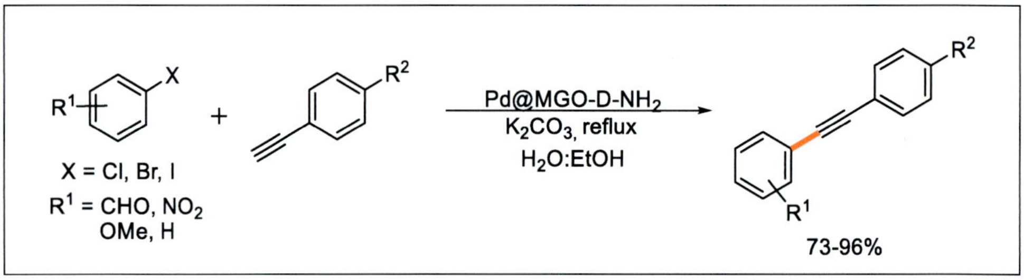

Scheme (22). Pd@MGO-D-NH_2 catalyzed Sonogashira coupling reaction.

CONCLUDING REMARKS

In this chapter, we focused on metal-supported graphene oxide for its catalytic potential. The prepared heterogeneous catalysts were beneficially used for C-C bond formation using different cross-coupling reactions such as Mizoroki-Heck, Suzuki-Miyaura, Kumada, Stille, Negishi, Hiyama, and Buchwald-Hartwig, *etc.* In a prevalent number of examples, the comparison of heterogeneous catalysts with corresponding homogeneous catalysts has been observed. The advantages of the discussed heterogeneous catalysts consisted of their simple preparation, inexpensive and easily available starting materials, their simple isolation, and the possibility of reuse. Easy performance of the reaction in aqueous media brings great benefits in the case of several cross-coupling reactions. Several of the described reactions possibly have a high potential for industrial usage, hence the catalytic systems presented in this chapter significantly contribute to the further development of green chemical processes.

LIST OF ABBREVIATIONS

AAPTMS	[3-(2-Aminoethylamino)propyl] trimethoxysilane
BET	Brunauer–Emmett–Teller
$ClPPh_2$	Chlorodiphenylphosphine
2D	Two dimensional
EPD	Electrophoretic deposition
ERGO-Pd	Electrochemically codeposited palladium nanoparticles supported on reduced graphene oxide
EDAX	Energy dispersive X-ray analysis
EDC2	ethyl(dimethylaminopropyl) carbodiimide
FESEM	Field emission scanning electron microscopy
FTIR	Fourier Transform Infrared Spectroscopy
GO	Graphene oxide
G-BI-Pd	Graphene-benzimidazole
GOH	Hydroxyl group on the surface of graphene
GNS	Graphite nanosheet
HRTEM	High-resolution transmission electron microscopy
ICP-AES	Nductively coupled plasma atomic emission spectroscopy
LAL	Laser ablation in liquids
NHS	N-Hydroxysuccinimide
PdNPs	Palladium nanoparticles
PdNPs-G	Palladium nanoparticles supported on graphene
PdNPs-rGO	Palladium nanoparticles reduced graphene oxide
Pd/graphene	Graphene supported Pd
PFG	Phosphine functionalized graphene
PFG	Phosphine grafted graphene oxide
Pd-slGO	Palladium nanoparticles supported on single layer oxide
Pd@GOF	Palladium nanoparticles catalyst supported on graphene oxide framework
PCA-GNS-Pd	Pyrenecarboxylic acid graphite nanosheet supported Palladium
PVP	Poly(N-vinyl-2-pyrrolidone)
PAMAM	Polyamidoamine
r-GO	Reduced graphene oxide
SEA	Strong electrostatic adsorption

SAED	Selected area diffraction
TEM	Transmission electron microscopy
XRD	Powder x-ray diffraction
XPS	X-ray photoelectron spectroscopy

CONSENT FOR PUBLICATION

Not applicable.

CONFLICT OF INTEREST

The authors declare no conflict of interest, financial or otherwise.

ACKNOWLEDGEMENTS

The authors thank the SERB SRG grant, grant no. SRG-0721 research grant for financial support.

REFERENCES

[1] (a) Ananikov, V.P.; Beletskaya, I.P. Toward the ideal catalyst: from atomic centers to a "cocktail" of catalysts. *Organometallics,* **2012**, *31*(5), 1595-1604. [http://dx.doi.org/10.1021/om201120n] (b) Negishi, E. Transition metal-catalyzed organometallic reactions that have revolutionized organic synthesis. *Bull. Chem. Soc. Jpn.,* **2007**, *80*(2), 233-257. [http://dx.doi.org/10.1246/bcsj.80.233] (c) Torborg, C.; Beller, M. Recent applications of palladium-catalyzed coupling reactions in the pharmaceutical, agrochemical, and fine chemical industries. *Adv. Synth. Catal.,* **2009**, *351*(18), 3027-3043. [http://dx.doi.org/10.1002/adsc.200900587]

[2] Magano, J.; Dunetz, J.R. Large-scale applications of transition metal-catalyzed couplings for the synthesis of pharmaceuticals. *Chem. Rev.,* **2011**, *111*(3), 2177-2250. [http://dx.doi.org/10.1021/cr100346g] [PMID: 21391570]

[3] Negishi, E.; Meijere, A. *Handbook of organo palladium chemistry for organic synthesis*; Wiley: New York, **2002**. Tsuji, J. *Perspective in Organo Palladium Chemistry for XXI Century*; Elsevier Science, **1999**. Malleron, J-L.; Fiaud, J-C.; Lagros, J.Y. *Handbook of Palladium Catalyzed Organic Reactions*; Academic Press: San Diego, **1997**.

[4] (a) Negishi, E. Magical power of transition metals: past, present, and future (Nobel Lecture). *Angew. Chem. Int. Ed.,* **2011**, *50*(30), 6738-6764. [http://dx.doi.org/10.1002/anie.201101380] [PMID: 21717531] (b) Suzuki, A. Cross-coupling reactions of organoboranes: an easy way to construct C-C bonds (Nobel Lecture). *Angew. Chem. Int. Ed.,* **2011**, *50*(30), 6722-6737. [http://dx.doi.org/10.1002/anie.201101379] [PMID: 21618370] (c) Nobelprize.org. **2013**. http://www.nobelprize. org/nobel_prizes/chemistry/laureates/2010/

[5] Astruc, D.; Lu, F.; Aranzaes, J.R. Nanoparticles as recyclable catalysts: the frontier between homogeneous and heterogeneous catalysis. *Angew. Chem. Int. Ed.,* **2005**, *44*(48), 7852-7872. [http://dx.doi.org/10.1002/anie.200500766] [PMID: 16304662]

[6] Yin, L.; Liebscher, J. Carbon-carbon coupling reactions catalyzed by heterogeneous palladium catalysts. *Chem. Rev.,* **2007**, *107*(1), 133-173. [http://dx.doi.org/10.1021/cr0505674] [PMID: 17212474]

[7] Bedford, R.B.; Singh, U.G.; Walton, R.I.; Williams, R.T.; Davis, S.A. Nanoparticulate palladium supported by covalently modified silicas: Synthesis, characterization, and application as catalysts for the Suzuki coupling of aryl halides. *Chem. Mater.*, **2005**, *17*(4), 701-707. [http://dx.doi.org/10.1021/cm048860s]

[8] Jana, S.; Dutta, B.; Bera, R.; Koner, S. Immobilization of palladium in mesoporous silica matrix: preparation, characterization, and its catalytic efficacy in carbon-carbon coupling reactions. *Inorg. Chem.*, **2008**, *47*(12), 5512-5520. [http://dx.doi.org/10.1021/ic8004294] [PMID: 18459724]

[9] Djakovitch, L.; Koehler, K. Heterogeneously catalysed Heck reaction using palladium modified zeolites. *J. Mol. Catal. Chem.*, **1999**, *142*(2), 275-284. [http://dx.doi.org/10.1016/S1381-1169(98)00292-1]

[10] Djakovitch, L.; Koehler, K. Heck reaction catalyzed by PD-modified zeolites. *J. Am. Chem. Soc.*, **2001**, *123*(25), 5990-5999. [http://dx.doi.org/10.1021/ja001087r] [PMID: 11414833]

[11] Zhao, D.; Fei, Z.; Geldbach, T.J.; Scopelliti, R.; Dyson, P.J. Nitrile-functionalized pyridinium ionic liquids: synthesis, characterization, and their application in carbon-carbon coupling reactions. *J. Am. Chem. Soc.*, **2004**, *126*(48), 15876-15882. [http://dx.doi.org/10.1021/ja0463482] [PMID: 15571412]

[12] Miao, T.; Wang, L.; Li, P.H.; Yan, J.C. A highly efficient and recyclable ionic liquid anchored pyrrolidine catalyst for enantioselective Michael additions. *Synthesis*, **2008**, 3828-3834.

[13] Lombardo, M.; Chiarucci, M.; Trombini, C. A recyclable triethylammonium ion-tagged diphenylphosphine palladium complex for the Suzuki–Miyaura reaction in ionic liquids. *Green Chem.*, **2009**, *11*(4), 574-579. [http://dx.doi.org/10.1039/b815568b]

[14] Biffis, A.; Zecca, M.; Basato, M. Metallic Palladium in the Heck Reaction: Active Catalyst or Convenient Precursor? *Eur. J. Inorg. Chem.*, **2001**, *2001*(5), 1131-1133. [http://dx.doi.org/10.1002/1099-0682(200105)2001:5<1131::AID-EJIC1131>3.0.CO;2-3]

[15] Ramchandani, R.K.; Vinod, M.P.; Wakharkar, R.D.; Choudhary, V.R.; Sudalai, A.; Sudalai, A. Pd–Cu–Exchanged montmorillonite K10 clay: an efficient and reusable heterogeneous catalyst for vinylation of aryl halides. *Chem. Commun. (Camb.)*, **1997**, (21), 2071-2072. [http://dx.doi.org/10.1039/a705870e]

[16] Singh, A.S.; Patil, U.B.; Nagarkar, J.M. Palladium supported on zinc ferrite: A highly active, magnetically separable catalyst for ligand free Suzuki and Heck coupling. *Catal. Commun.*, **2013**, *35*, 11-16. [http://dx.doi.org/10.1016/j.catcom.2013.02.003]

[17] Ogasawara, S.; Kato, S. Palladium nanoparticles captured in microporous polymers: a tailor-made catalyst for heterogeneous carbon cross-coupling reactions. *J. Am. Chem. Soc.*, **2010**, *132*(13), 4608-4613. [http://dx.doi.org/10.1021/ja9062053] [PMID: 20225817]

[18] Bai, L.; Wang, J.X. Reusable, polymer-supported, palladium-catalyzed, atom- efficient coupling reaction of aryl halides with sodium tetraphenylborate in water by focused microwave irradiation. *Adv. Synth. Catal.*, **2008**, *350*(2), 315-320. [http://dx.doi.org/10.1002/adsc.200700361]

[19] Alacid, E.; Nájera, C. Kaiser oxime resin-derived palladacycle: A recoverable polymeric precatalyst in Suzuki–Miyaura reactions in aqueous media. *J. Organomet. Chem.*, **2009**, *694*(11), 1658-1665. [http://dx.doi.org/10.1016/j.jorganchem.2009.01.016]

[20] Giacalone, F.; Gruttadauria, M. Covalently supported ionic liquid phases: An advanced class of recyclable catalytic systems. *ChemCatChem*, **2016**, *8*(4), 664-684.

[http://dx.doi.org/10.1002/cctc.201501086]

[21] Nasrollahzadeh, M.; Sajjadi, M.; Shokouhimehr, M.; Varma, R.S. Recent developments in palladium (nano)catalysts supported on polymers for selective and sustainable oxidation processes. *Coord. Chem. Rev.*, **2019**, *397*, 54-75.
[http://dx.doi.org/10.1016/j.ccr.2019.06.010]

[22] Nasrollahzadeh, M.; Motahharifar, N.; Ghorbannezhad, F.; Soheili Bidgoli, N.S.; Baran, T.; Varma, R.S. Recent advances in polymer supported palladium complexes as (nano)catalysts for Sonogashira coupling reaction. *Molecular Catalysis,* **2020**, *480*, 110645.
[http://dx.doi.org/10.1016/j.mcat.2019.110645]

[23] Ahmed, M.J. Application of agricultural based activated carbons by microwave and conventional activations for basic dye adsorption: Review. *J. Environ. Chem. Eng.*, **2016**, *4*(1), 89-99. [Review].
[http://dx.doi.org/10.1016/j.jece.2015.10.027]

[24] Harris, J.F. *Carbon Nanotubes Science: Synthesis, Properties and Applications Chap. 2*; Cambridge, **2009**.

[25] Wang, M.; Huang, Z-H.; Bai, Y.; Kang, F.; Inagaki, M. Porous Carbon Nanofibers: Preparation and Potential Applications. *Curr. Org. Chem.*, **2013**, *17*(13), 1434-1447.
[http://dx.doi.org/10.2174/13852728113179130010]

[26] Rao, C.N.R.; Sood, A.K. *Graphene: Synthesis, Properties and phenomena*; Wiley, **2012**.
[http://dx.doi.org/10.1002/9783527651122]

[27] Novoselov, K.S.; Geim, A.K.; Morozov, S.V.; Jiang, D.; Katsnelson, M.I.; Grigorieva, I.V.; Dubonos, S.V.; Firsov, A.A. Two-dimensional gas of massless Dirac fermions in graphene. *Nature,* **2005**, *438*(7065), 197-200.
[http://dx.doi.org/10.1038/nature04233] [PMID: 16281030]

[28] Geim, A.K. Graphene: Status and Prospects. *Science,* **2009**, *324*(5934), 1530-1534.
[http://dx.doi.org/10.1126/science.1158877] [PMID: 19541989]

[29] Geim, A.K.; Novoselov, K.S. The rise of graphene. *Nat. Mater.*, **2007**, *6*(3), 183-191.
[http://dx.doi.org/10.1038/nmat1849] [PMID: 17330084]

[30] Nasrollahzadeh, M.; Issaabadi, Z.; Tohidi, M.M.; Mohammad Sajadi, S. Recent progress in application of graphene supported metal nanoparticles in c-c and c-x coupling Reactions. *Chem. Rec.*, **2018**, *18*(2), 165-229.
[http://dx.doi.org/10.1002/tcr.201700022] [PMID: 28745452]

[31] Yang, Y.; Castano, C.E.; Gupton, B.F.; Reber, A.C.; Khanna, S.N. A fundamental analysis of enhanced cross-coupling catalytic activity for palladium clusters on graphene supports. *Nanoscale,* **2016**, *8*(47), 19564-19572.
[http://dx.doi.org/10.1039/C6NR06793J] [PMID: 27833943]

[32] Sakhaee-Pour, A.; Ahmadian, M.T.; Vafai, A. Potential application of single-layered graphene sheet as strain sensor. *Solid State Commun.*, **2008**, *147*(7-8), 336-340.
[http://dx.doi.org/10.1016/j.ssc.2008.04.016]

[33] Takamura, T.; Endo, K.; Fu, L.; Wu, Y.; Lee, K.J.; Matsumoto, T. Identification of nano-sized holes by TEM in the graphene layer of graphite and the high rate discharge capability of Li-ion battery anodes. *Electrochim. Acta,* **2007**, *53*(3), 1055-1061.
[http://dx.doi.org/10.1016/j.electacta.2007.03.052]

[34] Shendage, S.S.; Patil, U.B.; Nagarkar, J.M. Electrochemical deposition of highly dispersed palladium nanoparticles on nafion-graphene film in presence of ferrous ions for ethanol electrooxidation. *Fuel Cells (Weinh.),* **2013**, *13*(3), 364-370.
[http://dx.doi.org/10.1002/fuce.201300043]

[35] Groves, M.N.; Malardier-Jugroot, C.; Jugroot, M. Improving platinum catalyst durability with a doped graphene support. *J. Phys. Chem. C,* **2012**, *116*(19), 10548-10556.

[http://dx.doi.org/10.1021/jp203734d]

[36] Yamamoto, S.; Kinoshita, H.; Hashimoto, H.; Nishina, Y. Facile preparation of Pd nanoparticles supported on single-layer graphene oxide and application for the Suzuki–Miyaura cross-coupling reaction. *Nanoscale,* **2014**, *6*(12), 6501-6505. [http://dx.doi.org/10.1039/C4NR00715H] [PMID: 24838740]

[37] Hummers, W.S., Jr; Offeman, R.E. Preparation of graphitic oxide. *J. Am. Chem. Soc.,* **1958**, *80*(6), 1339-1339. [http://dx.doi.org/10.1021/ja01539a017]

[38] Djakovitch, L.; Koehler, K.; de Vries, J.G. *Nanoparticles and catalysis*; Wiley-VCH: Weinheim, **2009**.

[39] Narayanan, R. Recent advances in noble metal nanocatalysts for Suzuki and Heck cross-coupling reactions. *Molecules,* **2010**, *15*(4), 2124-2138. [http://dx.doi.org/10.3390/molecules15042124] [PMID: 20428032]

[40] Balanta, A.; Godard, C.; Claver, C. Pd nanoparticles for C–C coupling reactions. *Chem. Soc. Rev.,* **2011**, *40*(10), 4973-4985. [http://dx.doi.org/10.1039/c1cs15195a] [PMID: 21879073]

[41] Fihri, A.; Bouhrara, M.; Nekoueishahraki, B.; Basset, J.M.; Polshettiwar, V. Nanocatalysts for Suzuki cross-coupling reactions. *Chem. Soc. Rev.,* **2011**, *40*(10), 5181-5203. [http://dx.doi.org/10.1039/c1cs15079k] [PMID: 21804997]

[42] Molnár, Á. Efficient, selective, and recyclable palladium catalysts in carbon-carbon coupling reactions. *Chem. Rev.,* **2011**, *111*(3), 2251-2320. [http://dx.doi.org/10.1021/cr100355b] [PMID: 21391571]

[43] Taladriz-Blanco, P.; Hervés, P.; Pérez-Juste, J. Supported Pd nanoparticles for carbon–carbon coupling reactions. *Top. Catal.,* **2013**, *56*(13-14), 1154-1170. [http://dx.doi.org/10.1007/s11244-013-0082-6]

[44] Scheuermann, G.M.; Rumi, L.; Steurer, P.; Bannwarth, W.; Mülhaupt, R. Palladium nanoparticles on graphite oxide and its functionalized graphene derivatives as highly active catalysts for the Suzuki-Miyaura coupling reaction. *J. Am. Chem. Soc.,* **2009**, *131*(23), 8262-8270. [http://dx.doi.org/10.1021/ja901105a] [PMID: 19469566]

[45] Huang, Y.X.; Xie, J.F.; Zhang, X.; Xiong, L.; Yu, H.Q. Reduced graphene oxide supported palladium nanoparticles *via* photoassisted citrate reduction for enhanced electrocatalytic activities. *ACS Appl. Mater. Interfaces,* **2014**, *6*(18), 15795-15801. [http://dx.doi.org/10.1021/am504664r] [PMID: 25153308]

[46] Sun, T.; Zhang, Z.; Xiao, J.; Chen, C.; Xiao, F.; Wang, S.; Liu, Y. Facile and green synthesis of palladium nanoparticles-graphene-carbon nanotube material with high catalytic activity. *Sci. Rep.,* **2013**, *3*(1), 2527. [http://dx.doi.org/10.1038/srep02527] [PMID: 23982312]

[47] Scheuermann, G.M.; Rumi, L.; Steurer, P.; Bannwarth, W.; Mülhaupt, R. Palladium nanoparticles on graphite oxide and its functionalized graphene derivatives as highly active catalysts for the Suzuki-Miyaura coupling reaction. *J. Am. Chem. Soc.,* **2009**, *131*(23), 8262-8270. [http://dx.doi.org/10.1021/ja901105a] [PMID: 19469566]

[48] Hu, J.; Wang, Y.; Han, M.; Zhou, Y.; Jiang, X.; Sun, P. A facile preparation of palladium nanoparticles supported on magnetite/s-graphene and their catalytic application in Suzuki–Miyaura reaction. *Catal. Sci. Technol.,* **2012**, *2*(11), 2332-2340. [http://dx.doi.org/10.1039/c2cy20263h]

[49] Ioni, Y.V.; Lyubimov, S.E.; Korlyukov, A.A.; Antipin, M.Y.; Davankov, V.A.; Gubin, S.P. Activity of palladium nanoparticles on graphene oxide in the Suzuki—Miyaura reaction. *Russ. Chem. Bull.,* **2012**, *61*(9), 1825-1827.

[http://dx.doi.org/10.1007/s11172-012-0252-x]

[50] Morimoto, N.; Yamamoto, S.; Takeuchi, Y.; Nishina, Y. Palladium on graphene: the *in situ* generation of a catalyst for the chemoselective reduction of α,β-unsaturated carbonyl compounds. *RSC Advances,* **2013**, *3*(36), 15608-15612. [http://dx.doi.org/10.1039/c3ra42150c]

[51] Wang, Z.; Xu, C.; Gao, G.; Li, X. Facile synthesis of well-dispersed Pd–graphene nanohybrids and their catalytic properties in 4-nitrophenol reduction. *RSC Advances,* **2014**, *4*(26), 13644-13651. [http://dx.doi.org/10.1039/c3ra47721e]

[52] Yang, M.Q.; Pan, X.; Zhang, N.; Xu, Y.J. A facile one-step way to anchor noble metal (Au, Ag, Pd) nanoparticles on a reduced graphene oxide mat with catalytic activity for selective reduction of nitroaromatic compounds. *CrystEngComm,* **2013**, *15*(34), 6819-6828. [http://dx.doi.org/10.1039/c3ce40694f]

[53] Gómez-Martínez, M.; Buxaderas, E.; Pastor, I.M.; Alonso, D.A. Palladium nanoparticles supported on graphene and reduced graphene oxide as efficient recyclable catalyst for the Suzuki–Miyaura reaction of potassium aryltrifluoroborates. J. Mole. Catal. *Anal. Chem.,* **2015**, *404*, 1-7.

[54] Wang, P.; Zhang, G.; Jiao, H.; Liu, ; Deng, X.; Chen, Y.; Zheng, X. Pd/graphene nanocomposite as highly active catalyst for the Heck reactions. *Appl. Catal. A Gen.,* **2015**, *489*, 188-192. [http://dx.doi.org/10.1016/j.apcata.2014.10.044]

[55] Siamaki, A.R.; Khder, A.E.R.S.; Abdelsayed, V.; El-Shall, M.S.; Gupton, B.F. Microwave-assisted synthesis of palladium nanoparticles supported on graphene: A highly active and recyclable catalyst for carbon–carbon cross-coupling reactions. *J. Catal.,* **2011**, *279*(1), 1-11. [http://dx.doi.org/10.1016/j.jcat.2010.12.003]

[56] Gilliland, S.E., III; Tengco, J.M.M.; Yang, Y.; Regalbuto, J.R.; Castano, C.E.; Gupton, B.F. Electrostatic adsorption-microwave synthesis of palladium nanoparticles on graphene for improved cross-coupling activity. *Appl. Catal. A Gen.,* **2018**, *550*, 168-175. [http://dx.doi.org/10.1016/j.apcata.2017.11.007]

[57] Lee, K.H.; Han, S.W.; Kwon, K.Y.; Park, J.B. Systematic analysis of palladium–graphene nanocomposites and their catalytic applications in Sonogashira reaction. *J. Colloid Interface Sci.,* **2013**, *403*, 127-133. [http://dx.doi.org/10.1016/j.jcis.2013.04.006] [PMID: 23673006]

[58] Rana, S.; Maddila, S.; Yalagala, K.; Jonnalagadda, S.B. Organo functionalized graphene with Pd nanoparticles and its excellent catalytic activity for Suzuki coupling reaction. *Appl. Catal. A Gen.,* **2015**, *505*, 539-547. [http://dx.doi.org/10.1016/j.apcata.2015.07.018]

[59] Mahdavi, H.; Rahmani, O. Polyacrylamide-g-reduced graphene oxide supported pd nanoparticles as a highly efficient catalyst for Suzuki–Miyaura reactions in water. *Catal. Lett.,* **2016**, *146*(11), 2292-2305. [http://dx.doi.org/10.1007/s10562-016-1851-1]

[60] Sarvestani, M.; Azadi, R. Palladium nanoparticles deposited on a graphene-benzimidazole support as an efficient and recyclable catalyst for aqueous-phase Suzuki-Miyaura coupling reaction. *Appl. Organomet. Chem.,* **2017**, *31*(8), e3667. [http://dx.doi.org/10.1002/aoc.3667]

[61] Fareghi-Alamdari, R.; Haqiqi, M.G.; Zekri, N. Immobilized Pd(0) nanoparticles on phosphine-functionalized graphene as a highly active catalyst for Heck, Suzuki and N-arylation reactions. *New J. Chem.,* **2016**, *40*(2), 1287-1296. [http://dx.doi.org/10.1039/C5NJ02227D]

[62] Novoselov, K.S.; Geim, A.K.; Morozov, S.V.; Jiang, D.; Zhang, Y.; Dubonos, S.V.; Grigorieva, I.V.; Firsov, A.A. Electric field effect in atomically thin carbon films. *Science,* **2004**, *306*(5696), 666-669. [http://dx.doi.org/10.1126/science.1102896] [PMID: 15499015]

[63] Wu, Y.; Zhu, J.; Huang, L. A review of three-dimensional graphene-based materials: Synthesis and applications to energy conversion/storage and environment. *Carbon,* **2019**, *143*, 610-640. [http://dx.doi.org/10.1016/j.carbon.2018.11.053]

[64] Qiu, B.; Xing, M.; Zhang, J. Recent advances in three-dimensional graphene based materials for catalysis applications. *Chem. Soc. Rev.*, **2018**, *47*(6), 2165-2216. [http://dx.doi.org/10.1039/C7CS00904F] [PMID: 29412198]

[65] Tran, T.P.N.; Thakur, A.; Trinh, D.X.; Dao, A.T.N.; Taniike, T. Design of Pd@Graphene oxide framework nanocatalyst with improved activity and recyclability in Suzuki-Miyaura cross-coupling reaction. *Appl. Catal. A Gen.*, **2018**, *549*, 60-67. [http://dx.doi.org/10.1016/j.apcata.2017.09.026]

[66] Chua, C.K.; Pumera, M. Covalent chemistry on graphene. *Chem. Soc. Rev.*, **2013**, *42*(8), 3222-3233. [http://dx.doi.org/10.1039/c2cs35474h] [PMID: 23403465]

[67] Xiang, Q.; Cheng, B.; Yu, J. Graphene-based photocatalysts for solar-fuel generation. *Angew. Chem. Int. Ed.*, **2015**, *54*(39), 11350-11366. [http://dx.doi.org/10.1002/anie.201411096] [PMID: 26079429]

[68] Sharavath, V.; Ghosh, S. Palladium nanoparticles on noncovalently functionalized graphene-based heterogeneous catalyst for the Suzuki–Miyaura and Heck–Mizoroki reactions in water. *RSC Advances,* **2014**, *4*(89), 48322-48330. [http://dx.doi.org/10.1039/C4RA06868H]

[69] Feizi Mohazzab, B.; Jaleh, B.; Issaabadi, Z.; Nasrollahzadeh, M.; Varma, R.S. Stainless steel mesh-GO/Pd NPs: catalytic applications of Suzuki–Miyaura and Stille coupling reactions in eco-friendly media. *Green Chem.*, **2019**, *21*(12), 3319-3327. [http://dx.doi.org/10.1039/C9GC00889F]

[70] Premi, C.; Jain, N. A nitrogen and sulphur functionalized graphene oxide–palladium nanoparticle hybrid catalyst for an efficient Heck coupling. *RSC Advances,* **2016**, *6*(78), 74961-74967. [http://dx.doi.org/10.1039/C6RA09996C]

[71] Veisi, H.; Mirzaee, N. Ligand-free Mizoroki–heck reaction using reusable modified graphene oxide-supported Pd(0) nanoparticles. *Appl. Organomet. Chem.*, **2018**, *32*(2), 4067. [http://dx.doi.org/10.1002/aoc.4067]

[72] Nakhate, A.V.; Yadav, G.D. Palladium nanoparticles supported carbon based graphene oxide monolith as catalyst for Sonogashira coupling and hydrogenation of nitrobenzene and alkenes. *ChemistrySelect,* **2016**, *1*(13), 3954-3965. [http://dx.doi.org/10.1002/slct.201600819]

[73] Shendage, S.S.; Singh, A.S.; Nagarkar, J.M. Facile approach to the electrochemical synthesis of palladium-reduced graphene oxide and its application for Suzuki coupling reaction. *Tetrahedron Lett.*, **2014**, *55*(4), 857-860. [http://dx.doi.org/10.1016/j.tetlet.2013.12.022]

[74] Shendage, S.S.; Nagarkar, J.M. Electrochemically codeposited reduced graphene oxide and palladium nanoparticles: An efficient heterogeneous catalyst for Heck coupling reaction. *Colloid Interface Sci. Commun.*, **2014**, *1*, 47-49. [http://dx.doi.org/10.1016/j.colcom.2014.06.008]

[75] Liu, D.; Zhang, C.; Wang, F.; Huang, Z.; Zhang, N.; Zhou, H.; Kuang, Y. In situ preparation of graphene oxide supported Pd nanoparticles in an ionic liquid and the long-term catalytic stability for the Heck reaction. *J. Mater. Chem. A Mater. Energy Sustain.*, **2015**, *3*(32), 16583-16589. [http://dx.doi.org/10.1039/C5TA03767K]

[76] Liu, X.; Zhao, X.; Zhu, J.; Xu, J. One-pot synthesis of magnetic palladium-$NiFe_2O_4$-graphene oxide composite: an efficient and recyclable catalyst for Heck reaction. *Appl. Organomet. Chem.*, **2016**, *30*(5), 354-359.

[http://dx.doi.org/10.1002/aoc.3439]

[77] Elazab, H.A.; Siamaki, A.R.; Moussa, S.; Gupton, B.F.; El-Shall, M.S. Highly efficient and magnetically recyclable graphene-supported Pd/Fe3O4 nanoparticle catalysts for Suzuki and Heck cross-coupling reactions. *Appl. Catal. A Gen.,* **2015**, *491*, 58-69. [http://dx.doi.org/10.1016/j.apcata.2014.11.033]

[78] Fu, W.; Zhang, Z.; Zhuang, P.; Shen, J.; Ye, M. One-pot hydrothermal synthesis of magnetically recoverable palladium/reduced graphene oxide nanocomposites and its catalytic applications in cross-coupling reactions. *J. Colloid Interface Sci.,* **2017**, *497*, 83-92. [http://dx.doi.org/10.1016/j.jcis.2017.02.063] [PMID: 28273514]

[79] Hussain, N.; Borah, A.; Darabdhara, G.; Gogoi, P.; Azhagan, V.K.; Shelke, M.V.; Das, M.R. A green approach for the decoration of Pd nanoparticles on graphene nanosheets: An *in situ* process for the reduction of C–C double bonds and a reusable catalyst for the Suzuki cross-coupling reaction. *New J. Chem.,* **2015**, *39*(8), 6631-6641. [http://dx.doi.org/10.1039/C5NJ01221J]

[80] Anasdass, J.R.; Kannaiyan, P.; Raghavachary, R.; Gopinath, S.C.B.; Chen, Y. Palladium nanoparticle-decorated reduced graphene oxide sheets synthesized using Ficus carica fruit extract: A catalyst for Suzuki cross-coupling reactions. *PLoS One,* **2018**, *13*(2), e0193281. [http://dx.doi.org/10.1371/journal.pone.0193281] [PMID: 29466453]

[81] Kulkarni, P.A.; Shendage, S.S.; Awale, A.G. Carbon–carbon bond formation reaction with Pd/reduced graphene oxide composite. *Orient. J. Chem.,* **2018**, *34*, 881-886. [http://dx.doi.org/10.13005/ojc/340236]

[82] Mahanta, A.; Hussain, N.; Das, M.R.; Thakur, A.J.; Bora, U. Palladium nanoparticles decorated on reduced graphene oxide: An efficient catalyst for ligand- and copper-free Sonogashira reaction at room temperature. *Appl. Organomet. Chem.,* **2017**, *31*(8), e3679. [http://dx.doi.org/10.1002/aoc.3679]

[83] Tarahomi, M.; Alinezhad, H.; Maleki, B. Immobilizing Pd nanoparticles on the ternary hybrid system of graphene oxide, Fe_3O_4 nanoparticles, and PAMAM dendrimer as an efficient support for catalyzing sonogashira coupling reaction. *Appl. Organomet. Chem.,* **2019**, *33*(11), e5203. [http://dx.doi.org/10.1002/aoc.5203]

[84] Alinezhad, H.; Tarahomi, M.; Maleki, B.; Amiri, A. SO_3H-functionalized nano-MGO-D-NH_2: Synthesis, characterization and application for one-pot synthesis of pyrano[2,3- *d*]pyrimidinone and tetrahydrobenzo[*b*]pyran derivatives in aqueous media. *Appl. Organomet. Chem.,* **2019**, *33*(3), e4661. [http://dx.doi.org/10.1002/aoc.4661]

CHAPTER 8

Domino and Multicomponent Reactions by Graphene-Based Carbocatalysts – A Boon for Organic Transformations

S. Haripriya[1,ψ], K.M. Sreeranjan[1,ψ], Sagar Ulkesh Patil[1,ψ], G.S. Sahoo Navneet[1,ψ], H.S. Sushma[1], S. Ahallya[1], Rakesh R Panicker[1], Ivaturi Sai Vighnesh[1], Rajagopal Desikan[1], Vepa Jagannadha Rao[2] and **Akella Sivaramakrishna[1,*]**

[1] *Department of Chemistry, School of Advanced Sciences, Vellore Institute of Technology (VIT), Vellore 632014, Tamil Nadu, India*

[2] *Department of Chemistry, Anil Neerukonda Institute of Technology and Sciences, Affiliated to Andhra University, Sanghivalasa, Visakhapatnam 531162, Andhra Pradesh, India*

Abstract: The research on the development of high-performance metal-free carbocatalysis is one of the emerging areas in chemical sciences. The possible active sites of the surfaces of graphene, oxygenated, and reduced graphene oxide materials are responsible for diversified synthetic transformations, including redox reactions. At present, the advanced research is focused on scalable, sustainable, biocompatible, green, and cost-effective graphene-based carbocatalysts as wonderful alternatives over the noble metallic catalytic materials which are being employed despite their scarcity, high cost, and relative toxicity. The present chapter describes the utility of graphene-based carbocatalysts in multicomponent reactions.

Keywords: Catalytic Activity, Graphene Oxide Materials, Metal-Free Carbocatalysts, Multicomponent Reactions, Reduced Graphene Oxides.

1. INTRODUCTION

Numerous organic transformations have been reported by many homogeneous transition metal complexes including organometallics and heterogeneous catalysts [1 - 4]. Among these catalytic reactions, due to the significant medicinal properties of the reaction products, various multicomponent organic reactions yielding fused heterocyclic compounds have been designed employing a variety

* **Corresponding author Akella Sivaramakrishna:** Department of Chemistry, School of Advanced Sciences, Vellore Institute of Technology (VIT), Vellore 632014, Tamil Nadu, India; E-mail: asrkrishna@vit.ac.in

ψ All these authors have equal contributions to this Chapter.

Pinki Bala Punjabi, Rakshit Ameta & Sharoni Gupta (Eds.)

of metal and non-metal-based catalysts [5]. It would be appropriate to design alternative non-metal catalysts derived from renewable resources in terms of sustainability. Because of this, carbon materials resulting from biomass have been considered sustainable pathways [6, 7]. The economic burden and toxicity are the other specific factors besides sustainability, which favor the carbocatalysts since some of the precious metals (in particular noble metals) are very expensive over the carbon feedstock materials. Furthermore, it is sometimes a very difficult task to remove the residual catalyst from the final product, which can lead to undesired toxicity. These aspects are the driving force in developing metal-free catalysts by researchers. Among them, "carbocatalysis" is one of the promising sustainable routes to reduce metal-based reactions [8 - 10]. It was found that these carbocatalysts, especially the heteroatom doped graphene-based carbocatalysts are considered the best promising alternates of noble metals as an efficient and green catalyst.

In general, carbocatalysts are non-molecular entities and contain mostly carbon framework as an active component with a large surface area. But these carbonaceous catalytic materials are completely different from organocatalysts, which are active organic molecules [7, 11]. Furthermore, the incorporation of the active sites into the carbon materials by appropriate modifications is the key aspect for carbocatalysts, which makes the utilization of metals non-essential. Carbocatalyst is better suited for heterogeneous reactions over homogeneous systems when the particle sizes are large. However, the particle size of the graphene material can be reduced to the carbon dots having 2D morphology; carbocatalysts facilitate the single phased homogeneous catalysis [12]. The graphene derivatives can offer wonderful catalytic properties due to the presence of functional groups with heteroatoms on the G-sheet as active catalytic centers [13, 14]. The importance of graphene as catalytic material is highlighted based on the adsorption capacity, electron density, and morphology of the sheets and its efficiency is strongly dependent on the position and nature of active sites on the graphene sheets.

2. SYNTHESIS OF GRAPHENE-BASED CARBOCATALYSTS

A large number of research articles have been reported on the method of preparation on various Gmat [12 - 33] and characterization techniques [17, 18, 23, 34], including their important structural properties for the applications in electronics [34 - 40] and conductivity [38 - 41]. Furthermore, the studies on electrocatalysis [42 - 47] and photocatalytic applications [48] of graphene derivatives have been widely carried out; however, there is a growing interest in the graphene oxide (GO) materials as semiconductors [49] and the synthesis of organic heterocycles [50]. Notably, N and P double doped graphene carbon catalyst (NP-G) was employed for the reduction of nitrophenol over traditional graphene-based catalyst [14]. Nitrogen and sulfur co-doped graphene (NS-G) derived by one-pot thermal pyrolysis using heteroatom containing precursors like dopamine hydrochloride (DA) and 2-mercaptoethanol (ME) exhibits unusual catalytic activity due to its porous structure toward water splitting upon activation [51].

The synthesis of heteroatom donor graphene oxide (GO) is quite lengthy and cumbersome. Primarily, the modified Hummer's method [52] was employed to synthesize GO for the synthesis of N-doped rGO. By this process, GO was (1.0 g) dissolved in 50 mL ethanol, agitated for around 30 min, followed by sonication. After that, melamine (1.0 g) was incorporated into the mixture and the resulting suspension was heated to 50 °C on a hotplate while stirring (Scheme **1**). The desiccated mixture was heated at 450 °C and then calcination was done at varying temperatures of 600, 700, and 800 °C under an inert N_2 atmosphere for 1 h to obtain NG-600, NG-700, NG-800, respectively. Further washing with ultrapure water and ethanol and after overnight drying at 60 °C, the (N)G attains the catalytic activity [53].

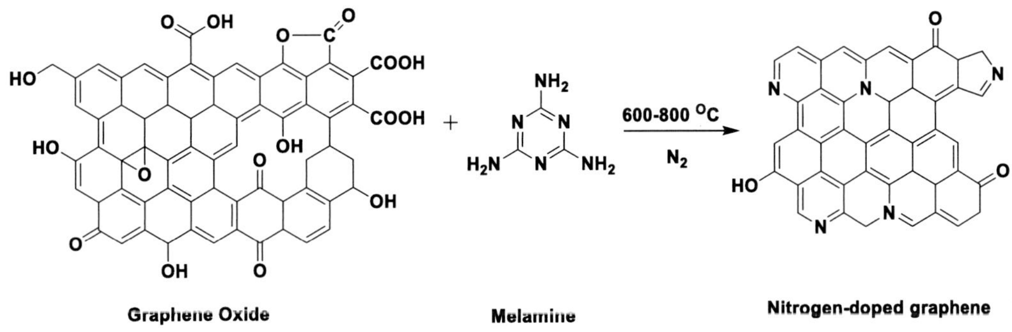

Scheme 1. Synthesis of N-doped graphene.

The graphene-based carbocatalytic reactions for various organic transformations are broadly classified into i) oxidation, ii) carbon-carbon bond formation, iii) hydrogenation, iv) aza-Michael addition, v) dehydration-hydrothiolation, and vi) Friedel-Crafts alkylation.

3. ORGANIC TRANSFORMATIONS BY GRAPHENE-BASED CARBOCATALYSIS

3.1. Oxidation Reactions

Similar to transition metal-based catalysts facilitating efficient incorporation of an oxygen atom into organic molecules by oxidation reactions, graphene oxide-based catalysts are also mild but effective carbocatalysts for oxidation reactions as given in Scheme **2** and Scheme **3** [54].

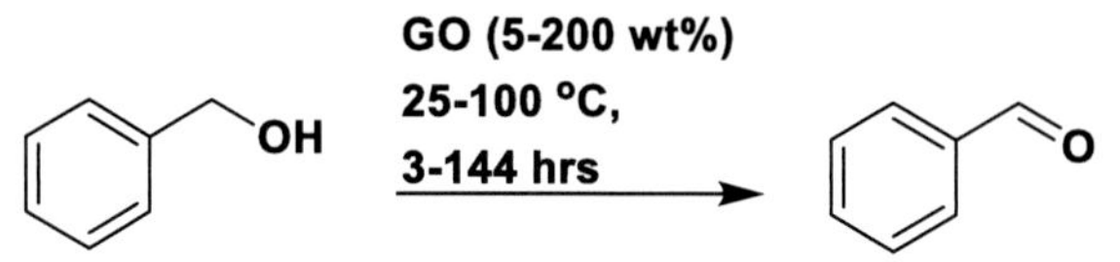

Scheme 2. Oxidation reaction of benzyl alcohol to benzaldehyde (92%) with GO catalyst.

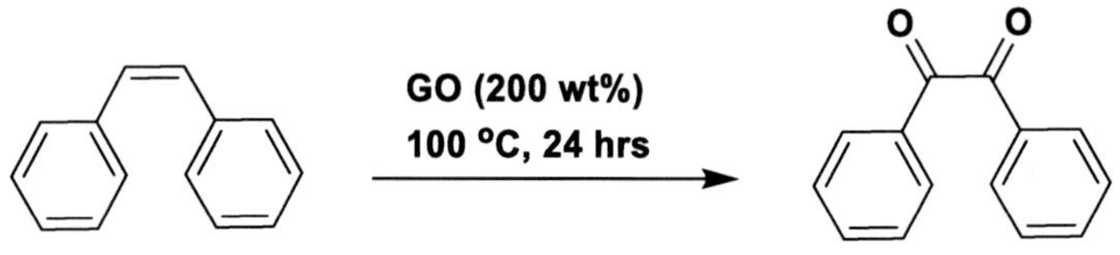

Scheme 3. Oxidation of *cis*-stilbene in to dione (49%) with GO catalyst.

The conversion of benzene to oxygen-containing aromatic compounds is one of the most active and still evolving fields of catalysis due to its industrial significance. But they are typically prepared by long processes that are both expensive and time-consuming (*e.g.* phenol produced by 3-step cumene processes as given in Scheme **4**). There are also other disadvantages, such as contamination and less selectivity towards compounds. A lot of changes and improvisations have been made to enhance this field of catalysis. Nano-sized carbon materials, such as graphene, have proven to be effective and selective catalysts for such reactions. The reason for the improved catalytic activity is due to the open π-electron system and the huge adsorption rate of graphene [55].

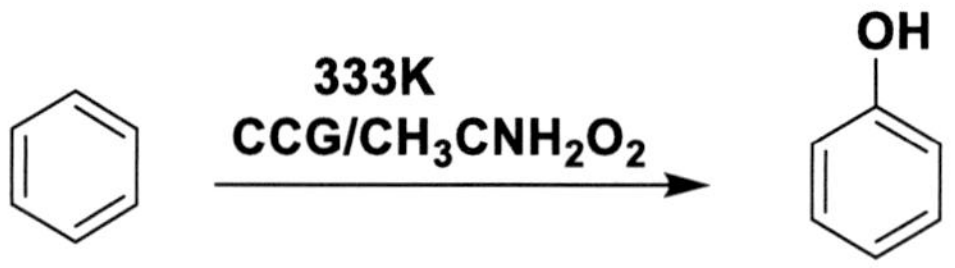

Scheme 4. Oxidation of benzene to phenol (> 99%) using CCG as a catalyst.

Traditionally the styrene was oxidized to benzaldehyde and benzoic acid by molecular oxygen, but it was investigated that the styrene oxide was also getting formed through epoxidation. So far it was extremely difficult to monitor the reaction so that it should not form unpleasant side products and undergo detrimental reactions. The interest of this (S)G synthesis arrived in minds of researchers from the fact that the oxidation of the alkene in aerobic conditions using transition or noble metals as the catalysts, and, hence it would be exceptionally profitable to formulate such catalysts which must be metal-free for such processes. The monitored reaction was done in the absence of catalyst which has led to only 4% product formation. This is attributed due to auto-oxidation. Moreover, the reaction was assessed by oxidation of styrene under aerobic and non-solvent reaction conditions in presence of (S)G as a catalyst. Promisingly, 13% of conversion was achieved after 7 h with 70 and 20% of benzaldehyde and styrene oxide respectively (Scheme **5**). The hallmark of (S)G catalyst is that it could be reused or recycled two times with no considerable changes in selectivity and efficiency [56].

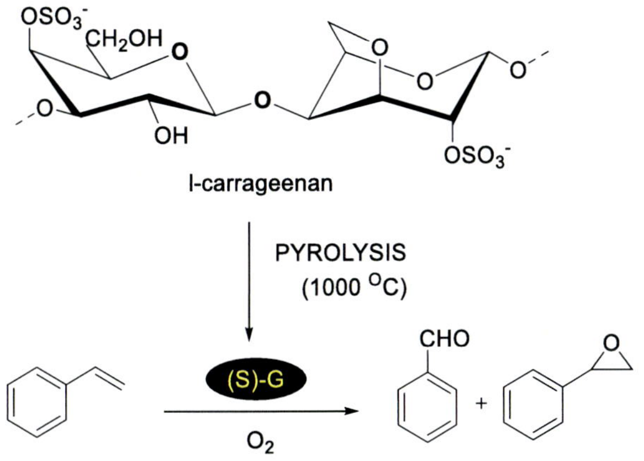

Scheme 5. Oxidation of styrene in the presence of graphene S-doped to benzaldehyde (100% selectivity at 100 °C) and to styrene oxide (maxmum 20% selectivity).

Further, the use of a metal-free sulfonic acid-functionalized reduced graphene oxide (SA-rGO) material for catalytic synthesis of carboxylic acids and its corresponding bio-isosteres, 5-substituted-1H-tetrazoles, has been reported. As a catalytic material, SA-rGO combines the intriguing properties of graphene oxide with the added benefit of highly acidic sites due to sulfonic acid groups. H_2O_2 as a green oxidant with high TOF values (9.06–9.89 h^{-1}) was used to oxidize aldehydes to carboxylic acids very efficiently. Similarly, the 5-substituted-1H-tetrazoles could also be synthesized efficiently with high TOF values (12.08–16.96 h^{-1}). By employing the SA-rGO catalyst, both reactions could be carried out efficiently at the gram scale level (10 g). SA-rGO has high reusability of up to eight runs without a significant decrease in productivity. As a result of these features, SA-rGO is an attractive alternative from the industrial point of view [57].

Notably, the graphene oxide reduction method influences the activity of the resulting reduced graphene oxide (rGO) for the aerobic oxidation of benzylamine. Beginning with GO obtained *via* the Hummers method, a series of rGO samples was prepared *via* chemical (hydroquinone (HQ), hydrazine (HZ), or ascorbic acid (ASC)) or thermal reduction. The analysis reveals that the chemical reducing agents reduce GO with different functional groups affecting the catalytic activity of the resulting rGO for molecular oxygen activation in the oxidation of benzylamine. The rGO-HQ1 sample prepared with HQ as a reducing agent exhibited the optimum activity in the aerobic oxidation of benzylamine at 80 °C. It has been hypothesized that HQ introduces hydroquinone/*p*-benzoquinone-like moieties onto the graphene sheet and these active sites are responsible for the oxidation reaction. By selective masking of oxygen functional groups present in the most active rGO sample, indicates the activity of HQ and/or *p*-benzoquinone as organocatalysts. In five consecutive uses, the most active rGO sample displayed good reusability and stability. Hydroperoxyl radicals were presumed as the primary reactive oxygen species formed in the system, according to selective quenching tests [58].

Anumula *et. al.* reported glutathione-protected Ag_6 clusters using a simple green reaction, transferred them to graphene oxide (GO), and used them as a catalyst for cyclohexane oxidation under ambient conditions. With *t*-butyl hydroperoxide (TBHP) as the oxidant in an oxygen-flowing atmosphere, high selectivity and yield of cyclohexanone is obtained. It is demonstrated that using density functional theory calculations that both TBHP and oxygen interact with the Ag6 clusters supported by GO, resulting in the peroxide structure and thus facilitating catalysis cyclohexane oxidation. It is interesting to note that hydrogen-added hydroperoxyl (-OOH) has a significantly lower activation barrier for O-O bond dissociation. The O-O dissociated intermediate then reacts with cyclohexane to form cyclohexanol, which is then dehydrogenated by successive oxidative reactions to form cyclohexanone. This finding validates the enhanced catalysis of TBHP and O_2 for the conversion of cyclohexane to cyclohexanone and contributes to the development of heterogeneous metal cluster catalysts for efficient C-H activation [59].

Reports show that by reacting respective ultra-small colloidal metal oxide/hydroxide with partially reduced colloidal graphene oxide, surfactant-free MGN for distinct noble metals can be prepared for high-performance fuel cell catalysis. The resulting MGN is made up of widely distributed $M-M^{+n}$ based nanoparticles with extremely small sizes that provide a strong and stable catalytic current for the oxidation of ethanol and formic acid [60].

3.2. Carbon-Carbon Bond Formation

Besides, *N*-doped defective graphene [(N)G] carbocatalyst was synthesized by pyrolysis of chitosan and employed as a basic catalyst for some C-C bond forming reactions which are nucleophilic Michael and Henry additions as shown in Scheme **6**. The most basic *N*-sites were detected by DFT calculations amongst pyridinic, pyrrolic, and graphitic residual nitrogen atoms of *N*-doped defective graphene [61].

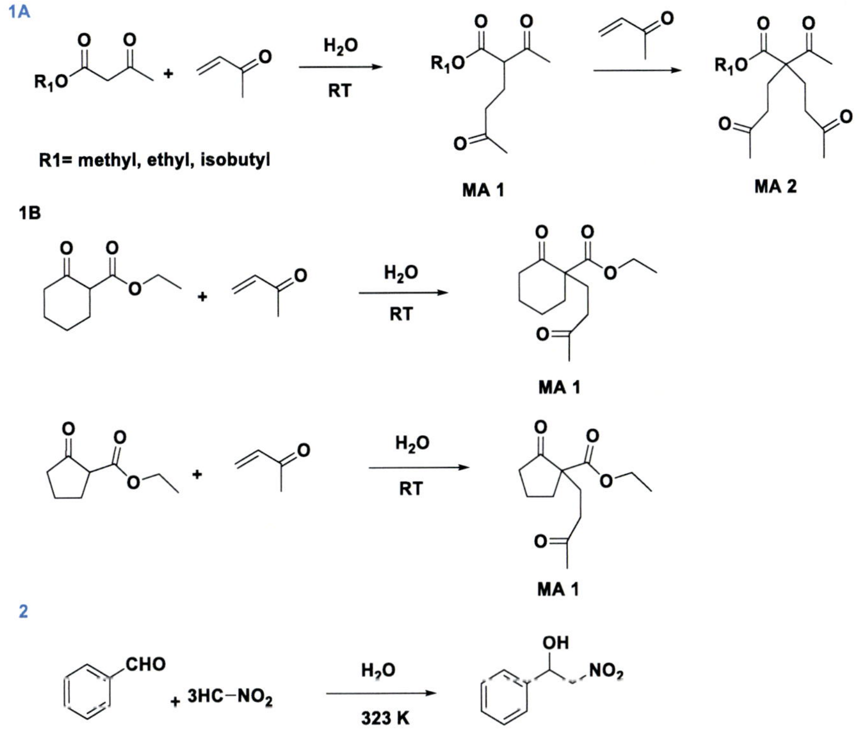

Scheme 6. Some C-C bond forming reactions through nucleophilic Michael and Henry additions by carbocatalysis.

Also, the fabrication of sulfur-doped graphene oxide (SA-GO) by the method of sulfuration and alkylation was reported [62].

Remarkably, GO-supported primary amine (AP-GO) facilitated one-pot Henry-Michael reactions provide synthetically profitable multi-functionalized nitroalkanes. Furthermore, the higher activity is recorded with GO-supported system more than those activated carbon-supported primary amine, and mesoporous silica-supported primary amine as catalysts. This predominant catalytic performance started from the interesting properties of AP-GO, which is based on the acid–base cooperative effect by the suitable primary amines on their basal planes and the carboxyl acids along their edges. The decreased diffusion resistance of the reactants and intermediates are responsible for the multistep catalytic cycles. Also, it may be promptly reused by simple filtration and consequently recycled without significant loss of its catalytic activity. The organoamines were anchored on the basal planes of GO through a basic and simple silane-coupling reaction to manage GO-supported organic amines, as depicted in Scheme **7** [63].

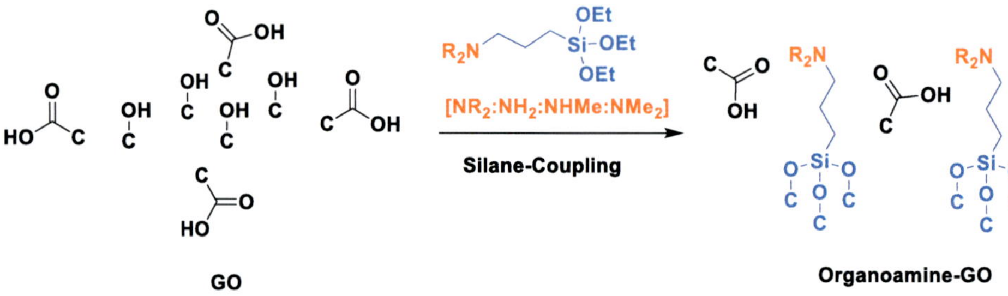

Scheme 7. Schematic illustration of organic amine-functionalized GO carbocatalysis.

The successful anchoring of organic amine groups onto GO surface was achieved by the Henry reaction between benzaldehyde and nitromethane to give the resulting products through nitroaldol addition and dehydration steps in the presence of different types of catalytic systems (Table **1**).

Table 1. Different catalysts in henry reaction.

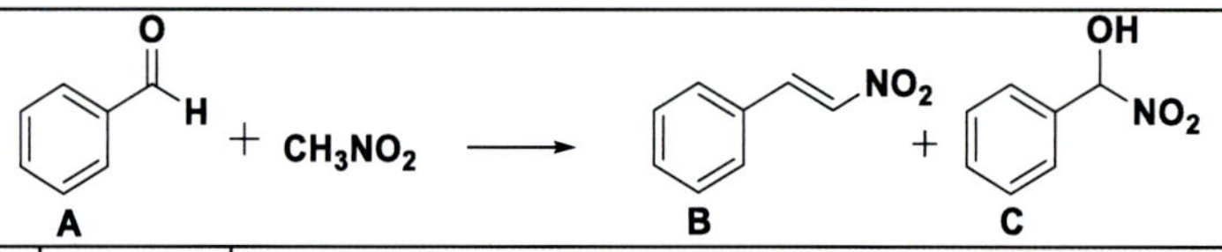

Carbocatalyst	Conversion of A (%)	Conversion to B (%)	Yield of C (%)
GO[2]	0	0	0
AP-GO	90	90	0
MAP-GO	91	76	15
DMAP-GO	84	53	31
PROPYLAMINE	39	39	0
GO+PROPYLAMINE	22	22	0
AP-GO-A[3]	13	13	0
AP-GO-B[4]	36	36	0

Interestingly, the catalytic reactions do not work with GO alone, but the GO-supported organic amines advanced to high conversion of products, showing that organic amines are important active species. The capability of the AP-GO catalyst among others to promote the one-pot Henry-Michael reaction is described in Table **2**. Initially, the carboxylic acids in the edges of GO activate the carbonyl group of benzaldehyde through hydrogen bonding, while the primary amines on the basal planes of GO attack the activated benzaldehyde to form the imine intermediates by the elimination of the water molecule. Eventually, nitromethane attacks this imine in the presence of carboxylic acids by following the elimination of amine species to generate *trans-β*-nitrostyrene. Finally, carboxylic acid activates *trans-β*-nitrostyrene again, to yield the targeted products as Michael adducts [63].

Table 2. Influence of different carbocatalysts in one-pot Henry-Michael reaction.

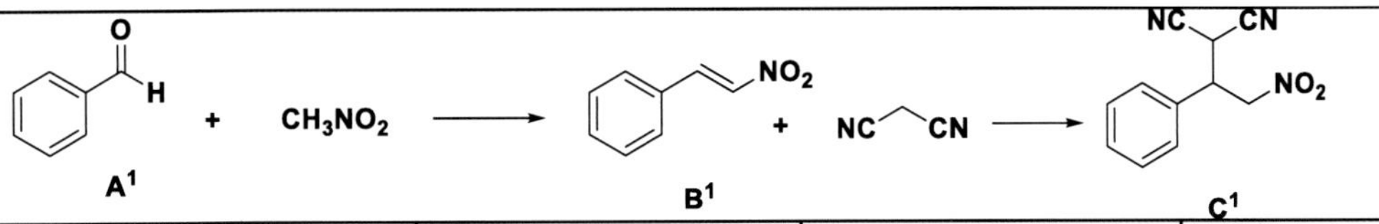

Carbocatalyst	Conversion of A^1 (%)	Conversion to B^1 (%)	Yield of C^1 (%)
AP-GO	90	97	87
Propylamine	39	49	19
AP-AC	53	58	31
AS-SBA-15	38	40	15
AP-GO-M[b]	34	39	13

Reaction conditions: 0.306 mL of benzaldehyde (3.0 mmol), 6.0 mL of nitromethane, 0.228 of malononitrile (3.6 mmol), a catalyst containing 0.075 mmol nitrogen species, 12 h. [b]Added 3.0 mL of methanol

Even though GO is inexpensive and has numerous remarkable mechanical and electronic properties, easy recovery, and metal-free reactivity from the reaction mixtures, the properties of GO have been relatively less explored as an oxidant or catalyst for facilitating various organic transformations. Eco-friendly synthesis of 2-amino-3-cyanopyridines *via* a GO catalyzed four-component condensation was studied [63] and the catalytic efficiency of GO over other catalysts is compared in Table **3**.

Table 3. A comparative analysis on the efficiency of carabocatalyst (GO) over other catalysts.

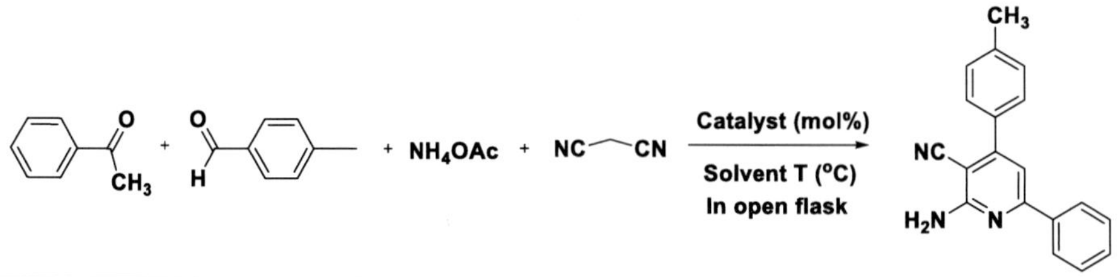

Entry	Catalyst (Mol %)	Reaction Conditions[1]	Time, h	Yield[2] (%)
1[3]	-	H_2O, 80°C	24	<10
2	GO (05)	H_2O, 80°C	05	81
3[4]	GO (10)	H_2O, 80°C	05	90
4	Graphite	H_2O, 80°C	24	20
5	r-GO	H_2O, 80°C	24	36
6	Activated carbon	H_2O, 80°C	24	57
7	Silica	H_2O, 80°C	05	49
8	H-beta Zeolite	H_2O, 80°C	05	66
9	TiO_2	H_2O, 80°C	05	40

[1]Reaction conditions: acetophenone (1 mmol), 4-methyl-benzaldehyde (1 mmol), ammonium acetate (1.5 mmol), malononitrile (1.3 mmol), catalyst (type indicated) in the presence of air
[2]Isolated yields
[3]r-GO-reduced graphene oxide

3.3. Hydrogenation Reaction

With regard to stability and safety issues associated with graphene oxide, when the compound is contaminated with potassium salts, its thermal stability is decreased. These potassium salts are obtained as GO synthesis products. These compounds are extremely flammable and unstable, therefore, a small spark is sufficient to burn down the entire GO film, resulting in a disastrous explosion. Since the compound's thermal instability in the presence of salts (by GO synthesis products) was observed, it was necessary to purify the GO material to prevent a fire hazard. In two phases, an acid wash proved to be a successful purification tool because it suppresses gelation, which is one of the major factors hindering purification. Furthermore, the energy produced from GO's self-propagating domino reactions can be used to generate other materials based on graphene that have high activity [64].

Among all the relevant graphene catalytic operations, its use as a composite photocatalyst for organic pollutant degradation is socially important. Graphene has a honeycomb-like structure with sp^2 hybridized carbon arranged like a sheet. By chemically modifying the graphene, its surface properties like chemical adsorption, transparency, and conductivity will improve, which enhances its use as a composite material. Photocatalysts like TiO_2 can be chemically bonded to graphene by a one-step hydrothermal reaction. P25 is a TiO_2 photocatalyst that is bonded to graphene and can be used for the photodegradation of dyes. P25-Graphene photocatalyst possesses good absorptivity of dyes, a wide range of light absorption, and also charge separation properties, making it different and more efficient than normal TiO_2. Due to its giant 2-D planar structure, the improved activity of the P25-Graphene photocatalyst over normal P25 and other photocatalysts provided a better surface for dye adsorption and also separation charge [65].

Studies show that peroxymonosulfate (PMS) can be effectively activated by chemically reduced graphene oxide to produce active sulfate radicals, which is a powerful oxidizing species with a high oxidizing potential having enough potential to decompose various aqueous contaminants. In comparison with graphite powder (GP), graphene oxide (GO), activated carbon (AC), and multiwall carbon nanotube (MWCNT), graphene exhibits a higher activity. It was demonstrated that graphene catalysis is better than that of Co_3O_4 in the degradation of 2,4-dichlorophenol (DCP), phenol, and methylene blue (MB) in water, which is an efficient system for ecological remediation.

The activation of peroxy-monosulfate by graphene is more like those in oxidative dehydrogenation (ODH) and dehydrogenation (DH) reactions, in the presence of nanocarbon (Lewis base). The reduction of GO hydrothermally tunes the surface acidity/alkalinity by removing the oxygen moieties. By modifying the surface acidity of GO to alkaline by ammonia solution, the activity of graphene in the initiation of PMS improves significantly [66].

The sp^2 carbon atoms are placed in a hexagonal cell as a conjugated device in the case of a basal plane in perfect graphene. The curvature of stacked graphene layers and edge defects create non-six carbon rings. Two kinds of edges, zigzag and armchair, are created when C-C σ bonds are broken. The localized state at the zigzag edge means that their π electrons are not limited to edge carbons, hence the chemical activity is increased. The zigzag edges of reduced graphene oxides derived from chemical treatment hold a small number of oxygen-containing groups, and the remaining groups have a higher intrinsic activity than those at the end of the flat surface. Chemically reduced graphene oxide at zigzag edges hold a very small number of oxygen groups, and the remaining groups have a significantly higher intrinsic activity than those at the end of the flat surface.

Oxygen functional surface species, such as ketone (C=O) groups, are abundant and have a great deal to organize the redox phase.

PMS (HSO_5^-) is capable of activating graphene to yield two different types of sulfate radicals ($SO_4^{\bullet-}$and $SO_5^{\bullet-}$). The former radical exhibits a higher oxidative potential value (2.5−3.1 V) and plays a vital role in phenol degradation. This reaction pathway is accompanied with the observations that the operation of different carbons follows the order: graphene > MWCNT > GP > GO, matching to the density of the graphite defects, (b) the action is closely related to the loading of the catalyst, (c) the influence of temperature is less which reflects the character of the graphene carrier, and (d) the similar catalysis analogs applicable to nanocarbons. The activity of graphene is higher than other carbon allotropes (GP, AC, MWCNT, and GO), and then it is on par with the activity of popular transition metal oxides.

Moreover, hydrogenation reactions of nitroarenes by reduced graphene oxide (RGO)-supported platinum (Pt) catalyst derived from the reduction of ethylene glycol (EG) and its characterization study demonstrated that EG as a reductant exhibits better advantages than the extensively used hydrated hydrazine to produce Pt nanoparticles that are small-sized and monodispersed on the entire surface of RGO. The results reveal that the yields of aniline over the Pt/RGO-EG catalyst are much higher than that of active carbon-supported and multi-walled carbon nanotube-based Pt catalysts. When the reaction temperature was changed to 20°C, the catalytic activity of Pt/RGO-EG was enhanced to 1138.3 mol (mol-Pt min), which proved to be extremely fruitful for the hydrogenation of nitroarene derivatives. The peculiar catalytic behavior of Pt/RGO-EG can be attributed to the well-scattered Pt clusters on the RGO sheets and the homogeneously dispersed Pt/RGO-EG in the reaction mixture. Thus, it is evident that RGO plays a vital role as a support to the catalyst and it is a better promoter than AC and MWCNT. Furthermore, Pt/RGO-EG appears to be a promisingly active and stable catalyst for the hydrogenation of NB (Scheme **8** and Scheme **9**).

The distinctive catalytic operation and simple fabrication process of GO and GO-based composites have the notable ability as heterogeneous catalysis [67].

The hydrothermal treatment of a mixed aqueous precursor solution of $PdCl_4^{2-}$, carbon nanotube (CNT), and GO, as well as the simultaneous spontaneous redox reaction between $PdCl_4^{2-}$ and CNTGO hybrid, resulted in the formation of a 3D macroscopic Pd-CNT-rGO hydrogel. The resulting Pd-CNT-rGO bifunctional nanocatalyst has a large surface area and excellent durability, and thus exhibits remarkably high and stable catalytic activity toward p-nitrophenol reduction, with 100 percent conversion in only 30s. The proposed simple synthetic route and designed bifunctional catalysts will serve as a universal platform for the fabrication of nano-catalysts that can be used in a variety of catalysis, environmental protection, energy storage, drug delivery, chemical and biological sensing, and other applications [68].

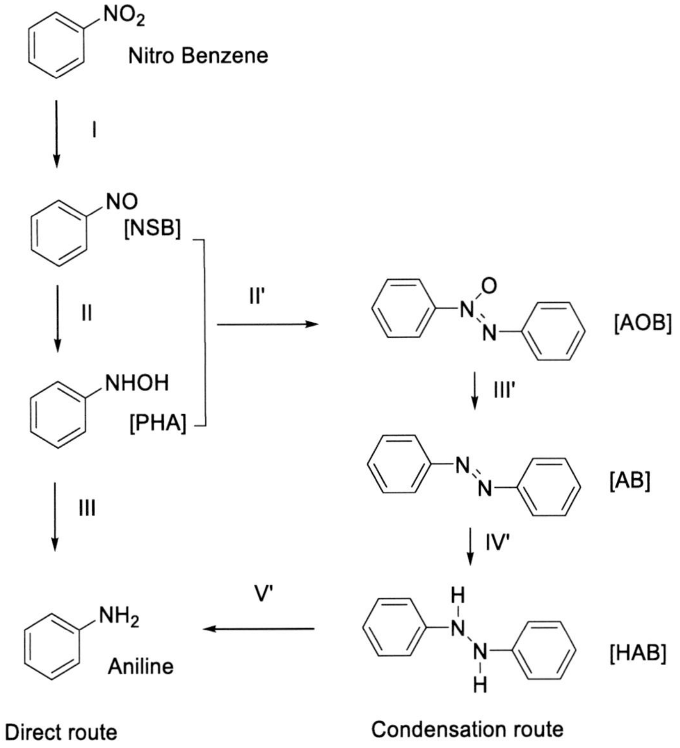

Scheme 8. Mechanistic steps for the hydrogenation of NB.

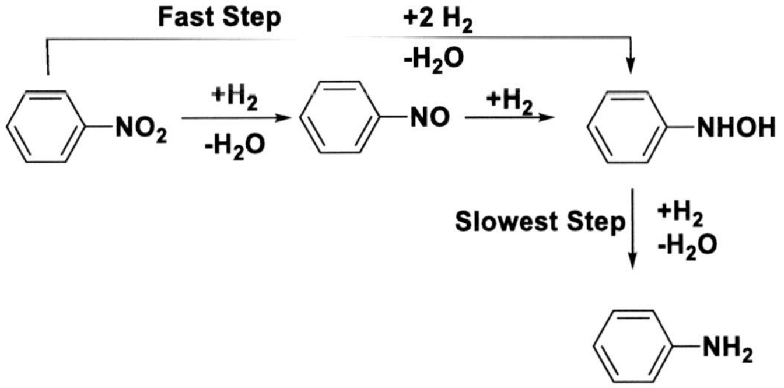

Scheme 9. Proposed mechanism for the hydrogenation of NB over Pt/RGO-EG catalyst.

Song *et.al.* reported the effect of capping agents on the catalytic function of rGO for redox reactions using nitrogen-containing surface modifiers with distinctly different chemical structures, such as poly (diallyl dimethylammonium chloride), cetyltrimethylammonium chloride, and poly (allylamine hydrochloride), which can endow rGO with improved suspension stability. The reduction of graphene oxide with hydrazine in the presence of capping agents resulted in the facile preparation of functionalized rGOs. The results of model redox reactions, such as 4-nitrophenol and ferricyanide reduction reactions, catalyzed by functionalized rGOs, confirmed that the way capping agents functionalize rGO, which is highly correlated with their chemical structure, has a significant impact on overall reaction kinetics, such as induction time, reduction rate, total reaction time, and reaction order. This strongly shows that careful capping agent selection is essential for fully exploiting rGO's catalytic function and therefore designing innovative rGO-based non-metallic catalysts with controlled reaction kinetics [69].

GO and MoS_2 solution in H_2O

↓ **Freeze drying and annealing**

Porous rGO + MoS_2 Foam

↓ **Sulfurization**

rGO + MoS_2 Foam Composite

Scheme 10. Procedure of *in situ* growth of rGO supported MoS_2 with high defect density by sulfurization.

Joyner *et. al.* outlined a chemical vapor deposition (CVD) method for growing MoS_2 structures by sulfurizing rGO supported MoO_2 particles. On the basal plane of rGO supported MoS_2 structures, this approach created various morphologies of MoS_2 with large edges and high defect densities, which are considered active sites for hydrogen evolution reactions (HER) catalysis. Furthermore, MoS_2 nanostructures on the surface of the porous rGO network exhibit strong physical interactions, such as van der Waals and π–π interactions between MoS_2 and rGO. As a result of these features, the HER catalyst has been enhanced. With the duration of the sulfurization was 30 min, the 1rGO- 2MoS_2 composition provided the lowest onset potential of -197 mV. Theoretical calculations show that defects on MoS_2's basal plane do indeed increase catalytic activity, resulting in more active sites. Due to alternative geometries other than triangle structures take longer to saturate the interactions with hydrogen atoms, having alternative MoS_2 geometries allows the system to be active for longer. According to the calculations, having MoS_2 crystals with diverse geometries and defects is favorable for hydrogen catalytic activity because the varied geometries allow the system to remain active for longer periods and the defects increase the number of active sites accessible for hydrogen interactions (Scheme **10**) [70].

Graphene doped with a heteroatom such as oxygen, nitrogen, boron, sulfur, and phosphorus yields a wide range of metal-free organocatalysts carrying improved catalytic performance. In this effort, Jiangbo Xi *et al.* synthesized an N and P-double-doped graphene (NPG) carbocatalyst by microwave-exfoliated graphite (MEG) in presence of hexachlorocyclotriphosphazene (HCCP) in a facile way, which in turn was used as N as well as a P source. This NPG showed a wonderful catalytic activity towards reduction reactions of nitroarenes, furthermore, excellent stability, selectivity, and recyclability [71].

3.4. Aza-Michael Addition

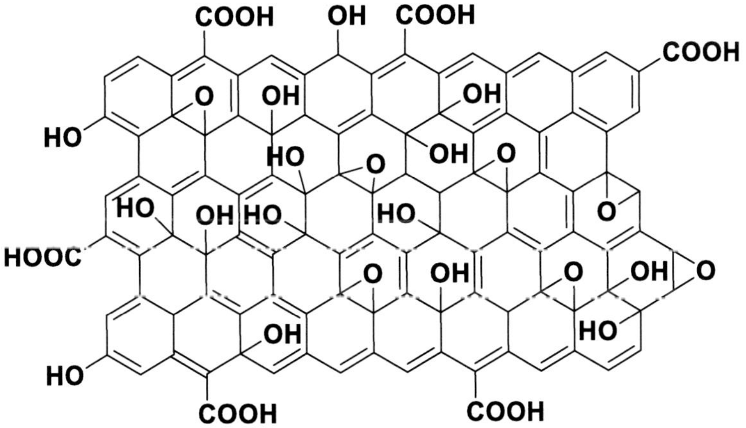

Fig. (1). A schematic design of graphene oxide (single sheet) showing active oxygen groups.

Oxygen-containing graphene oxide (GO) mainly functions as a precursor to the reduced graphene oxide. Various solvents including ionic liquids, water, and aprotic organic solvents, like dimethyl sulfoxide (DMSO), have been used as reaction media in the dehydration of carbohydrates to HMF. Further, ionic liquids are capable of solvents in terms of high selectivity in biomass conversion. For the dehydration of fructose, graphene oxide can be an efficient carbocatalyst in the presence of isopropanol. Also, GO has been employed under various experimental conditions including strongly acidic and ultrasonic conditions [72]. The arrangement of new carbon-heteroatom bonds through *aza*-Michael expansion is a notable change in organic chemistry, which is of biological significance through active nitrogen-containing molecules, chiral auxiliaries, natural products, and antibiotics. The presence of oxygen-carrying functionalities in graphene oxide (Fig. **1**) facilitates hydrophilicity and allows dispersion in water and further polar solvents. The catalytic potential of graphene oxide nano-sheets was examined for the Aza-Michael addition reaction, as shown in Scheme **11**.

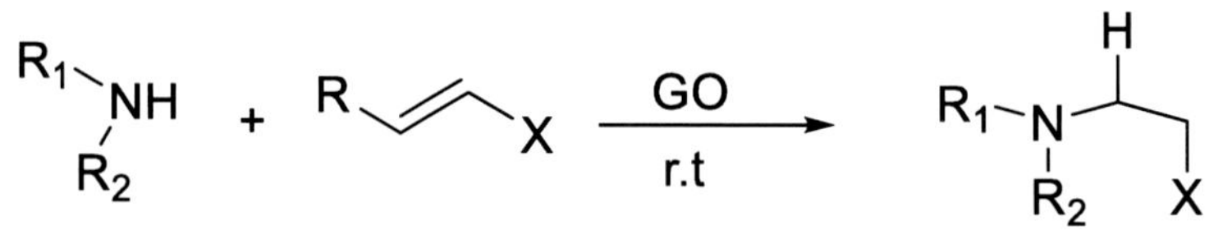

R= H, CH_3
R_1= Et, n-Bu, n-Pr, Bz
R_2= Et, n-Bu, n-Pr, Bz
X = CN, $CONH_2$, $COOCH_3$

Scheme 11. The catalytic Aza-Michael addition reaction by GO.

Further reaction scope was extended towards aromatic amines, such as aniline and *p*-anisidine, and was found to be less reactive and afforded moderate product yield in longer reaction times as given in Table **4** [73].

Table 4. Graphene oxide catalyzed aza-Michael addition of amines to α,β-unsaturated compounds.

Entry	Amine	Michael Acceptor	Reaction Time, Min	Yield (%)
1	Et(Et)NH	CN	35,[c] 150[d]	96, 94
		CN	30.0,[e] 5.0	95, 97
		$CONH_2$	5.0	97
		$COOCH_3$	7.0	96
		$COOCH_3$ (methyl-substituted)	7.0	95
2	NH_2 (cyclohexyl)	CN	5.0	95
		$CONH_2$	7.0	92
		$COOCH_3$	5.0	92
		$COOCH_3$ (methyl-substituted)	5.0	94

(Table 4) cont.....

3	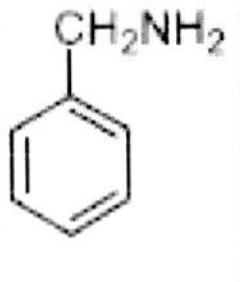	CN	7.0	90
		$CONH_2$	10.0	89
		$COOCH_3$	10.0	92
		$COOCH_3$	10.0	92
4	*n*-Bu NH *n*-Bu	CN	5.0	96
		$CONH_2$	10.0	85
		$COOCH_3$	7.0	94
5	*n*-$BuNH_2$	CN	5.0	95
		$CONH_2$	5.0	94
6	*n*-Pr NH *n*-Pr	CN	5.0	95
		$CONH_2$	5.0	96
7		CN	5.0	94
		$CONH_2$	5.0	95
8	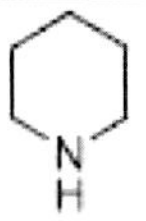	CN	5.0	95
		$CONH_2$	5.0	94
9	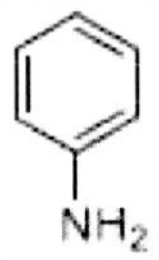	CN	45	65
10	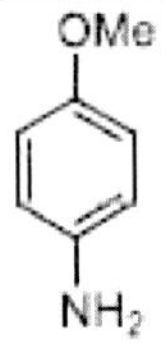	CN	30	70

3.5. Dehydration-Hydrothiolation Reactions

To optimize the catalytic activity and observing the yields of the products, various combinations of the reactants were used under varied conditions like changing the amount of catalyst, time, temperature, and different aryl and alky groups. Thioethers can be utilized for synthesizing various biologically vital molecules. It was observed that the reactions with both long-chain and alicyclic thiols were reported (Scheme **12**). The electron-withdrawing group enhances dehydration-hydrothiolation reactions.

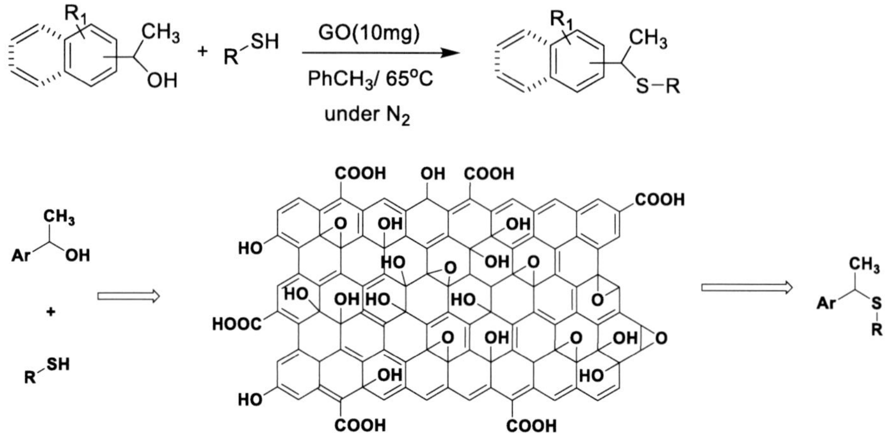

Scheme 12. Graphene oxide-catalyzed dehydration and hydrothiolation.

By this method, the catalyst can be retrieved from the product and can be reused at least 4-5 times without losing the catalytic activity. It was observed from the FT-IR spectrum of the GO catalyst and the GO's functional groups were not at all affected even after several cycles of catalytic reactions and they retained their functionality [74].

Peng *et. al.* developed a three-dimensional monolith of neat GO aerogel as a fixed-bed carbocatalyst for the synthesis of thiol compounds (Scheme **13**), demonstrating the advantages of ultrafast catalytic speed (5-8 h), high selectivity (100%), high yields (near 100%), easy product isolation, and long-life recyclability (> 18 times). For the first time, thiol compounds with halogen and hydroxyl functional groups that cannot be produced using conventional catalysts have been reported. The effective catalysis mechanism is primarily attributed to the protonic functional groups, ultra large size, and unpaired electrons of GO, as well as the "cage effect" in nanoscale constrained areas of aerogel cells, according to control experiments [75].

3.6. Friedel-Crafts Alkylation

The carbonyl functional sites in graphene can also be used as active sites for chemical reactions. The developed carbocatalysts are one without pH adjusted film (G1 film) and another pH adjusted film using ammonia (G2 film). Also, it was noticed that AA (L-ascorbic acid) plays a vital role in the preparation of carbocatalyst. The pH-mediated carbocatalyst enhances electrostatic repulsive interactions and it was observed that the pH and its effect can be seen only during carbocatalyst formation with AA and after the reaction, the active edges are exposed (Scheme **14**) [76].

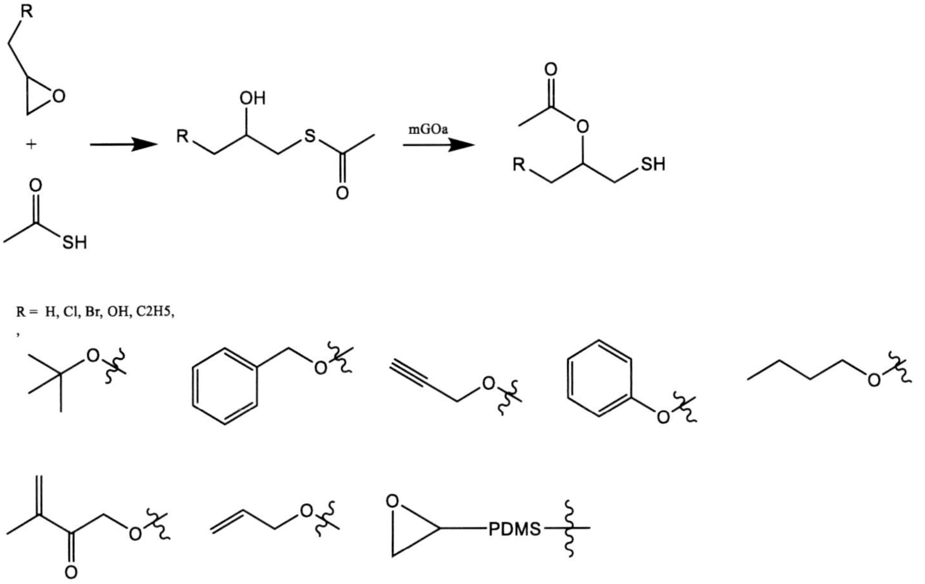

Scheme 13. Acetyl migration reaction of latent thiol compounds into thiol compounds catalyzed by mGO fixed-bed.

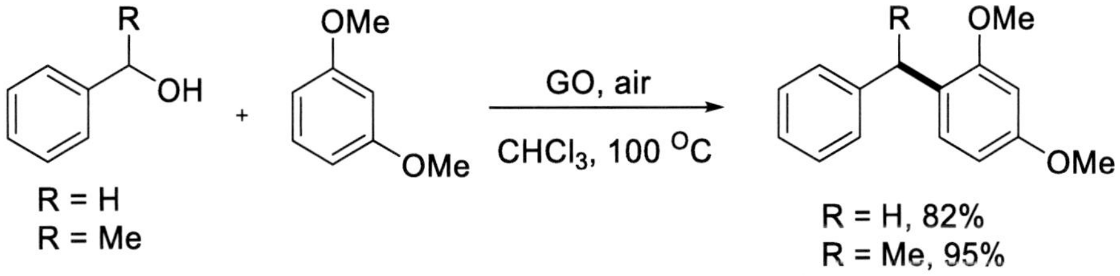

Scheme 14. GO-catalyzed direct Friedel-Crafts alkylation of arenes to alcohols.

Alkylation of arenes to diarylmethanes catalyzed by graphenes has been reported as shown in Fig. (**2**). The activation of the reactants on the catalyst surface is due to the active edges of aromatic polar functional groups on the surface of graphene.

It was observed that RGO can be used as a very effective catalyst in electrochemical and in-situ polymerization reactions of ABA whereas the GO can be used as an active catalyst for in-situ polymerization reaction of ABA only and in the case of electrochemical polymerization RGO, it enhances the formation of ABA rather than GO. Whereas in the case of *in-situ* both the carbocatalysts exhibits good efficacy and the reaction of polymcrization depends on the concentration of monomer and oxidants on the surfaces (Scheme **15**) [77].

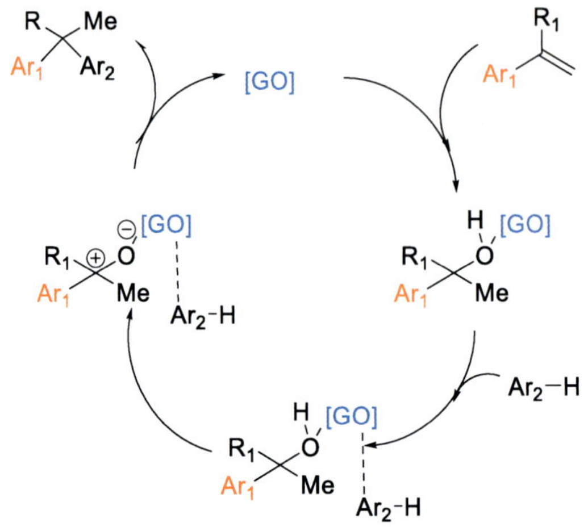

Fig. (2). Reaction pathway for the GO-catalysed alkylation of arenes.

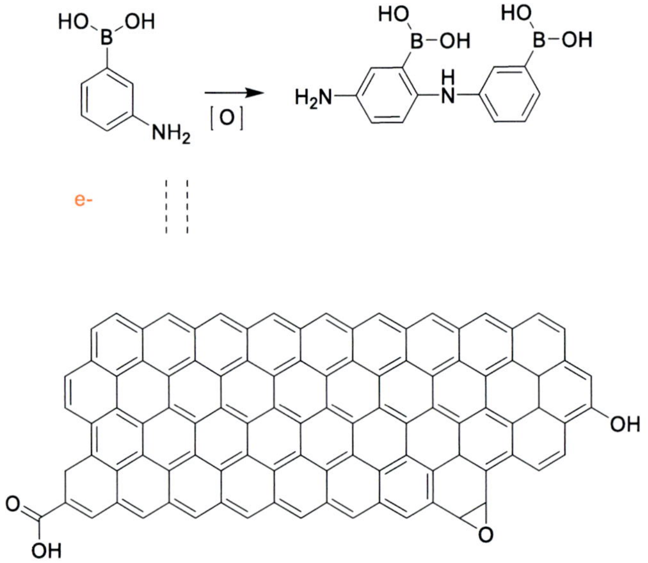

Scheme 15. Schematic illustration of the electron-withdrawing process from the RGO Substrate to ABA.

The preparation of graphene oxide (GO) carbocatalyst by the Hummers method is well-known. A graphite's oxidation in presence of strong acidic conditions and strong oxidizing agents gives GO. This method enhances the oxygen-functionalized groups on the GO surface. GO is can be retrieved back effectively after the reaction process and it is also be used as an acid catalyst for opening the ring of epoxides. The conversion of carbohydrates to 5-ethoxy methyl furfural (EMF) has been reported. The formation of 5-hydroxymethylfurfural (HMF) on etherification with ethanol and GO catalyst gives EMF. The yield of EMF is affected by the loading capacity on the carbocatalyst surface and the temperature. This conversion of bio-mass compounds to useful chemicals using GO as a catalyst would open up a wide area of research (Scheme **16**) [78].

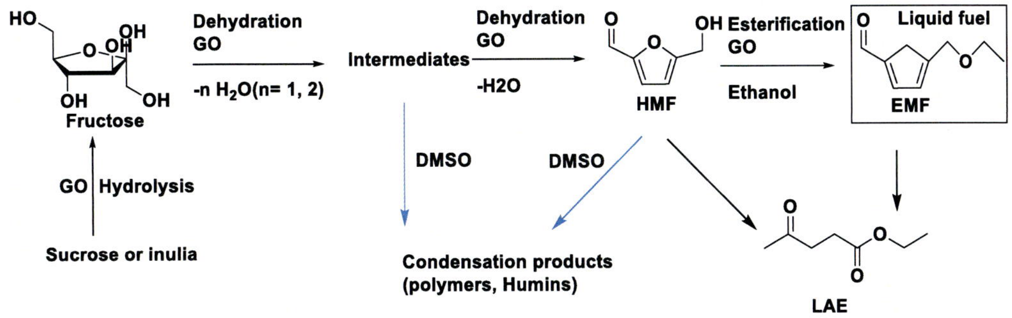

Scheme 16. The conversion of carbohydrates to EMF catalyzed by GO.

3.7. Other Useful Reactions

In a range of applications, graphene-based materials with hierarchical structures and multifunctionality have roused attention. Yang *et al.* reported a simple, yet universal, the method for making graphene-based composite cellular foams (GCCFs) using a mixture of redox reactions. Cationic polymers facilitate assembly and biomimetic mineralization. Cationic polymers (*e.g.*, polyethyleneimine, lysozyme, *etc.*) could not only decrease and assemble graphene oxide (GO) into cellular foams, but they could also induce mineralization, allowing the formation of inorganic nanoparticles (silica, titania, silver, and so on). The GCCFs have a highly porous structure and adequate structural stability, with nanoparticles evenly distributed on the surface of the reduced GO. A series of GCCFs are synthesized by varying polymer/inorganic pairs, which exhibit significantly improved catalytic performance in enzyme catalysis, heterogeneous chemical catalysis, and photocatalysis when compared to nanoparticulate catalysts [79].

Furthermore, graphene oxide is good support for immobilizing lipase in an organic solvent for catalysis. Lipase was immobilized using carboxyl functionalized graphene oxide (GO-COOH). To characterize lipase immobilization, FTIR, UV-vis, and XPS spectra were measured. The immobilized lipase retains 80% of the hydrolytic activity of the native lipase at the optimum temperature of 40 °C. The catalysis efficiency of the immobilized lipase is 1.6 times that of native lipase. when catalyzing the enantioselective reaction in the organic solvent heptane at 50 °C (optimal), and the immobilized lipase retains the native lipase's selectivity [80].

Using a spray-drying approach, Z. Xu *et. al.* developed hollow reduced graphene oxide microspheres containing well-distributed Fe_2O_3 nanoparticles (Fe-rGOS). L-ascorbic acid was used to decrease graphite oxide (GO) and increase electron transfer velocity. The *in situ* Fe_2O_3 nanoparticles with a mean size of 5-10 nm were uniformly deposited on rGO support due to the L-ascorbic acid and spray-drying procedure, and the rGO migrated to the drop's surface to form microspheres. The well-dispersed nanoparticles not only generated more active sites and interface contact, which improved catalyst stability but also acted as pillars between the rGO layers, resulting in a mesoporous structure. The formed mesoporous frameworks significantly improved mass transfer, resulting in significantly higher catalytic efficiency. As a result, the prepared Fe-rGOS demonstrated remarkable photocatalytic activity across a wide pH range, as well as superior recyclability with low iron ion leaching [81].

The detailed syntheses of miscellaneous heterocyclic compounds catalyzed by graphene-based carbocatalysts were very well-reviewed [76]. The Cu–Cu_2O@RGO nanocomposite was studied as a promising reusable catalyst in a multicomponent reaction of halides and sodium azide with terminal alkynes for the synthesis of 1,4-disubstituted 1,2,3-triazoles [82] (Scheme **17**). Strikingly, the merging of RGO and Cu–Cu_2O nanocomposite prevents the aggregation of RGO but also enhances the improved stability and catalytic performance of the metal nanoparticles loaded at RGO.

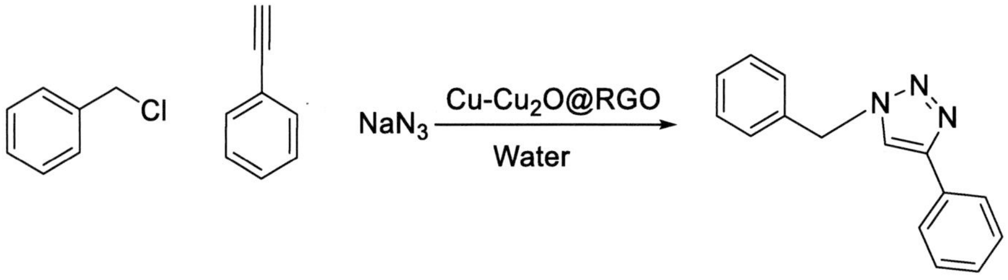

Scheme 17. Cu-Cu_2O@RGO catalyzed synthesis of triazoles by a facile cycloaddition.

Gupta *et al.* [83] described an effective and viable click reaction for the one-pot synthesis of hybrid molecules containing flavanone with triazole by amalgamating Michael addition using copper oxide/reduced graphene oxide (CuO/rGO) nanocomposite (Scheme **18**). A comparison of CuO/rGO with other catalysts was carried out.

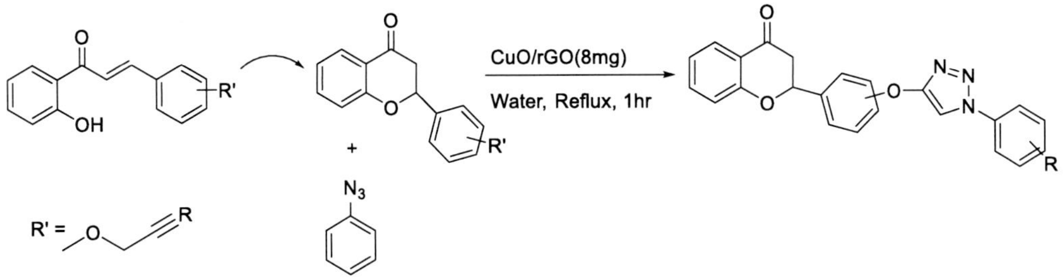

Scheme 18. CuO/rGO catalyzed synthesis of flavanone containing triazole derivatives.

A three-component reaction of sodium azide and alkyl halides in the presence of terminal alkynes was reported [84] by using functionalized graphene oxide copper(I) complex (GO–NH–IA–Cu(I)) under ultrasonic irradiation at room temperature (Scheme **19**).

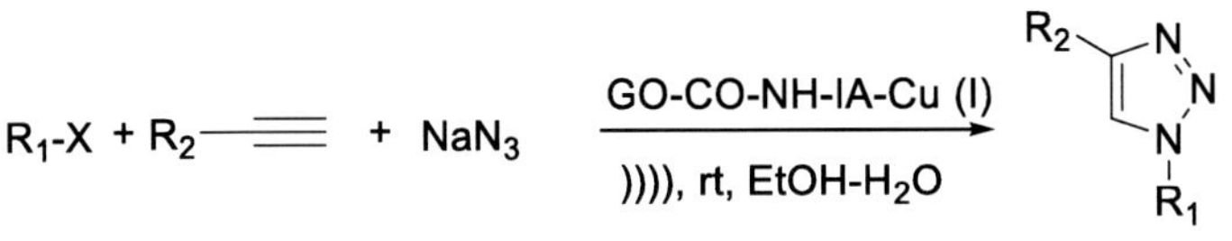

Where X = Cl, Br, I; R_1 = alkyl, benzyl, cinnamyl and R_2 = Ph, CH_3-Ph, $CH_3CH_2CH_3$

Scheme 19. Catalytic synthesis of triazole derivatives by functionalized graphene oxide-Cu(I) complex.

In addition to this, studies on the catalytic synthesis of β-hydroxy-1,2,3-triazole derivatives from aryl-substituted epoxide, phenylacetylene, and sodium azide by GO@polytriazole–Cu (GO@PTA–Cu) mediated by ultrasound irradiation were carried out [85] (Scheme **20**).

Where Ph = C_6H_5, 4-CH_3-C_6H_5 and $C_6H_5OCH_2$

Scheme 20. Catalytic synthesis of β-hydroxy-1,2,3-triazole derivatives using GO@polytriazole-Cu-catalyst.

Furthermore, Xiong *et al.* [86] used GO/Fe_3O_4–CuBr catalyst was employed as a special semi-heterogeneous catalyst for the multicomponent reaction of azide, alkyl halide, and alkyne through 1,3-dipolar cycloaddition by microwave irradiation to yield mono/bis-1,2,3-triazoles (Scheme **21**).

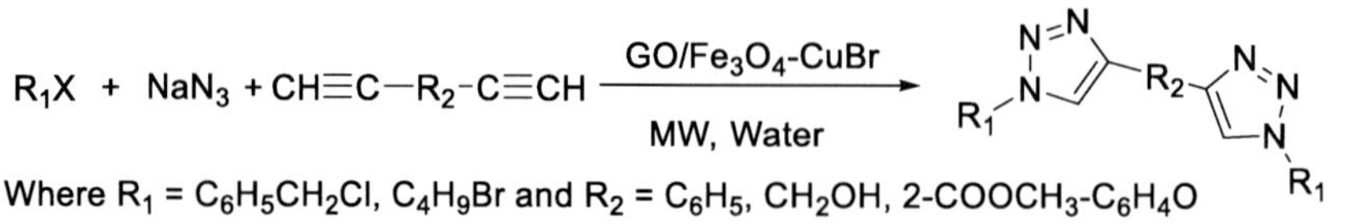

Scheme 21. Microwave-assisted catalytic synthesis of substituted-1,2,3-triazoles using GO/Fe_3O_4-CuBr.

Narayanan *et al.* [87] reported neat, effective, and simple Biginelli reactions of aryl aldehyde, ethyl acetoacetate, and urea in the presence of clay-graphene oxide as recyclable nanocatalyst (montmorillonite clay–GO) (Scheme **22**).

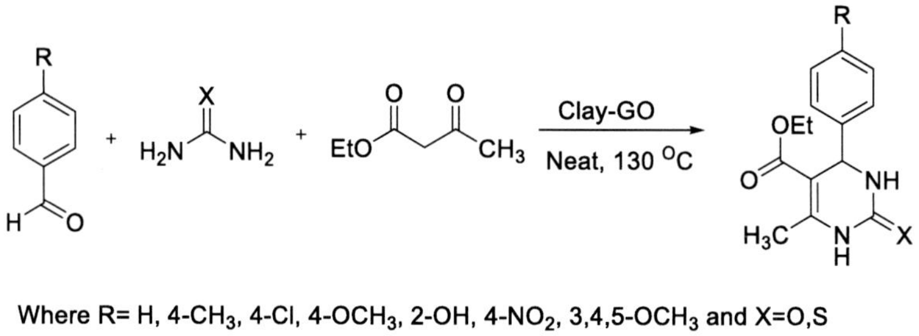

Scheme 22. Clay-GO nanocomposite catalyzed synthesis of 3,4-dihydropyrimidinone.

Similarly, Moitra *et al.* [88] reported a solvent-free preparation of 3,4-dihydropyrimidin-2(1H)-ones (Scheme **23**) in presence of $BiFeO_3$ nanowire-reduced graphene oxide (BFO-RGO) nanocatalyst. Furthermore, Akocak *et al.* [89] established an eco-friendly method for the synthesis of 2-amino--H-chromene derivatives *via* condensation of resorcinol, malononitrile, and diverse aromatic aldehydes in the presence of palladium nanoparticles loaded on graphene oxide (Pd@GO) (Scheme **24**).

Investigations were carried out on the usage of magnetically separable nanocatalyst, Fe_3O_4-supported-N-pyridin-4-amine-grafted graphene oxide for the synthesis of chromenes from a multicomponent reaction of malononitrile, dimedone, and aryl aldehydes [90] (Scheme **25**).

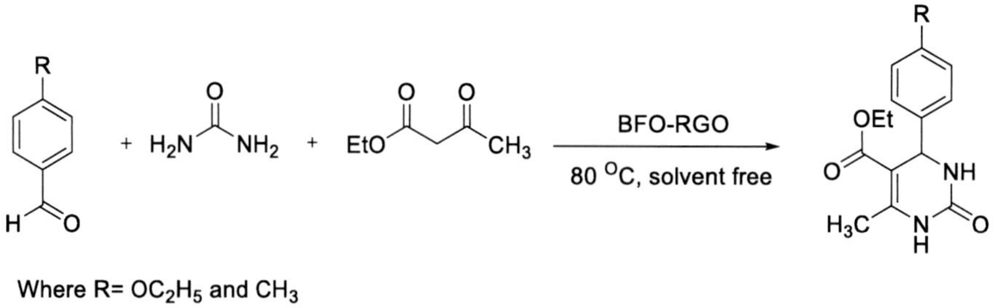

Scheme 23. Biginelli reaction catalyzed by $BiFeO_3$-reduced graphene oxide nanocatalyst.

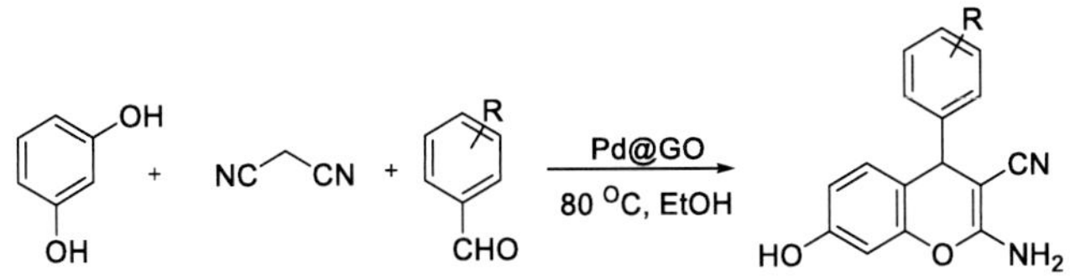

Where R= H, 4-CH_3, 4-$CH(CH_3)_2$, 4-CN, 4-OCH_3, 4-$N(CH_3)_2$, 3,4-OCH_3, 3,4,5-OCH_3, 3,5-CH_3 and 2,3,5,6-F

Scheme 24. Catalytic synthesis of 2-amino-4*H*-chromene by Pd@GO nanoparticles.

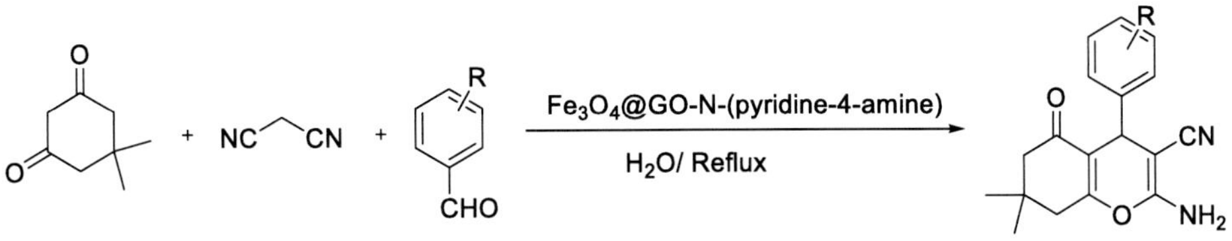

Where R= H, 4-F, 4-CH_3, 4-Cl, 2-Cl, 3-NO_2, 4-NO_2 and 4-OH

Scheme 25. Fe_3O_4@GO-*N*-(Pyridin-4-amine) catalytic synthesis of chromenes.

A multi-component reaction of aryl glyoxals with 4-hydroxycoumarin, and malononitrile was reported using metal-free graphene oxide (GO) nanosheets [91] (Scheme **26**).

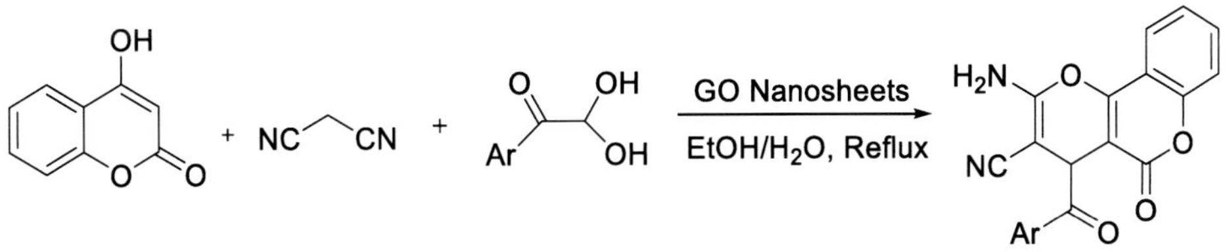

Where Ar= C_6H_5, 1-naphthyl, 2-Cl-C_6H_4, 4-NO_2-C_6H_4, 4-Br-C_6H_4 and 4-CH_3O-C_6H_4

Scheme 26. GO nanosheets catalyzed preparation of pyranocoumarins.

A four-component reaction of phenylhydrazine, aldehydes, ethyl acetoacetate, and 4-hydroxycoumarin was demonstrated for the preparation of highly functionalized benzylpyrazolyl coumarins employing mesoporous/biogenic graphene oxide nanoparticles by Siddiqui *et al.* [92] (Scheme **27**).

The catalytic preparation of a variety of 1-H spiro[isoindoline-1,2- quinazoline]-3,4-(3-H)-diones was reported by the reaction of aliphatic, or aromatic primary amines with isatoic anhydride in the presence of isatin using acidic magnetic graphene oxide nanosheets (Fe_3O_4–GO SO$_3$H) by Abdi *et al.* [93] (Scheme **28**).

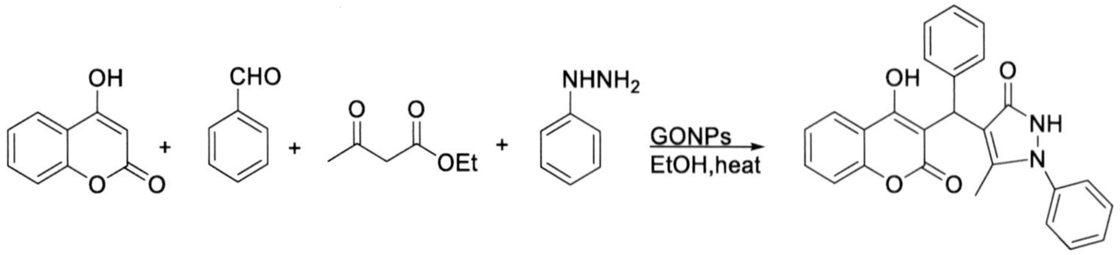

Scheme 27. GONPs catalyzed synthesis of benzyl pyrazolyl coumarins.

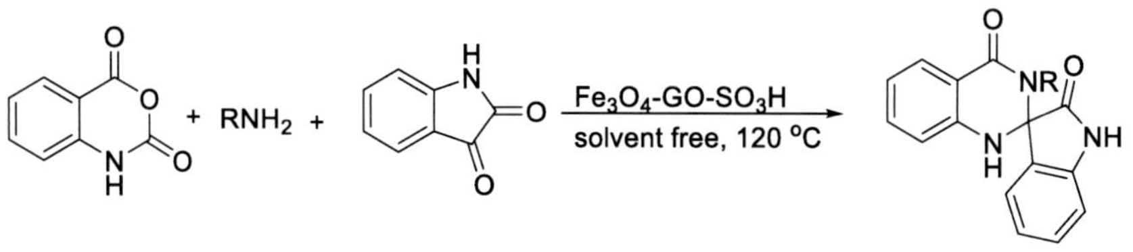

Where R= H, cyclohexyl, Ph, n-Bu, 4-CH_3-C_6H_4, 4-CH_{30}-C_6H_4, 3-CH_3O-C_6H_4, 4-Br-C_6H_4, 3-Cl-C_6H_4, 4-NO_2-C_6H_4

Scheme 28. Catalytic synthesis of 1-H spiro[isoindoline-1,2-quinazoline]-3,4-(3-*H*)-diones by Fe_3O_4-GO-SO_3H.

A tandem synthesis of benzo[b] [1, 4]diazepine derivatives was performed [94] by the reaction of ethyl-3-oxobutanoate,2-nitroaniline, isocyanide, and $TMSN_3$ as substrates using the bimetallic nanoparticles (AuCu and AgCu) supported on guanidine-grafted-rGO nanosheets (AuCu@G-rGO and AgCu@G-rGO) as catalysts and (Scheme **29**).

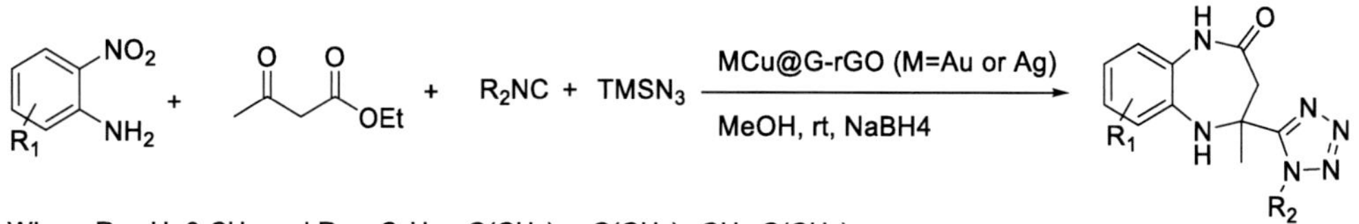

Where R_1= H, 3-CH_3 and R_2 = C_6H_{11}, $C(CH_3)_3$, -$C(CH_3)_2$-CH_2-$C(CH_3)_3$

Scheme 29. AuCu@G-rGO/AgCu@G-rGOcatalyzedsynthesis of benzo[b] [1, 4]diazepine derivatives.

Kausar *et al.* [95] reported a metal-free carbocatalyst to prepare a variety of dibenzo [1, 4] diazepine scaffolds from *o*-phenylene diamine, aldehydes, or ketones, and 1,3-diketo compounds (Scheme **30**) employing graphene oxide (GO) nanosheets as carbocatalyst to producedibenzo [1, 4]diazepines.

Several carbocatalysts have been examined for PMS activation for the oxidation of phenol in an aqueous medium. In absence of carbocatalyst, the PMS could unable to generate activated radicals for phenol degradation. In parallel, without PMS addition, hardly 4% of phenol was removed because of adsorption on the NG-700 surface.

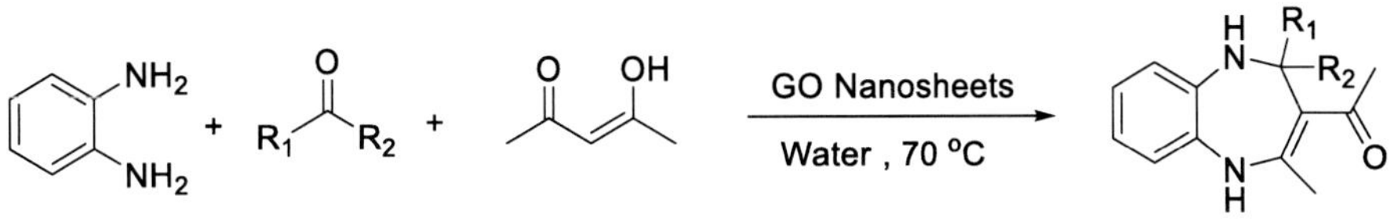

Where R_1 = H, CH_3 and R_2 = aryl, heteroaryl, alkyl and benzoyl

Scheme 30. GO nanosheet catalyzed preparation of dibenzo [1, 4]diazepines.

The GO and RGO catalysts usually are not that stable or soluble in an aqueous medium and there is a need for optimization of this catalytic activity of GO and RGO by incorporating a polymer PSSS (poly(sodium 4-styrene sulfonate)) which enhances the dispersion of this catalyst in an aqueous medium. PSSS is water-soluble which is incorporated into the graphene surface through a radical initiator (ATRP). ATRP use is that it facilitates the incorporation of PSSS on the graphene surface. Once this GO-g-PSSS catalyst is ready, it can be used as the catalyst for various chemical reactions. In the same way, RGO-*g*-PSSS is also improvised which also helps in the synthesis of isoamyl benzoate (Scheme **31**). Also, the PSSS is pH-sensitive, such that the introduction of any drugs to the surface can be released concerning the drug's response to the change of pH. This method is used to improvise the dispersion of graphene oxide and reduced graphene oxide in the aqueous solution [96].

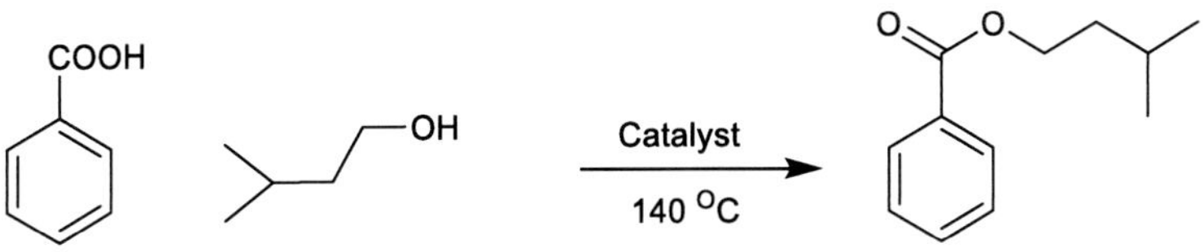

Scheme 31. Preparation of isoamyl benzoate.

CONCLUSION AND FUTURE PROSPECTS

With the development of several advanced catalytic systems for diverse chemical processes, carbocatalysts are one of the efficient classes of heterogeneous materials of high industrial significance. Several GO based catalysts employed for a wide range of organic transformations reported in the past are summarised in Table **5**. Owing to their remarkable properties, including high intrinsic mobility, large specific surface area, excellent electrical conductivity, high thermal conductivity, ease of preparation, chemical inertness, and facile functionalization, graphene-based materials have become promising recyclable and sustainable catalysts. In the light of this, the present chapter highlights the preparation of different types of carbocatalysts, the possible reaction mechanisms, and multicomponent organic transformations. Various graphene-based materials (graphene oxide/ reduced graphene oxide/ their nanosheets/metal-nanoparticles grafted on graphene/ functionalized graphene oxide/ quantum dots/ graphene oxide-spinel composites) have been employed for the preparation of a variety of heterocyclic molecules. Relatively faster synthesis of thiophenes, triazoles, aziridines, tetrazoles, pyrazoles, imidazoles, furans, oxindoles, pyrimidines, chromenes, coumarins, quinazolines, oxazoles, quinoxalines, xanthenes, pyridines, and diazepines was reported in good to excellent yields. The high catalytic activity of graphene-based catalysts is mainly attributed to the incorporation of Brønsted acid sites and/ or Lewis acid-base centers in these carbon materials, accompanied by π–π stacking and electrostatic interactions through the increased adsorption of reactants. Additionally, the advantage of these carbocatalysts is that they are very easy to recover by filtration/centrifugation and/or magnetic decantation. Moreover, the low catalyst loading, renewability, and solvent-free conditions make them very special catalytic systems concerning sustainability and environmentally benign processes. The major limitations of these graphene-based carbocatalysts are the identification of the actual reactive sites of graphene oxide, functionalities, defects, and edges. These problems can be overcome when suitable models are designed describing the structure of graphene-based materials at the molecular level. Still, better graphene-based carbocatalysts need to be developed with less complexity and cost-effectiveness by doping/ double doping with B/ B-N/ P-N and other appropriate dopants, as evident in the facile activation of molecular hydrogen at the periphery and holes with zigzag edges. This will be a promising field for future applications of catalysis by graphene-based materials.

Table 5. Previously reported graphene oxide catalyzed reactions.

S.no	Catalyst	Reaction	Reference
1	rGO	Oxidation of methylene blue, phenol, 2,4-dichlorophenol by H_2O_2	[97]
2	(B, N) doped rGO	Oxidation of phenol by peroxymonosulfate or H_2O_2	[98]
3	rGO	Hydrogenation	[99-100]
4	rGO or N-rGO	Reduction of 4-nitrophenol to 4-aminophenol by $NaBH_4$	[101-102]
5	rGO	Synthesis of dipyrromethene, calix [4] pyrrole and 5-substituted 1H-tetrazoles	[103-104]
6	rGO	Thermal dehydrogenation and re-hydrogenation of $LiBH_4$	[105]
7	rGO-NEt_3	Hydrolysis of ethyl acetate	[106]
8	rGO-PAMAM	Knoevenagel condensation	[107]
9	rGO-SO_3H	Dehydration of xylose to furfural, Esterification and transesterification, pechmann reaction, hydration of propylene oxide	[108-110]

(Table 5) cont.....

10	rGO-poly(styrene sulfonic acid)	Synthesis of isoamyl benzoate	[111]
11	rGO-Ar-SO_3H	Hydrolysis of ethyl acetate, synthesis of xanthenes and benzoxanthenes	[112-113]
12	GO	Aerobic oxidation of alcohols	[114-115]
13	GO	Dehydrogenation of propane to propene	[116]
14	GO	Aerobic oxidation of SO_2	[117]
15	GO or rGO	Polymerization of 3-aminophenylboronic acid	[118]
16	GO	Ring opening of epoxides	[119]
17	GO	Michael addition reaction	[120]
18	GO	Aza-michael addition of amines to alkenes	[121]
19	GO	Dehydrative esterification of benzyl alcohol	[122]

LIST OF ABBREVIATIONS

AC Activated Carbon

AA L-Ascorbic Acid

ABA Azo Benzoic Acid

ATRP Atom Transfer Radical Polymerization

ASC Ascorbic acid

CDG Chemically derived Graphene

CVD Chemical Vapour Deposition

DH Dopamine Hydrochloride

DCP Dichlorophenol

DH Dehydrogenation

DMSO Dimethyl Sulfoxide

EG Ethyleneglycol

EMF 5-Ethoxy Methyl Furfural

GO Graphene Oxide

GP Graphite Powder

GCCF Graphene Composite Cellular Foam

HMF 5-Hydroxy Methyl Furfural

HCCP Hexachlorocyclotriphosphazene

HQ Hydroquinone

HZ Hydrazine

HER Hydrogen Evolution Reaction

ME	2-Mercapto Ethanol
MWCNT	Multi Wall Carbon Nanotube
MB	Methylene Blue
MEG	Microwave Exfoliated Graphite
MGN	Metal Graphene Nanocomposite
NPG	N&P double doped Graphene
[(N)G]	N-doped defective Graphene
ODH	Oxidative Dehydrogenation
PMS	Peroxymonosulphate
PSSS	[Poly(Sodium-4-Styrene Sulfonate)]
rGO	Reduced Graphene Oxide
rGO-PAMAM	Poly(amidoamine)-modified Reduced Graphene Oxide
SA-GO	Sulfur Graphene Oxide
TBHP	Tert-Butyl Hydroperoxide
$TMSN_3$	Trimethylsilylazide
TOF	Turn Over Frequency

CONSENT FOR PUBLICATION

Not applicable.

CONFLICT OF INTEREST

The author declares no conflict of interest, financial or otherwise.

ACKNOWLEDGEMENTS

Declared none.

REFERENCES

[1] a) Stratakis, M.; Garcia, H. Catalysis by supported gold nanoparticles: beyond aerobic oxidative processes. *Chem. Rev.*, **2012**, *112*(8), 4469-4506. [http://dx.doi.org/10.1021/cr3000785] [PMID: 22690711] b) Sivaramakrishna, A.; Clayton, H.S.; Mogorosi, M.M.; Moss, J.R. Hydrocarbon (π- and σ-) complexes of nickel, palladium and platinum: Synthesis, reactivity and applications. *Coord. Chem. Rev.*, **2010**, *254*(23-24), 2904-2932. [http://dx.doi.org/10.1016/j.ccr.2010.05.021] c) Sivaramakrishna, A.; Suman, P.; Veerashekhar Goud, E.; Janardan, S.; Sravani, C.; Sandeep, T.; Vijayakrishna, K.; Clayton, H.S. Review: active homogeneous reagents and catalysts in *n* -alkane activation. *J. Coord. Chem.*, **2013**, *66*(12), 2091-2109. [http://dx.doi.org/10.1080/00958972.2013.797572]

[2] Sherry, B.D.; Fürstner, A. The promise and challenge of iron-catalyzed cross coupling. *Acc. Chem. Res.*, **2008**, *41*(11), 1500-1511. [http://dx.doi.org/10.1021/ar800039x] [PMID: 18588321]

[3] Jana, R.; Pathak, T.P.; Sigman, M.S. Advances in transition metal (Pd, Ni, Fe)-catalyzed cross-coupling reactions using alkyl-organometallics as reaction partners. *Chem. Rev.*, **2011**, *111*(3), 1417-1492.
[http://dx.doi.org/10.1021/cr100327p] [PMID: 21319862]

[4] a) Punniyamurthy, T.; Velusamy, S.; Iqbal, J. Recent advances in transition metal catalyzed oxidation of organic substrates with molecular oxygen. *Chem. Rev.*, **2005**, *105*(6), 2329-2364.
[http://dx.doi.org/10.1021/cr050523v] [PMID: 15941216] b) Yan, Y.; Xia, B.Y.; Zhao, B.; Wang, X. A review on noble-metal-free bifunctional heterogeneous catalysts for overall electrochemical water splitting. *J. Mater. Chem. A Mater. Energy Sustain.*, **2016**, *4*(45), 17587-17603.
[http://dx.doi.org/10.1039/C6TA08075H]

[5] a) Dömling, A.; Wang, W.; Wang, K. Chemistry and biology of multicomponent reactions. *Chem. Rev.*, **2012**, *112*(6), 3083-3135.
[http://dx.doi.org/10.1021/cr100233r] [PMID: 22435608] b) Reddy, M.M.; Sivaramakrishna, A. Remarkably flexible quinazolinones—synthesis and biological applications. *J. Heterocycl. Chem.*, **2020**, *57*(3), 942-954.
[http://dx.doi.org/10.1002/jhet.3844] c) Rocha, R.O.; Rodrigues, M.O.; Neto, B.A.D. Review on the Ugi Multicomponent Reaction Mechanism and the Use of Fluorescent Derivatives as Functional Chromophores. *ACS Omega*, **2020**, *5*(2), 972-979.
[http://dx.doi.org/10.1021/acsomega.9b03684] [PMID: 31984252] d) José Climent, M.; Corma, A.; Iborra, S. Homogeneous and heterogeneous catalysts for multicomponent reactions. *RSC Advances*, **2012**, *2*(1), 16-58.
[http://dx.doi.org/10.1039/C1RA00807B] e) Saranya, S.; Rohit, K.R.; Radhika, S.; Anilkumar, G. Palladium-catalyzed multicomponent reactions: an overview. *Org. Biomol. Chem.*, **2019**, *17*(35), 8048-8061.
[http://dx.doi.org/10.1039/C9OB01538H] [PMID: 31410440]

[6] Titirici, M.M.; Antonietti, M. Chemistry and materials options of sustainable carbon materials made by hydrothermal carbonization. *Chem. Soc. Rev.*, **2010**, *39*(1), 103-116.
[http://dx.doi.org/10.1039/B819318P] [PMID: 20023841]

[7] Schaetz, A.; Zeltner, M.; Stark, W.J. Carbon Modifications and Surfaces for Catalytic Organic Transformations. *ACS Catal.*, **2012**, *2*(6), 1267-1284.
[http://dx.doi.org/10.1021/cs300014k]

[8] Dreyer, D.R.; Bielawski, C.W. Carbocatalysis: Heterogeneous carbons finding utility in synthetic chemistry. *Chem. Sci. (Camb.)*, **2011**, *2*(7), 1233-1240.
[http://dx.doi.org/10.1039/c1sc00035g]

[9] Su, C.; Loh, K.P. Carbocatalysts: graphene oxide and its derivatives. *Acc. Chem. Res.*, **2013**, *46*(10), 2275-2285.
[http://dx.doi.org/10.1021/ar300118v] [PMID: 23270430]

[10] Pyun, J. Graphene oxide as catalyst: application of carbon materials beyond nanotechnology. *Angew. Chem. Int. Ed.*, **2011**, *50*(1), 46-48.
[http://dx.doi.org/10.1002/anie.201003897] [PMID: 20979072]

[11] Navalon, S.; Dhakshinamoorthy, A.; Alvaro, M.; Garcia, H. Heterogeneous fenton catalysts based on activated carbon and related materials. *ChemSusChem*, **2011**, *4*(12), 1712-1730.
[http://dx.doi.org/10.1002/cssc.201100216] [PMID: 22162405]

[12] Su, D.S.; Perathoner, S.; Centi, G. Nanocarbons for the development of advanced catalysts. *Chem. Rev.*, **2013**, *113*(8), 5782-5816.
[http://dx.doi.org/10.1021/cr300367d] [PMID: 23721498]

[13] Rao, C.N.R.; Sood, A.K.; Subrahmanyam, K.S.; Govindaraj, A. Graphene: the new two-dimensional nanomaterial. *Angew. Chem. Int. Ed.*, **2009**, *48*(42), 7752-7777.
[http://dx.doi.org/10.1002/anie.200901678] [PMID: 19784976]

[14] a) Blanita, G.; Lazar, M.D. Mater. Review of graphene-supported metal nanoparticles as new and efficient heterogeneous catalysts. *Micro Nanosyst.,* **2013**, *5*(2), 138-146. [http://dx.doi.org/10.2174/1876402911305020009] b) Zhu, Y.; Murali, S.; Stoller, M.D.; Velamakanni, A.; Piner, R.D.; Ruoff, R.S. Microwave assisted exfoliation and reduction of graphite oxide for ultracapacitors. *Carbon,* **2010**, *48*(7), 2118-2122. [http://dx.doi.org/10.1016/j.carbon.2010.02.001] c) Xi, J.; Wang, Q.; Liu, J.; Huan, L.; He, Z.; Qiu, Y.; Zhang, J.; Tang, C.; Xiao, J.; Wang, S. N,P-dual-doped multilayer graphene as an efficient carbocatalyst for nitroarene reduction: A mechanistic study of metal-free catalysis. *J. Catal.,* **2018**, *359*, 233-241. [http://dx.doi.org/10.1016/j.jcat.2018.01.003]

[15] Dreyer, D.R.; Park, S.; Bielawski, C.W.; Ruoff, R.S. The chemistry of graphene oxide. *Chem. Soc. Rev.,* **2010**, *39*(1), 228-240. [http://dx.doi.org/10.1039/B917103G] [PMID: 20023850]

[16] Jia, X.; Campos-Delgado, J.; Terrones, M.; Meunier, V.; Dresselhaus, M.S. Graphene edges: a review of their fabrication and characterization. *Nanoscale,* **2011**, *3*(1), 86-95. [http://dx.doi.org/10.1039/C0NR00600A] [PMID: 21103548]

[17] Kuila, T.; Bose, S.; Mishra, A.K.; Khanra, P.; Kim, N.H.; Lee, J.H. Chemical functionalization of graphene and its applications. *Prog. Mater. Sci.,* **2012**, *57*(7), 1061-1105. [http://dx.doi.org/10.1016/j.pmatsci.2012.03.002]

[18] Soldano, C.; Mahmood, A.; Dujardin, E. Production, properties and potential of graphene. *Carbon,* **2010**, *48*(8), 2127-2150. [http://dx.doi.org/10.1016/j.carbon.2010.01.058]

[19] Chua, C.K.; Pumera, M. Covalent chemistry on graphene. *Chem. Soc. Rev.,* **2013**, *42*(8), 3222-3233. [http://dx.doi.org/10.1039/c2cs35474h] [PMID: 23403465]

[20] Georgakilas, V.; Otyepka, M.; Bourlinos, A.B.; Chandra, V.; Kim, N.; Kemp, K.C.; Hobza, P.; Zboril, R.; Kim, K.S. Functionalization of graphene: covalent and non-covalent approaches, derivatives and applications. *Chem. Rev.,* **2012**, *112*(11), 6156-6214. [http://dx.doi.org/10.1021/cr3000412] [PMID: 23009634]

[21] Chen, L.; Hernandez, Y.; Feng, X.; Müllen, K. From nanographene and graphene nanoribbons to graphene sheets: chemical synthesis. *Angew. Chem. Int. Ed.,* **2012**, *51*(31), 7640-7654. [http://dx.doi.org/10.1002/anie.201201084] [PMID: 22777811]

[22] Chua, C.K.; Pumera, M. Chemical reduction of graphene oxide: a synthetic chemistry viewpoint. *Chem. Soc. Rev.,* **2014**, *43*(1), 291-312. [http://dx.doi.org/10.1039/C3CS60303B] [PMID: 24121318]

[23] Singh, V.; Joung, D.; Zhai, L.; Das, S.; Khondaker, S.I.; Seal, S. Graphene based materials: Past, present and future. *Prog. Mater. Sci.,* **2011**, *56*(8), 1178-1271. [http://dx.doi.org/10.1016/j.pmatsci.2011.03.003]

[24] Tunckol, M.; Durand, J.; Serp, P. Carbon nanomaterial–ionic liquid hybrids. *Carbon,* **2012**, *50*(12), 4303-4334. [http://dx.doi.org/10.1016/j.carbon.2012.05.017]

[25] Krishnan, D.; Kim, F.; Luo, J.; Cruz-Silva, R.; Cote, L.J.; Jang, H.D.; Huang, J. Energetic graphene oxide: Challenges and opportunities. *Nano Today,* **2012**, *7*(2), 137-152. [http://dx.doi.org/10.1016/j.nantod.2012.02.003]

[26] Hur, S.H.; Park, J.N. Graphene and its application in fuel cell catalysis: a review. *Asia-Pac. J. Chem. Eng.,* **2013**, *8*(2), 218-233. [http://dx.doi.org/10.1002/apj.1676]

[27] Malig, J.; Jux, N.; Guldi, D.M. Toward multifunctional wet chemically functionalized graphene-integration of oligomeric, molecular, and particulate building blocks that reveal photoactivity and

redox activity. *Acc. Chem. Res.*, **2013**, *46*(1), 53-64.
[http://dx.doi.org/10.1021/ar300124z] [PMID: 22916796]

[28] Rodríguez-Pérez, L.; Herranz, M.Á.; Martín, N. The chemistry of pristine graphene. *Chem. Commun. (Camb.)*, **2013**, *49*(36), 3721-3735.
[http://dx.doi.org/10.1039/c3cc38950b] [PMID: 23539514]

[29] James, D.K.; Tour, J.M. Graphene: powder, flakes, ribbons, and sheets. *Acc. Chem. Res.*, **2013**, *46*(10), 2307-2318.
[http://dx.doi.org/10.1021/ar300127r] [PMID: 23276286]

[30] Zhang, Y.; Zhang, L.; Zhou, C. Review of chemical vapor deposition of graphene and related applications. *Acc. Chem. Res.*, **2013**, *46*(10), 2329-2339.
[http://dx.doi.org/10.1021/ar300203n] [PMID: 23480816]

[31] Wang, H.; Yuan, X.; Wu, Y.; Huang, H.; Peng, X.; Zeng, G.; Zhong, H.; Liang, J.; Ren, M. M. Graphene-based materials: fabrication, characterization and application for the decontamination of wastewater and wastegas and hydrogen storage/generation. *Adv. Colloid Interface Sci*, **2013**, 195–196-19-40.

[32] Ciesielski, A.; Samorì, P. Graphene *via* sonication assisted liquid-phase exfoliation. *Chem. Soc. Rev.*, **2014**, *43*(1), 381-398.
[http://dx.doi.org/10.1039/C3CS60217F] [PMID: 24002478]

[33] Loh, K.P.; Bao, Q.; Ang, P.K.; Yang, J. The chemistry of graphene. *J. Mater. Chem.*, **2010**, *20*(12), 2277-2289.
[http://dx.doi.org/10.1039/b920539j]

[34] Brownson, D.A.C.; Kampouris, D.K.; Banks, C.E. Graphene electrochemistry: fundamental concepts through to prominent applications. *Chem. Soc. Rev.*, **2012**, *41*(21), 6944-6976.
[http://dx.doi.org/10.1039/c2cs35105f] [PMID: 22850696]

[35] Zhu, Y.; Murali, S.; Cai, W.; Li, X.; Suk, J.W.; Potts, J.R.; Ruoff, R.S. Graphene and graphene oxide: synthesis, properties, and applications. *Adv. Mater.*, **2010**, *22*(35), 3906-3924.
[http://dx.doi.org/10.1002/adma.201001068] [PMID: 20706983]

[36] Shen, J.; Zhu, Y.; Yang, X.; Li, C. Graphene quantum dots: emergent nanolights for bioimaging, sensors, catalysis and photovoltaic devices. *Chem. Commun. (Camb.)*, **2012**, *48*(31), 3686-3699.
[http://dx.doi.org/10.1039/c2cc00110a] [PMID: 22410424]

[37] Dirian, K.; Herranz, M.Á.; Katsukis, G.; Malig, J.; Rodríguez-Pérez, L.; Romero-Nieto, C.; Strauss, V., Martín, N.; Guldi, D.M. Low dimensional nanocarbons – chemistry and energy/electron transfer reactions. *Chem. Sci. (Camb.)*, **2013**, *4*(12), 4335-4353.
[http://dx.doi.org/10.1039/c3sc51100f]

[38] Agnoli, S.; Granozzi, G. Second generation graphene: Opportunities and challenges for surface science. *Surf. Sci.*, **2013**, *609*, 1-5.
[http://dx.doi.org/10.1016/j.susc.2012.11.016]

[39] Li, N.; Cao, M.; Hu, C. Review on the latest design of graphene-based inorganic materials. *Nanoscale*, **2012**, *4*(20), 6205-6218.
[http://dx.doi.org/10.1039/c2nr31750h] [PMID: 22961110]

[40] Wang, H.X.; Wang, Q.; Zhou, K.G.; Zhang, H.L. Graphene in light: design, synthesis and applications of photo-active graphene and graphene-like materials. *Small*, **2013**, *9*(8), 1266-1283.
[http://dx.doi.org/10.1002/smll.201203040] [PMID: 23554268]

[41] Zheng, Y.; Liu, J.; Liang, J.; Jaroniec, M.; Qiao, S.Z. Graphitic carbon nitride materials: controllable synthesis and applications in fuel cells and photocatalysis. *Energy Environ. Sci.*, **2012**, *5*(5), 6717-6731.
[http://dx.doi.org/10.1039/c2ee03479d]

[42] Sharma, S.; Pollet, B.G. Support materials for PEMFC and DMFC electrocatalysts—A review. *J.*

Power Sources, **2012**, *208*, 96-119.
[http://dx.doi.org/10.1016/j.jpowsour.2012.02.011]

[43] Candelaria, S.L.; Shao, Y.; Zhou, W.; Li, X.; Xiao, J.; Zhang, J.G.; Wang, Y.; Liu, J.; Li, J.; Cao, G. Nanostructured carbon for energy storage and conversion. *Nano Energy,* **2012**, *1*(2), 195-220.
[http://dx.doi.org/10.1016/j.nanoen.2011.11.006]

[44] Cheng, F.; Chen, J. Metal–air batteries: from oxygen reduction electrochemistry to cathode catalysts. *Chem. Soc. Rev.,* **2012**, *41*(6), 2172-2192.
[http://dx.doi.org/10.1039/c1cs15228a] [PMID: 22254234]

[45] Pumera, M. Graphene-based nanomaterials and their electrochemistry. *Chem. Soc. Rev.,* **2010**, *39*(11), 4146-4157.
[http://dx.doi.org/10.1039/c002690p] [PMID: 20623061]

[46] Chen, D.; Tang, L.; Li, J. Graphene-based materials in electrochemistry. *Chem. Soc. Rev.,* **2010**, *39*(8), 3157-3180.
[http://dx.doi.org/10.1039/b923596e] [PMID: 20589275]

[47] Chen, D.; Feng, H.; Li, J. Graphene oxide: preparation, functionalization, and electrochemical applications. *Chem. Rev.,* **2012**, *112*(11), 6027-6053.
[http://dx.doi.org/10.1021/cr300115g] [PMID: 22889102]

[48] a) Xiang, Q.; Yu, J.; Jaroniec, M. Graphene-based semiconductor photocatalysts. *Chem. Soc. Rev.,* **2012**, *41*(2), 782-796.
[http://dx.doi.org/10.1039/C1CS15172J] [PMID: 21853184] b) Sun, Z.; Ye, Q.; Chi, C.; Wu, J. Low band gap polycyclic hydrocarbons: from closed-shell near infrared dyes and semiconductors to open-shell radicals. *Chem. Soc. Rev.,* **2012**, *41*(23), 7857-7889.
[http://dx.doi.org/10.1039/c2cs35211g] [PMID: 22918297]

[49] Al Kausor, M.; Chakrabortty, D. Graphene oxide based semiconductor photocatalysts for degradation of organic dye in waste water: A review on fabrication, performance enhancement and challenges. *Inorg. Chem. Commun.,* **2021**, *129*, 108630.
[http://dx.doi.org/10.1016/j.inoche.2021.108630]

[50] Gupta, S.; Banu, R.; Ameta, C.; Ameta, R.; Punjabi, P.B. Emerging Trends in the Syntheses of Heterocycles Using Graphene-based Carbocatalysts: An Update. *Top. Curr. Chem. (Cham),* **2019**, *377*(3), 13.
[http://dx.doi.org/10.1007/s41061-019-0238-3] [PMID: 31054016]

[51] Li, X.; Duan, X.; Han, C.; Fan, X.; Li, Y.; Zhang, F.; Zhang, G.; Peng, W.; Wang, S. Chemical activation of nitrogen and sulfur co-doped graphene as defect-rich carbocatalyst for electrochemical water splitting. *Carbon,* **2019**, *148*, 540-549.
[http://dx.doi.org/10.1016/j.carbon.2019.04.021]

[52] Hummers, W.S., Jr; Offeman, R.E. Preparation of Graphitic Oxide. *J. Am. Chem. Soc.,* **1958**, *80*(6), 1339-1339.
[http://dx.doi.org/10.1021/ja01539a017]

[53] Duan, X.; Ao, Z.; Sun, H.; Indrawirawan, S.; Wang, Y.; Kang, J.; Liang, F.; Zhu, Z.H.; Wang, S. Nitrogen-doped graphene for generation and evolution of reactive radicals by metal-free catalysis. *ACS Appl. Mater. Interfaces,* **2015**, *7*(7), 4169-4178.
[http://dx.doi.org/10.1021/am508416n] [PMID: 25632991]

[54] Dreyer, D.R.; Jia, H.P.; Bielawski, C.W. Graphene Oxide: A Convenient Carbocatalyst for Facilitating Oxidation and Hydration Reactions. Angew. Chem. Intern. Ed. , **2010**; pp. 6813-6815.

[55] Yang, J.H.; Sun, G.; Gao, Y.; Zhao, H.; Tang, P.; Tan, J.; Lu, A.H.; Ma, D. Direct catalytic oxidation of benzene to phenol over metal-free graphene-based catalyst. *Energy Environ. Sci.,* **2013**, *6*(3), 793-798.
[http://dx.doi.org/10.1039/c3ee23623d]

[56] Dhakshinamoorthy, A.; Latorre-Sanchez, M.; Asiri, A.M.; Primo, A.; Garcia, H. Sulphur-doped graphene as metal-free carbocatalysts for the solventless aerobic oxidation of styrenes. *Catal. Commun.*, **2015**, *65*, 10-13.
[http://dx.doi.org/10.1016/j.catcom.2015.02.018]

[57] Mittal, R.; Kumar, A.; Awasthi, S.K. Practical scale up synthesis of carboxylic acids and their bioisosteres 5-substituted-1 *H*-tetrazoles catalyzed by a graphene oxide-based solid acid carbocatalyst. *RSC Advances,* **2021**, *11*(19), 11166-11176.
[http://dx.doi.org/10.1039/D1RA01053K] [PMID: 35423636]

[58] a) Espinosa, J.C.; Álvaro, M.; Dhakshinamoorthy, A.; Navalón, S.; García, H. Engineering active sites in reduced graphene oxide: Tuning the catalytic activity for aerobic oxidation. *ACS Sustain. Chem.& Eng.,* **2019**, *7*(19), 15948-15956.
[http://dx.doi.org/10.1021/acssuschemeng.9b02237] b) Kitano, M.; Yamaguchi, D.; Suganuma, S.; Nakajima, K.; Kato, H.; Hayashi, S.; Hara, M. Adsorption-enhanced hydrolysis of β-1,4-glucan on graphene-based amorphous carbon bearing SO3H, COOH, and OH groups. *Langmuir,* **2009**, *25*(9), 5068-5075.
[http://dx.doi.org/10.1021/la8040506] [PMID: 19397353]

[59] Anumula, R.; Cui, C.; Yang, M.; Li, J.; Luo, Z. Catalytic Oxidation of Cyclohexane on Small Silver Clusters Supported by Graphene Oxide. *J. Phys. Chem. C,* **2019**, *123*(35), 21504-21512.
[http://dx.doi.org/10.1021/acs.jpcc.9b04405]

[60] Mondal, A.; Jana, N.R. Surfactant-free, stable noble metal–graphene nanocomposite as high-performance electrocatalyst. *ACS Catal.,* **2014**, *4*(2), 593-599.
[http://dx.doi.org/10.1021/cs401032p]

[61] Candu, N.; Man, I.; Simion, A.; Cojocaru, B.; Coman, S.M.; Bucur, C.; Primo, A.; Garcia, H.; Parvulescu, V.I. Nitrogen-doped graphene as metal free basic catalyst for coupling reactions. *J. Catal.,* **2019**, *376*, 238-247.
[http://dx.doi.org/10.1016/j.jcat.2019.07.011]

[62] Ma, J.; Xiao, Y.; Sun, Y.; Hu, J.; Wang, Y. Sulfur-Doped Alkylated Graphene Oxide as High-Performance Lubricant Additive. *Nanoscale Res. Lett.,* **2020**, *15*(1), 26.
[http://dx.doi.org/10.1186/s11671-020-3257-7] [PMID: 32002704]

[63] a) Khalili, D. Graphene oxide: a reusable and metal-free carbocatalyst for the one-pot synthesis of 2-amino-3-cyanopyridines in water. *Tetrahedron Lett.,* **2016**, *57*(15), 1721-1723.
[http://dx.doi.org/10.1016/j.tetlet.2016.03.020] b) Zhang, F.; Jiang, H.; Wu, X.; Mao, Z.; Li, H. Organoamine-functionalized graphene oxide as a bifunctional carbocatalyst with remarkable acceleration in a one-pot multistep reaction. *ACS Appl. Mater. Interfaces,* **2015**, *7*(3), 1669-1677.
[http://dx.doi.org/10.1021/am507221a] [PMID: 25556875] c) Scheuermann, G.M.; Rumi, L.; Steurer, P.; Bannwarth, W.; Mülhaupt, R. Palladium nanoparticles on graphite oxide and its functionalized graphene derivatives as highly active catalysts for the Suzuki-Miyaura coupling reaction. *J. Am. Chem. Soc.,* **2009**, *131*(23), 8262-8270.
[http://dx.doi.org/10.1021/ja901105a] [PMID: 19469566]

[64] Kim, F.; Luo, J.; Cruz-Silva, R.; Cote, L.J.; Sohn, K.; Huang, J. Self-propagating domino-like reactions in oxidized graphite. *Adv. Funct. Mater.,* **2010**, *20*(17), 2867-2873.
[http://dx.doi.org/10.1002/adfm.201000736]

[65] Zhang, H.; Lv, X.; Li, Y.; Wang, Y.; Li, J. P25-graphene composite as a high performance photocatalyst. *ACS Nano,* **2010**, *4*(1), 380-386.
[http://dx.doi.org/10.1021/nn901221k] [PMID: 20041631]

[66] Sun, H.; Liu, S.; Zhou, G.; Ang, H.M.; Tadé, M.O.; Wang, S. Reduced graphene oxide for catalytic oxidation of aqueous organic pollutants. *ACS Appl. Mater. Interfaces,* **2012**, *4*(10), 5466-5471.
[http://dx.doi.org/10.1021/am301372d] [PMID: 22967012]

[67] Nie, R.; Wang, J.; Wang, L.; Qin, Y.; Chen, P.; Hou, Z. Platinum supported on reduced graphene

oxide as a catalyst for hydrogenation of nitroarenes. *Carbon,* **2012**, *50*(2), 586-596. [http://dx.doi.org/10.1016/j.carbon.2011.09.017]

[68] Zhang, Z.; Sun, T.; Chen, C.; Xiao, F.; Gong, Z.; Wang, S. Bifunctional nanocatalyst based on three-dimensional carbon nanotube-graphene hydrogel supported Pd nanoparticles: one-pot synthesis and its catalytic properties. *ACS Appl. Mater. Interfaces,* **2014**, *6*(23), 21035-21040. [http://dx.doi.org/10.1021/am505911h] [PMID: 25375195]

[69] Song, J.; Kang, S.W.; Lee, Y.W.; Park, Y.; Kim, J.H.; Han, S.W. Regulating the catalytic function of reduced graphene oxides using capping agents for metal-free catalysis. *ACS Appl. Mater. Interfaces,* **2017**, *9*(2), 1692-1701. [http://dx.doi.org/10.1021/acsami.6b13970] [PMID: 27991751]

[70] Joyner, J.; Oliveira, E.F.; Yamaguchi, H.; Kato, K.; Vinod, S.; Galvao, D.S.; Salpekar, D.; Roy, S.; Martinez, U.; Tiwary, C.S.; Ozden, S.; Ajayan, P.M. Graphene supported MoS2 structures with high defect density for an efficient HER electrocatalyst. *ACS Appl. Mater. Interfaces,* **2020**, *12*(11), 12629-12638. [http://dx.doi.org/10.1021/acsami.9b17713] [PMID: 32045208]

[71] Xi, J.; Wang, Q.; Liu, J.; Huan, L.; He, Z.; Qiu, Y.; Zhang, J.; Tang, C.; Xiao, J.; Wang, S. N,P-dua--doped multilayer graphene as an efficient carbocatalyst for nitroarene reduction: A mechanistic study of metal-free catalysis. *J. Catal.,* **2018**, *359*, 233-241. [http://dx.doi.org/10.1016/j.jcat.2018.01.003]

[72] Wang, H.; Kong, Q.; Wang, Y.; Deng, T.; Chen, C.; Hou, X.; Zhu, Y. Graphene Oxide Catalyzed Dehydration of Fructose into 5-Hydroxymethylfurfural with Isopropanol as Cosolvent. *ChemCatChem,* **2014**, *6*(3), 728-732. [http://dx.doi.org/10.1002/cctc.201301067]

[73] Basu, B.; Kundu, S.; Sengupta, D. Graphene oxide as a carbocatalyst: the first example of a one-pot sequential dehydration–hydrothiolation of secondary aryl alcohols. *RSC Advances,* **2013**, *3*(44), 22130-22134. [http://dx.doi.org/10.1039/c3ra44712j]

[74] Hu, H.; Wang, X.; Miao, D.; Wang, Y.; Lai, C.; Guo, Y.; Wang, W.; Xin, J.H.; Hu, H. A pH-mediated enhancement of the graphene carbocatalyst activity for the reduction of 4-nitrophenol. *Chem. Commun. (Camb.),* **2015**, *51*(93), 16699-16702. [http://dx.doi.org/10.1039/C5CC05826K] [PMID: 26434405]

[75] Peng, L.; Zheng, Y.; Li, J.; Jin, Y.; Gao, C. Monolithic neat graphene oxide aerogel for efficient catalysis of S→ O acetyl migration. *ACS Catal.,* **2015**, *5*(6), 3387-3392. [http://dx.doi.org/10.1021/acscatal.5b00233]

[76] Hu, F.; Patel, M.; Luo, F.; Flach, C.; Mendelsohn, R.; Garfunkel, E.; He, H.; Szostak, M. Graphene-catalyzed direct Friedel–Crafts alkylation reactions: mechanism, selectivity, and synthetic utility. *J. Am. Chem. Soc.,* **2015**, *137*(45), 14473-14480. [http://dx.doi.org/10.1021/jacs.5b09636] [PMID: 26496423]

[77] Tan, L.; Wang, B.; Feng, H. Comparative studies of graphene oxide and reduced graphene oxide as carbocatalysts for polymerization of 3-aminophenylboronic acid. *RSC Advances,* **2013**, *3*(8), 2561-2565. [http://dx.doi.org/10.1039/c2ra22516f]

[78] Wang, H.; Deng, T.; Wang, Y.; Cui, X.; Qi, Y.; Mu, X.; Hou, X.; Zhu, Y. Graphene oxide as a facile acid catalyst for the one-pot conversion of carbohydrates into 5-ethoxymethylfurfural. *Green Chem.,* **2013**, *15*(9), 2379-2383. [http://dx.doi.org/10.1039/c3gc41109e]

[79] Yang, D.; Zhao, J.; Shi, J.; Wang, X.; Zhang, S.; Jiang, Z. Combination of redox assembly and biomimetic mineralization to prepare graphene-based composite cellular foams for versatile catalysis. *ACS Appl. Mater. Interfaces,* **2017**, *9*(50), 43950-43958.

[http://dx.doi.org/10.1021/acsami.7b11601] [PMID: 29171256]

[80] Li, Q.; Fan, F.; Wang, Y.; Feng, W.; Ji, P. Enzyme immobilization on carboxyl-functionalized graphene oxide for catalysis in organic solvent. *Ind. Eng. Chem. Res.*, **2013**, *52*(19), 6343-6348. [http://dx.doi.org/10.1021/ie400558u]

[81] Xu, Z.; Li, W.; Zhang, Y.; Xue, Z.; Guo, X.; Zhang, G. Facile synthesis of mesoporous reduced graphene oxide microspheres with well-distributed Fe_2O_3 nanoparticles for photochemical catalysis. *Ind. Eng. Chem. Res.*, **2016**, *55*(40), 10591-10599. [http://dx.doi.org/10.1021/acs.iecr.6b01004]

[82] Li, Z.; Zhao, H.; Han, H.; Song, J.; Liu, Y.; Guo, W.; Sun, Z.; Chu, W. A one-pot method for synthesis of reduced graphene oxide-supported Cu-Cu_2O and catalytic application in tandem reaction of halides and sodium azide with terminal alkynes. *Appl. Organomet. Chem.*, **2018**, *32*(4), e4301. [http://dx.doi.org/10.1002/aoc.4301]

[83] Gupta, A.; Jamatia, R.; Patil, R.A.; Ma, Y.R.; Pal, A.K. Copper Oxide/Reduced Graphene Oxide Nanocomposite-Catalyzed Synthesis of Flavanones and Flavanones with Triazole Hybrid Molecules in One Pot: A Green and Sustainable Approach. *ACS Omega*, **2018**, *3*(7), 7288-7299. [http://dx.doi.org/10.1021/acsomega.8b00334] [PMID: 31458889]

[84] Naeimi, H.; Shaabani, R. Ultrasound promoted facile one pot synthesis of triazole derivatives catalyzed by functionalized graphene oxide Cu(I) complex under mild conditions. *Ultrason. Sonochem.*, **2017**, *34*, 246-254. [http://dx.doi.org/10.1016/j.ultsonch.2016.05.043] [PMID: 27773242]

[85] Naeimi, H.; Ansarian, Z. Functionalized polytriazoles on graphene oxide-supported copper(I) complex as an effective reusable catalyst for sonochemical click synthesis of triazoles in aqueous media. *Inorg. Chim. Acta*, **2017**, *466*, 417-425. [http://dx.doi.org/10.1016/j.ica.2017.06.057]

[86] Xiong, X.; Chen, H.; Tang, Z.; Jiang, Y. Supported CuBr on graphene oxide/Fe_3O_4: a highly efficient, magnetically separable catalyst for the multi-gram scale synthesis of 1,2,3-triazoles. *RSC Advances*, **2014**, *4*(19), 9830-9837. [http://dx.doi.org/10.1039/c3ra45994b]

[87] Narayanan, D.P.; Gopalakrishnan, A.; Yaakob, Z.; Sugunan, S.; Narayanan, B.N. A facile synthesis of clay – graphene oxide nanocomposite catalysts for solvent free multicomponent Biginelli reaction. *Arab. J. Chem.*, **2020**, *13*(1), 318-334. [http://dx.doi.org/10.1016/j.arabjc.2017.04.011]

[88] Moitra, D.; Ghosh, B.K.; Chandel, M.; Ghosh, N.N. Synthesis of a $BiFeO_3$ nanowire-reduced graphene oxide based magnetically separable nanocatalyst and its versatile catalytic activity towards multiple organic reactions. *RSC Advances*, **2016**, *6*(100), 97941-97952. [http://dx.doi.org/10.1039/C6RA22077K]

[89] Akocak, S.; Şen, B.; Lolak, N.; Şavk, A.; Koca, M.; Kuzu, S.; Şen, F. One-pot three-component synthesis of 2-Amino-4H-Chromene derivatives by using monodisperse Pd nanomaterials anchored graphene oxide as highly efficient and recyclable catalyst. *Nano-Structures & Nano-Objects*, **2017**, *11*, 25-31. [http://dx.doi.org/10.1016/j.nanoso.2017.06.002]

[90] Azarifar, D.; Khaleghi-Abbasabadi, M. Fe_3O_4-supported N-pyridin-4-amine-grafted graphene oxide as efficient and magnetically separable novel nanocatalyst for green synthesis of 4H-chromenes and dihydropyrano[2,3-c]pyrazole derivatives in water. *Res. Chem. Intermed.*, **2019**, *45*(2), 199-222. [http://dx.doi.org/10.1007/s11164-018-3597-4]

[91] Khodabakhshi, S.; Karami, B. Graphene oxide nanosheets as metal-free catalysts in the three-component reactions based on aryl glyoxals to generate novel pyranocoumarins. *New J. Chem.*, **2014**, *38*(8), 3586-3590. [http://dx.doi.org/10.1039/C4NJ00228H]

[92] Siddiqui, T.A.J.; Ghule, B.G.; Shaikh, S.; Shinde, P.V.; Gunturu, K.C.; Zubaidha, P.K.; Yun, J.M.; O'Dwyer, C.; Mane, R.S.; Kim, K.H. Metal-free heterogeneous and mesoporous biogenic graphene-oxide nanoparticle-catalyzed synthesis of bioactive benzylpyrazolyl coumarin derivatives. *RSC Advances,* **2018**, *8*(31), 17373-17379.
[http://dx.doi.org/10.1039/C7RA12550J] [PMID: 35539260]

[93] Abdi, M.; Rostamizadeh, S.; Zekri, N. An efficient and green synthesis of 1′H spiro [isoindoline-1, 2′-quinazoline]-3, 4′ (3′H)-Dione derivatives in the presence of nano Fe_3O_4–GO–SO_3H. *Polycycl. Aromat. Compd.,* **2019**, *39*(5), 413-424.
[http://dx.doi.org/10.1080/10406638.2017.1340313]

[94] Shaabani, A.; Hezarkhani, Z.; Nejad, M.K. AuCu and AgCu bimetallic nanoparticles supported on guanidine-modified reduced graphene oxide nanosheets as catalysts in the reduction of nitroarenes: tandem synthesis of benzo[b][1,4]diazepine derivatives. *RSC Advances,* **2016**, *6*(36), 30247-30257.
[http://dx.doi.org/10.1039/C6RA03132C]

[95] Kausar, N.; Mukherjee, P.; Das, A.R. Practical carbocatalysis by graphene oxide nanosheets in aqueous medium towards the synthesis of diversified dibenzo[1,4]diazepine scaffolds. *RSC Advances,* **2016**, *6*(91), 88904-88910.
[http://dx.doi.org/10.1039/C6RA17520A]

[96] Sun, X.; Wang, W.; Wu, T.; Qiu, H.; Wang, X.; Gao, J. Grafting of graphene oxide with poly(sodium 4-styrenesulfonate) by atom transfer radical polymerization. *Mater. Chem. Phys.,* **2013**, *138*(2-3), 434-439.
[http://dx.doi.org/10.1016/j.matchemphys.2012.12.022]

[97] Gao, Y.; Hu, G.; Zhong, J.; Shi, Z.; Zhu, Y.; Su, Sheng Gao, Y.; Hu, G.; Zhong, J.; Shi, Z.; Zhu, Y.; Su, D.S.; Wang, J.; Bao, X.; Ma, D. Nitrogen-doped sp2-hybridized carbon as a superior catalyst for selective oxidation. *Angew. Chem. Int. Ed.,* **2013**, *52*(7), 2109-2113.
[http://dx.doi.org/10.1002/anie.201207918] [PMID: 23307693]

[98] Sun, H.; Wang, Y.; Liu, S.; Ge, L.; Wang, L.; Zhu, Z.; Wang, S. Facile synthesis of nitrogen doped reduced graphene oxide as a superior metal-free catalyst for oxidation. *Chem. Commun. (Camb.),* **2013**, *49*(85), 9914-9916.
[http://dx.doi.org/10.1039/c3cc43401j] [PMID: 23945634]

[99] Gao, Y.; Ma, D.; Wang, C.; Guan, J.; Bao, X. Reduced graphene oxide as a catalyst for hydrogenation of nitrobenzene at room temperature. *Chem. Commun. (Camb.),* **2011**, *47*(8), 2432-2434.
[http://dx.doi.org/10.1039/C0CC04420B] [PMID: 21170437]

[100] Perhun, T.I.; Bychko, I.B.; Trypolsky, A.I.; Strizhak, P.E. Catalytic properties of graphene material in the hydrogenation of ethylene. *Theor. Exp. Chem.,* **2013**, *48*(6), 367-370.
[http://dx.doi.org/10.1007/s11237-013-9282-1]

[101] Kong, X.; Chen, Q.; Lun, Z. Probing the influence of different oxygenated groups on graphene oxide's catalytic performance. *J. Mater. Chem. A Mater. Energy Sustain.,* **2014**, *2*(3), 610-613.
[http://dx.doi.org/10.1039/C3TA13946H]

[102] Kong, X.; Sun, Z.; Chen, M.; Chen, C.; Chen, Q. Metal-free catalytic reduction of 4-nitrophenol to 4-aminophenol by N-doped graphene. *Energy Environ. Sci.,* **2013**, *6*(11), 3260.
[http://dx.doi.org/10.1039/c3ee40918j]

[103] Singh Chauhan, S.M.; Mishra, S. Use of graphite oxide and graphene oxide as catalysts in the synthesis of dipyrromethane and calix[4]pyrrole. *Molecules,* **2011**, *16*(9), 7256-7266.
[http://dx.doi.org/10.3390/molecules16097256] [PMID: 21869753]

[104] Qi, G.; Zhang, W.; Dai, Y. An efficient synthesis of 5-substituted 1H-tetrazoles catalyzed by graphene. *Res. Chem. Intermed.,* **2015**, *41*(2), 1149-1155.
[http://dx.doi.org/10.1007/s11164-013-1260-7]

[105] Xu, J.; Meng, R.; Cao, J.; Gu, X.; Qi, Z.; Wang, W.; Chen, Z. Enhanced dehydrogenation and

rehydrogenation properties of LiBH4 catalyzed by graphene. *Int. J. Hydrogen Energy,* **2013**, *38*(6), 2796-2803.
[http://dx.doi.org/10.1016/j.ijhydene.2012.12.046]

[106] Yuan, C.; Chen, W.; Yan, L. Amino-grafted graphene as a stable and metal-free solid basic catalyst. *J. Mater. Chem.,* **2012**, *22*(15), 7456.
[http://dx.doi.org/10.1039/c2jm30442b]

[107] Wu, T.; Wang, X.; Qiu, H.; Gao, J.; Wang, W.; Liu, Y. Amino-grafted graphene as a stable and metal-free solid basic catalyst. *J. Mater. Chem.,* **2012**, *22*, 4772.
[http://dx.doi.org/10.1039/c2jm15311d]

[108] Lam, E.; Chong, J.H.; Majid, E.; Liu, Y.; Hrapovic, S.; Leung, A.C.W.; Luong, J.H.T. Carbocatalytic dehydration of xylose to furfural in water. *Carbon,* **2012**, *50*(3), 1033-1043.
[http://dx.doi.org/10.1016/j.carbon.2011.10.007]

[109] Liu, F.; Sun, J.; Zhu, L.; Meng, X.; Qi, C.; Xiao, F.S. Sulfated graphene as an efficient solid catalyst for acid-catalyzed liquid reactions. *J. Mater. Chem.,* **2012**, *22*(12), 5495.
[http://dx.doi.org/10.1039/c2jm16608a]

[110] Wang, L.; Wang, D.; Zhang, S.; Tian, H. Synthesis and characterization of sulfonated graphene as a highly active solid acid catalyst for the ester-exchange reaction. *Catal. Sci. Technol.,* **2013**, *3*(5), 1194.
[http://dx.doi.org/10.1039/c3cy20646g]

[111] Sun, X.; Wang, W.; Wu, T.; Qiu, H.; Wang, X.; Gao, J. Grafting of graphene oxide with poly(sodium 4-styrenesulfonate) by atom transfer radical polymerization. *Mater. Chem. Phys.,* **2013**, *138*(2-3), 434-439.
[http://dx.doi.org/10.1016/j.matchemphys.2012.12.022]

[112] Ji, J.; Zhang, G.; Chen, H.; Wang, S.; Zhang, G.; Zhang, F.; Fan, X. Sulfonated graphene as water-tolerant solid acid catalyst. *Chem. Sci. (Camb.),* **2011**, *2*(3), 484-487.
[http://dx.doi.org/10.1039/C0SC00484G]

[113] Shaabani, A.; Mahyari, M.; Hajishaabanha, F. The synthesis of xanthenes and benzoxanthenes on graphene oxide and sulfated graphene nanosheets in water. *Res. Chem. Intermed.,* **2014**, *40*(8), 2799-2810.
[http://dx.doi.org/10.1007/s11164-013-1129-9]

[114] Ramirez-Barria, C.S.; Isaacs, M.; Parlett, C.; Wilson, K.; Guerrero-Ruiz, A.; Rodríguez-Ramos, I. Ru nanoparticles supported on N-doped reduced graphene oxide as valuable catalyst for the selective aerobic oxidation of benzyl alcohol. *Catal. Today,* **2020**, *357*, 8-14.
[http://dx.doi.org/10.1016/j.cattod.2019.05.057]

[115] Boukhvalov, D.W.; Dreyer, D.R.; Bielawski, C.W.; Son, Y.W. A Computational Investigation of the Catalytic Properties of Graphene Oxide: Exploring Mechanisms by using DFT Methods. *ChemCatChem,* **2012**, *4*(11), 1844-1849.
[http://dx.doi.org/10.1002/cctc.201200210]

[116] Tang, S.; Cao, Z. Site-dependent catalytic activity of graphene oxides towards oxidative dehydrogenation of propane. *Phys. Chem. Chem. Phys.,* **2012**, *14*(48), 16558-16565.
[http://dx.doi.org/10.1039/c2cp41343d] [PMID: 22801590]

[117] Long, Y.; Zhang, C.; Wang, X.; Gao, J.; Wang, W.; Liu, Y. Oxidation of SO_2 to SO_3 catalyzed by graphene oxide foams. *J. Mater. Chem.,* **2011**, *21*(36), 13934.
[http://dx.doi.org/10.1039/c1jm12031j]

[118] Tan, L.; Wang, B.; Feng, H. Comparative studies of graphene oxide and reduced graphene oxide as carbocatalysts for polymerization of 3-aminophenylboronic acid. *RSC Advances,* **2013**, *3*(8), 2561.
[http://dx.doi.org/10.1039/c2ra22516f]

[119] Sreeprasad, T.S.; Berry, V. How do the electrical properties of graphene change with its functionalization? *Small,* **2013**, *9*(3), 341-350.

[http://dx.doi.org/10.1002/smll.201202196] [PMID: 23169614]

[120] Kim, Y.; Some, S.; Lee, H. Graphene oxide as a recyclable phase transfer catalyst. *Chem. Commun. (Camb.),* **2013**, *49*(50), 5702-5704.
[http://dx.doi.org/10.1039/c3cc42787k] [PMID: 23689290]

[121] Verma, S.; Mungse, H. P.; Kumar, N.; Choudhary, S.; Jain, S. L.; Sain, B., ; Khatri, O. P, Graphene oxide: an efficient and reusable carbocatalyst for aza-Michael addition of amines to activated alkenes. *Chem. Commun. (Camb.),* **2011**, *47*(47), 12673-12675.
[http://dx.doi.org/10.1039/c1cc15230k] [PMID: 22039588]

[122] Yu, H.; Wang, X.; Zhu, Y.; Zhuang, G.; Zhong, X.; Wang, J. Solvent-free catalytic dehydrative etherification of benzyl alcohol over graphene oxide. *Chem. Phys. Lett.,* **2013**, *583*, 146-150.
[http://dx.doi.org/10.1016/j.cplett.2013.08.011]

CHAPTER 9

Oxidation and Reduction Reactions Catalyzed by Graphene Based Catalysts

Chetna Ameta[1,*], **Priyanka Chundawat**[1], **Yogeshwari Vyas**[1], **Reema Agarwal**[1] and **Dharmendra Dharmendra**[1]

[1] *Photochemistry Laboratory, Department of Chemistry University College of Science, Mohanlal Sukhadia University, Udaipur (Raj.), India*

Abstract: Graphene has several features, such as charge mobility, high conductivity, and a large specific surface area with a two-dimensional structure. It also has exceptional electrical, mechanical, and thermal properties. Graphene has exceptional mechanical, physical, and chemical properties, which are responsible for the development of efficient graphene-based catalysts for selective organic synthesis. This chapter recapitulates the various applications of graphene-based catalysts in oxidation and reduction reactions. It gives a complete overview of graphene, reduced graphene oxide, functionalized graphene oxide, N-doped graphene oxide, and their catalytic applications in various oxidation and reduction reactions. The previous studies reveal that metal-free GO has many catalytic applications. Owing to its high surface area, graphene oxide has a high adsorption tendency for hydrocarbons, gases, and ions.When reacting with oxygenating functionalities, GO provides several paths for introducing and modifying various functional groups. The large potential is employed for the use of metal-free carbon catalysts to resolve the industrial problems arising from traditional catalysts. Since graphene/GO catalysts are synthesized from environmental-friendly material, their applications in green synthesis should be discovered vigorously. The graphene-based catalysts have several applications. They allow only selective, mild, and highly effective transformations and undergo the synthesis and synthesis in an easy, recyclable, regenerable, and environmentally friendly manner.

Keywords: Graphene, Reduced Graphene Oxide, Functionalized Graphene Oxide, Metal-free Structured Carbon Catalysts, Oxidation and Reduction Reactions, Green Method.

INTRODUCTION

In the last decade, the utilization of nanomaterials as catalysts and supporting catalysts has propagated immensely. More exciting is the fast development in the research on the application of structured carbon nanomaterial in catalysis.

* **Corresponding author Chetna Ameta:** Photochemistry Laboratory, Department of Chemistry University College of Science, Mohanlal Sukhadia University, Udaipur (Raj.), India; E-mail: chetna.ameta@yahoo.com

Pinki Bala Punjabi, Rakshit Ameta & Sharoni Gupta (Eds.)

Graphene is composed of a two-dimensional carbon network and it is also used for the synthesis of other carbon materials such as single and multiwalled carbon tubes (1D), fullerenes (0D), and graphite (3D) [1]. All the above-mentioned carbon materials are used as a support system for catalytic active phase and metals, it can also be used as a metal-free catalyst.

Carbon nanomaterials own unique properties, such as electrochemical and chemical stability, the tendency for a high degree of functionalization, and high surface area. In the last few years, extensive studies have been done on single and multi-walled carbon nanotubes to explore their use in catalysis, energy and electronic devices, polymer fillers, biological applications, and drug delivery. In 2004 mechanically exfoliated single-layer graphene sheets were reported.

Graphene-based materials have a huge application in energy storage, photocatalysis, catalysis, solar cells, memory devices, drug delivery, nano circuitry, and sensing platforms [2]. Several unique thermal, electronic, mechanical, and optical properties are owned by pristine graphene, these above properties are different from other forms of carbon. But, pristine graphene has some limitations, such as complex synthesis procedures and poor processability. Graphene oxide, reduced graphene oxide, and chemically modified graphene can be used as alternatives to pristine graphene because they can be easily prepared and processed in large quantity and also they can be easily synthesized, processed and dispersed in water and other organic solvents. However, they have some mechanical and electrical properties which are somehow different from graphene and responsible for disturbance in few applications. Yet, they are the appropriate starting materials for several applications [3].

Pristine graphene has zero bandgap and is a stable and inert material, whereas graphene oxide and reduced graphene oxides are either insulators or semiconductors. Graphene consists of only carbon material, while graphene oxide and reduced graphene oxide contain several organic functional groups which can further undergo modifications. At present main function of GO, graphene, and CMGs (chemically modified Graphene) are to act as supporting materials for catalytically active transition metals. So, we need to find more striking alternatives concerning extended stability, low cost, flexibility, and ecological influence.

The biological grade synthesis processes are highly toxic processes which is a matter of great concern, particularly the leakage of metal from their supports. To solve these problems, heterogenous metal-free catalysts made up of carbon materials [4] and noncarbon materials [5] can be used as an alternative. Over the past two decades, innovation and growth in new forms of carbon like nanotubes, fullerenes, graphene, and nanofibers have resulted in an emphasis on the development and use of metal-free carbon catalysts. Important applications of CMGs and metal-free graphene oxides are their use in various chemical reactions as green catalysts.

The application of graphene and graphene oxide in catalysis has been discovered recently. Because of their use in battery and fuel cells, these materials have gained much attention already, specially N-doped graphene shows excellent activity for electrocatalytic oxygen reduction reaction (ORR). In this chapter, we will emphasize recent development and synthesis and oxidation and reduction reactions catalyzed by GO, FGO (Functionalized graphene oxide), and heteroatom doped graphene, which does not gain much consideration.

SYNTHESIS OF GRAPHENE/GRAPHENE OXIDE

Graphene

The exfoliation method was used first time for the experimental research of a single atomic layer of graphene. But this synthesis method is not adaptable for producing high-quality graphene on a large scale. The development of graphene sheets from graphite is done by the use of various chemical methods. The solvent used in the chemical method can interrupt the p–p interaction between sheets and result in individual stabilization of graphene [6], another alternative method is the bottom-up method in which, at high-temperature organic precursors are used for the direct growth of graphene [7].

The above methods have a large number of limitations, such as poor solubility, which results in graphitization and an increase in defect sites, and the number of unwanted side reactions. Even with the lots of development in the above alternative techniques, the mechanical exfoliation route is still the best method to produce graphene sheets of good quality.

Graphene Oxide and Reduced Graphene Oxide

Brodie [8], Staudenmeier [9], and Hummers [10] proposed the most common method to prepare to GO. In this method, graphite is oxidized with a strong oxidizing agent and results in samples having the same degree of oxidation. In the last decade, a modified Hummers method has become much more popular because it does not produce hazardous ClO_2 gas, has a short reaction time, and peroxidation steps promote further oxidation.

In Modified Hummers method, oxidization of flake graphite powder takes place, in these method flake graphite powder is mixed with phosphorus peroxide (P_2O_5)

and potassium persulphate ($K_2S_2O_8$) and stirred after the above mixture is dissolved in concentrated sulfuric acid (H_2SO_4). Afterward, the further oxidation of graphite is done by adding a mixture of potassium permanganate ($KMnO_4$) and H_2SO_4 and it is cool down to remove the reaction heat. After that, exfoliation of oxidized graphene is done by sonication, or mechanically it is done by polymers or surface stabilizing solvents [11].

GO is made up of a carbon framework containing a variety of oxygenating functionalities. GO lacks well-defined stoichiometry or uniquely defined structure, and distribution, location, and density of different oxygenate functionalities are defined by the synthesis method and the starting graphite material. Lerf and Klinowski [12] proposed the most acceptable structural model of GO, as shown in Fig. (**1**). This model indicates that and epoxy and hydroxyl groups are present at the basal plane of a GO sheet, and carbonyl groups and carboxylic acid are present at the edges of GO sheets. The above groups disturb the delocalization of p electrons and make graphene oxide electrically insulating. The polarity of the above group and their capability to contribute to hydrogen bonding results in the hydrophilic nature of GO and make it easily soluble in polar solvents or water. The carboxylic acid groups present at the edges of graphene oxide sheet are ionizable in less acidic aqueous solution, and the suspension of GO sheets in polar and aqueous solvents are stabilized by electrostatic repulsion between above-charged groups, and this repulsion is also responsible for the separation of GO sheets.

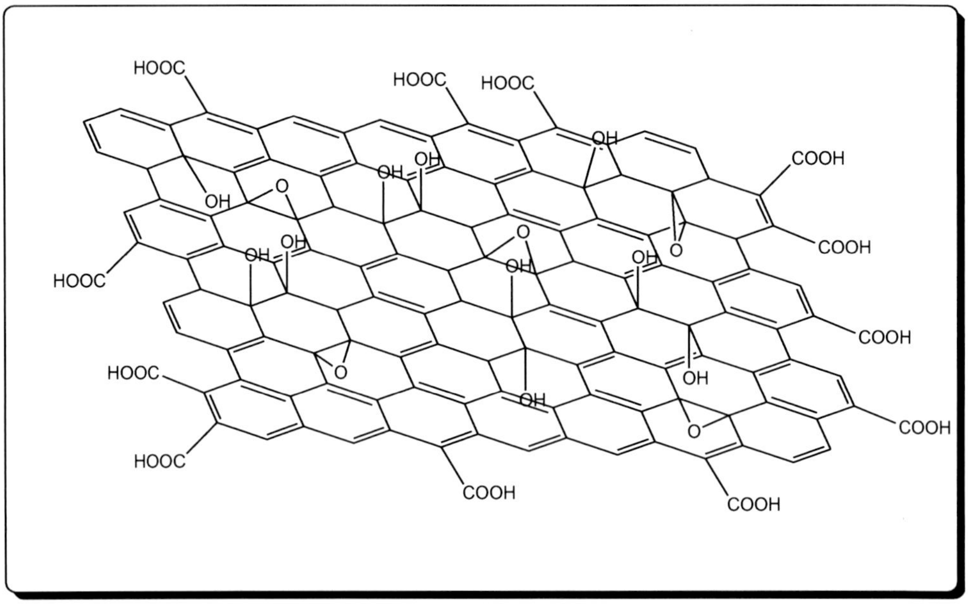

Fig. (1). Proposed structure of graphene oxide (honeycomb graphene lattice).

Few exceptional properties of graphene are retained by the reduction in oxygen functionalities. After reduction GO is signified as reduced graphene oxide, indicating the incomplete reduction. However, at a present complete reduction has not been achieved by researchers. Chemical, thermal and electrochemical methods are the commonly used method for the synthesis of rGO, and all the above methods retain the conjugated p-p bond up to some extent. For the chemical reduction of GO, lithium aluminum hydride is used as a strong reducing agent.

However, lithium aluminum hydride gives dangerous side reactions, so it cannot be used for the reduction of GO. So to avoid the above problems, less reactive chemicals such as hydrazine monohydrate are used for reduction of GO and reduction is also possible in aqueous solvents. Sodium borohydride, hydroquinone, and strong alkaline solutions are some other effective reducing agents employed for the reduction of GO. The main disadvantage associated with the chemical reduction is the incorporation of heteroatom impurities into the basal plane. The heteroatom nitrogen is incorporated into the basal plane by the reacting carbonyl group with hydrazine. This contamination affects the resulting electronic properties.

The ideal condition for effective reduction is achieved at high temperatures especially between 500-1,000°C using a dilute hydrogen atmosphere. Thermal reduction of graphene oxide and reduced graphene oxide is made in an inert atmosphere at a temperature above 500°C. More complete reduction is obtained at very high temperatures up to 1,000°C. This is because of the thermal cracking of C-C and C-O bonds in decarbonylation, dehydration, dehydrogenation, and decarboxylation reactions. Gas produced in reduction reaction is considered a byproduct, consisting of carbon dioxide, carbon monoxide, water, and perhaps hydrogen. Fast production of these gases creates pressure in the graphene oxide layers, which exfoliates the sample.

As the exfoliation process results in a higher surface area of the sample as compared to the area obtained by chemical reduction, it also has some drawbacks, such as it results in huge mass losses (30%) and it produces significant defects, these defects affect the electronic and chemical properties of rGO. These defects formation is controlled by employing successive reduction of chemical and thermal reduction. The third commonly used method for reduction is electrochemical reduction. In the above method, the electrode is prepared by placing a graphene oxide sample on the substrate, and this electrode is immersed in an electrolyte solution. The reduction process is completed by passing a current among a counter electrode and substrate [13]. The major drawback associated with this method is that the large-scale production of rGO is not possible.

Functionalized Graphene Oxide

To get a more adaptable material, various routes have been employed to change the GO surface through a change in a covalent bond. The various route which was employed to discover the functionalized carbon nanotubes can be used for graphene. For GO, numerous information is available that explains the functional group's nature, so the chemical reactions used to convert them into other functional groups have been employed.

Unfortunately, there are no particular techniques available for successful selective transformations, so effective transformation is not possible. However, these transformations are broadly used transformations, and the extensively used conversion reaction of an epoxy group and carboxylic acid are discussed below and also, several effective methods used for the synthesis of catalysts are discussed below. Unlike another organic synthesis, activation of surface carboxylic acid groups is necessary for their transformation. Once the carboxylic acid group gets stimulated, the carbon present in carboxylic acid becomes more susceptible to nucleophilic attack forming the covalent bond with nucleophiles.

Thionyl chloride ($SOCl_2$) [14, 15], 1-ethyl-3-(3-dimethyl aminopropyl) carbodiimide (EDC) or N-(3-dimethyl aminopropyl)-N-ethyl carbodiimide hydrochloride (EDCI) [16 - 19], and N,N-dicyclohexylcarbodiimide (DCC) are the commonly used activation or coupling reagents. Commonly used nucleophiles have a hydroxyl group or an amine that forms an ester and amide bond with graphene oxide and results in the formation of targeted entities on the graphene oxide surface. A ring-opening mechanism is used for the transformation of graphene oxide surface by its reaction with an epoxy group [20], but the above reaction is not much selective. For example, in the reaction of an amine with carboxylic acid, simultaneous aminolysis of epoxide may occur. Another method used for the activation of a carboxylic acid is the solvothermal method [21].

Other methods used to functionalize GO surface are the condensation of methoxy and ethoxy groups with surface hydroxyl group [22]; this is done by using the same method which is employed for fixing molecules on usual heterogeneous supports like Al_2O_3 and SiO_2. The rarely used method, which is the covalent functionalization of the graphitic area of reduced graphene oxide with diazonium salt is done by using a radical mechanism, and is also employed for catalyst synthesis.

N-Doped (Heteroatom-Doped) Graphene

Post-treatment of graphene or GO or in situ growth are generally used methods for the synthesis of heteroatom-doped graphene catalysts. In the last few years, N-doping has become the most popular subject [23]. In the in-situ growth method, heteroatoms are integrated through the growth procedure by adding in an N-containing reagent, by CVD takes place. The key advantage of the in situ method is that it offers new catalytic properties and it also allows retains the intrinsic physical and chemical properties of pristine graphene. N-doped graphene can be synthesized by CVD by passing a gaseous mixture of CH_4, NH_3, and H_2 in an Ar carrier over a substrate.

The incorporation of nitrogen into graphene aromatic structures can be done by using various forms such as pyrrolic and pyridine nitrogen (Fig. **2**). Solvothermal synthesis processes are another type of in situ method. For example, during the pyrolysis process reaction of lithium nitride undergoes the reaction with tetrachloromethane by using stainless steel autoclave. In the post-treatment method heteroatoms are incorporated by indulging graphene oxide with various compounds by using a physical and chemical method like bypassing NH_3 over graphene oxide at high temperatures or combining graphene oxide with nitrogen sources such as dicyandiamide, melamine, and pyrrole, and then post-treatment method like thermal and chemical reduction is employed. Depending on the experimental details, nitrogen atoms can be merged into graphene oxide in various forms.

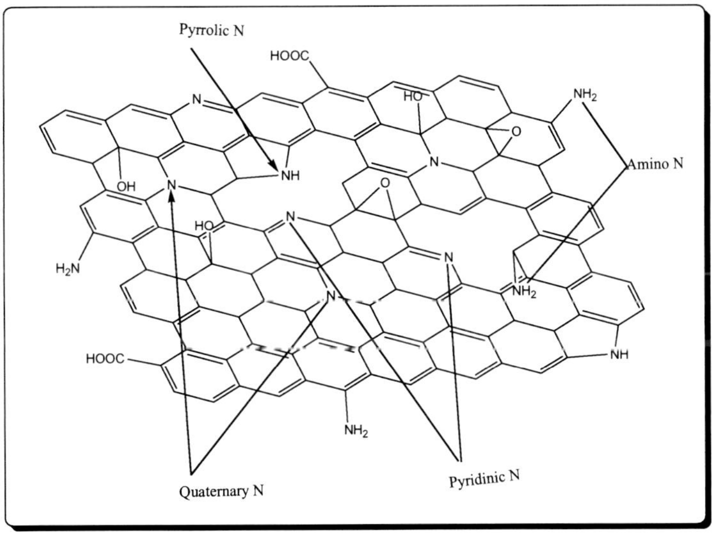

Fig. (2). Schematic illustration of the various forms of 'N' atom incorporation into the graphene structure.

CATALYTIC APPLICATIONS

As we already know that carbon and carbon-based materials like graphitic carbon and activated carbon play a very important role in catalysis and the functionalized carbons are generally more selective and active as compared to their unfunctionalized counterparts. Since, because of heterogeneity of functionalized carbons, the determination of a detailed reaction mechanism is found to be very difficult, and their characterization at an atomic level is also, although functionalized carbons are found to be very active for several reactions such as oxidation of aromatic compounds, decarboxylation of fatty acids, and selective oxidation dehydrogenation of hydrocarbons [24]. Here we emphasize graphene and graphene oxide-based materials.

Oxidation Reactions

Catalysis by Graphene Oxide and Graphite Oxide

Dreyer and Bielawski [25] explained that synthesized graphene oxide and graphite oxide, which is exfoliated version of graphite oxide give a highly disrupted graphene network. GO has a huge application in catalysis like easy synthesis procedures, low cost, it can be easily recovered by filtration, and it also has a surface to volume ratio. As it is already mentioned that the edges and basal plane of GO contain numerous oxygen functional groups like carbonyl, epoxy, carboxylic acids, and hydroxyl, all these groups are responsible for the acidic character, chemical activity, oxidative nature, and catalytic activity of GO.

However, GO can be used as a catalyst in numerous reactions. Bielawski and his coworkers [26] reported the effect of GO on the oxidation of alkenes to diketones, alcohols to aldehyde and diketones, and oxidation of alkynes to hydrates at 75-150 °C in air. Above oxidation reactions are highly selective. For example, the yield in the oxidation of cis1,2-diphenylmethane-1,2-diol to 1,2-diphenylethan--1,2dione was 96%, and 98% for diphenylmethanol to benzophenone. Interestingly, whereas GO oxidized 1,2diphenylethene to 1,2-diphenylethane-1-2-dione with 56% conversion, it was either inactive or could only catalyze the isomerization of other alkene systems.

So, it is found that cis-1,2-diphenylethene is reactive because of a combination of cis geometry and aromatic substituents. To determine the catalytic activity of GO researchers performed several experiments inert nitrogen environment and found that absence of oxygen results in very little oxidation, and also GO can be extracted from the reaction mixture and reused up to several cycles. But, GO catalyst has very low activity and to obtain a reasonable yield in 24 h high carbon to substrate ratio up to 200% is required. Although GO catalyst requires high loading and has low activity, it has the main advantage that it has a high turnover frequency per dollar as compared to tetrapropyl ammonium per ruthenate oxidation catalyst.

GO was found to catalyze other reactions, such as oxidation of sulfides to sulfoxides and the coupling of thiols to disulfides [27], dehydrogenation of various hydrocarbons and oxidation of methylbenzenes to their respective aldehydes, diarylmethanes to their respective ketones, and other hydrocarbons with sp3-hybridized C–H bonds [28]. It has also shown high activities for the Claisen–Schmidt coupling reactions [29], Friedel-Crafts reactions [30], Aza-Michael additions [31], and ring-opening polymerization [32].

Oxidation of Various Alcohols

In the initial experiment, under ambient conditions and in the presence of 20 wt% of GO, benzyl alcohol ($PhCH_2OH$) is heated at 100 °C for 14 h (Scheme **1**). After that, it is filtered several times and gives the mixture of benzaldehyde (PhCHO)/$PhCH_2OH$ in the molar ratio of 1:3, as verified by 1H NMR spectroscopy. Under these conditions, the formation of any other by-products and benzoic acid ($PhCO_2H$) were not observed.

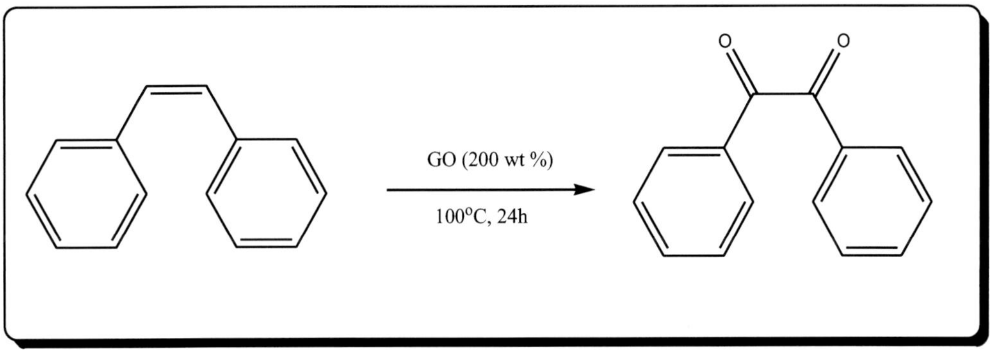

Scheme 1. Oxidation of benzyl alcohol using graphene oxide (GO).

The reactivity of the above carbocatalyst is studied by using various primary and secondary aliphatic and benzylic alcohols under ambient conditions such as 100°C, 24 h and 200 wt% GO, which results in the highest conversion of $PhCH_2OH$ into its oxidized products. Table **1**, entries 1–3 reveals that benzylic alcohols have moderate to high product conversion efficiency. Table **1**, entry 4 reveals the high catalytic power of GO to oxidize cyclohexanol. Table **1**, entry 5 explains the conversion of 2-thiophenemethanol into aldehyde, and the sulfur oxidation does not take place. The latter results explain that GO exhibit broad chemoselectivity and have a high tolerance towards functional groups, so GO can be used to overcome the challenges given by transition metal catalysts.

Table 1. Oxidation of Various Alcohols Using GO[a].

Entry	Alcohol	Product	Conversion (%)
1[b]	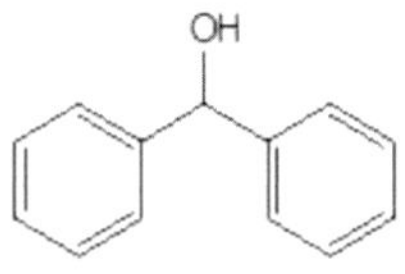	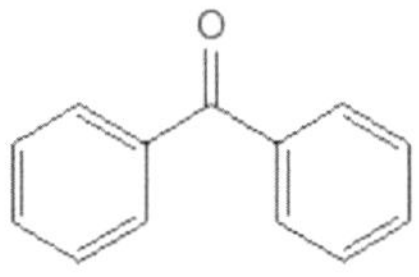	> 98
2[b]	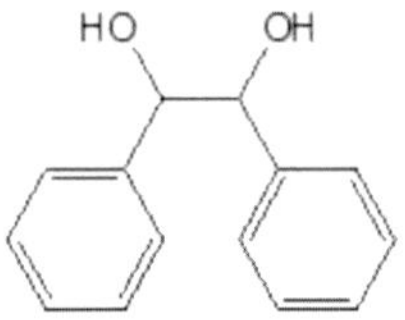	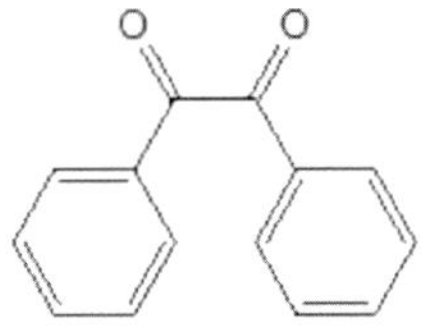	96
3[d]	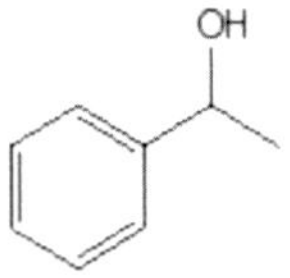	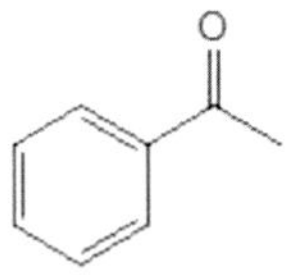	26
4	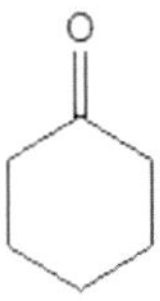	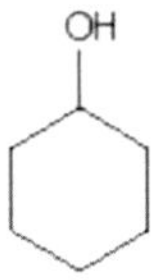	> 98
5[d]	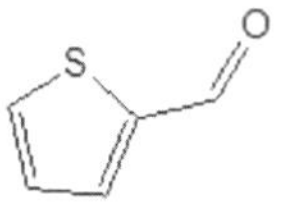	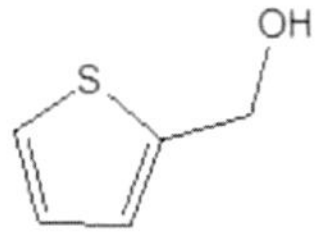	18

Oxidation of Cis-Stilbene

As mentioned above, GO can undergo the oxidation of various alcohols, so nowadays various efforts are used to explore the ability of GO to catalyze the oxidation of unsaturated hydrocarbons. As illustrated in Scheme **2**, under the optimized conditions on treatment of cis-stilbene at 100°C for 24 hours 49% benzil is isolated (56% conversion by NMR spectroscopy); in the crude reaction mixture, benzene or any oxidation product is not found.

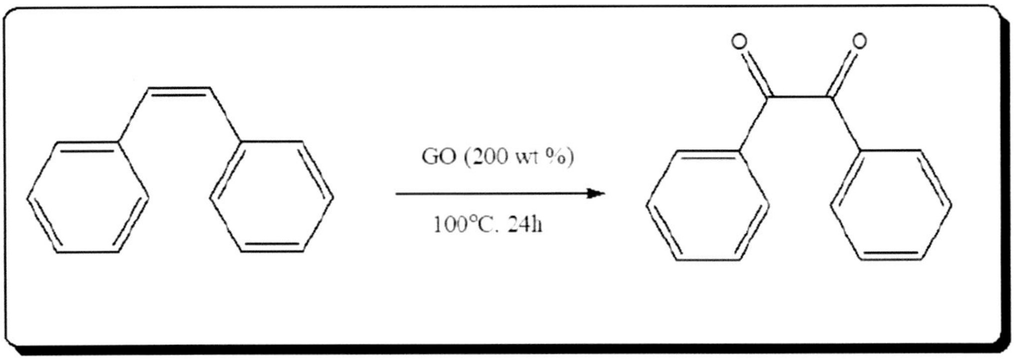

Scheme 2. Oxidation of cis-stilbene using GO.

To facilitate the oxidation process, it is necessary to combine cis geometry with aromatic substituents because when other alkenes (trans-stilbene, 3-hexene, cyclohexene, and b-methyl styrene) are used as substrates under the same environment, no reaction or isomerization is observed these may be due to orientation of the interaction tangled between catalyst surface and olefins or may be due to steric hindrance.

Oxidation of Organosulfur Compounds

The oxidation of sulfides and thiols resulted in various applications in drug metabolism, synthesis, and preparation of bioconjugates. Several methods have been investigated for the oxidation of organosulfur compounds but all the methods are associated with a major drawback of over-oxidation. For example, dimerization of thiols resulted in thiols and then it further undergoes the oxidation of disulfinates, thiosulfinates, disulfones, and sulfinylsulfones. Similarly, the oxidation of sulfides resulted in sulfoxides along with sulfones and sulfonic acid.

To solve this problem of overoxidation several advanced catalysts such as transition metal with active species are investigated. Among all the metals, catalyst based on palladium, iron, and molybdenum decreases the reaction time and increases the selectivity for the desired product. Despite this advantage, because of toxicity and separation issues, numerous metal catalysts cannot be used in biotechnological and pharmaceutical applications.

So for the organosulfur compounds, it is necessary to develop highly active metal-free oxidants. Furthermore, the heterogeneous catalysts allow not only the easy and facile separation of the target compounds by distillation and filtration but also reuse and recovery of the catalyst itself. We found that many of the above-mentioned challenges to ing organosulfur oxidants can be solved using a graphene oxide catalyst.

Oxidation of Thiols

In an initial experiment, 25 mg of thiophenol is reacted with 75 mg of GO by using 0.3 ml of $CHCl_3$ at 100 ◦C for 10 min in a closed vessel (Scheme **3**). After that, the above reaction mixture is dissolved in 50 ml of $CHCl_3$ and then it undergoes filtration to remove the volatile solvent, the thiol is determined by 1H NMR spectroscopy ($CDCl_3$) to convert them into diphenyl disulfide.

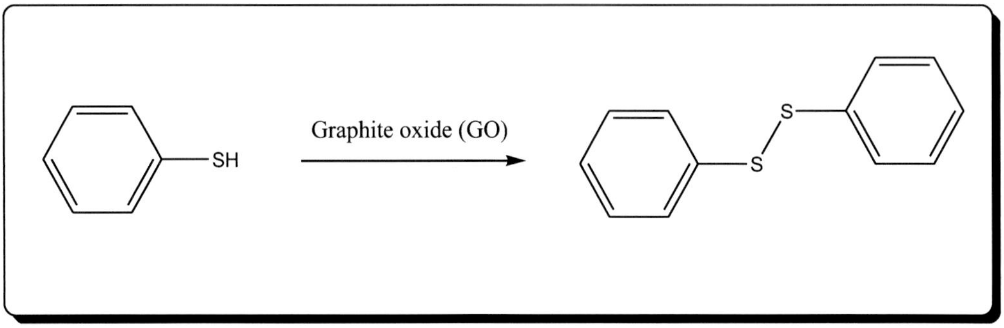

Scheme 3. Oxidation of thiol using graphene oxide (GO).

By using the above-mentioned experimental conditions (25 mg thiol, 15 mg GO, 0.3 mL $CHCl_3$, 100°C, 10 min), after that, we determined the substrate scope of this method. As presented in Table **2**, the reaction of various thiols with GO resulted in exceptional conversion to their disulfides, and by dissolving the reaction mixture in 50 ml $CHCl_{3,}$ filtering the remaining carbon, removing the volatile solvent under close atmosphere, in some cases purification by using column chromatography resulted in the pure products.

Entries 1, 4–9 of Table **2** show that the aryl thiols have a longer reaction time of 30 min as compared to alkyl thiols which are 10 min (entries 2–3, 10), but the reactivity of aryl thiol is higher than alkyl thiol these is because of higher activity of GO towards arene functionalized substrate. To further indicate the superior reactivity of aryl thiol over alkyl thiols, 25 mg of thiophenol and 1-butanethiol in 1:1 molar ratio is reacted with 60 wt% of GO at 100°C for 10 min. The 1HNMR spectroscopy revealed the 63% conversion of reactant into product and the product mixture contains a 1.0:5.0:2.9 molar ratio of dibutyl disulfide: butyl phenyl disulfide: diphenyl disulfide along with 34% unreacted 1-butylsulfide and a <5% unreacted thiophenol.

The incorporation of electron-donating or withdrawing substituents to the aromatic thiols does not influence the quantity of desired product. For example, electron-rich diols as given in entries 4–7 and electron deficient thiols as presented in entries 8–9 of Table **2** gives the exceptional 94-99% yield of their respective disulfides, in all the above reaction, no over-oxidation and nitrogen oxidation in case of 2-amino thiophenol is detected, which has been occurred by use of chromic potassium sulfate ($CrK(SO_2)_2.12H_2O$) and other transition metal-based oxidation catalysts [33]. Although these challenges are overcome by the use of more advanced metal species such as cobalt porphyrins and imidazole [34], MOFs, and various bismuth species [35] but these catalysts do not show the extent of reactivity as shown by GO.

Table 2. Oxidation of various thiols using GO[a].

Entry	Starting material	Product	Reaction Time	Yield[b] (%)
1	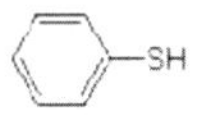	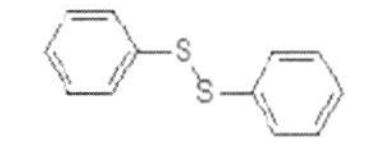	10 min	100
2		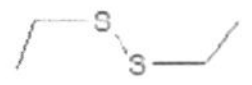	30 min[c]	77
3		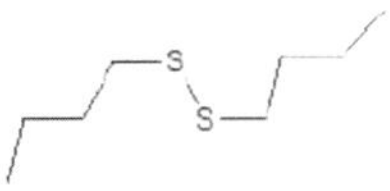	30 min[c]	90
4	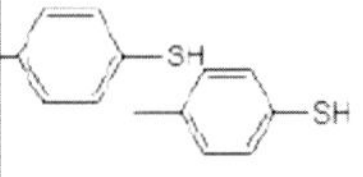	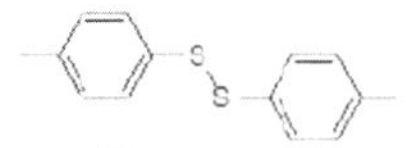	10 min	95
5	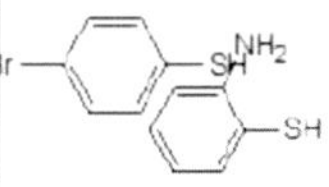	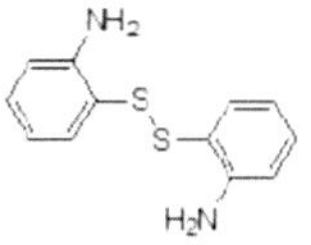	10 min	96
6	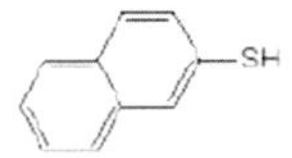	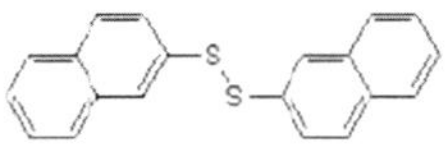	10 min	97
7	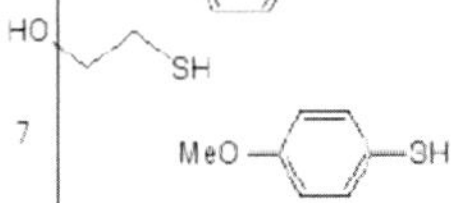	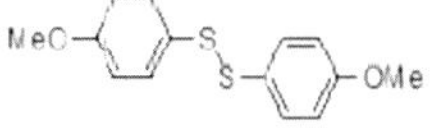	10 min	97
8	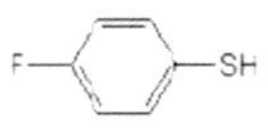	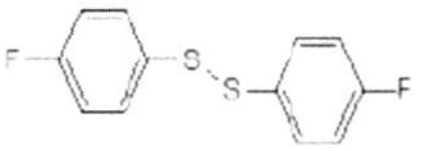	10 min	99
9	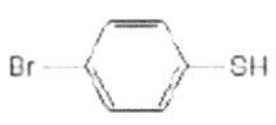	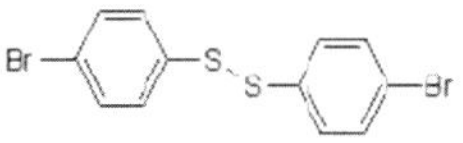	10 min	94
10		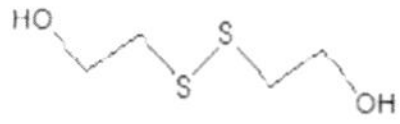	30 min[c]	75

[a]All reactions were performed at 100°C in sealed 7.5 mL vials using 25 mg thiol, 15 mg GO (60 wt%) and 0.3 mL $CHCl_3$ for the indicated reaction time. [b]Isolated yield after purification by column chromatography. [c]Relatively low yields (40–65%) were observed after 10 min; higher yields were obtained after 30 min.

Oxidation of Sulfides

In an initial experiment, diphenyl sulfide is oxidized to diphenyl sulfoxide at 100°C (300 wt% GO) after 24 h 86% isolated yield was obtained (Scheme **4**), and in the reaction mixture, no other oxidation product is observed. In the way illustrated above, the reaction is improved concerning reaction time, reaction temperature, and GO loading. It is found that GO concentration below 300 wt% does not give the desired product, yet with an increase in GO loading, no enhancement in isolated yield is observed.

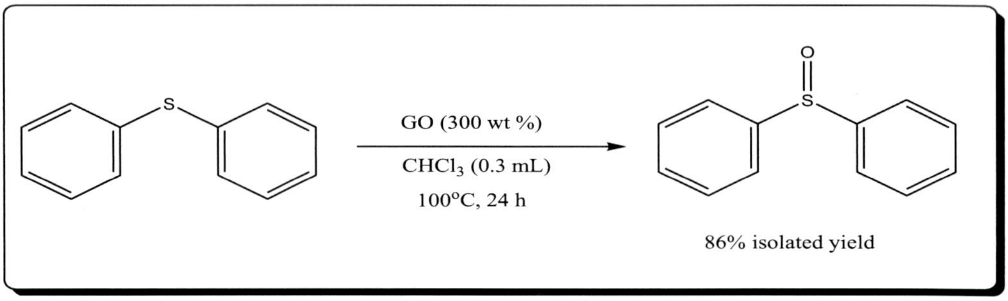

Scheme 4. Oxidation of diphenyl sulfide using GO.

Here the GO loadings required are much higher than that of metal catalysts but their heterogeneity and low cost make them more feasible to use. It is found that the used GO loadings which undergo the partial deoxygenation as given below, can be again reoxidized to GO by using the oxidation method, which was used by the graphite starting material. The above mention thiol coupling shows that when carbons other than GO are used, such as activated carbon, graphite, and hydrazine-reduced graphene oxide minimum reactivity is observed.

As shown in Table **3**, by using the optimum conditions like 75 mg GO, 25 mg sulfide, 0.3 mL $CHCl_3$, 24 h, and 100°C a large number of sulfides can be converted into their respective sulfoxides. In most cases, after purification by column chromatography exceptional yields of their subsequent products are gained. Differences in the electronic structure of the arene substituted sulfides slightly affect sulfoxide product yield. For example, an electron-rich sulfide (*p*-methoxythianisole, entry8) on oxidation gives a 96% product yield, while an electron deficient sulfides (*p*-chlorothianisole, entry 9) gives the 81% product yield. As Compared to Ru and Fe catalysts (using H_2O_2 or O_2 as the oxidant), GO resulted in an increase in selectivity and product yield for the sulfoxide over the sulfone.

Table 3. Oxidation of Various Sulfdies by Using GO[a].

Entry	Starting material	Product	Yield[b]
1	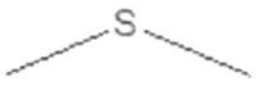	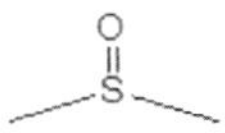	89%
2	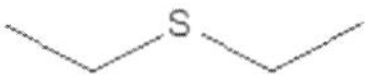	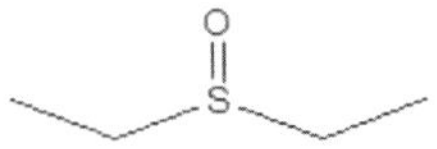	92%
3	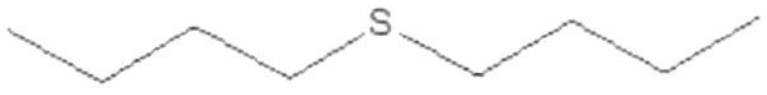	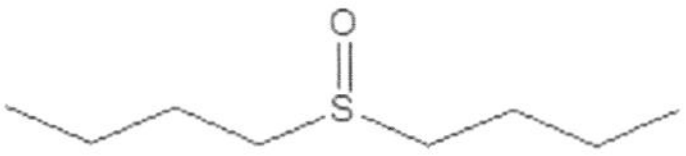	90%
4	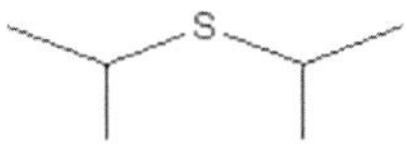	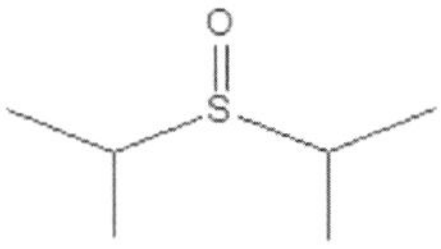	51%
5	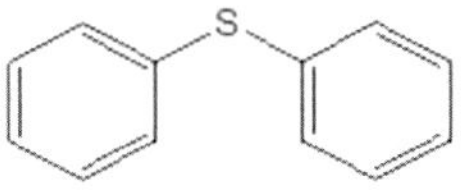	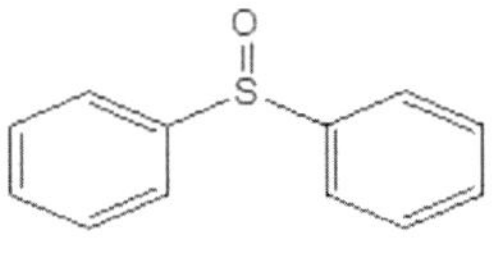	86%
6	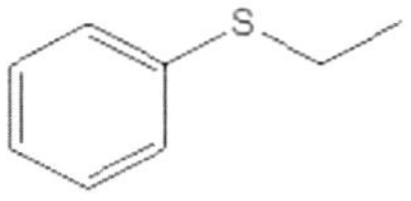	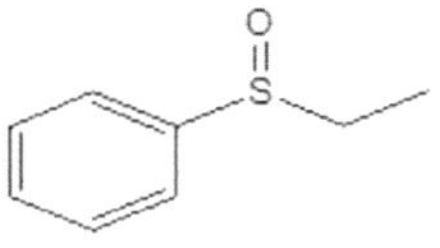	85%
7	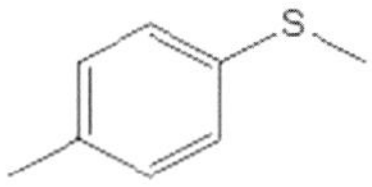	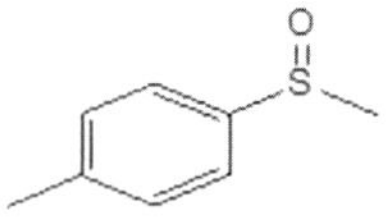	75%
8	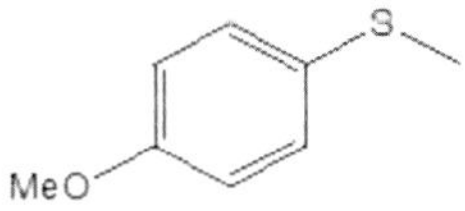	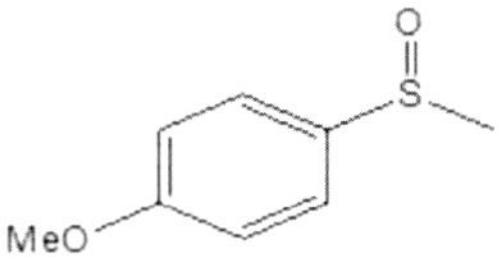	96%
9	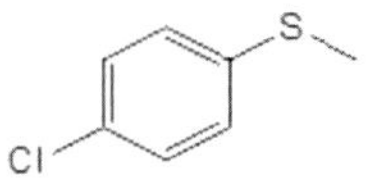	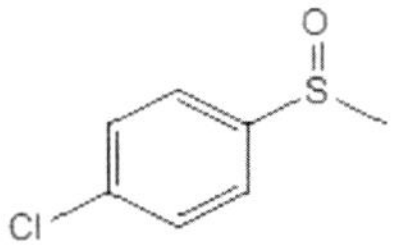	81%

[a]All reactions were performed at 100°C in sealed 7.5 mL vials using 25 mg sulfide, 75 mg GO (300 wt%), and 0.3 mL $CHCl_3$ for 24 h. [b]Isolated yield after purification by column chromatography.

Oxidation of Amines

A large number of efficient methods have been for the oxidation of alcohols into their respective carbonyl groups but the methods for the catalytic oxidation of amines into imines are still not much studied, although imines are a good source of nitrogen in many chemical reactions [36 - 40]. In the conventional method, imine is prepared by condensation of primary amines and aldehyde and ketones [41], but this method requires some critical conditions like longer reaction time and catalyst Lewis as the catalyst [42]. In the last few years, to optimize these reaction conditions, many methods have been developed such as self-condensation of primary amines, oxidation of secondary amines, and direct coupling of amines and alcohols [43 - 47].

In recent years, a large number of catalysts have been explored for the oxidation of amines [48, 49]. But these catalysts still have some drawbacks, such as high toxicity, high cost, difficult preparation methods, and critical reaction conditions. Therefore, a catalyst with high selectivity, separability, and easy preparation procedure is required.

Recently, Go has drawn widespread attention [50 - 52]. Go along with its derivatives are widely used as a metal-free catalyst [53 - 58]. Because of the presence of a large number of oxygenated functional groups in GO, GO has high catalytic and oxidation capabilities [59]. Several advantages of GO like low toxicity, simple preparation method, and tendency to get separate makes it a good catalyst for oxidation of amine.

It is found that GO is an efficient catalyst for the oxidation of amines into imines but the effect of various functional groups on Go and their decreased activity in recycling experiments are not clarified [60]. Bermudez and co-workers [61] reported the use of hydrogen peroxide (H_2O_2) as the oxidant in GO-catalyzed oxidation of amines and disclose the effect of GO and H_2O_2 and GO on the selectivity of products.

But, in several reactions, the effect of free radical is uncertain so because of the above reason the mechanism of oxidation of amines catalyzed by carbocatalysts, which is necessary for efficiency and design of catalyst, is still in doubt. It is observed that GO and its derivative such as improved graphene oxide (IGO), can oxidize the amines into their respective imines in the absence of oxygen and catalyst and improved graphene oxide has higher reactivity for amine oxidation. GO can function as both catalyst and oxidant, the epoxy and hydroxyl groups present in GO act as an oxidant, and Carboxyl groups along with defects and edges play the role of catalyst.

This remarkable metal-free reactivity of graphene oxide has encouraged researchers to explore the probabilities offered for catalytic oxidation of amines into imines. Go is reported as an easily available, easily applicable, cheap, recyclable, and efficient catalyst for the direct conversion of amines into imines. It is the first suitable metal-free catalytic method that is used for the synthesis of imine from various amines by using molecular oxygen as the oxidant under neat conditions [62].

Benzylamine acts as an intermediate for several polymers, pharmaceuticals, and dyes, and it is also used to study the use of graphene oxide in the oxidative coupling reaction of amines. Entry 1 of Table **4** shows that the reaction of GO and benzylamine with p-xylene at 130°C in the air for 8 h resulted in 87.39% of N-benzylidene benzylamine product. Entry 2 of Table **4** shows that when a similar experiment is performed in the absence of GO, only 9.33% of product yield is obtained, which indicates that GO can catalyze the aerobic oxidation of benzylamine.

Table 4. Oxidation of Benzylamine to N-Benzylidene Benzylamine under various conditions[a].

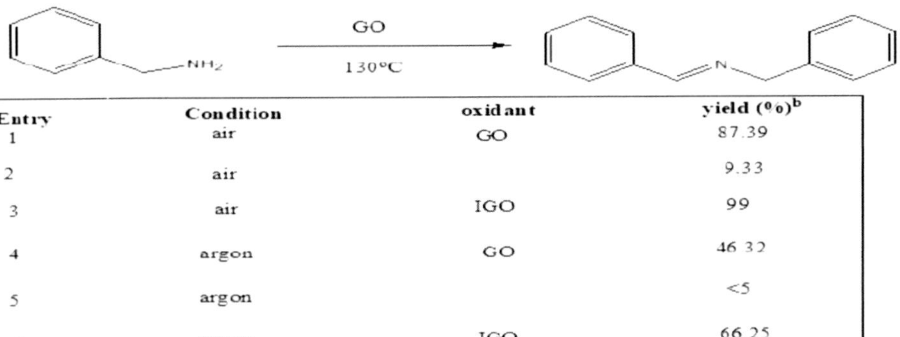

Entry	Condition	oxidant	yield (%)[b]
1	air	GO	87.39
2	air		9.33
3	air	IGO	99
4	argon	GO	46.32
5	argon		<5
6	argon	IGO	66.25

[a]Reaction conditions: benzylamine (2.5 mmol), GO (26 mg), and p xylene (5 mL) at 130°C for 8 h, entry 1–3 in air condition, entry 4– 6 in argon condition. [b]Isolated yield.

Entry 4 of Table **4** states that when the above reaction is carried out in the same condition in the presence of argon results in 46.32% of product yield. Entry 5 of Table **4** shows when the above reaction is carried out with argon in the absence of GO then less than 5% of the product is obtained which indicates that GO also plays the role of oxidant in this reaction.

Nitrogen-Doped Graphene Catalysts

Nowadays, N-doped graphene has become brilliant catalysts for oxygen reduction reaction (ORR) battery and fuel cell applications. The inclusion of nitrogen atom changes the charge distribution of surrounding carbon, and spin density and broadens the bandgap of GO. Generally, the nitrogen can be incorporated into carbon lattice in three forms: pyridinic N, quaternary (graphitic) N and pyrrolic N. Since all the three forms of nitrogen affect the oxygen reduction reaction and the improvement in the reactivity is depends on the amount of quaternary and pyridinic nitrogen. Despite the above advantage use of N-doped graphene in various reactions is not much studied only a few studies like oxidation of benzene into phenol [63] and of a range of benzylic alcohols [64] are available. N-doped graphene is synthesized by flowing NH_3 over GO and it is found that the quaternary nitrogen increases the catalytic activity of oxidation of benzylic alcohol into their respective aldehyde. The incorporation of quaternary nitrogen into GO can be increased by increasing the temperature up to 900°C.

Oxidation of Alkyl Arenes and Alcohols

In organic synthesis, the selective oxidation of organic compounds at laboratory and industrial scales is a major challenge. In the aerobic oxidation reaction, there are generally three major considerable aspects are eradicating the requirement of the activators like inorganic bases and accomplishment of the reaction without activation in a green solvent at room temperature.

At an industrial scale, the grouping of these characteristics is highly challenging and this grouping provides benefits for chemical reactions. To solve the above challenge catalyst is the key, several studies are ongoing to get a suitable catalyst having all aforementioned aspects. Recently the researcher emphasis on the development of the catalyst by using the green and biometric principal. Because of this reason metal phthalocyanines and metal porphyrins are used as biomimetic catalysts in various oxidation reactions.

Zhang *et al.* [65] reported that the oxidation of alcohols and alkyl arenes in the solvent-free aerobic atmosphere with Pd@ N-doped carbon at 120°C resulted in very low product yield and the use of Pd in the catalyst as an active site also restrict its application. The aforementioned drawbacks lead to the development of new catalysts such as copper(II) tetrasulfo phthalocyanine reinforced on three-dimensional nitrogen-doped graphene-based frameworks (CuTSPc@3D-(N)GFs). The above catalyst is the first catalyst that was synthesized by using 3D-(N)GF--supported phthalocyanine and it has the unique property of selective aerobic oxidation of alcohols and alkyl arenes into their respective carbonyl groups. The selective aerobic oxidation reactions are conducted at room temperature in an ecofriendly environment for a short reaction time and resulted in excellent product yield.

To optimize the required amount of solvent and catalyst, the oxidation of 2,3-dihydro-1H-indene is carried out by using 0.4 mol% of H_2O and Cu as a solvent for both the reactions at room temperature and the results are given in Table **5**. The important advantage of 3D-(N)GFs support are: (i) They have high surface area, macroporous structures, good physical and chemical stability (ii) simple preparation method, (iii) the presence of a bifunctional heterogeneous catalyst (copper(II) as the active site and N as the base) eliminates the requirement of activators like inorganic bases, (iv) chemical modification of the surface with nitrogen doping increases the reactivity and in catalytic oxidation, they also improves the oxygen interaction with substrate and, and (v) because of its hydrophilic nature it is easily soluble in water which is used as a reaction solvent.

Entries 2-4 of Table **5** present the catalytic oxidation of benzaldehyde and 2,3-dihydro-1H-indene using GO, 3D-(N)GFs, and CuTSPc [66]. Entry 2 of Table **5** shows that when GO (0.005 g) is used as a catalyst under a mild condition in oxidation reaction results in a very low product yield.

Table 5. Comparison of the Results Obtained From Cutspc@3D-(N)Gfs with GO, Cutspc, and 3D-(N)Gfs for the Oxidation Reactions.

Entry	Catalyst	Product	Yield[b] (%)	Selectivity[b] (%)
1	CuTSPc@3D-(N)GFs	2,3-Dihydro-1*H*-indene	98	>99
		Benzaldehyde	98	>99
2	GO	2,3-Dihydro-1*H*-indene	24	>99
		Benzaldehyde	30	>99
3	3D-(N)GFs	2,3-Dihydro-1*H*-indene	33	>99
		Benzaldehyde	35	>99
4	CuTSPc	2,3-Dihydro-1*H*-indene	86	97
		Benzaldehyde	88	98
5	CuTSPc@MCM-41	2,3-Dihydro-1*H*-indene	75	98
		Benzaldehyde	81	99

[a]Reaction conditions: 2,3-dihydro-1H-indene or benzyl alcohol (1.0 mmol), catalyst, H2O (4.0 mL), air (1 atm), r.t., 3 h. [b]Yield and selectivity determined by GC analysis.

It is found that the use of 3D-(N)GFs support increases the adsorption of aromatic reactants these may be due to the p-p interaction between graphene and benzene skeleton of reactant. Furthermore, the interaction of 3D-(N)GFs with nitrogen-doped support resulted in the adsorption of oxygen on the bridge sites of graphene by using migration energy of more than 1.0 eV. Graphene contains mobile oxygen, which endorses hydrogen elimination resulting in the enhancement of the aerobic oxidation of organic compounds. Entries 1 and 4 of Table **5** show the increase in the product yield as compared to the homogenous catalyst, several factors improve the yield, such as the presence of basic sites on solid and p-p interaction between support and aromatic groups favor the proximity of reagents with the catalytic site.

Selective Oxidation of Olefins and Alcohols

Although the selective oxidation of alcohols and olefins with molecular oxygen is very much important for the environment, life, and synthetic chemistry, it is still a very difficult task. Several reagents and methods have been discovered to accomplish these oxidation reactions. Most of the oxidation reaction uses the fixed amounts of heavy metal ions such as Mn, Cr, Co, and Fe and explosive oxidants such as hydrogen peroxide and peroxy acids. Because of environmental issues, there is a need to develop a safer and cheaper heterogeneous catalyst for oxidation. A large number of noble metal nanoparticles like Pd, Ru, Pt, and Au have been studied for use in the oxidation of alcohols and olefins.

Graphene has several advantages, such as it has a high surface area which helps in the transfer of reactant and product by scattering the metal nanoparticles micropore-free structure with one or a few sheet layers, it also has appropriate electrical conductivity, which helps in the transfer of electrons during catalysis process, and also have high chemical reactivity, and it is also found that the graphene has p character which is responsible for interaction between aromatic derivatives and graphene and this interaction makes the reactants available for active sites. Because of the large availability and low cost of graphite, it can be easily broken down into individual graphene by using chemical and physical methods. Traditionally it can be broken down into graphene using several methods such as liquid-phase exfoliation, use of graphite intercalation compound, oxidation exfoliation, and electrochemical exfoliation. In the last few years, it has become a great challenge to modify graphene's mechanical, chemical, and electronic properties by doping using N, B, and P elements.

It is found that N-dopants can decline the formation of "re-graphitized" carbon regions in the carbon skeleton and also, they are favorable sites to enhance the nucleation of particle and decreases particle size. be usually observed as favorable anchor sites or defects for enhancing particle nucleation and reducing the particle size. N dopants have several advantages, such as they can change the electronic structure of the carbon matrix, improve the interaction between the guest molecule and carbon, and also be responsible for the giving metal like d-band to carbon which hosts the reactive species. The effective heterogeneous oxidation reaction of both alcohol and olefins is carried out by using molecular oxygen over a Co_3O_4/N-doped graphene hybrid (Co_3O_4/RGO-N) catalyst. It is observed that the dispersed Co_3O_4 NPs and well-exfoliated N-doped graphene can be grown simultaneously by employing a one-pot hydrothermal reaction using NH_3 solution. Here we discuss the structural characteristics of Co_3O_4/RGO-N and its performance in the green oxidation of olefins and alcohols [67].

Co_3O_4/RGO-N hybrid is used in the epoxidation of styrene. Entries 1 and 2 of Table **6** reveal that amides like DMF and DMA can be used to get the high conversion of styrene up to 97.7 and 86.2% and high epoxide selectivity of 72.7 and 63.8%. along with the desired product small amount of cleavage product like benzaldehyde is also observed.

Table 6. Epoxidation of Styrene by O_2 in Different Solvents Over Co_3O_4/RGO-N[a] in different solvents over.

Entry	Solvent	Conversion (%)	Epoxide selectivity (%)
1	DMF	97.7	72.7
2	DMA	86.2	63.8
3	Dioxane	77.7	55.7
4	DMSO	26.3	15.2
5	Toluene	0	0
6	n-Heptane	0	0
7	1-Butanol	0	0
8[b]	DMF	22.8	49.1
9[c]	DMF	79.9	66.0
10[d]	DMF	98.7	69.9

[a]Reaction conditions: temperature (100○C), reaction time (3 h), Co_3O_4/ RGO-N (20 mg), styrene (8 mmol), DMF (25 mL), low rate of O_2 (20 mL min-1). [b]The reaction was carried out at 80○C. [c]The reaction was carried out at 90○C. dThe reaction was carried out at 110○C.

Entries 3 and 4 of Table **6** show that by using DMSO and dioxane solvent certain amount of styrene conversion is obtained but their epoxide selectivity is very low. Entries 5-7 of Table **6** reveals that by using toluene, n-heptane, or 1-butanol solvent, no products are obtained. Entries 8-10 of Table **6** present the temperature dependency of oxidation reaction and we found that at the temperature 80 -90°C both conversion and selectivity decrease and a small decrease in selectivity is observed at higher temperature (110°C).

Entry 2 of Table **7** shows that Co_3O_4/ RGO gives the relatively low activity (57.3%) and lower epoxide selectivity (70.3%) as compared to Co_3O_4/ RGO-N. Entry 3 of Table **7** shows that Co_3O_4 seems to be poorer in catalyzing the epoxidation of styrene. Entry 4 shows that significant catalytic activity can be observed over pure RGO-N alone. Entries 5-7 reveal that the use of commonly used oxidation catalysts such as MnO_2, Fe_3O_4, and CuO results in low activity and low epoxide selectivity. The above results show that separated Co_3O_4or RGO-N gives the lower catalytic activity while hybrids result in the superior catalytic activity suggests that Co_3O_4 and N-doped graphene play synergistic roles in the Co_3O_4/RGO-N hybrid for the selective epoxidation of styrene.

Table 7. Catalyst Screening in the Epoxidation of Styrene by O_2[a].

Entry	Catalyst	Conversion (%)	Epoxide selectivity (%)
1	Co_3O_4/RGO-N	97.7	72.7
2	Co_3O_4/RGO	57.3	70.9
3	Co_3O_4	38.6	67.6
4[b]	RGO-N	7.2	58.3
5	MnO_2	5.2	57.7
6	Fe_3O_4	32.2	63.4
7	CuO	3.5	42.9

[a]Reaction conditions: temperature (100°C), reaction time (3 h), the amount of M (0.15 mmol), styrene (8 mmol), DMF (25 mL), low rate of O_2 (20 mL min-1). [b]Reaction conditions: 7.8 mg RGO-N was added.

The flexibility of the Co_3O_4/RGO-N catalyst is further explained by using different types of olefins used as substrates under optimized conditions (Table **8**). Entry 2 reveals that using styrene with an electron-donating group such as 4-methylphenylene results in a similar conversion to styrene (entry 1), but it has high epoxide selectivity of 80.6%. Entry 3 shows that the presence of electron-withdrawing group on styrene results in lower conversion and lower epoxide selectivity of 70.7% as compared to unsubstituted styrene. Entry 4 shows that the use of ethyl cinnamate results gives a lower conversion of 55.6% but gives a high epoxide selectivity of 83.9%.

Table 8. Selective Oxidation of Alkenes and Alcohols Over the Co3O4/RGO-N Catalyst[a].

Entry	Substrates	T (h)	Conversion (%)	Product	Selectivity (%)
1[b]	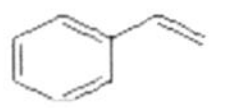	3	97.7	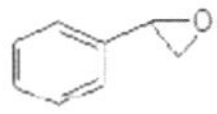	72.7
2	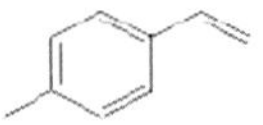	3	97.5	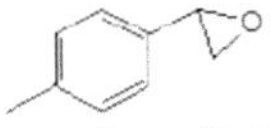	80.6
3		4	96.5	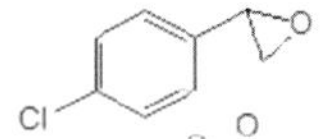	70.7
4	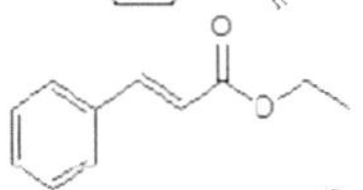	30	55.6	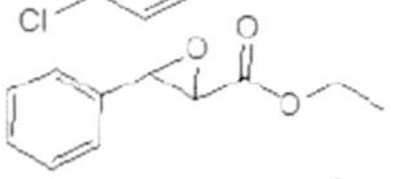	83.9
5	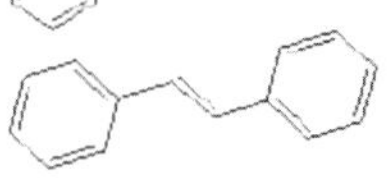	10	95.9	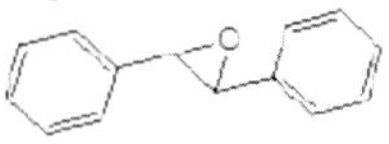	98.3
6[c]		2	>99.9	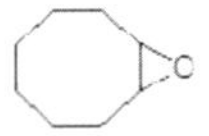	94.9
7[c]	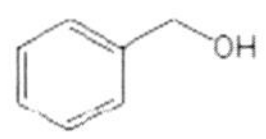	7	93.9	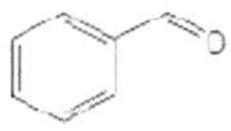	>99.9
8[c]	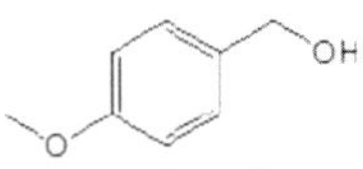	2	98.1	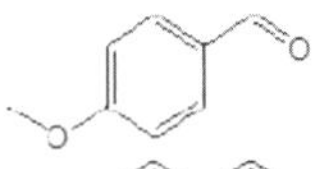	>99.9
9[c]	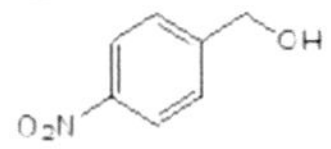	2	99.8	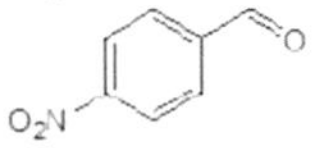	>99.9
10	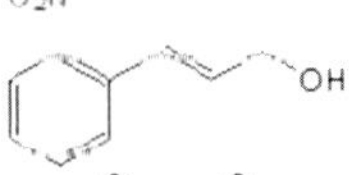	2.5	>99.9	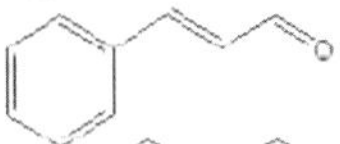	94.4
11[c]	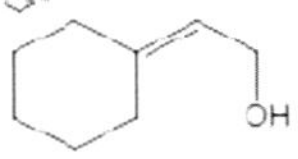	1	92.5	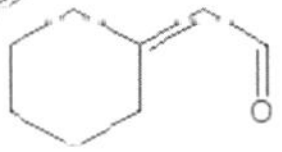	91.5
12[c]	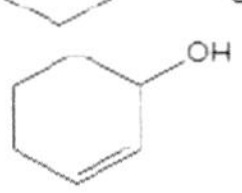	2	98.2	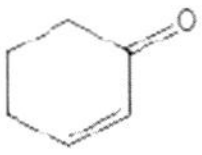	>99.9
13[c]	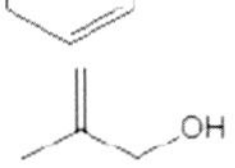	2	95.4	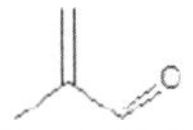	>99.9

[a]Reaction conditions: substrate (1.5 mmol), Co3O4/RGO-N (20 mg), DMF (25 mL), temperature (100◦C), low rate of O_2 (20 mL min-1). [b]8 mmol of the substrate was added. [c]sup> The reaction was carried out at 120◦C.

Entries 5 and 6 revels that by using trans-stilbene and cyclooctene higher selectivity (98.3% and 94.9%) and conversion (95.9% and >99.9%) is obtained.

Entries 7-9 show that the use of both electron-donating and electron-withdrawing substituents results in high conversion (>93%) and high selectivity (>99%). Entries 10-13 show that oxidation of allylic alcohols or other aliphatic unsaturated alcohols results in the respective unsaturated aldehydes/ketones without significant oxidation of the double bond.

REDUCTION REACTIONS

Reduction of Nitro Compounds

Nitro compounds such as 4-nitrophenol (4-NP) and 2-nitrophenol (2-NP) are highly soluble in water and are responsible for water pollution and they are referred to as priority pollutants. To date, several methods have been used to convert nitro pollutants into amines [68 - 72]. To remove the water pollutants generally used, three methods are photocatalytic degradation, catalytic reduction, and adsorption [73].

The catalytic reduction method is used to convert toxic nitrophenols into their corresponding amino phenols. The reduction process is an important method to remove toxic pollutants such as nitroarenes and convert them into amino arenes. The amino arenes play an important role in the synthesis of dyes, pesticides, pharmaceuticals, and have several advantages in the chemical industries [74 - 79]. Catalytic reduction is one of the most efficient methods used for the reduction of nitroarenes, by using the heterogenous catalysts the 4NP and 2-NP are converted into 4aminophenol (4-AP) and 2-aminophenol (2- respectively. Because simple processing, high efficiency, and economic nature of heterogenous catalyst make it a suitable reagent for the catalytic reduction of nitroarenes. Subsequently, the appropriate design and careful selection of the heterogeneous (nano) catalysts results in the enhancement of the reduction process.

At present several heterogeneous catalytic reduction processes have been discovered which does not use water-soluble toxic solvents [80 - 84]. Because of the presence of sp^2 hybridized carbon atomic layer in graphene, it becomes an appropriate candidate for catalytic reduction of nitroarenes. Geim and colleagues in 2004 reported that by using a peeling method having a scotch tape, single-layer graphene can be prepared [85]. Hexagonal flat layer carbon sheets of graphene have the special properties of material science, physics, mechanics, and chemistry [86 - 90] and it also has several advantages in hydrogen storage, catalysis, field emitters, and hydrogen storage. To prepare the heterogeneous catalyst graphene has a high tendency of immobilization of metal oxide between the layers or on the surface, because of above mention property graphene becomes an excellent (nano) catalyst support.

Graphene has very good conductivity, which is responsible for electron transfer during various transformations. In between the reaction steps, electric polarization is developed between the substrates and graphene, which is responsible for the generation of thermal and electrical conductivity that is useful in various processes [91 - 96]. In recent years several graphene derivatives such as functionalized graphene, reduced graphene oxide, and graphene oxide have gained significant attention [97 - 101]. Considering their excellent properties, graphene-based nanostructures with high catalytic activity are one of the outstanding selections for the synthesis of diverse organic compounds [102, 103]. With regards to the greener aspect of the catalytic (nano) systems, developing the most facile and environmentally benign strategies in the efficient remediation of toxic/hazardous nitroarenes is of specific significance worldwide [104].

Graphene-Based (nano) Catalysts in the Reduction of Nitro Compounds

GO and RGO

In 2012, Chen *et al.* reported that nitrogen-doped graphene (NG) can be used as a catalyst for electrochemical reduction of aromatic nitro compounds such as 2,4dinitrotoluene (2,4-DNT), 1,3-dinitrobenzene (1,3-DNB), 2,4,6-trinitrotoluene (TNT) and 1,3,5-trinitrobenzene (TNB) [105].

Colunga and his co-workers stated that using biogenic sulfide graphene oxide can undergo the catalytic reduction of various pollutants like 3-chloronitrobenzene (3CNB) and reactive red 2(RR2) [106]. It is found that the use of GO nanocatalyst in the reduction of 3-chloronitrobenzene results in a high product yield. Mahata and his co-workers, by using cashew leaf extract, discovered an economical, efficient, and described an efficient, economical, and green route for the synthesis of graphene oxide [108].

Au/Graphene-Based Nanocomposites

Adhikari and colleagues stated that GO-based supramolecular hydrogels can be prepared by using biologically active polyamines such as tris (aminoethyl) amine, spermine, and spermidine. Moreover, by using vitamin C reductant and greener route they synthesize the Ag, Au, Pt nanoparticles, and RGO-based hybrid hydrogels *via* the reduction of the metal ions and GO The obtained results reveal that metal nanoparticles are uniformly distributed on reduced graphene oxide nanosheet.

Fakhri and colleagues reported an easy and facile route for the synthesis of GO/Au NPs [109]. As shown in Scheme **5** that prepared GO/Au NPs undergo the catalytic reduction of nitro compounds at ambient temperature and catalyst can be reused several times.

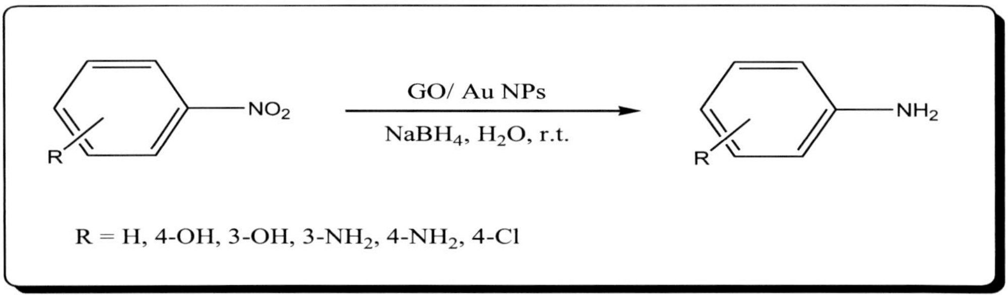

Scheme 5. Catalytic reduction of the nitroarenes using GO/AuNPs.

Ag/Graphene-based Nanocomposites

In 2013, Jeon and his co-workers suggested an easy and facile method for the synthesis of Ag nanoparticles decorated on GO surface-functionalized with them mussel inspired dopamine (GO-Dopa) [110]. It is found that the green synthesized Ag/GO- Dopa hybrid by using $NaBH_4$ undergoes the catalytic reduction of 4-NP to 4-AP. Paul and colleagues in 2016 reported a novel method for the hydrothermal synthesis of Ag/ Fe_2O_3 present on rGO surface by using urea as a hydroxylating agent and polyethylene glycol (PEG4000) as a surfactant [111]. As shown in Scheme **6** synthesized Ag/ Fe_2O_3 -rGO nanocomposite undergoes the effective catalytic reduction of nitro compounds into their respective amines and by using an external magnet catalyst can be reused five times.

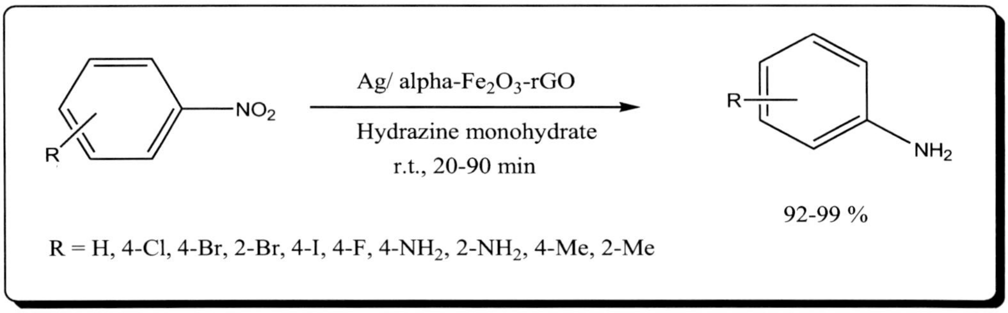

Scheme 6. Catalytic reduction of nitroarenes using Ag/Fe_2O_3-rGO nanocomposite.

Pd/Graphene-Based Nanocomposites

Li and his co-workers developed a synthesis method for the preparation of ultrafine Pd nanocrystals (PdNCs) immobilized on the SnO_2-decorated graphene nanosheets *via* immobilization of Pd NPs with 3.4 nm average size on the PDDA functionalized SnO_2-GNS (SnO_2-GNS) [112]. As shown in Scheme **7** that by using $NaBH_4$.Pd NPs/ RGO the Pd NCs/ SnO_2 -GNS undergoes the extraordinary reduction of 4-nitrophenols by $NaBH_4$ in an EtOH and H_2O (1:2) mixture and the catalyst can be reused for 5 times.

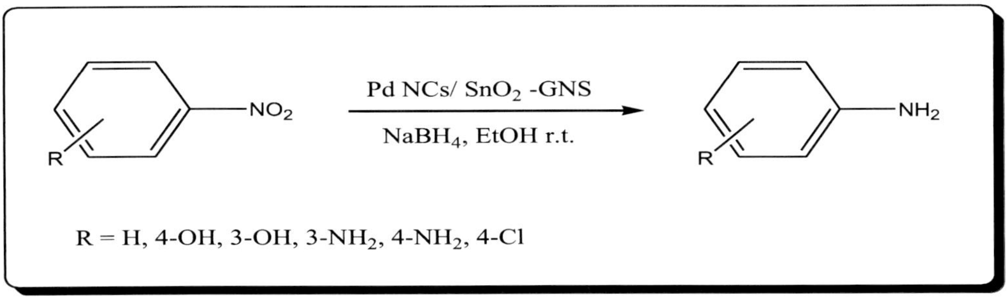

Scheme 7. The reduction of nitroarenes using Pd NPs/RGO.

Zhang *et al.* in 2014 reported an easy and facile route for the synthesis of Go nanocomposite decorated by Pd nanoparticles [113]. It is evident for Scheme **8** that Pd/GO nanocatalyst gives that exceptional catalytic activity during the reduction of aromatic nitro compounds by using $NaBH_4$. The catalyst can be reused 6 times.

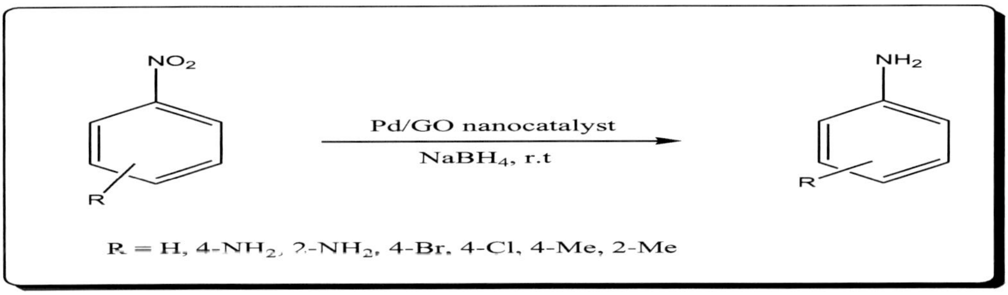

Scheme 8. The catalytic activity of Pd/GO nanocatalyst in nitroarenes reduction by $NaBH_4$.

Ni/Graphene-based Nanocomposites

Kubendhiran *et al.* in 2017 by employing reduction of GO with caffeic acid(CA) and the non-covalent π-π stacking interaction between Ni-TPP and RGO fabricated an RGO/nickel tetraphenyl porphyrin (GRGO/Ni-TPP) nanocomposite modified electrode [114]. The above-prepared nanocomposite GRGO/Ni-TPP is used for the reduction of nitrobenzene and it is found that GRGO/Ni-TPP modified electrode results in high product yield and high selectivity.

Karthik and his co-workers reported RGO-Ni nanocomposite is synthesized by using L-ascorbic acid as a reducing and capping agent by reduction of GO and $NiCl_2.6H_2O$ under hydrothermal conditions [115]. From Scheme **9** it is concluded that by using $NaBH_4$ the RGO-Ni heterogeneous catalyst results in high catalytic activity during the catalytic reduction of aromatic nitro compounds. By using the magnetic separation method, the above catalyst can be recovered and reused 5 times.

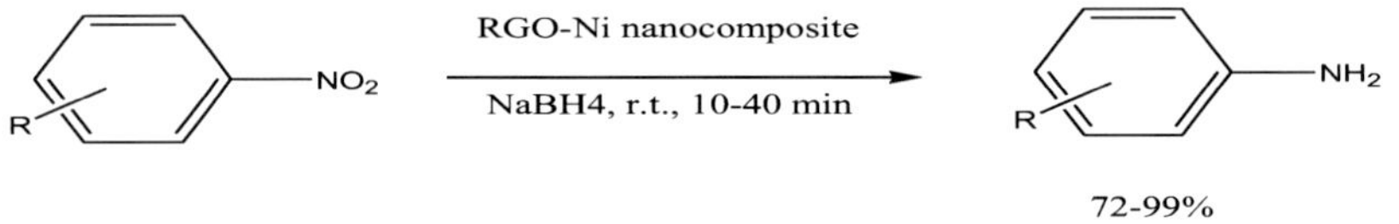

R= H, 4-Me, 4-F, 4-Cl, 4-I, 4-Br, 2-Br, 4-OH, 2-OH, 2-NH_2, 4-COOH, 4-COMe, 4-CN, 3-CN, 5- chloro-3-methyl-2-nitropyridine, 4-Cl-2-OH, 3-Br-5-OMe, 2-NH_2-4-Br, 1-nitrobutane, 1-nitrohexan, 2-methyl-4-nitroquinoline

Scheme 9. Catalytic reduction of nitroarenes using RGO-N in a nanocomposite.

Pt/Graphene-based Nanocomposites

Li and his colleagues in 2015 reported a simple method for the fabrication of (Pt NCs/rGO) decorated on rGO by using a water-based, co-chemical reduction process using polyallylamine hydrochloride as a multi-functional agent in the functionalization of GO [116]. It is found that by using $NaBH_4$ the Pt-NCs/rGO composite results in high catalytic activity during the reduction of 4-Nitrophenols and the synthesized catalyst can be used 5 times.

In 2017, Sahoo and his co-workers by using graphene oxide and metal precursor, reported an efficient and green method for the synthesis of d metals (Au, Pd, and Pt)/ 3D-graphene nanocomposites [117]. The metals/3D-graphene nano-composites by using $NaBH_4$ at ambient temperature result in a high catalytic reduction of 4-NP and MB. The catalytic activity rate of metals is as Pt/>Pd/ >Au/3D-graphene for both the pollutants, showing a minimum number of metal nanoparticles results in excellent catalytic activity. The prepared catalyst can be reused 5 times.

Cu/Graphene-based Nanocomposites

In 2018, Liu and Guo developed a simple approach to fabricate the SH-β--D-RGO/Cu NSs nanocomposite by depositing the Cu nanospheres on thio---cyclodextrin (SH-βCD) functionalized RGO (SH-β-CD-RGO) [118]. The aforementioned nanocomposite plays an important role in the catalytic reduction of 4-Nitrophenol.

Bimetallic/Graphene-based Nanocomposites

Feng and colleagues fabricated the graphene decorated Pd-Cu NPs by employing co-reduction of $CuCl_2.H_2O$ and $PdCl_2$ [119]. As shown in Scheme **10** that Pd Cu-graphene undergoes the effective nitroarene reduction by using $NaBH_4$ in EtOH: H_2O (1:2) at 50 °C. The obtained results reveal that the Pd Cu-graphene can be reused up to six times.

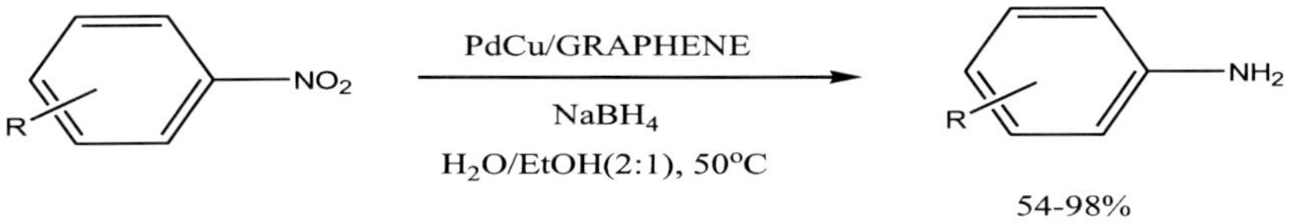

R= H, 4-Me, 3-Me, 2-Me, 4-OMe, 4-OH, 3-OH, 4-CH_2OH, 2-Me^{-4}-Cl, 4-Cl, 4-Br, 4-NH_2, 4-$CONH_2$, 4-CN, 4-COOMe, [1,1'-biphenyl]$^{-2}$-amine, 1H-indol^{-5}-amine

Scheme 10. Catalytic reduction of nitroarenes using Pd Cu-graphene.

In 2013, Goksu and co-workers reported an efficient method for the fabrication of rGO decorated $Co_{30}Pd_{70}$ alloy (rGO- $Co_{30}Pd_{70}$)NPs by co-reduction of cobalt(II) and Pd(II) acetyl acetates in borane-tertbutyl amine (BTB) and oleyl amine (OAm) mixture [120]. It is clear from Scheme **11** that rGO- $Co_{30}Pd_{70}$ catalyst by using ammonia borane in H_2O/methanol mixture undergoes the catalytic reduction of aromatic nitro compounds into their respective amines and results in good product yield.

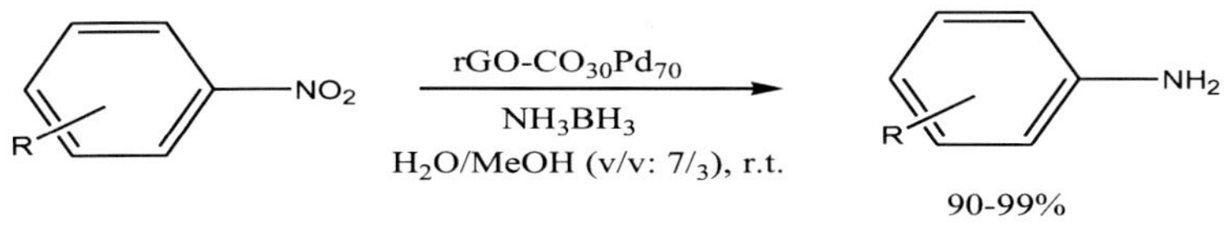

R= H, 4-Me, 2-OMe, 3-OH, 4-OH, 1-naphthalene, 3-nitro^{-9}H-fluorene, 2-choloro^{-5}-nitropyridine, 3-NO_2, 2-NH_2, 4-NH_2, 4-CN

Scheme 11. Catalytic reduction of nitroarenes using rGO-Co30Pd70.

Fang and his coworkers developed an easy and eco-friendly method for the synthesis of graphene/CuNi nanocomposite. They synthesized the two-dimensional graphene decorated Cu Ni with 2D nanoparticles of particle size 2 nm. From Scheme **12** it is clear that the graphene/CuNi nanocomposite undergoes the selective catalytic reduction of aromatic nitro compounds by using $NaBH_4$ and the catalyst can be reused up to 25 times.

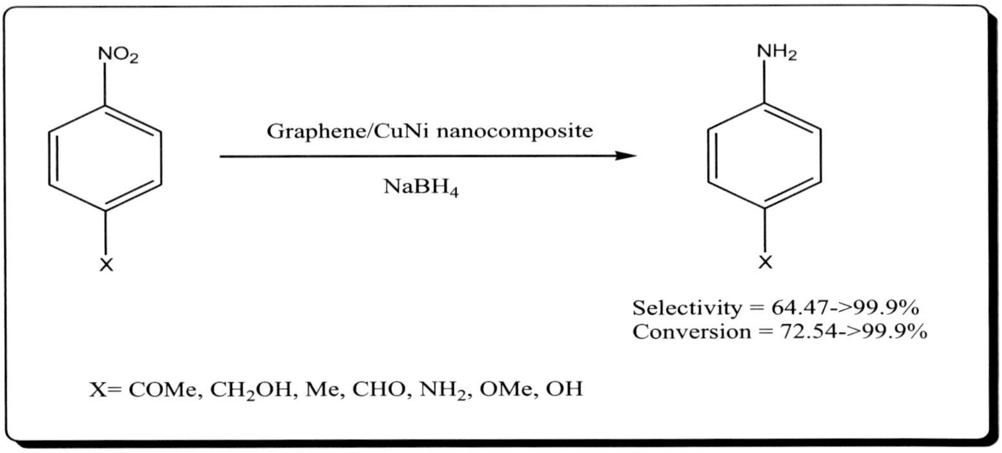

Scheme 12. Reduction of nitroarenes by graphene/CuNi nanocomposite.

CONCLUDING REMARKS

The previous studies reveal that the metal-free GO has many catalytic applications. The surface area of GO is high, the adsorption capacity of GO towards hydrocarbons, gases, and ions is very high, and also the reaction of GO with oxygenating functionalities resulted in various paths for the introduction and modification of various functional groups.

Although lots of studies are done on the numerous catalytic application of GO, still there are lots of challenges that should be examined and resolved. Among them, the three main important challenges are: (1) As we know that the performance of the GO catalytic system is mainly dependent on composition and structure, so there is a need to develop such methods which undergo the precise characterization for the identification of active sites. (2) Much care should be taken to eliminate the probability that the catalytic activity could be recognized to metal contaminants, even at a trace level. (3) Further understanding and more research are needed to describe the potential cooperative effect of Go on the catalytic activity. To design the balanced metal-free catalyst systems it is necessary to understand the advantages of graphene oxide over the activated carbon and traditional metal oxide. However, the metal-free carbon catalyst has the above drawbacks. It can be used to solve the various industrial problems related to the traditional catalysts. Since graphene oxide catalysts are synthesized environmentally friendly material, their applications should be explored in green synthesis.

LIST OF ABBREVIATIONS

0D	Zero dimensional
1D	One dimensional
2D	Two dimensional
AP	Aminophenol
BTB	Borane-tertbutyl amine
CVD	Chemical vapor deposition
CuTSPc@3D-(N)GFs	Copper(II) tetrasulfo phthalocyanine reinforced on three-dimensional nitrogen-doped graphene-based frameworks
CNB	Chloronitrobenzene
Cu NSs	Copper nanospheres
CMGs	Chemically modified Graphenes
DCC	N,N-dicyclohexylcarbodiimide
DMF	Dimethylformamide
DMA	Dimethylacetamide
DMSO	Dimethyl sulfoxide
DNT	Dinitrotoluene
DNB	Dinitrobenzene
EDC	1-ethyl-3-(3-dimethyl aminopropyl) carbodiimide
EDCI	N-(3-dimethyl aminopropyl)-N-ethyl carbodiimide hydrochloride
FGO	Functionalized graphene oxide
GO	Graphene oxide
GNS	Graphene nanosheets
IGO	Improved graphene oxide
MOFs	Metal–organic frameworks
MB	Methylene Blue
NMR	Nuclear magnetic resonance
NP	Nitrophenol
NG	Nitrogen-doped graphene
NPs	Nanoparticles
NCs	Nanocrystals

ORR	Oxygen reduction reaction
OAm	Oleyl amine
rGO	Reduced graphene oxide
RR	Reactive red
R.t.	Room temperature
SH-βCD	Thio-β-cyclodextrin
TNT	Trinitrotoluene
TNB	Trinitrobenzene
TPP	Tetraphenyl porphyrin

CONSENT FOR PUBLICATION

Not applicable.

CONFLICT OF INTEREST

The author declares no conflict of interest, financial or otherwise.

ACKNOWLEDGEMENTS

Declared none.

REFERENCES

[1] Geim, A.K.; Novoselov, K.S. The rise of graphene. *Nat. Mater.*, **2007**, *6*(3), 183-191. [http://dx.doi.org/10.1038/nmat1849] [PMID: 17330084]

[2] Georgakilas, V.; Otyepka, M.; Bourlinos, A.B.; Chandra, V.; Kim, N.; Kemp, K.C.; Hobza, P.; Zboril, R.; Kim, K.S. Functionalization of graphene: covalent and non-covalent approaches, derivatives and applications. *Chem. Rev.*, **2012**, *112*(11), 6156-6214. [http://dx.doi.org/10.1021/cr3000412] [PMID: 23009634]

[3] Huang, X.; Yin, Z.; Wu, S.; Qi, X.; He, Q.; Zhang, Q.; Yan, Q.; Boey, F.; Zhang, H. Graphene-based materials: synthesis, characterization, properties, and applications. *Small,* **2011**, *7*(14), 1876-1902. [http://dx.doi.org/10.1002/smll.201002009] [PMID: 21630440]

[4] Sun, X.; Wang, R.; Su, D. Research progress in metal-free carbon-based catalysts. *Chin. J. Catal.*, **2013**, *34*(3), 508-523. [http://dx.doi.org/10.1016/S1872-2067(11)60515-9]

[5] Yu, D.; Nagelli, E.; Du, F.; Dai, L. Metal-free carbon nanomaterials become more active than metal catalysts and last longer. *J. Phys. Chem. Lett.*, **2010**, *1*(14), 2165-2173. [http://dx.doi.org/10.1021/jz100533t]

[6] Lotya, M.; Hernandez, Y.; King, P.J.; Smith, R.J.; Nicolosi, V.; Karlsson, L.S.; Blighe, F.M.; De, S.; Wang, Z.; McGovern, I.T.; Duesberg, G.S.; Coleman, J.N. Liquid phase production of graphene by exfoliation of graphite in surfactant/water solutions. *J. Am. Chem. Soc.*, **2009**, *131*(10), 3611-3620. [http://dx.doi.org/10.1021/ja807449u] [PMID: 19227978]

[7] Kim, K.S.; Zhao, Y.; Jang, H.; Lee, S.Y.; Kim, J.M.; Kim, K.S.; Ahn, J.H.; Kim, P.; Choi, J.Y.; Hong, B.H. Large-scale pattern growth of graphene films for stretchable transparent electrodes. *Nature,* **2009**,

457(7230), 706-710.
[http://dx.doi.org/10.1038/nature07719] [PMID: 19145232]

[8] Brodie, B.C. XIII. On the atomic weight of graphite. *Philos. Trans. R. Soc. Lond.,* **1859**, *149*, 249-259. [http://dx.doi.org/10.1098/rstl.1859.0013]

[9] Staudenmaier, L. Verfahren zur Darstellung der Graphitsäure. *Ber. Dtsch. Chem. Ges.,* **1898**, *31*(2), 1481-1487.
[http://dx.doi.org/10.1002/cber.18980310237]

[10] Hummers, W.S., Jr; Offeman, R.E. Preparation of graphitic oxide. *J. Am. Chem. Soc.,* **1958**, *80*(6), 1339-1339.
[http://dx.doi.org/10.1021/ja01539a017]

[11] Stankovich, S.; Dikin, D.A.; Dommett, G.H.B.; Kohlhaas, K.M.; Zimney, E.J.; Stach, E.A.; Piner, R.D.; Nguyen, S.T.; Ruoff, R.S. Graphene-based composite materials. *Nature,* **2006**, *442*(7100), 282-286.
[http://dx.doi.org/10.1038/nature04969] [PMID: 16855586]

[12] Lerf, A.; He, H.; Forster, M.; Klinowski, J. Structure of graphite oxide revisited. *J. Phys. Chem. B,* **1998**, *102*(23), 4477-4482.
[http://dx.doi.org/10.1021/jp9731821]

[13] Zhou, M.; Wang, Y.; Zhai, Y.; Zhai, J.; Ren, W.; Wang, F.; Dong, S. Controlled synthesis of large-area and patterned electrochemically reduced graphene oxide films. *Chemistry,* **2009**, *15*(25), 6116-6120.
[http://dx.doi.org/10.1002/chem.200900596] [PMID: 19444826]

[14] Park, J.S.; Cho, S.M.; Kim, W.J.; Park, J.; Yoo, P.J. Fabrication of graphene thin films based on layer-by-layer self-assembly of functionalized graphene nanosheets. *ACS Appl. Mater. Interfaces,* **2011**, *3*(2), 360-368.
[http://dx.doi.org/10.1021/am100977p] [PMID: 21207942]

[15] Cao, H.; Wu, X.; Yin, G.; Warner, J.H. Synthesis of adenine-modified reduced graphene oxide nanosheets. *Inorg. Chem.,* **2012**, *51*(5), 2954-2960.
[http://dx.doi.org/10.1021/ic2022402] [PMID: 22356685]

[16] Liu, Q.; Shi, J.; Sun, J.; Wang, T.; Zeng, L.; Jiang, G. Graphene and graphene oxide sheets supported on silica as versatile and high-performance adsorbents for solid-phase extraction. *Angew. Chem. Int. Ed.,* **2011**, *50*(26), 5913-5917.
[http://dx.doi.org/10.1002/anie.201007138] [PMID: 21567667]

[17] Sun, X.; Liu, Z.; Welsher, K.; Robinson, J.T.; Goodwin, A.; Zaric, S.; Dai, H. Nano-graphene oxide for cellular imaging and drug delivery. *Nano Res.,* **2008**, *1*(3), 203-212.
[http://dx.doi.org/10.1007/s12274-008-8021-8] [PMID: 20216934]

[18] Jin, L.; Yang, K.; Yao, K.; Zhang, S.; Tao, H.; Lee, S.T.; Liu, Z.; Peng, R. Functionalized graphene oxide in enzyme engineering: a selective modulator for enzyme activity and thermostability. *ACS Nano,* **2012**, *6*(6), 4864-4875.
[http://dx.doi.org/10.1021/nn300217z] [PMID: 22574614]

[19] Kou, L.; He, H.; Gao, C. Click chemistry approach to functionalize two-dimensional macromolecules of graphene oxide nanosheets. *Nano-Micro Lett.,* **2010**, *2*(3), 177-183.
[http://dx.doi.org/10.1007/BF03353638]

[20] Lee, D.; Choi, M.C.; Ha, C.S. Polynorbornene dicarboximide/amine functionalized graphene hybrids for potential oxygen barrier films. *J. Polym. Sci. A Polym. Chem.,* **2012**, *50*(8), 1611-1621.
[http://dx.doi.org/10.1002/pola.25932]

[21] Wang, B.; Luo, B.; Liang, M.; Wang, A.; Wang, J.; Fang, Y.; Chang, Y.; Zhi, L. Chemical amination of graphene oxides and their extraordinary properties in the detection of lead ions. *Nanoscale,* **2011**, *3*(12), 5059-5066.

[http://dx.doi.org/10.1039/c1nr10901d] [PMID: 22041992]

[22] Zhang, W.; Wang, S.; Ji, J.; Li, Y.; Zhang, G.; Zhang, F.; Fan, X. Primary and tertiary amines bifunctional graphene oxide for cooperative catalysis. *Nanoscale,* **2013**, *5*(13), 6030-6033. [http://dx.doi.org/10.1039/c3nr01323e] [PMID: 23714770]

[23] Wang, H.; Maiyalagan, T.; Wang, X. Review on recent progress in nitrogen-doped graphene: synthesis, characterization, and Its potential applications. *ACS Catal.,* **2012**, *2*(5), 781-794. [http://dx.doi.org/10.1021/cs200652y]

[24] Fu, J.; Shi, F.; Thompson, L.T., Jr; Lu, X.; Savage, P.E. Activated carbons for hydrothermal decarboxylation of fatty acids. *ACS Catal.,* **2011**, *1*(3), 227-231. [http://dx.doi.org/10.1021/cs1001306]

[25] Dreyer, D.R.; Bielawski, C.W. Carbocatalysis: Heterogeneous carbons finding utility in synthetic chemistry. *Chem. Sci. (Camb.),* **2011**, *2*(7), 1233-1240. [http://dx.doi.org/10.1039/c1sc00035g]

[26] Dreyer, D.R.; Jia, H.P.; Bielawski, C.W. Graphene oxide: a convenient carbocatalyst for facilitating oxidation and hydration reactions. *Angew. Chem. Int. Ed.,* **2010**, *49*(38), 6686. [http://dx.doi.org/10.1002/anie.201003238] [PMID: 20602388]

[27] Dreyer, D.R.; Jia, H.P.; Todd, A.D.; Geng, J.; Bielawski, C.W. Graphite oxide: a selective and highly efficient oxidant of thiols and sulfides. *Org. Biomol. Chem.,* **2011**, *9*(21), 7292-7295. [http://dx.doi.org/10.1039/c1ob06102j] [PMID: 21909587]

[28] Jia, H.P.; Dreyer, D.R.; Bielawski, C.W. C–H oxidation using graphite oxide. *Tetrahedron,* **2011**, *67*(24), 4431-4434. [http://dx.doi.org/10.1016/j.tet.2011.02.065]

[29] Jia, H.P.; Dreyer, D.R.; Bielawski, C.W. Graphite oxide as an auto-tandem oxidation-hydration-aldol coupling catalyst. *Adv. Synth. Catal.,* **2011**, *353*(4), 528-532. [http://dx.doi.org/10.1002/adsc.201000748]

[30] Vijay Kumar, A.; Rama Rao, K. Recyclable graphite oxide catalyzed Friedel–Crafts addition of indoles to α,β-unsaturated ketones. *Tetrahedron Lett.,* **2011**, *52*(40), 5188-5191. [http://dx.doi.org/10.1016/j.tetlet.2011.08.002]

[31] Verma, S.; Mungse, H.P.; Kumar, N.; Choudhary, S.; Jain, S.L.; Sain, B.; Khatri, O.P. Graphene oxide: an efficient and reusable carbocatalyst for aza-Michael addition of amines to activated alkenes. *Chem. Commun. (Camb.),* **2011**, *47*(47), 12673-12675. [http://dx.doi.org/10.1039/c1cc15230k] [PMID: 22039588]

[32] Dreyer, D.R.; Jarvis, K.A.; Ferreira, P.J.; Bielawski, C.W. Graphite oxide as a carbocatalyst for the preparation of fullerene-reinforced polyester and polyamide nanocomposites. *Polym. Chem.,* **2012**, *3*(3), 757-766. [http://dx.doi.org/10.1039/c2py00545j]

[33] Supale, A.; Gokavi, G. Oxidation of thiols to disulfides using H_2O_2 catalyzed by recyclable chromic potassium sulphate at room temperature. *React. Kinet. Catal. Lett.,* **2008**, *93*(1), 141-148. [http://dx.doi.org/10.1007/s11144-008-5223-8]

[34] Goifman, A.; Gun, J.; Gelman, F.; Ekeltchik, I.; Worch, E.; Lev, O. Catalytic oxidation of thiol compounds by novel fuel cell□inspired Co□porphyrin and Co□imidazole catalysts. *Isr. J. Chem.,* **2006**, *46*(1), 17-26. [http://dx.doi.org/10.1560/8J05-UHG8-K6DN-VVB2]

[35] Postel, M.; Duñach, E. Bismuth derivatives for the oxidation of organic compounds. *Coord. Chem. Rev.,* **1996**, *155*, 127-144. [http://dx.doi.org/10.1016/0010-8545(96)90179-4]

[36] Qiu, X.; Len, C.; Luque, R.; Li, Y. Solventless oxidative coupling of amines to imines by using transition-metal-free metal-organic frameworks. *ChemSusChem,* **2014**, *7*(6), 1684-1688.

[http://dx.doi.org/10.1002/cssc.201301340] [PMID: 24801486]

[37] Jaiswal, G.; Landge, V.G.; Jagadeesan, D.; Balaraman, E. Sustainable iron-catalyzed direct imine formation by acceptorless dehydrogenative coupling of alcohols with amines. *Green Chem.*, **2016**, *18*(11), 3232-3238.
[http://dx.doi.org/10.1039/C6GC00565A]

[38] Zhang, B.; Guo, X.W.; Liang, H.; Ge, H.; Gu, X.; Chen, S.; Yang, H.; Qin, Y. Tailoring Pt–Fe_2O_3 Interfaces for Selective Reductive Coupling Reaction To Synthesize Imine. *ACS Catal.*, **2016**, *6*(10), 6560-6566.
[http://dx.doi.org/10.1021/acscatal.6b01756]

[39] Xu, Y.; Chen, Y.; Fu, W.F. Visible-light driven oxidative coupling of amines to imines with high selectivity in air over core-shell structured CdS@C_3N_4. *Appl. Catal. B*, **2018**, *236*, 176-183.
[http://dx.doi.org/10.1016/j.apcatb.2018.03.098]

[40] Chen, B.; Shang, S.; Wang, L.; Zhang, Y.; Gao, S. Mesoporous carbon derived from vitamin B_{12} : a high-performance bifunctional catalyst for imine formation. *Chem. Commun. (Camb.)*, **2016**, *52*(3), 481-484.
[http://dx.doi.org/10.1039/C5CC06179B] [PMID: 26529323]

[41] Naya, S.; Kimura, K.; Tada, H. One-step selective aerobic oxidation of amines to imines by gold nanoparticle-loaded rutile titanium (iv) oxide plasmon photocatalyst. *ACS Catal.*, **2013**, *3*(1), 10-13.
[http://dx.doi.org/10.1021/cs300682d]

[42] Biswas, S.; Dutta, B.; Mullick, K.; Kuo, C.H.; Poyraz, A.S.; Suib, S.L. Aerobic oxidation of amines to imines by cesium-promoted mesoporous manganese oxide. *ACS Catal.*, **2015**, *5*(7), 4394-4403.
[http://dx.doi.org/10.1021/acscatal.5b00325]

[43] Tang, L.; Sun, H.; Li, Y.; Zha, Z.; Wang, Z. Highly active and selective synthesis of imines from alcohols and amines or nitroarenes catalyzed by Pd/DNA in water with dehydrogenation. *Green Chem.*, **2012**, *14*(12), 3423-3428.
[http://dx.doi.org/10.1039/c2gc36312g]

[44] Chen, B.; Wang, L.; Gao, S. Recent advances in aerobic oxidation of alcohols and amines to imines. *ACS Catal.*, **2015**, *5*(10), 5851-5876.
[http://dx.doi.org/10.1021/acscatal.5b01479]

[45] Largeron, M.; Fleury, M.B. A biologically inspired Cu(I)/topaquinone-like co-catalytic system for the highly atom-economical aerobic oxidation of primary amines to imines. *Angew. Chem. Int. Ed.*, **2012**, *51*(22), 5409-5412.
[http://dx.doi.org/10.1002/anie.201200587] [PMID: 22499249]

[46] Xu, J.; Zhuang, R.; Bao, L.; Tang, G.; Zhao, Y. KOH-mediated transition metal-free synthesis of imines from alcohols and amines. *Green Chem.*, **2012**, *14*(9), 2384-2387.
[http://dx.doi.org/10.1039/c2gc35714c]

[47] Zhang, H.; Wu, C.; Wang, W.; Bu, J.; Zhou, F.; Zhang, B.; Zhang, Q. Effect of Ceria on redox-catalytic property in mild condition: A solvent-free route for imine synthesis at low temperature. *Appl. Catal. B*, **2018**, *227*, 209-217.
[http://dx.doi.org/10.1016/j.apcatb.2018.01.012]

[48] Chen, B.; Wang, L.; Dai, W.; Shang, S.; Lv, Y.; Gao, S. Metal free and solvent-free oxidative coupling of amines to imines with mesoporous carbon from macrocyclic compounds. *ACS Catal.*, **2015**, *5*(5), 2788-2794.
[http://dx.doi.org/10.1021/acscatal.5b00244]

[49] Hirakawa, H.; Katayama, M.; Shiraishi, Y.; Sakamoto, H.; Wang, K.; Ohtani, B.; Ichikawa, S.; Tanaka, S.; Hirai, T. One-pot synthesis of imines from nitroaromatics and alcohols by tandem photocatalytic and catalytic reactions on Degussa (Evonik) P25 titanium dioxide. *ACS Appl. Mater. Interfaces*, **2015**, *7*(6), 3797-3806.
[http://dx.doi.org/10.1021/am508769x] [PMID: 25621386]

[50] Su, C.; Loh, K.P. Carbocatalysts: graphene oxide and its derivatives. *Acc. Chem. Res.*, **2013**, *46*(10), 2275-2285.
[http://dx.doi.org/10.1021/ar300118v] [PMID: 23270430]

[51] Xie, J.; Li, L.; Sun, L.; Pei, Z.; Wen, B.; Xing, B. Reduced graphene oxide-catalyzed oxidative coupling reaction of 4-methoxyphenol in aerobic aqueous solution. *Carbon,* **2017**, *121*, 418-425.
[http://dx.doi.org/10.1016/j.carbon.2017.06.016]

[52] Fan, X.; Peng, W.; Li, Y.; Li, X.; Wang, S.; Zhang, G.; Zhang, F. Deoxygenation of exfoliated graphite oxide under alkaline conditions: a green route to graphene preparation. *Adv. Mater.*, **2008**, *20*(23), 4490-4493.
[http://dx.doi.org/10.1002/adma.200801306]

[53] Pedrosa, M.; Pastrana-Martínez, L.M.; Pereira, M.F.R.; Faria, J.L.; Figueiredo, J.L.; Silva, A.M.T. N/S-doped graphene derivatives and TiO_2 for catalytic ozonation and photocatalysis of water pollutants. *Chem. Eng. J.*, **2018**, *348*, 888-897.
[http://dx.doi.org/10.1016/j.cej.2018.04.214]

[54] Marinescu, C.; Ben Ali, M.; Hamdi, A.; Cherifi, Y.; Barras, A.; Coffinier, Y.; Somacescu, S.; Raditoiu, V.; Szunerits, S.; Boukherroub, R. Cobalt phthalocyanine-supported reduced graphene oxide: A highly efficient catalyst for heterogeneous activation of peroxymonosulfate for rhodamine B and pentachlorophenol degradation. *Chem. Eng. J.*, **2018**, *336*, 465-475.
[http://dx.doi.org/10.1016/j.cej.2017.12.009]

[55] Su, C.; Acik, M.; Takai, K.; Lu, J.; Hao, S.; Zheng, Y.; Wu, P.; Bao, Q.; Enoki, T.; Chabal, Y.J.; Ping Loh, K. Probing the catalytic activity of porous graphene oxide and the origin of this behaviour. *Nat. Commun.*, **2012**, *3*(1), 1298-1306.
[http://dx.doi.org/10.1038/ncomms2315] [PMID: 23250428]

[56] Yang, F.; Fan, X.; Wang, C.; Yang, W.; Hou, L.; Xu, X.; Feng, A.; Dong, S.; Chen, K.; Wang, Y.; Li, Y. P-doped nanomesh graphene with high-surface-area as an efficient metal-free catalyst for aerobic oxidative coupling of amines. *Carbon,* **2017**, *121*, 443-451.
[http://dx.doi.org/10.1016/j.carbon.2017.05.101]

[57] Su, C.; Tandiana, R.; Balapanuru, J.; Tang, W.; Pareek, K.; Nai, C.T.; Hayashi, T.; Loh, K.P. Tandem catalysis of amines using porous graphene oxide. *J. Am. Chem. Soc.*, **2015**, *137*(2), 685-690.
[http://dx.doi.org/10.1021/ja512470t] [PMID: 25551171]

[58] Zhao, Q.; Chen, D.; Li, Y.; Zhang, G.; Zhang, F.; Fan, X. Rhodium complex immobilized on graphene oxide as an efficient and recyclable catalyst for hydrogenation of cyclohexene. *Nanoscale,* **2013**, *5*(3), 882-885.
[http://dx.doi.org/10.1039/C2NR33290F] [PMID: 23238302]

[59] Li, Y.; Chen, H.; Voo, L.Y.; Ji, J.; Zhang, G.; Zhang, G.; Zhang, F.; Fan, X. Synthesis of partially hydrogenated graphene and brominated graphene. *J. Mater. Chem.*, **2012**, *22*(30), 15021-15024.
[http://dx.doi.org/10.1039/c2jm32307a]

[60] Huang, H.; Huang, J.; Liu, Y.M.; He, H.Y.; Cao, Y.; Fan, K.N. Graphite oxide as an efficient and durable metal-free catalyst for aerobic oxidative coupling of amines to imines. *Green Chem.*, **2012**, *14*(4), 930-934.
[http://dx.doi.org/10.1039/c2gc16681j]

[61] Bermudez, J.M.; Menendez, J.A.; Arenillas, A.; Martinez-Palou, R.; Romero, A.A.; Luque, R. Selectivity matters: Graphene oxide-mediated oxidative coupling of benzylamine to N□benzylidine---phenylmethanamine under microwave irradiation. *J. Mol. Catal. Chem.*, **2015**, *406*, 19-22.
[http://dx.doi.org/10.1016/j.molcata.2015.05.010]

[62] He, H.; Li, Z.; Li, K.; Lei, G.; Guan, X.; Zhang, G.; Zhang, F.; Fan, X.; Peng, W.; Li, Y. Bifunctional Graphene-Based Metal-Free Catalysts for Oxidative Coupling of Amines. *ACS Appl. Mater. Interfaces,* **2019**, *11*(35), 31844-31850.
[http://dx.doi.org/10.1021/acsami.9b08741] [PMID: 31382740]

[63] Yang, J.H.; Sun, G.; Gao, Y.; Zhao, H.; Tang, P.; Tan, J.; Lu, A.H.; Ma, D. Direct catalytic oxidation of benzene to phenol over metal-free graphene-based catalyst. *Energy Environ. Sci.*, **2013**, *6*(3), 793-800.
[http://dx.doi.org/10.1039/c3ee23623d]

[64] Long, J.; Xie, X.; Xu, J.; Gu, Q.; Chen, L.; Wang, X. Nitrogendoped graphene nanosheets as metal-free catalysts for aerobic selective oxidation of benzylic alcohols. *ACS Catal.*, **2012**, *2*(4), 622-631.
[http://dx.doi.org/10.1021/cs3000396]

[65] Zhang, P.; Gong, Y.; Li, H.; Chen, Z.; Wang, Y. Solvent-free aerobic oxidation of hydrocarbons and alcohols with Pd@N-doped carbon from glucose. *Nat. Commun.*, **2013**, *4*(1), 1593-1604.
[http://dx.doi.org/10.1038/ncomms2586] [PMID: 23481401]

[66] Mahyari, M.; Laeini, M.S.; Shaabani, A. Aqueous aerobic oxidation of alkyl arenes and alcohols catalyzed by copper(II) phthalocyanine supported on three-dimensional nitrogen-doped graphene at room temperature. *Chem. Commun. (Camb.)*, **2014**, *50*(58), 7855-7857.
[http://dx.doi.org/10.1039/C4CC01406E] [PMID: 24912023]

[67] Nie, R.; Shi, J.; Du, W.; Ning, W.; Hou, Z.; Xiao, F.S. A sandwich N-doped graphene/Co_3O_4 hybrid: an efficient catalyst for selective oxidation of olefins and alcohols. *J. Mater. Chem. A Mater. Energy Sustain.*, **2013**, *1*(32), 9037-9045.
[http://dx.doi.org/10.1039/c3ta11672g]

[68] Vellaichamy, B.; Prakash, P.; Thomas, J. Synthesis of AuNPs@RGO nanosheets for sustainable catalysis toward nitrophenols reduction. *Ultrason. Sonochem.*, **2018**, *48*, 362-369.
[http://dx.doi.org/10.1016/j.ultsonch.2018.05.012] [PMID: 30080561]

[69] Sarvestani, M.; Azadi, R. Efficient reduction of nitroarenes in water catalyzed by reusable Pd nanoparticles immobilized on chitosan-functionalized graphene oxide. *Can. J. Chem.*, **2019**, *97*(3), 191-196.
[http://dx.doi.org/10.1139/cjc-2018-0011]

[70] Ansar, S.M.; Kitchens, C.L. Impact of gold nanoparticle stabilizing ligands on the colloidal catalytic reduction of 4-nitrophenol. *ACS Catal.*, **2016**, *6*(8), 5553-5560.
[http://dx.doi.org/10.1021/acscatal.6b00635]

[71] Ayad, M.M.; Amer, W.A.; Kotp, M.G. Magnetic polyaniline-chitosan nanocomposite decorated with palladium nanoparticles for enhanced catalytic reduction of 4-nitrophenol. *Molecular Catalysis*, **2017**, *439*, 72-80.
[http://dx.doi.org/10.1016/j.mcat.2017.06.023]

[72] Naghdi, S.; Sajjadi, M.; Nasrollahzadeh, M.; Rhee, K.Y.; Sajadi, S.M.; Jaleh, B. Cuscuta reflexa leaf extract mediated green synthesis of the Cu nanoparticles on graphene oxide/manganese dioxide nanocomposite and its catalytic activity toward reduction of nitroarenes and organic dyes. *J. Taiwan Inst. Chem. Eng.*, **2018**, *86*, 158-173.
[http://dx.doi.org/10.1016/j.jtice.2017.12.017]

[73] Hu, H.; Xin, J.H.; Hu, H.; Wang, X.; Miao, D.; Liu, Y. Synthesis and stabilization of metal nanocatalysts for reduction reactions – a review. *J. Mater. Chem. A Mater. Energy Sustain.*, **2015**, *3*(21), 11157-11182.
[http://dx.doi.org/10.1039/C5TA00753D]

[74] Kariper, İ.A.; Çağlayan, M.O.; Üstündağ, Z. Heterogeneous Au/Ru hybrid nanoparticle decorated graphene oxide nanosheet catalyst for the catalytic reduction of nitroaromatics. *Res. Chem. Intermed.*, **2019**, *45*(2), 801-813.
[http://dx.doi.org/10.1007/s11164-018-3644-1]

[75] Lin, S-C.A.; Liu, Y.H.; Peng, S.M.; Liu, S.T. Diruthenium complex catalyzed reduction of nitroarenes-investigation of reaction pathway. *Molecular Catalysis*, **2019**, *466*, 46-51.
[http://dx.doi.org/10.1016/j.mcat.2019.01.005]

[76] Fiore, A.M.; Romanazzi, G.; Dell'Anna, M.M.; Latronico, M.; Leonelli, C.; Mali, M.; Rizzuti, A.; Mastrorilli, P. Mild and efficient synthesis of secondary aromatic amines by one-pot stepwise reductive amination of arylaldehydes with nitroarenes promoted by reusable nickel nanoparticles. *Molecular Catalysis,* **2019**, *476*, 110507-110516. [http://dx.doi.org/10.1016/j.mcat.2019.110507]

[77] Romanazzi, G.; Fiore, A.M.; Mali, M.; Rizzuti, A.; Leonelli, C.; Nacci, A.; Mastrorilli, P.; Dell'Anna, M.M. Catalytic activities of heterogeneous catalysts obtained by copolymerization of metal-containing 2-(acetoacetoxy)ethyl methacrylate. *Mol. Catal.,* **2018**, *446*, 31-38. [http://dx.doi.org/10.1016/j.mcat.2017.12.015]

[78] Metin, Ö.; Can, H.; Şendil, K.; Gültekin, M.S. Monodisperse Ag/Pd core/shell nanoparticles assembled on reduced graphene oxide as highly efficient catalysts for the transfer hydrogenation of nitroarenes. *J. Colloid Interface Sci.,* **2017**, *498*, 378-386. [http://dx.doi.org/10.1016/j.jcis.2017.03.066] [PMID: 28343135]

[79] Goksu, H.; Sert, H.; Kilbas, B.; Sen, F. Recent advances in the reduction of nitro compounds by heterogenous catalysts. *Curr. Org. Chem.,* **2017**, *21*(9), 794-820. [http://dx.doi.org/10.2174/1385272820666160525123907]

[80] Bustamante, T.M.; Dinamarca, R.; Torres, C.C.; Pecchi, G.; Campos, C.H. Pd-Co catalysts prepared from palladium-doped cobalt titanate precursors for chemoselective hydrogenation of halonitroarenes. *Molecular Catalysis,* **2020**, *482*, 110702-110715. [http://dx.doi.org/10.1016/j.mcat.2019.110702]

[81] Jin, Q.; Shen, Y.; Li, X.; Zeng, Y. Resource utilization of waste deNOx catalyst for continuous-flow catalysis by supported metal reactors. *Molecular Catalysis,* **2020**, *480*, 110634-110640. [http://dx.doi.org/10.1016/j.mcat.2019.110634]

[82] Hu, A.; Lu, X.; Cai, D.; Pan, H.; Jing, R.; Xia, Q.; Zhou, D.; Xia, Y. Selective hydrogenation of nitroarenes over MOF-derived Co@CN catalysts at mild conditions. *Molecular Catalysis,* **2019**, *472*, 27-36. [http://dx.doi.org/10.1016/j.mcat.2019.04.008]

[83] Zhang, K.; Suh, J.M.; Choi, J.W.; Jang, H.W.; Shokouhimehr, M.; Varma, R.S. Recent advances in the nanocatalyst-assisted $NaBH_4$ reduction of nitroaromatics in water. *ACS Omega,* **2019**, *4*(1), 483-495. [http://dx.doi.org/10.1021/acsomega.8b03051] [PMID: 31032469]

[84] Iijima, S.; Ichihashi, T. Single-shell carbon nanotubes of 1-nm diameter. *Nature,* **1993**, *363*(6430), 603-605. [http://dx.doi.org/10.1038/363603a0]

[85] Castro Neto, A.H.; Guinea, F.; Peres, N.M.R.; Novoselov, K.S.; Geim, A.K. The electronic properties of graphene. *Rev. Mod. Phys.,* **2009**, *81*(1), 109-162. [http://dx.doi.org/10.1103/RevModPhys.81.109]

[86] Shi, R.; Ren, M.; Li, H.; Zhao, J.; Liu, S.; Li, Z.; Ren, J. Graphene supported Cu nanoparticles as catalysts for the synthesis of dimethyl carbonate: Effect of carbon black intercalation. *Molecular Catalysis,* **2018**, *445*, 257-268. [http://dx.doi.org/10.1016/j.mcat.2017.12.002]

[87] Gao, Z.; Liu, X.; Li, A.; Li, X.; Ding, X.; Yang, W. Bimetallic sites supported on N-doped graphene ((Fe,Co)/N-GN) as a new catalyst for NO oxidation: A theoretical investigation. *Molecular Catalysis,* **2019**, *470*, 56-66. [http://dx.doi.org/10.1016/j.mcat.2019.03.024]

[88] Wang, R.; Liu, H.; Fan, C.; Gao, J.; Chen, C.; Zheng, Z. Selective oxidative esterification of alcohols over Au-Pd/graphene. *Molecular Catalysis,* **2020**, *484*, 110687-110699. [http://dx.doi.org/10.1016/j.mcat.2019.110687]

[89] Li, D.; Li, W.; Zhang, J. Catalytic CO oxidation by Fe doped penta-graphene: A density functional

study. *Molecular Catalysis,* **2019**, *470*, 48-55. [http://dx.doi.org/10.1016/j.mcat.2019.03.023]

[90] Behjatmanesh-Ardakani, R. Catalytic CO oxidation by Fe doped penta-graphene: A density functional study. *Mol. Catal.,* **2018**, *455*, 239-249. [http://dx.doi.org/10.1016/j.mcat.2018.06.008]

[91] Kumar, K.V.A.; Chandana, L.; Ghosal, P.; Subrahmanyam, C. Simultaneous photocatalytic degradation of p -cresol and Cr (VI) by metal oxides supported reduced graphene oxide. *Molecular Catalysis,* **2018**, *451*, 87-95. [http://dx.doi.org/10.1016/j.mcat.2017.11.014]

[92] Sharma, H.; Sharma, S.; Sharma, C.; Paul, S.; Clark, J.H. Magnetically recoverable graphene oxide supported Co@Fe_3O_4/L-dopa for C-C cross-coupling and oxidation reactions in aqueous medium. *Molecular Catalysis,* **2019**, *469*, 27-39. [http://dx.doi.org/10.1016/j.mcat.2019.02.023]

[93] Yang, J.; Shen, X.; Ji, Z.; Zhou, H.; Zhu, G.; Chen, K. In situ growth of hollow CuNi alloy nanoparticles on reduced graphene oxide nanosheets and their magnetic and catalytic properties. *Appl. Surf. Sci.,* **2014**, *316*, 575-581. [http://dx.doi.org/10.1016/j.apsusc.2014.08.097]

[94] Cheng, Y.; Fan, Y.; Pei, Y.; Qiao, M. Graphene-supported metal/metal oxide nanohybrids: synthesis and applications in heterogeneous catalysis. *Catal. Sci. Technol.,* **2015**, *5*(8), 3903-3916. [http://dx.doi.org/10.1039/C5CY00630A]

[95] Samak, N.A.; Tan, Y.; Sui, K.; Xia, T.T.; Wang, K.; Guo, C.; Liu, C. CotA laccase immobilized on functionalized magnetic graphene oxide nano-sheets for efficient biocatalysis. *Molecular Catalysis,* **2018**, *445*, 269-278. [http://dx.doi.org/10.1016/j.mcat.2017.12.004]

[96] Nayak, P.S.; Barik, B.; Achary, L.S.K.; Kumar, A.; Dash, P. Gold nanoparticles deposited on MnO2 nanorods modified graphene oxide composite: A potential ternary nanocatalyst for efficient synthesis of betti bases and bisamides. *Molecular Catalysis,* **2019**, *474*, 110415-110430. [http://dx.doi.org/10.1016/j.mcat.2019.110415]

[97] Julkapli, N.M.; Bagheri, S. Graphene supported heterogeneous catalysts: An overview. *Int. J. Hydrogen Energy,* **2015**, *40*(2), 948-979. [http://dx.doi.org/10.1016/j.ijhydene.2014.10.129]

[98] Fu, J.Y.; Chen, L.W.; Dai, Y.M.; Liu, F.Y.; Huang, S.T.; Chen, C.C. BiOmFn/BiOxIy/GO Nanocomposites: Synthesis, characterization, and photocatalytic activity. *Molecular Catalysis,* **2018**, *455*, 214-223. [http://dx.doi.org/10.1016/j.mcat.2018.06.014]

[99] Zhu, Y.; Lin, B.; Hu, Y.; Cai, Z.; Xie, H.; Wang, K.; Zhou, Y. Enhanced stability of Pd/SPS catalyst over the one-pot liquid-phase synthesis of methyl isobutyl ketone by adding GO. *Molecular Catalysis,* **2019**, *478*, 110609-110617. [http://dx.doi.org/10.1016/j.mcat.2019.110609]

[100] Tang, Y.; Zhou, J.; Chen, W.; Chai, H.; Li, Y.; Feng, Z.; Dai, X. Theoretical evaluation on single-atom Fe doped divacancy graphene for catalytic CO and NO oxidation by O_2 molecules. *Molecular Catalysis,* **2019**, *476*, 110524-110537. [http://dx.doi.org/10.1016/j.mcat.2019.110524]

[101] Wu, C.; Gates, I.D. Methane activation by a single iron atom supported on graphene: Impact of substrates. *Molecular Catalysis,* **2019**, *469*, 40-47. [http://dx.doi.org/10.1016/j.mcat.2019.03.002]

[102] Mohazzab, B.F.; Jaleh, B.; Nasrollahzadeh, M.; Issaabadi, Z.; Varma, R.S. Laser ablation-assisted synthesis of GO/TiO2/Au nanocomposite: Applications in $K_3[Fe(CN)_6]$ and Nigrosin reduction. *Molecular Catalysis,* **2019**, *473*, 110401-110411.

[http://dx.doi.org/10.1016/j.mcat.2019.110401]

[103] Lv, T.; Wang, H.; Hong, W.; Wang, P.; Jia, L. In situ self-assembly synthesis of sandwich-like TiO2/reduced graphene oxide/LaFeO_3 Z-scheme ternary heterostructure towards enhanced photocatalytic hydrogen production. *Molecular Catalysis,* **2019**, *475*, 110497-110509. [http://dx.doi.org/10.1016/j.mcat.2019.110497]

[104] Li, X.; Qi, T.; Wang, J.; She, W.; Mao, G.; Yan, P.; Li, W.; Li, G. Enhanced catalytic performance of nitrogen-doped carbon supported FeOx-based catalyst derived from electrospun nanofiber crosslinked N, Fe-containing MOFs for efficient hydrogenation of nitroarenes. *Molecular Catalysis,* **2019**, *477*, 110544-110549. [http://dx.doi.org/10.1016/j.mcat.2019.110544]

[105] Chen, T.W.; Xu, J.Y.; Sheng, Z.H.; Wang, K.; Wang, F.B.; Liang, T.M.; Xia, X.H. Enhanced electrocatalytic activity of nitrogen-doped graphene for the reduction of nitro explosives. *Electrochem. Commun.,* **2012**, *16*(1), 30-33. [http://dx.doi.org/10.1016/j.elecom.2011.12.016]

[106] Colunga, A.; Rangel-Mendez, J.R.; Celis, L.B.; Cervantes, F.J. Graphene oxide as electron shuttle for increased redox conversion of contaminants under methanogenic and sulfate-reducing conditions. *Bioresour. Technol.,* **2015**, *175*, 309-314. [http://dx.doi.org/10.1016/j.biortech.2014.10.101] [PMID: 25459837]

[107] Trandafir, M.M.; Florea, M.; Neaţu, F.; Primo, A.; Parvulescu, V.I.; García, H. Graphene from Alginate Pyrolysis as a Metal-Free Catalyst for Hydrogenation of Nitro Compounds. *ChemSusChem,* **2016**, *9*(13), 1565-1569. [http://dx.doi.org/10.1002/cssc.201600197] [PMID: 27246529]

[108] Mahata, S.; Sahu, A.; Shukla, P.; Rai, A.; Singh, M.; Rai, V.K. Bio-inspired unprecedented synthesis of reduced graphene oxide: a catalytic probe for electro-/chemical reduction of nitro groups in an aqueous medium. *New J. Chem.,* **2018**, *42*(3), 2067-2073. [http://dx.doi.org/10.1039/C7NJ04732K]

[109] Fakhri, P.; Nasrollahzadeh, M.; Jaleh, B. Graphene oxide supported Au nanoparticles as an efficient catalyst for reduction of nitro compounds and Suzuki–Miyaura coupling in water. *RSC Advances,* **2014**, *4*(89), 48691-48697. [http://dx.doi.org/10.1039/C4RA06562J]

[110] Jeon, E.K.; Seo, E.; Lee, E.; Lee, W.; Um, M.K.; Kim, B.S. Mussel-inspired green synthesis of silver nanoparticles on graphene oxide nanosheets for enhanced catalytic applications. *Chem. Commun. (Camb.),* **2013**, *49*(33), 3392-3394. [http://dx.doi.org/10.1039/c3cc00115f] [PMID: 23435315]

[111] Paul, B.; Purkayastha, D.D.; Dhar, S.S.; Das, S.; Haldar, S. Facile one-pot strategy to prepare Ag/Fe_2O_3 decorated reduced graphene oxide nanocomposite and its catalytic application in chemoselective reduction of nitroarenes. *J. Alloys Compd.,* **2016**, *681*, 316-323. [http://dx.doi.org/10.1016/j.jallcom.2016.04.229]

[112] Li, H.; Gan, S.; Han, D.; Ma, W.; Cai, B.; Zhang, W.; Zhang, Q.; Niu, L. High performance Pd nanocrystals supported on SnO_2 -decorated graphene for aromatic nitro compound reduction. *J. Mater. Chem. A Mater. Energy Sustain.,* **2014**, *2*(10), 3461-3467. [http://dx.doi.org/10.1039/C3TA14506A]

[113] Zhang, K.; Hong, K.; Suh, J.M.; Lee, T.H.; Kwon, O.; Shokouhimehr, M.; Jang, H.W. Facile synthesis of monodispersed Pd nanocatalysts decorated on graphene oxide for reduction of nitroaromatics in aqueous solution. *Res. Chem. Intermed.,* **2019**, *45*(2), 599-611. [http://dx.doi.org/10.1007/s11164-018-3621-8]

[114] Kubendhiran, S.; Sakthinathan, S.; Chen, S.M.; Tamizhdurai, P.; Shanthi, K.; Karuppiah, C. Green reduction of reduced graphene oxide with nickel tetraphenyl porphyrin nanocomposite modified electrode for enhanced electrochemical determination of environmentally pollutant nitrobenzene. *J.*

Colloid Interface Sci., **2017**, *497*, 207-216.
[http://dx.doi.org/10.1016/j.jcis.2017.03.003] [PMID: 28285048]

[115] Karthik, M.; Suresh, P. Greener Synthesis of Reduced Graphene Oxide-Nickel Nanocomposite: Rapid and Sustainable Catalyst for the Reduction of Nitroaromatics. *ChemistrySelect,* **2017**, *2*(23), 6916-6928.
[http://dx.doi.org/10.1002/slct.201701314]

[116] Li, F.; Gao, X.; Xue, Q.; Li, S.; Chen, Y.; Lee, J.M. Reduced graphene oxide supported platinum nanocubes composites: one-pot hydrothermal synthesis and enhanced catalytic activity. *Nanotechnology,* **2015**, *26*(6), 065603.
[http://dx.doi.org/10.1088/0957-4484/26/6/065603] [PMID: 25612090]

[117] Kumar Sahoo, P.; Panigrahy, B.; Thakur, D.; Bahadur, D. Ice-templating synthesis of macroporous noble metal/3D-graphene nanocomposites: their fluorescence lifetimes and catalytic study. *New J. Chem.,* **2017**, *41*(16), 7861-7869.
[http://dx.doi.org/10.1039/C7NJ00128B]

[118] Liu, Z.; Guo, Y. Sensitive determination of trace 4-nitrophenol in water based on thio-β-cyclodextrin functionalized graphene/copper nanospheres. *J. Electroanal. Chem. (Lausanne),* **2018**, *825*, 57-64.
[http://dx.doi.org/10.1016/j.jelechem.2018.08.005]

[119] Feng, Y.S.; Ma, J-J.; Kang, Y-M.; Xu, H-J. PdCu nanoparticles supported on graphene: an efficient and recyclable catalyst for reduction of nitroarenes. *Tetrahedron,* **2014**, *70*(36), 6100-6105.
[http://dx.doi.org/10.1016/j.tet.2014.04.034]

[120] Fang, H.; Wen, M.; Chen, H.; Wu, Q.; Li, W. Graphene stabilized ultra-small CuNi nanocomposite with high activity and recyclability toward catalysing the reduction of aromatic nitro-compounds. *Nanoscale,* **2016**, *8*(1), 536-542.
[http://dx.doi.org/10.1039/C5NR05016B] [PMID: 26646949]

CHAPTER 10

Graphene Based Biocatalysts

Monika Jangid[1], **Sunita Panchawat**[2] and **Abhilasha Jain**[3,*]

[1] *Department of Chemistry, PAHER University, Udaipur, Rajasthan, 313001, India*

[2] *Department of Phatmaceutical Sciences, Mohanlal Sukhadia University, Udaipur, Rajasthan, 313001, India*

[3] *Department of Chemistry, St. Xavier's College (Autonomus), Mumbai, Maharashtra, 400001, India*

Abstract: Enzymes have catalytic properties and can be used for different purposes as biocatalysts in some industrial processes. However, their applications are limited due to some drawbacks, such as lack of long-term stability and recovery under conditions of any particular process. Enzymes can improve their catalytic activity, stability, reusability, and half life, if these are immobilized on some support. Graphene and graphene based nanomaterials are good supports for enzymes as they are also non-toxic materials in nature. Such materials can also find applications in the fields of medical diagnostics, biofuel cells, biosensors, *etc*. These particular aspects have been discussed in this chapter.

Keywords: Biocatalyst, Biofuel Cell, Biosensors, Graphene, Graphene Oxide, Immobilization, Reduced Graphene Oxide.

INTRODUCTION

Nowadays, graphene and its derivatives are attracting the attention of scientists all over the globe due to their unique properties such as chemical, optical, mechanical, electrical, *etc*. There could be multivalent functionalized and surface loaded efficiently with the number of biomolecules. Apart from it, abundance, easy synthesis, excellent solubility and processability, high surface area, conductivity, and biocompatibility are the added advantages of graphene-derivatives over other carbon-based materials like diamond, carbon nanotubes, fullerene, *etc*. Recent advancement in the synthesis and surface functionalization of this graphene or graphene-based nanostructured materials have opened up new horizons as a host material for polymers grafting and the designing of state-of-the-art nano-biocatalytic systems. Graphene-immobilized nanocatalyst varies from

* **Corresponding author Abhilasha Jain:** Department of Chemistry, St. Xavier's College (Autonomus), Mumbai, Maharashtra,400001, India; E-mail: jainabhilasha5@gmail.com

conventional nano-bioconjugates with respect to functional stability, catalytic efficacy, and applicability. Graphene-assisted biocatalysts find numerous applications in various fields such as enzyme engineering, biotransformation, biosensing, biofuel & energy production, and designing highly biocompatible, non-toxic (Arvidsson *et al.* [4]), and environmentally-friendly nanomaterials and assemblies.

The substances, which can catalyse different chemical reactions in living systems are called biocatalysts. Best examples are enzymes, which are biological compounds (derived from micro- organisms, plants or animal tissue) worked as nature's catalysts and promoted chemical reactions in biological as well as natural systems. Microbial lipases are a class of enzymes, well known as biocatalysts and these are used in a broad range of chemical and biological reactions. These enzymes find numerous applications such as pharmaceuticals, cosmetics, waste water treatments, foods, tanning, detergents, *etc.*

Various enzymes can be used as biocatalysts for different analytical purposes in medical diagnostics and manufacturing purposes in large-scale industrial processes. They have catalytic properties, but their industrial applications are limited due to a lack of long-term stability under process conditions. There is a demand to resolve the difficulties associated with recycling before the enzyme is implemented on an industrial scale. Here, immobilization of enzymes is a novel approach, which can improve half-life, stability, catalytic activity, and reusability of enzymes. Graphene-based nanomaterials are nanoscale and thermo stable inorganic carriers. There are also nontoxic materials and selective modulators for enzyme activity.

APPLICATIONS OF GRAPHENE BASED BIOCATALYSTS

Immobilization of Enzymes

Enzyme immobilization is an interesting field of research and many enzymes are useful in medical fields and food processing industries and this immobilization can improve their activity and stability. Immobilization has certain advantages such as increasing catalytic activity, improvement of stability, easy recovery of enzyme, reuse, *etc.*

A green and efficient Rassi has developed one-pot procedure and Baharfar [1] to achieve indole based amides using Passerini reaction, which involves indole---carboxylic acids, aromatic aldehydes and alkyl isocyanides in the presence of immobilized lipase on graphene oxide. This method has many advantages such as simple operation, mild reaction conditions, good to excellent yields, environmental compatibility and no side products.

The Passerini reaction was performed in the presence of GO@lipase as a green heterogeneous catalyst for the synthesis of indole based a-acyloxycarboxamide derivatives. Activity, efficiency and reusability of *Aspergillus Niger* Lipase (ANL) immobilized on graphene oxide (GO@lipase) were demonstrated. As-prepared nanobiocatalyst was found applicable in multi-component reactions (MCRs) as a heterogenous biocatalyst.

Adeel *et al.* [2] reported that graphene-based nanomaterials had attracted research interest in different fields related to proteins. They have exceptional physiochemical properties such as optical, thermal, electrical, and mechanical strength, so graphene-based nanostructured materials are considered suitable for many applications in varried fields such as electrochemical sensors, electroanalytical chemistry and immobilization of biomolecules and enzymes. Graphene oxide has different functionalities, such as epoxy, hydroxyl, and carboxylic group, which has opened new avenues to modify GO to achieve materials with desired properties.

Ariaeenejad *et al.* [3] reported bifunctional applications of a novel immobilized enzyme on the modified magnetic graphene oxide (GO) for the effective removal of dyes from water. The amine functionalized GO nano-carrier was covalently attached to a model enzyme (PersiManXyn1). It was observed that the specific activities of the free and immobilized enzymes were 856.05 and 1141.1 $\mu molmin^{-1}mg^{-1}$, respectively based an enzyme assays. The free enzyme showed only 5% of its maximum activity, while the immobilized PersiManXyn1 preserved more than 35% of its activity at 90°C. It was revealed that the free enzyme was deactivated after four weeks of storage, but the immobilized enzyme retained 54% of its initial activity. ImmobilizedPersiManXyn1 was applied for the removal of dye from water. The immobilized PersiManXyn1 could catalytically reduce the dye solutions (94% dye removal after 15th cycle), while pristine nano-carrier and free enzyme showed no considerable catalytic ability within 150s. It retained its superior reusability. This immobilized PersiManXyn1 finds its practical applications in the treatment of real textile effluent.

Graphene oxide is a potential candidate for enzyme immobilization, because it has large specific surface area. GO was first functionalized with L-aspartic acid (GO-Asp) by Monajati *et al.* [5]. Then L-asparaginase was immobilized on the GO-Asp either physically or through chemical conjugation. It was reported that a significant enzyme loading could be achieved through covalent immobilization (almost 100% immobilization efficiency). They also examined stability of free and immobilized L-asparaginase in a wide range of temperatures (20–60°C) and pH (5–9). It was revealed that covalently immobilized L-asparaginase exhibited higher enzyme activity than the free enzyme at pH 8, with the maximum recovered activity of 100, 90.5 and 40.6% after 24h of incubation at 20, 40 and 60°C, respectively. The covalently immobilized L-asparaginase on GO-Asp was found to have 42% recovered activity even after eight continuous reaction cycles at 60°C. It was also indicated that no significant changes were there in the enzyme affinity through covalent conjugation. It was suggested that GO-Asp is suitable as a nanosheet support for L-asparaginase loading and later, it can find some industrial applications in future.

They synthesized a graphene-based matrix for immobilization of L-asparaginase. Asp was used for GO functionalization so that its dispersion and stability could be improved. Here, Asp acts as a spacer between graphene and immobilized enzyme to avoid any extensive structural damage of the enzyme. It was observed that thermal and pH stability of L-asparaginase was improved on immobilization. The covalent attachment of L-asparaginase on GO-Asp was found to retain the enzyme activity more at high temperature and pH 40.6% after 24 h heating at 60°C and pH 8 as compared to physical adsorption (25.1%) and free L-asparaginase (12.9%). It was reported that recovered activity of L-asparaginase retained about 29% and 42% of the initial activity for physically adsorbed and chemically conjugated L-asparaginase after 8 cycles at 60°C.

Dutta *et al.* [6] reported that immobilization conferred psychrostability to a cold active α-amylase (amy) enzyme using magnesium nanoparticle (Mg NP) and graphene oxide. It was revealed that GO-Mg NP-amy nanocomposite showed enhanced enzymatic activity and thermo stability at both; the temperature extremes; upper (90°C) and lower (8°C). The GO-Mg NP-amy exhibited increased affinity towards the substrate, which is reflected in the decline in its Km by 2.35 and 14.9-fold at 8° and 90°C, respectively, as-compared to untreated enzyme. They also found that GO-Mg NP-amy showed 2.34-fold and 4.29-fold increase in Vmax at 8° and 90°C, respectively, compared to the untreated enzyme. The half life of GO-Mg NP-amy was increased by 44-fold with simultaneous increase in E_d by 1.9-fold as compared to the native enzyme at 90°C. The same is true for8°C, half-life of GO-Mg NP-amy increased by 6.48-fold with the simultaneous increase in E_d by 2.21-fold as compared to the native enzyme. It was observed that enzymatic activity of GO-Mg NP-amy was retained even after 12 consecutive uses and had storage stability for more than 120 days at 4°C. The ability of GO-Mg NP to sustain and aggravate enzyme activity and stability at temperatures in a wide range can make it a better candidate in bioprocessing industries, which requires functioning at these extreme ranges of temperature.

Gür *et al.* [7] prepared graphene oxide and reduced graphene oxide (rGO) Langmuir-Blodgett (LB) films with specific surface properties and functional groups as negative ideal substrates. They immobilized polyphenol oxidase (PPO) enzyme on these films by electrostatic interaction without any requirement of chemical bonding or surface modification. It was reported that catalytic performances of the enzyme immobilized on GO and rGO substrates were higher in GO-PPO LB films as compared to rGO-PPO LB films *in-vitro* media. It was revealed that the immobilized PPO enzyme on biodegradable GO was superior to the inorganic materials in terms of its use in biological systems. They prepared regular film structures with a single molecule thickness by LB film technique, which contributed to the stability and recycling of immobilization of PPO enzyme on GO and rGO. It was found that *in-vitro* storage stability of immobilization on both film structures was about one month at +4°C.

Li *et al.* [8] prepared three-dimensional magnetic graphene oxide-magnetite polyvinyl alcohol (3D-GO/PVA/Fe_3O_4) nanocomposites. It was reported that the specific surface and the average pore size of 3D-GO/PVA/Fe_3O_4 nanocomposites were 388.87 m^2g^{-1} and 9.6nm, respectively. The higher specific surface indicated that the three-dimensional structure avoided any aggregation for GO sheets. The saturation magnetization (Ms) of the nanocomposites was found to be large (30.5 $emug^{-1}$), which enabled the easy cycling of such nanocomposites. It was revealed that 3D-GO/PVA/Fe_3O_4 nanocomposites exhibited better performance for porcine pancreatic lipase (PPL) enzyme immobilization and it could reach the efficiency of 91%, and the enzyme immobilized nanocomposites reached up to 90% of their activities. It was observed that the activity of the immobilized enzyme retained 70.8% of its initial activity after 10 cycles. The activity of the immobilized enzyme remained up to 71.1% at 4°C for 56 days. Hydroxyl-rich PVA provided more binding sites for immobilizing enzyme and best-immobilized efficiency for (3D-GO/PVA/Fe_3O_4)-PPL (91%) was much more as compared to (GO/Fe_3O_4)-PPL (76%).

Immobilization of lipase was successfully achieved by Heidarizadeh *et al.* [9] on the surface of magnetically separable Fe_3O_4/graphene oxide through a post-modification. This modification was achieved in alternation to glutar aldehyde post-modification. It was revealed that the activity of immobilized lipase had no significant loss in the activity, and it can be simply separated (by keeping its major activity) from reaction crude using a magnet.

Dithiocarbamate can modify magnetic graphene oxide nanocomposite (Fe_3O_4-GO) for immobilization of covalent enzyme (lipase).Fabricated Fe_3O_4/GO-NHCS-Lip acts as a nanobiocatalyst for enzymatic hydrolysis of PNPD. It was reported that the recyclability of immobilized lipase is due to the fact that it is covalently attached to the surface and the leaching issue was not observed in this protocol.

The synthesis, characterization and biological application of reduced graphene oxide–chitosan (GC) based benign supramolecular scaffold (SMS) has been reported by Mondal *et al.* [10]. It was observed that immobilization of polyphenol oxidase (PPO) onto GC lowers the Michaelis-Menten constant (Km) value, which helps in achieving maximum velocity at lower substrate concentrations. In corporation of gold nanoparticles in GC (GCG) further strengthens the microbial resistance properties of GC and also improves its biocompatibility by reducing cytotoxicity.

A new rGO-chitosan (GC) based supramolecular scaffold (SMS) has desired thickness (55 nm), surface morphology and functionality for immobilization of enzyme and AuNP implantation. Its bio fabrication properties revealed that-

- Synthesized GC serves as an excellent enzyme (PPO) immobilizing matrix with low Km value and hence, a lesser amount of substrate is required to achieve maximum velocity,
- GC and GCG both possess antibacterial and antifungal activity and the presence of Au NP in GC (GCG) improved this antimicrobial activity as compared to GC alone,
- Biocompatibility study (MTT assay) on macrophage culture indicated that 91 and 98% cells are viable for GC and GCG, respectively and therefore, both the composite possesses no noticeable adverse effect on the cells and,
- It was also found that as-synthesized materials are suitable for *in vitro* cell growth.

Hence, these environmentally benign and biocompatible materials can be used in industries for the removal of phenolic, food packaging, biomedical coating and tissue engineering.

Magnetic graphene oxide (GO-MNP) was synthesized by Paz-Cedeno *et al.* [11] and used as the immobilization support for an industrial cellulase- and xylanase-containing preparation. The enzymatic preparation was immobilized utilizing carbodiimide cross-linking chemistry using 1-ethyl-3-(3-dimethylaminopropyl) carbodiimide and *N*-hydroxysuccinimide (NHS). It was revealed that the supported final biocatalyst (GO-MNP-Enz) exhibited higher activity for the hydrolysis of pretreated sugarcane bagasse (PSB). It presented relative β-glucosidase, endoglucanase, β-xylosidase and xylanase activities of 70, 66, 88, and 70%, respectively, after 10 cycles of hydrolysis of their respective substrates. It was also revealed that biocatalyst also maintained approximately 50 and 80% of its efficiency for cellulose and xylan hydrolysis, respectively. It was suggested that GO-MNP-Enz may be a prospective candidate for industrial applications such as second-generation ethanol production.

The yield, efficiency, and relative activity of the immobilization process ranked from 27–97, 37–113, and 24–110%, respectively, for the different enzymatic activities assessed. It was observed that cellulose conversion to glucose was diminished in comparison to the initial cycles, when GO-MNP-Enz was used multiple times in the hydrolysis of PSB, where xylan conversion to xylose remained high and stable at levels greater than 60% up to the tenth cycle. It was proposed that the use of GO-MNP-Enz is a cost-effective strategy for more competitive hydrolysis of PSB, and it has a potential for application in the production of second-generation ethanol, where the cost of enzymes is considered the bottleneck that prevents its economic viability.

Ameri *et al.* [12] immobilized thermo alkalophilic lipase from *Bacillus atrophaeus* (*Ba*L) onto amine-functionalized graphene oxide nanosheets, which were coated with the poly(maleic anhydride-alt-1-octadecene) copolymer (GO-NH_2-PMAO) and activated with glutaraldehyde as spacer arm through interfacial activation and subsequent multipoint covalent attachment. Immobilization conditions were optimized, such as GO-NH_2- concentration of PMAO, buffer & enzyme glutaraldehyde concentration, pH, time of sonication, concentration, time, and temperature. It was reported that specific activity of the immobilized *Ba*L (105.95 ± 2.37 Umg^{-1}) reached at 5 mg mL^{-1} for GO-NH_2-PMAO, 25 mM of buffer, pH 6.0, 60 min sonication time, 100 mM glutaraldehyde, 60 U mL^{-1} of the enzyme, and 4 h of immobilization time at 25°C, and it was very close to the predicted amount (106.08 ± 1.42 U mg^{-1}). Maximum immobilization yield and efficiency were found to be 81.35 and 277.63% in optimal immobilization conditions. It was indicated that the immobilized *Ba*L showed better stability at extreme temperature and pH than the free *Ba*L and it could retain more than 90% of the initial activity at a temperature of 90°C and pH 11 even after 10 days of storage and 6 cycles of application. It was revealed that maximum bioconversion of valeric acid to pentyl valerate was there in the xylene medium after 48 h of incubation at 60°C using the free *Ba*L (34.5%) and the immobilized *Ba*L (96.3%). This, *Ba*L immobilized on GO-NH_2-PMAO can be used as an effective biocatalyst to synthesize ester with the flavour of green apple.

Asmat *et al.* [13] treated *Candida rugosa* lipase (CRL) with surfactants and immobilized it onto the novel formulated magnetic graphene anchored silica nanocomposite (Fe_3O_4/SiO_2/Gr NC). The surface of lipase was initially coated with Triton-X 100 and cetyltrimethylammonium bromide (CTAB) surfactants to stabilize the enzyme in its open form and it was then adsorbed onto aminated Fe_3O_4/SiO_2/Gr NC. Here, glutaraldehyde (GA) was used to cross-link the adsorbed lipase onto the NC. It was reported that surfactant stabilized immobilized lipase exhibited significantly enhanced thermo stability, and also tolerance to various metal ions and inhibitors. The immobilization yield was remarkably enhanced by

6-folds and 3-folds using Triton X-100 and CTAB, respectively. As-fabricated nanobiocatalysts were used to synthesise green apple flavour ester, ethyl valerate *via* esterification reaction. It was found that Triton X-100 stabilized immobilized lipase performed better, giving about 90% ester yield as compared to 78% yield obtained with CTAB stabilized immobilized lipase.

A simple and effective method for lipase immobilization onto nanocellulose fused Ppy functionalized GO composite was developed by Asmat and Husain [14]. As-prepared nanobioconjugate exhibited significant catalytic performance, stability and solvent tolerance. This is an easy method of preparation and high stability nanobiocatalyst CRL@ Ppy/NCe/GO CRL immobilized on Ppy/NCe/GO, which exhibited enhanced catalytic performance and 160% of activity yield percentage was obtained. It was reported that immobilized lipase formulation (nanobiocatalyst) retained 85% of its original catalytic activity even after 10 cycles of application. It was employed for the synthesis of ethyl acetoacetate. The use of immobilized lipase gave flavour compounds with 27.5 and 75.5% ester yields in solvent-free media and n-hexane, respectively.

Jafarian *et al.* [15] synthesised graphene oxide nanosheets (GON) and its functionalization with 2, 4, 6-trichloro-1, 3, 5-triazine (TCT) through two different routes:

- Directly reaction of GON with TCT (GON-1), and
- Reaction of GON with pre-functionalized TCT with 3-aminopropyl-triethoxysilane (APTS) (GON-2).

Then GON, GON-1 and GON-2 were used as supports for immobilization of *Candida rugosa* lipase (CRL). The values of loading capacity, immobilization yield and leaching of CRL were found in the following order:

GON>GON-1>GON-2

But this order is reversed for a specific activity and recovered activity. It was found that storage stability was high with reasonable reusability for all the immobilized CRL, while the pH and thermal stability of CRL@GON-2 were higher than the remaining two. The advantages of these supports include biocompatibility, low toxicity, high enzyme stability, low cost production, convenient separation of the supports from the solution and relatively high reusability of the enzyme, which make them promising candidates for practical applications.

Kashefi *et al.* [16] synthesized graphene oxide by modified Hummer's method and used it as an ideal enzyme immobilization support because of its specific chemical and structural features. Then they covalently immobilized laccase from genetically modified Aspergillusonto GO (nanobiocatalyst). It exhibited interesting results such as laccase loading (156.5mgg^{-1}) and immobilization yield (64.6%) at laccase concentration (0.9mgmL^{-1}). Bioconversion of two anionic dyes; Direct red 23 (DR23) and acid blue 92 (AB92) was studied using nanobiocatalyst and optimized. It was found that average decolorization effectiveness of the nanobiocatalyst was more than 75% for both these dyes after six cycles, which indicated its excellent operational stability and good reusability.

Gao *et al.* [17] fabricated and characterized glucose oxidase (GOx) and chloroperoxidase (CPO) on the surface of MGO. The performance of as-synthesized immobilized CPO was enhanced by coupling with GOx providing required H_2O_2*in situ* through oxidation of glucose. The activity of MGO-GO--CPO for the decolorization of orange G was much higher (96.6%) than MGO-GOx+MGO-CPO (86.2%). It may be due to reduced mass transfer resistance between CPO and H_2O_2 generated from GOx molecules. It was found that MGO-GOx-CPO reached its highest activity (95.2%) in the temperature range 42-50°C as compared to 35°C for the free enzyme. It was attributed to dispersion of the otherwise stacked MGO below 42°C. This can be conveniently reused as it can be easily recovered in the presence of an external magnetic field, and retained ~38.5% activity even after 6 cycles of applications. Thus, as-synthesized MGO-GOx-CPO has potential applications in the environmental field.

Xylanaseenzyme was immobilized by Mehnati-Najafabadi*et al* [18]. on superparamagnetic garphene oxide nanosheets (GOMNP). It is a hydrolytic enzyme of xylan, which is widely used in industry. Asxylan is bulk, polyethylene glycol bisamine (PEGA) and, it was used as a spacer for functionalization of GOMNP to enhance the accessibility of active sites of the immobilized xylanase. It was reported that xylanase was attached to the functionalized nanocomposite with a yield of 273 mg enzyme g^{-1} PEGA-GOMNP. It was found that free and immobilized xylanase displayed an optimal enzymatic activity and pH 6.5 and 7.5, respectively, at 60°C. The immobilized enzyme retained about 40% of its initial activity even after 8 cycles. It was observed that immobilized and free enzymes retained 35 and 20% of initial catalytic activity after 90 days storage at 4 °C, respectively. The PEGA- modified GOMNP can be used as biodegradable and suitable support for bioengineering.

Desai *et al.* [19] immobilizedamylase on graphene oxide (GO)–magnetite (Fe_3O_4) nanoparticles through covalent bonding. Different operating parameters were optimized as concentrations of GO (1.3 mg), Fe_3O_4 (58 μg), and amylase (4.5 mg).

High loading capacity of 77.58 μg amylase over 1 μg GO-magnetite nanoparticles could be achieved under optimum conditions. The optimum pH remained the same (7), but alkalitolerance was increased by ~20% in relative activities upon immobilization. It was observed that half-life of soluble amylase was found to be 13 h, which was enhanced to 20 h upon immobilization in 20 mM phosphate buffer, pH 7 at 50°C. The immobilized amylase could be used for 11 consecutive runs.

Pectinase was immobilized by Dai *et al.* [20] onto sodium alginate/graphene oxide beads *via* amide bonds. They used *N,N'*-dicyclohexylcarbodiimide/*N*-hydroxysuccinimide as the activating agent. The activity of the as-prepared immobilized pectinase (1236.86±40.21 U g^{-1}) at optimal conditions had enzyme activity recovery 83.5%. The optimal pH of free pectinase was 4.5, which was shifted to 4.0 in the case of immobilized pectinase. The optimal temperature of immobilized pectinase was also increased by 10°C (60°C) as compared to the free form (50°C). It was observed that the immobilized pectinase had a superior thermal and storage stability than that of free pectinase. It was indicated that immobilized pectinase retained 73% of initial activity even after six cycles. Such immobilized pectinase may find various applications in the food industry.

Bifunctional acid urease is useful in reducing harmful substances in rice wine and it also effectively removes ethyl carbamate (EC). The acid urease was immobilized by Yang *et al.* [21] on graphene oxide/chitosan beads. It was observed that immobilized urease can be recovered to 77% from the graphene oxide/chitosan composite when it was coupled for 8 h with 5% glutaraldehyde. The optimum temperature for urease activity was found to be 37°C. The stability of the immobilized urease was improved at 37°C and the half-life of decay in urease activity was only 12 h, whereas it was 43 days for soluble and chitosan/graphene oxide immobilized urease at 4°C. The Km value for urea and EC in the free state was 9.14 mM and 386 mM, respectively, which were 5.28 mM and 180 mM, respectively, relatively lower in the immobilized state. The values of Vmax for immobilized EC and urea were observed to be 1.08 and 1.16 μmol min^{-1}, respectively, which were higher for soluble EC and urea 0.782 and 0.931 μmol min^{-1}, respectively. Furthermore, 90% of the initial activity of graphene oxide/chitosan–urease beads was retained after using the beads 10 times; thus, indicating its excellent reusability.

Dutta *et al.* purified an extracellular phytase from a strain of *Bacillus subtilis* obtained from cattle sheds in 24 Parganas (Kolkata, West Bengal) by Dutta *et al.* [22]. The native enzyme showed optimum activity at 42°C and pH 8.5. but it showed 2.40 fold increased activity at 42°C and more than 20-fold activity increment at 85°C in the presence of magnesium nanoparticle (Mg NP) graphene oxide immobilized enzyme, respectively. The GO-Mg NP-phy complex (phytase coupled to Mg NP immobilized on graphene oxide) had a decrease in Km by 3.29 folds at 42°C and 6 folds at 85°C, and the enzyme system showed 3 fold and 5.5 fold increase in Vmax at 42°C and 85°C, respectively than the untreated enzyme. The energy of activation (Ea) was also lowered by 2.5 and 6 folds, respectively at 42°C and 85°C for GO-MgN-phy as compared to the untreated enzyme. The half life for GO-Mg NP-phy was found to be 29.5 times of that of the untreated enzyme. It was reported that enzymatic activity of GO-Mg NP-phy was retained by 86.6%, when it was tested for storage stability at 4°C after 140 days. This immobilization technique may be used in developing a thermostable phytase enzyme,which could withstand high temperatures during the pelleting of animal feeds and increased activity of the enzyme would also be beneficial to break down phytic acid during storage of feeds.

Different kinds of amino acids were used by Zhuang *et al.* [23] to reduce and modify GO. It was found that reduction and adsorption disparities of amino acids were responsible for the difference of structural heterogeneity such as density, thermal tolerance and disorder of the oxygen-containing groups of GO. This modification also resulted in changes in surface properties of the GO samples, such as hydrophobicity and surface charge. They functionalized GO as a carrier for the immobilization of lipase from *Thermomyces lanuginosus* (lipase TL). It was found that lipase immobilized on functionalized GO exhibited superior enzymatic loading and relative activity compared to GO, with a maximum increase of 70 and 47%, respectively. The relative activity of lipase immobilized on functionalized GO was reported to be 19.21 times more than commercial immobilized lipase TLIM.

Zhuang *et al.* [24] also used disodium guanosine 5′-monophosphate (GMP-2Na) as a reducing agent for the synthesis of functionalized GO (FGO) for lipase immobilization. This method could solve the toxic and hazardous issues of the agents in the chemical reduction process of graphene oxide.The degree of hydrophobicity of FGO was conducive to the formation of optimal conformation of lipase, when the concentration of GMP-2Na is 0.20mg mL^{-1}.The activity of the immobilized enzyme was enhanced by two folds of free enzyme after immobilization. Here, the protein loading reached 600mg g^{-1}. The immobilized enzyme has much durability as no significant decrease was observed after incubation for 8h at pH6.0 and 7.0. The activity of the GO-G@ enzyme was found to be 54.47% higher than that of free enzyme after storage for 35days.

Ariaeenejad *et al.* [25] presented a dual crosslinking (DC) strategy to synthesize novel 3D networks of carboxymethyl cellulose grafted copolymers of 2-acrylamido-2-methyl propane sulfonate and acrylamide (CMC-g-poly(AMPS-co-AAm)) hydrogels. They utilized graphene oxide nano-sheets as nano-fillers and physical cross-linkers, making H-bondings between polymeric chains to prepare GO@CMC-g-poly(AMPS-co-AAm) networks. The effects of GO content on the performance of as-synthesized material was examined with a model enzyme (PersiCel1). It was revealed that persiCel1 immobilization on the GO reinforced hydrogels retained almost 60% of its maximum activity at 90°C. Its specific activity and storage stability were remarkably enhanced. The immobilized enzyme on the GO containing hydrogels showed a 154.8% increase in conversion of alkaline-treated sugar beet pulp compared with the free PersiCel1, while the PersiCel1/neat-hydrogel indicated only an increment of 66.7% under the same conditions.

Ranjbari *et al.* [26] reported the synthesis of graphene oxide nanosheets (GON) by a modified Hummer's method. Then they immobilized a protease purified from the *Penaeus*vannamei shrimp on the GON activated with glutaraldehyde. It was observed that immobilization of the protease on the GON activated with glutaraldehyde did not affect the optimum pH much, but it significantly improved thermal stability and stability at extreme pH values, as well as activity at 90°C. It was interesting to note that the free enzyme retained less than 10% of the activity, but immobilized enzyme could retain more than 90% of its original activity after 24h of incubation at 90°C. The immobilized enzyme exhibited higher activity for hydrolysis of casein at 70°C and in the presence of 2M urethane than that of the free enzyme.

The effect of the hydrophobicity of reduced graphene oxide (rGO) on the activity of candida rugose lipase (CRL) was studied by Chen *et al.* [27]. They observed that the higher hydrophobicity of rGO surface could increase CRL activity, but it decreased the activity. It was revealed that the rGOB showed a three-dimensional micro-nanoporous structure and surface hydrophobicity with improved adsorption properties and catalytic activity of CRL. It was reported that CRL was immobilized onto the beads by physical adsorption with the maximum adsorption capacity of 117.67 mg g^{-1}. The immobilized lipase (rGOB@CRL) exhibited good temperature, thermal, and reuse stability as compared to free CRL. The conversion rate of triglycerides (TG) in the free enzyme reaction system in the glycerolysis reaction of olive oil was found to be only 67.83%, while the conversion rate of TG in the $rGOB_4$@CRL reaction system could reach 74.87%. As-synthesised rGOB@CRL can be used to adsorb emulsified oil due to the excellent oil-absorbing ability of rGOB, producing monoglyceride (MG) and diglyceride (DG) *in situ* by catalyzing the glycerolysis reaction. The conversion rate of triglyceride (TG) could reach 89.70%, when the oil concentration was kept 4 mg mL^{-1}.

Agrawal *et al.* [28] immobilized *β*-amylase onto GQDs using 3-aminopropyltriethoxysilane and glutaraldehyde. It was reported that the predicted optimum immobilization efficiency (88.64%) was quite close to the actual value (87.98%), and it confirmed that the immobilization process was successful. The immobilized enzyme exhibited maximum activity at pH 5.0 and 57°C, The K_m and V_{max} were found to be 6.40 mg mL^{-1} and 714.28 μmol^{-1} min^{-1} mg, respectively. The enzyme could retain 75% activity even after 12 uses at 30°C. An increase in thermostability of immobilized fenugreek *β*-amylase and non-toxic nature of GQDs could be used for maltose production in beverage, food and pharmaceutical industries.

Laccase enzymes of *Aspergillus oryzae* were immobilized on graphene nanosheets by Skoronski *et al.* [29] *via* physical adsorption and covalent bonding. It was carried out through contact between graphene and a solution of laccase enzyme, which was dissolved in deionized water. It was reported that this adsorption process followed a Freundlich model, while the process of immobilization by covalent bonding was carried out by nitration of graphene. It was followed by a reduction of sodium borohydride and crosslinking with glutaraldehyde. It was observed that process of immobilization by both these techniques increased the pH range of its activity as compared to the free enzyme and also its operating temperature. It was found that the enzyme quickly lost its activity after the second reaction cycle, when it was immobilized *via* physical adsorption. It retained about 80% of it activity after six cycles.

Wang *et al.* [30] fabricated hydrophilic polyethylenimine (PEI) modified magnetic graphene oxide composite (Fe_3O_4@PDA@GO-PEI) and used it as efficient solid support for the immobilization of dextranase, so as to improve stability and recyclable performance. It was reported that GO nanosheet component not only provided abundant binding sites for decorating PEI molecules, but it increased the loading capacity of immobilized dextranase further and acted as a soft matrix to enhance its stability. A combination of the magnetic separation property and the recyclable ability of immobilized dextranase was found to be further improved. The loading amount of immobilized dextranase on as-developed Fe_3O_4@PDA@GO-PEI was determined to be 305.1 mg g^{-1}, and it was about 1.61 times that without GO component. It was revealed that as-developed Fe_3O_4@PDA@GO-PEI-dextranase exhibited high immobilization efficiency (76.27%) enzymatic activity (83.67%), improved stabilities in a wide range of temperature and pH conditions.It retained 85.78% of its initial activity even after 20 cycles. It was suggested that the immobilized dextranase has a potential for various applications in biocatalytic engineering fields.

Alpha-amylase was immobilized by Mulko *et al.* [31] inside three different polymeric matrices (polyacrylamide hydrogel) PAAm; polyacrylamide-graphene oxide nanocomposite, PAAm-GO; and alginate so that effect of the matrix on the catalytic performance can be compared. They also studied morphology, swelling, mechanical properties, retention efficiency, and the catalytic behaviour of supported biocatalysts. It was found that nanocomposite made of PAAm-GO matrix incorporated 98% of the enzyme, which may be due to cooperative effect, while alginate gels incorporated only 30%. The enzyme retention using PAAm-GO could achieve a value of 97.5%. It was reported that PAAm matrices protect the biocatalyst so that it can be reused. An improvement in catalytic properties of enzyme *via* immobilization makes it an excellent candidate in bioindustrial applications. It will therefore be possible to produce sequential or simultaneous saccharification and fermentation using it.

Atiroğlu [32] made some efforts to develop stability in *lipase B Candida antarctica* as Cross-Linked Enzyme Aggregates (CLEAs). Grapheneoxide nanoparticles (GO-MNPs) were prepared for this purpose. Different operating conditions were compared, such as temperature, pH, storage stability, reusability, and organic solvents. It was reported that saturated ammonium sulfate (55%) and cross-linked with 125mM glutaraldehyde for 10h at 4°C were found to optimize the synthesis of lipase-GO-MNPs-CLEAs with the maximum activity recovery of 96.5%. Hyaluronicacid (HA) was used to modify MNPs, which improved the stability and nano-biocatalytic performance so that it can be used in industrial applications.

D-Psicose (D-ribo-2-hexulose or D-allulose), (an epimer of D-fructose) is considered as a rare low-calorie sugar and its enzymatic production using ketose 3-epimerases is a feasible process for the production of D-Psicose. But there are some drawbacks in the application of ketose 3-epimerases, such as bioconversion efficiency and reusability of the enzyme. Dedania *et al.* [33] immobilized ketose 3-epimerases from *Agrobacterium tumefaciens (agtu)* D-psicose 3-epimerase (DPEase) on graphene oxide. It was reported that graphene oxide immobilized *agtu*-DPEase (GO-*agtu*-DPEase) had optimal pH 7.5 and temperature 60°C. A significant improvement in thermal stability was observed, which showed a half-life of 720 min at 60°C, while *Agrobacterium tumefaciens (agtu)* DPEase displayed only 3.99 min. The bioconversion efficiency accounted for graphene oxide immobilized DPEase at equilibrium, 40:60 (D-psicose: D-fructose), which is higher than the *agtu*-DPEase. It was observed that graphene oxide immobilized DPEase exhibited bioconversion efficiency up to 10 cycles of consecutive uses.

A simple, and efficient approach has been reported by Lin *et al.* [34] to fabricate a biocatalytic system by immobilizing the enzyme onto graphene oxide nanosheets by two methods for the practical application of Bacillus licheniformis γ-glutamyltranspeptidase (BlGGT) *via* both; non-covalent (GO-BlGGT) and covalent (GO/GA-BlGGT) bonds. It was found that the enzyme-loading capacity for the prepared GO/GA nanomaterial was 3.47mg mg^{-1} support, corresponding to a 68.7% recovery of the initial activity. The immobilized enzymes exhibited significantly higher activity as compared to the free form of BlGGT, may be due to the beneficial effect of the layered GO carrier. It was revealed that covalently immobilized enzyme had comparable stability with respect to free enzyme during a storage period of 30days and it could be recycled nine times, retaining about 45.3% of its initial activity. Apart from it, the biocatalytic synthesis of γ---glutamyl-phenylalanine and γ-l-glutamyl-leucine by immobilized enzymes resulted in a yield > 31%.

Orfanakis *et al.* [35] synthesized hybrid nanostructures of magnetic iron nanoparticles and graphene oxide. They used these as nano supports for the covalent immobilization of β-glucosidase. It was revealed that the immobilization efficiency depends on the structure as well as the surface chemistry of nanostructures used. As-obtained hybrid nanostructure-based biocatalysts exhibited a two to four-fold higher thermostability as compared to the free enzyme, and also an enhanced performance at higher temperatures (up to 70°C); that too in a wide pH range. It was claimed that these biocatalysts retained up to 40% of their bioactivity after twelve cycles.

Biofunctionalization of Protein

As biological macromolecules have an essential role in our daily life; thus, it is important to control the activity and stability of such macromolecules. The most promising mothed for enhancing in activity and stability is to immobilize proteins on different support materials. Furthermore, large surface area and surface functional groups are the important features that are required for a better support system. Chaudhary *et al.* [36] opined that these features of graphene oxide and reduced graphene oxide make them ideal support materials for protein immobilization. The successful formation of GO/RGO-protein complexes with enhancement in structural/thermal stability may be due to various interactions at the nano-bio interface, making their utilization in various functional applications. It was emphasized that mechanism and interactions (hydrophilic, hydrophobic, electrostatic, local protein-protein, hydrogen bonding and van der Walls) between protein and GO/RGO influence structural stability and activity of enzymes/proteins. Furthermore, GO/RGO-protein complexes are utilized in various applications such as biosensors, bioimaging and theranostic agent, targeted drug delivery agents, and nanovectors for drug and protein delivery.

Wang *et al.* [37] constructed a hybrid composite of gold nanoparticles, phospholipids bilayer membrane, and graphene and there used this composite as matrices for microperoxidase-11 (MP11) immobilization. They also investigated direct electrochemistry and corresponding bioelectrocatalysis of the enzyme electrode. Phospholipid bilayer membrane was found to protect gold nanoparticles (Au NPs) assembled on polyelectrolyte functionalized graphene sheets may be due to electrostatic attraction to form a hybrid bionanocomposite. Due to the biocompatible microenvironment, which was (provided by the mimetic biomembrane) microperoxidase-11 entrapped in this matrix retained its original structure well and thus, exhibited higher bioactivity. The Au NPs–graphene assemblies, could also promote the direct electron transfer efficiently between the immobilized MP11 and the substrate electrode. As-prepared enzyme electrodes exhibited good direct electrochemistry and electrocatalytic responses to the reduction of hydrogen peroxide. The resulting H_2O_2 biosensor showed a low detection limit (2.6×10^{-6} M), broad linear range (2.0×10^{-5}–2.8×10^{-4} M), stability and good reproducibility. This sensor was used for real-time detection of H_2O_2 dynamically released from the tumor cells MCF-7 in response to a pro-inflammatory stimulant.

Biosensors

Narwal and Pundir [38] fabricated an improved amperometric biosensor by immobilizing glycerol kinase (GK) and glycerol-3-phosphate oxidase (GPO) nanoparticles (NPs) onto graphene oxide nanoparticles (GrO NPs) modified pencil graphite (PG) electrode. It was reported that this biosensor exhibited optimal current response at an applied potential of 0.45V and pH8.0 at 35°C. It displayed a wide linear response for glycerol concentration 0.001-60mM with a detection limit of 0.002μM. A very high sensitivity ($121.45 \mu A \cdot mM^{-1} \cdot cm^{-2}$) and rapid response time (2s) were there. It was revealed that 98.5% analytical recovery was there on adding glycerol to the sera samples. The biosensor was used to measure blood serum glycerol levels in patients suffering from hyperglyceridemia. It was indicated that 25% of its initial activity was lost after its regular use for a period of 210 days or more at 4°C under storage conditions.

Two-dimensional GO nanosheets and one-dimensional lysozyme nanofibrils were hybridized by Wu *et al.* [39] through electrostatic interaction to obtain a novel amyloid-GO composite. It provides a biocompatible immobilization platform for Au nanocatalysts as well as enzymes, which could load a large and tunable amount of Au NPs and its high catalytic activity was also maintained. These immobilized catalysts showed high electrochemical behaviors, which is ideal as glucose sensing systems. It was reported that enzymes could also be immobilized on the residual bare surfaces of amyloid-GO, which could serve as a sensitive and selective analytical glucose-detecting platform.

A competitive-displacement reaction strategy based on target-induced dissociation of gold nanoparticle coated graphene nanosheet (Au NPs/GN) from CdS quantum dot functionalized mesoporous titanium dioxide (CdS QDs/TiO_2) was designed by Cai *et al.* [40] for the sensitive photoelectrochemical (PEC) aptasensing of prostate-specific antigen (PSA) through the exciton–plasmon interaction (EPI) between CdS QDs and AuNPs. To construct such an aptasensing system, capture DNA was initially conjugated covalently onto CdS QDs/TiO_2-modified electrode, and then Au NPs/GN-labeled PSA aptamer was bound onto biofunctionalized CdS QDs/TiO_2 *via* hybridization chain reaction of partial bases with capture DNA. The introduction of AuNPs/GN efficiently quenched the photocurrent of CdS QDs/TiO_2 due to energy transfer. Upon addition of target PSA, the sandwiched aptamer between CdS QDs/TiO_2 and AuNPs/GN reacted with the analyte; thus, resulting in the dissociation of Au NPs/GN from the CdS QDs/TiO_2 to increase the photocurrent. Under optimum conditions, the aptasensing platform exhibited a high sensitivity for PSA detection within a dynamic linear range of 1.0 pg mL^{-1} to 8.0 ng mL^{-1} at a low limit of detection of 0.52 pg mL^{-1}. The interparticle distance of exciton–plasmon interaction and contents of AuNPs corresponding to the EPI effect in this system were also studied. Good selectivity and high reproducibility were obtained for the analysis of the target PSA. Importantly, the accuracy and matrix effect of the PEC aptasensor was evaluated for the determination of human serum specimens and newborn calf serum-diluted PSA standards, giving a well-matched result with the referenced PSA ELISA kit.

A simple and versatile method has been described by Liu *et al.* [41] to construct high-performance three-dimensional (3D) graphene-based enzyme biosensor. They used carbon nanotubes (CNTs) as carriers of methylene blue molecules to immobilize them on 3D graphene surface through strong π-π stacking force. The polydopamine (PDA) layer could also effectively prevent the leakage of inner electron mediators. The as-fabricated biosensor exhibited an outstanding performance for reagentless detection of hydrogen peroxide in terms of wide linear range (0.2 M-1.1 mM), high sensitivity (227.8 μA mM^{-1} cm^{-2}), low detection limit (58.0 nM), and fast response (reaching 95% of the steady current within 3 s). It was also revealed that biosensors exhibited high reproducibility and stability also.

Au nanoparticles and TiO_2 nanosheets co-modified reduced graphene oxide nanocomposite (Au-TiO_2NSs-rGO) were synthesized by Liu *et al.* [42] *via* a simple synthetic route and then used to immobilize hemoglobin (Hb) in order to fabricate a mediator-free biosensor. It was revealed that the Au-TiO_2NSs-rGO would act as immobilization support for Hb, which afforded good protein stability and bioactivity. It was reported that direct electron transfer of Hb was facilitated with the special structure of Au-TiO_2NSs-rGO. This biosensor displayed good performance for the detection of H_2O_2 in a wide linear range of 7.5~250 μM with a low detection limit of 3 μM, supported by fast response and good stability.

Various multi-tapered (four, five, and eight tapered) optical fiber sensors (OFSs) have been developed by Zhu *et al.* [43]. They immobilized different nanomaterials, such as gold nanoparticles (Au NPs) and graphene oxide (GO) over the bare probe. These nanomaterials immobilized probes were then functionalized with ascorbate oxidase. The Au NPs were found to be well-suited for biosensing applications. GO-based materials were also equally promising in nanostructure-based optical sensing due to their good chemical stability,broad surface area, high electrical conductivity, and excellent mechanical behavior. The performance parameters of this sensor were found to be greatly improved such as sensitivity (8.3 nmmM^{-1}), correlation correlator (0.9724), and limit of detection (51.94 μM).

A novel and sensitive electrochemical biosensor were constructed by He and Liu [44] for detecting phosphate using AuNRs@Cu_2O-NDs loaded poly (diallyldimethylammonium chloride) (PDDA) functionalized graphene immobilized pyruvate oxidase (PyO). PDDA-Gr (which has good conductive ability) was used to increase the active site and supporting substrate of biosensor. Then AuNRs@Cu_2O-NDs composite material was prepared to enhance the capability for PyO immobilization and achieved fast electron transfer between hydrogen peroxide and Cu_2O-NDs to amplify the electrical signal. The proposed biosensor showed a wide linearity range from 0.01-80μM and a low detection limit of 0.4nM under optimal conditions. This biosensor exhibited satisfactory anti-interference, good stability, and excellent reproducibility when it was applied to the monitoring of phosphate content in food.

Centeno *et al* [45]. isolated a new plant peroxidase from the leaves of guinea grass (Panicum maximum). This peroxidase had a specific activity of 408 U mg^{-1} and a molecular weight of 30 kDa. It showed optimum activity at pH 8.0 and high thermostability at 66°C. The best substrates for this peroxidase from guinea grass are o-dianisidine and 2,2′-azino-bis-(3-ethylbenzthiazoline-6-sulfonic acid). The graphene-modified electrode exhibited excellent electrocatalytic activity for reduction of H_2O_2, with a linear response in the range (100 μM-3.5 mM)with detection limit of 150 μM.

Liu *et al.* [46] reported a hydrothermal method to prepare reduced graphene oxide (rGO)-based hydrogel (GelrGONR), using neutral red (NR) to mediate the assembly of rGO sheets and tune the pore size of GelrGONR. A large pore size of up to 20 μm and interconnected porous structure of GelrGONR were observed. The GelrGONR-M was used as a biocatalyst for establishing a BOD biosensor and it exhibited a linear range of 2–64 mg O L^{-1} and a detection limit 0.4 mg O^{-1} for glucose–glutamic acid (GGA). This BOD detection strategy had long-term viability over one year and stability up to 2 months with a relative standard deviation of 2.1%.

A novel nanocomposite material prepared by Povedano *et al.* [47], which consists of reduced graphene oxide/Rh nanoparticles through a one-pot reaction process. It involved simultaneous reduction of both; $RhCl_3$ and graphene oxide with $NaBH_4$ as well as in situ deposition of the metal nanoparticles on the 2D carbon nanomaterial planar sheets. Glassy carbon electrode was coated with this nanocomposite and it was used as nanostructured support for the cross-linking of the enzyme laccase with glutaraldehyde to provide a voltammperometric biosensor for 17β-estradiol in the 0.9^{-11}pM range. This biosensor showed excellent analytical performance with high sensitivity of 25.7 A μM^{-1} $cm^{-1}$with a very low detection limit of 0.54 pM and high selectivity. It was applied to fast determination of the hormone in spiked synthetic and real human urine samples.

Amperometric biosensors based on glassy carbon electrodes (GCE) were modified with graphene oxide (GO) and multi-walled carbon nanotubes (MWCNTs) by Vlamidis *et al.* [48]. Two types of biosensors were realized using tyrosinase (Tyr) or laccase (Lac). The enzyme immobilisation on the modified GC was optimised employing different agents (bovine serum albumin and glutaraldehyde as crosslinking agent, Nafion, and chitosan). The conditions for the fabrication and the storage of the biosensors were established in order to obtain high sensitivities, good enzyme retention, and long-life devices. This biosensor was used for the determination of catechol, pyrogallol, epicatechin, gallic acid, 1,2-dihydroxybenzoic acid, caffeic acid, chlorogenic acid, rutin, catechin and dopamine. These were also used for the determination of total polyphenols concentration in juice samples.

Hashim *et al.* [49] used the spin coating technique for the preparation of tyrosinase immobilized graphene oxide thin film on a gold surface. It was revealed that the thin film is homogeneous, where the surface of graphene oxide is covered by tyrosinase as evident for AFM studies. The thin film has high absorbance value with a band gap of 4.082 eV. This modified thin film could sense the sensing presenceof phenol with a sensitivity of 0.00234° μM-1 and the ability to detect phenol down to 1 μM.

Zrinski *et al.* [50] developed a simple and highly sensitive electrochemical biosensor based on laccase immobilized onto a gold nanoparticles/graphene nanoplatelets-modified screen-printed carbon electrode (LACC/AuNP/GN-Pl/SPCE), and then it was used for the determination of hydroquinone (HQ) and other phenolic compounds. This biosensor shows excellent electrocatalytic activity towards the oxidation of hydroquinone at a potential of -0.05V (vs Ag/AgCl, 3M KCl) in phosphate buffer as supporting electrolyte (0.1M, pH 7.0) using hydrodynamic amperometry.

The biosensor was then applied to the determination of phenolic antioxidant capacity (AOC) in wine and blueberry syrup based on Trolox (6-hydroxy-2,5,7-8-tetramethylchromane-2-carboxylic acid) and hydroquinone equivalents. The results were found to be satisfactory and agreed well with the results of the reference method.

Enzymatic Biofuel Cells

Biofuel cell is a bioenergy device, which is useful in areas of biomedicine, bioelectronics, *etc.* Enzymatic biofuel cells are a class of biofuel cells which utilizes enzymes to generate energy from renewable sources.

Graphene has excellent electrical conductivity and ease of functionalization. As a result, it is considered a promising material for use in enzymatic bioelectrodes and biofuel cells. Such biofuel cells have attracted the attention of scientists due to their potential to harvest energy from organic materials.

Xie and Huang [51] encapsulated magnetic Fe_3O_4 nanoparticles in graphene oxides, and then these were employed as a magnetic carrier for the immobilization of lipase from *Candida rugosavia* interfacial activation of enzyme on hydrophobic surfaces. It was reported that activity recovery of 64.9% and enzyme immobilization efficiency of 85.5% could be achieved using the immobilization strategy. It was observed that Fe_3O_4 nanoparticles were successfully encapsulated into GO, and the lipase was tethered on this magnetic support. The immobilized lipase could efficiently catalyze the transesterification of soybean oil with methanol and produce biodiesel and the yield was 92.8% at 40°C. It was claimed that immobilized lipase could be easily recovered and recycling of the biocatalyst was possible five times without any significant loss of its activity.

Babadi *et al.* [52] fabricated 3-dimensional graphene (3DG)/glucose oxidase (GOx) bio-nanocomposite just to improve enzyme immobilisation and enzyme lifetime with an enhanced electron transfer rate. These enhancements were attributed to the unique physical properties of 3DG, (high porosity, large surface area, and excellent electrical conductivity). They could achieve a power density of

164 μW cm^{-2} at 0.4 V from this enzymatic biofuel cell (EBFC), which is comparable to that of the other glucose biofuel cells (GBFCs). This 3DG enhanced the enzyme lifetime, but decreased enzyme leaking and as a result, its good conductivity facilitates the electron harvest and transfer from the enzyme active site to the electrode.

Microfluidic microbial fuel cells (μMFC) combine the microfluidic technology and biocatalysts anchored at the electrode, representing an emerging bio-power generator. However, a major challenge towards further development is their low bio-catalysts loading on the electrodes. It has already been revealed that bioelectrode structure and nutrient-feeding type played a significant role in the power generation of μMFC. Yang *et al.* [53] fabricated a membrane-free μMFC with a piece of graphene-decorated nickel mesh as bio-anode. It was indicated that mixed bacteria grow densely on the surface of the electrode. Due to enriched bacterial colonization and sufficient nutrient supply, μMFC produces a remarkably high areal power density of 1.11 W m^{-2} at a current density of 2.15 A m^{-2}, which was higher than those obtained with conventional two-dimensional electrodes and flow-over architectures.

A nanocomposite comprised of graphene (G) supported manganese dioxide (MnO_2) incorporated into the network of polythioaniline (MnO_2-G/PTA) was synthesized by Perveen *et al.* [54]. This hybrid composite was applied as an electrode material for the development of a bioanode, which was fabricated by the electrochemical entrapment of ferritin (Frt) as mediator and glucose oxidase (GOx) enzyme in the matrix of the as-synthesized MnO_2-G/PTA deposited on glassy carbon electrode (GCE) surface. It was revealed that this hybrid electroactive support (MnO_2-G/PTA) used for the immobilization of the enzyme (GOx) provides appropriate electrical cabling between the redox enzyme (GOx) and the electrode surface. The as-developed GCE/MnO_2-G/PTA/Frt/GOx bioanode attained a maximum current density of 3.68 $mAcm^{-2}$ at 35 mM glucose concentration at a scan rate of 100 mVs^{-1}. This modified electrode possesses a high potential and good biocompatibility for production of bio-electricity from glucose.

Yaqoob *et al.* [55] used oil palm biomass waste-derived graphene derivatives (L-GO) as an anode in a double chamber of MFCs. This composite based anode (L-GO/ZnO) was fabricated to generate energy and remove Pb^{2+} from wastewater. It was reported that L-GO/ZnO anode showed 91.07% removal efficiency of lead ions, while only 85% removal efficiency was exhibited by L-GO anode. It was found that the composite anode (L-GO/ZnO) showed maximum power density (1350×10^{-3} mW m^{-2}) and current density (142.98 mA m^{-2}), which was relatively many times higher than the L-GO anode (power density = 20×10^{-3}mW m^{-2} and current density = 17.54 mA m^{-2}).

There are a number of reactions, which involve biotransformations/organic transformations by the biocatalysts. Normally, the activity of an enzyme is limited because of its mobility; therefore, immobilization of such entity is necessary to enhance its activity. Graphene and graphene based materials such as graphene oxide and reduced graphene oxide have a good potential in this respect. These biocatalysts can effectively promote some bioreactions. Other applications of such biocatalyst are biosensors, biofuels, *etc*. It will prove to be a promising area of research in coming decades.

Some recent development in the fields of immobilization of enzymes, biofunctionalization of protein, biosensors, enzymatic biofuel cells, *etc*. have been summarized in Table **1**.

Table 1. Recent applications of graphene- based biocatalayst.

SUPPORT	APPLICATION	REFERENCE
Graphene oxide	Immobilization of Adenylate cyclase	[56]
Reduced graphene oxide	Immobilization of Geotrichumcandidum acetophenone reductase	[57]
Graphene oxide	Functionalized by loading laccase	[58]
Reduced graphene oxide	Biosensor for detection of sarcosine	[59]
N-Doped graphene	High-energy enzyme biofuel cells	[60]
Graphene	Xylose-fueled microbial fuel cells	[61]
Graphene oxide	Adsorption of dyes through immobilized lipase	[62]
Graphene oxide	Biosensor for detection of doxorubicin hydrochloride	[63]

CONCLUSION

There are a number of reactions that involve biotransformations/organic transformations by the biocatalyst. Normally, the activity of an enzyme is limited because of its mobility; therefore, immobilization of such entity is necessary to enhance its activity. Graphene and -based material such as graphene oxide and reduced graphene oxide have a good potential in this respect. These biocatalysts can effectively promote some bioreactions. Other applications such as biosensors, biofuels *etc*. will surface as a promising area of research in the coming decades.

CONSENT FOR PUBLICATION

Not applicable.

CONFLICT OF INTEREST

The author declares no conflict of interest, financial or otherwise.

ACKNOWLEDGEMENTS

Declared none.

REFERENCES

[1] Rassi, S.; Baharfar, R. Synthesis of indole based amides using immobilized lipase on graphene oxide (GO@lipase) as a retrievable heterogeneous nano biocatalyst. *Polyhedron,* **2019**, *174*, 114153. [http://dx.doi.org/10.1016/j.poly.2019.114153]

[2] Adeel, M.; Bilal, M.; Rasheed, T.; Sharma, A.; Iqbal, H.M.N. Graphene and graphene oxide: Functionalization and nano-bio-catalytic system for enzyme immobilization and biotechnological perspective. *Int. J. Biol. Macromol.,* **2018**, *120*(Pt B), 1430-1440. [http://dx.doi.org/10.1016/j.ijbiomac.2018.09.144] [PMID: 30261251]

[3] Ariaeenejad, S.; Motamedi, E.; Hosseini Salekdeh, G. Application of the immobilized enzyme on magnetic graphene oxide nano-carrier as a versatile bi-functional tool for efficient removal of dye from water. *Bioresour. Technol.,* **2021**, *319*, 124228. [http://dx.doi.org/10.1016/j.biortech.2020.124228] [PMID: 33254455]

[4] Arvidsson, R.; Boholm, M.; Johansson, M.; de Montoya, M.L. "Just carbon": ideas about graphene risks by graphene researchers and innovation advisors. *NanoEthics,* **2018**, *12*(3), 199-210. [http://dx.doi.org/10.1007/s11569-018-0324-y] [PMID: 30546498]

[5] Monajati, M.; Borandeh, S.; Hesami, A.; Mansouri, D.; Tamaddon, A.M. Immobilization of -asparaginase on aspartic acid functionalized graphene oxide nanosheet: Enzyme kinetics and stability studies. *Chem. Eng. J.,* **2018**, *354*, 1153-1163. [http://dx.doi.org/10.1016/j.cej.2018.08.058]

[6] Dutta, N.; Biswas, S.; Saha, M.K. Nano-magnesium aided activity enhancement and biophysical characterization of a psychrophilic α-amylase immobilized on graphene oxide nanosupport. *J. Biosci. Bioeng.,* **2017**, *124*(1), 15-22. [http://dx.doi.org/10.1016/j.jbiosc.2017.02.002] [PMID: 28268061]

[7] Gür, B.; Ayhan, M.E.; Türkhan, A.; Gür, F.; Kaya, E.D. A facile immobilization of polyphenol oxidase enzyme on graphene oxide and reduced graphene oxide thin films: An insight into in-vitro activity measurements and characterization. *Colloids Surf. A Physicochem. Eng. Asp.,* **2019**, *562*, 179-185. [http://dx.doi.org/10.1016/j.colsurfa.2018.11.041]

[8] Li, Y.; Jing, T.; Xu, G.; Tian, J.; Dong, M.; Shao, Q.; Wang, B.; Wang, Z.; Zheng, Y.; Yang, C.; Guo, Z. 3-D magnetic graphene oxide-magnetite poly(vinyl alcohol) nanocomposite substrates for immobilizing enzyme. *Polymer (Guildf.),* **2018**, *149*, 13-22. [http://dx.doi.org/10.1016/j.polymer.2018.06.046]

[9] Heidarizadeh, M.; Doustkhah, E.; Rostamnia, S.; Rezaei, P.F.; Harzevili, F.D.; Zeynizadeh, B. Dithiocarbamate to modify magnetic graphene oxide nanocomposite (Fe_3O_4-GO): A new strategy for covalent enzyme (lipase) immobilization to fabrication a new nanobiocatalyst for enzymatic hydrolysis of PNPD. *Int. J. Biol. Macromol.,* **2017**, *101*, 696-702. [http://dx.doi.org/10.1016/j.ijbiomac.2017.03.152] [PMID: 28363653]

[10] Mondal, M.K.; Mukherjee, S.; Saha, S.K.; Chowdhury, P.; Sinha Babu, S.P. Design and synthesis of reduced graphene oxide based supramolecular scaffold: A benign microbial resistant network for enzyme immobilization and cell growth. *Mater. Sci. Eng. C,* **2017**, *75*, 1168-1177. [http://dx.doi.org/10.1016/j.msec.2017.02.136] [PMID: 28415403]

[11] Paz-Cedeno, F.R.; Carceller, J.M.; Iborra, S.; Donato, R.K.; Godoy, A.P.; Veloso de Paula, A.; Monti, R.; Corma, A.; Masarin, F. Magnetic graphene oxide as a platform for the immobilization of cellulases and xylanases: Ultrastructural characterization and assessment of lignocellulosic biomass hydrolysis. *Renew. Energy,* **2021**, *164*, 491-501. [http://dx.doi.org/10.1016/j.renene.2020.09.059]

[12] Ameri, A.; Shakibaie, M.; Khoobi, M.; Faramarzi, M.A.; Gholibegloo, E.; Ameri, A.; Forootanfar, H. Optimization of immobilization conditions of Bacillus atrophaeus FSHM2 lipase on maleic copolymer coated amine-modified graphene oxide nanosheets and its application for valeric acid esterification. *Int. J. Biol. Macromol.,* **2020**, *162*, 1790-1806. [http://dx.doi.org/10.1016/j.ijbiomac.2020.08.101] [PMID: 32814102]

[13] Asmat, S.; Husain, Q.; Shoeb, M.; Mobin, M. Tailoring a robust nanozyme formulation based on surfactant stabilized lipase immobilized onto newly fabricated magnetic silica anchored graphene nanocomposite: Aggrandized stability and application. *Mater. Sci. Eng. C,* **2020**, *112*, 110883. [http://dx.doi.org/10.1016/j.msec.2020.110883] [PMID: 32409040]

[14] Asmat, S.; Husain, Q. Exquisite stability and catalytic performance of immobilized lipase on novel fabricated nanocellulose fused polypyrrole/graphene oxide nanocomposite: Characterization and application. *Int. J. Biol. Macromol.,* **2018**, *117*, 331-341. [http://dx.doi.org/10.1016/j.ijbiomac.2018.05.216] [PMID: 29857098]

[15] Jafarian, F.; Bordbar, A.K.; Zare, A.; Khosropour, A. The performance of immobilized *Candida rugosa* lipase on various surface modified graphene oxide nanosheets. *Int. J. Biol. Macromol.,* **2018**, *111*, 1166-1174. [http://dx.doi.org/10.1016/j.ijbiomac.2018.01.133] [PMID: 29371152]

[16] Kashefi, S.; Borghei, S.M.; Mahmoodi, N.M. Covalently immobilized laccase onto graphene oxide nanosheets: Preparation, characterization, and biodegradation of azo dyes in colored wastewater. *J. Mol. Liq.,* **2019**, *276*, 153-162. [http://dx.doi.org/10.1016/j.molliq.2018.11.156]

[17] Gao, F.; Guo, Y.; Fan, X.; Hu, M.; Li, S.; Zhai, Q.; Jiang, Y.; Wang, X. Enhancing the catalytic performance of chloroperoxidase by co-immobilization with glucose oxidase on magnetic graphene oxide. *Biochem. Eng. J.,* **2019**, *143*, 101-109. [http://dx.doi.org/10.1016/j.bej.2018.12.013]

[18] Mehnati-Najafabadi, V.; Taheri-Kafrani, A.; Bordbar, A.K. Xylanase immobilization on modified superparamagnetic graphene oxide nanocomposite: Effect of PEGylation on activity and stability. *Int. J. Biol. Macromol.,* **2018**, *107*(Pt A), 418-425. [http://dx.doi.org/10.1016/j.ijbiomac.2017.09.013] [PMID: 28888544]

[19] Desai, R.P.; Dave, D.; Suthar, S.A.; Shah, S.; Ruparelia, N.; Kikani, B.A. Immobilization of α-amylase on GO-magnetite nanoparticles for the production of high maltose containing syrup. *Int. J. Biol. Macromol.,* **2021**, *169*, 228-238. [http://dx.doi.org/10.1016/j.ijbiomac.2020.12.101] [PMID: 33338531]

[20] Dai, X.Y.; Kong, L.M.; Wang, X.L.; Zhu, Q.; Chen, K.; Zhou, T. Preparation, characterization and catalytic behavior of pectinase covalently immobilized onto sodium alginate/graphene oxide composite beads. *Food Chem.,* **2018**, *253*, 185-193. [http://dx.doi.org/10.1016/j.foodchem.2018.01.157] [PMID: 29502820]

[21] Yang, L.; Liu, X.; Zhou, N.; Tian, Y. Characteristics of refold acid urease immobilized covalently by graphene oxide-chitosan composite beads. *J. Biosci. Bioeng.,* **2019**, *127*(1), 16-22. [http://dx.doi.org/10.1016/j.jbiosc.2018.06.012] [PMID: 30139654]

[22] Dutta, N.; Raj, D.; Biswas, N.; Mallick, M.; Omesh, S. Nanoparticle assisted activity optimization and characterization of a bacterial phytase immobilized on single layer graphene oxide. *Biocatal. Agric. Biotechnol.*, **2017**, *9*, 240-247. [http://dx.doi.org/10.1016/j.bcab.2017.01.008]

[23] Zhuang, W.; Quan, X.; Wang, Z.; Zhou, W.; Yang, P.; Ge, L.; Villacorta Hernandez, B.; Wu, J.; Li, M.; Zhou, J.; Zhu, C.; Ying, H. Interfacial microenvironment for lipase immobilization: Regulating the heterogeneity of graphene oxide. *Chem. Eng. J.*, **2020**, *394*, 125038. [http://dx.doi.org/10.1016/j.cej.2020.125038]

[24] Zhuang, W.; Gu, W.; Zhu, Q.; Zhu, J.; Wang, Z.; Niu, H.; Liu, D.; Wu, J.; Chen, Y.; Li, M.; Zhu, C.; Ying, H. Surface functionalization of graphene oxide by disodium guanosine 5′-monophosphate and its excellent performance for lipase immobilization. *Appl. Surf. Sci.*, **2019**, *492*, 27-36. [http://dx.doi.org/10.1016/j.apsusc.2019.05.166]

[25] Ariaeenejad, S.; Motamedi, E.; Hosseini Salekdeh, G. Stable cellulase immobilized on graphene oxide@CMC-g-poly(AMPS-co-AAm) hydrogel for enhanced enzymatic hydrolysis of lignocellulosic biomass. *Carbohydr. Polym.*, **2020**, *230*, 115661. [http://dx.doi.org/10.1016/j.carbpol.2019.115661] [PMID: 31887893]

[26] Ranjbari, N.; Razzaghi, M.; Fernandez-Lafuente, R.; Shojaei, F.; Satari, M.; Homaei, A. Improved features of a highly stable protease from Penaeus vannamei by immobilization on glutaraldehyde activated graphene oxide nanosheets. *Int. J. Biol. Macromol.*, **2019**, *130*, 564-572. [http://dx.doi.org/10.1016/j.ijbiomac.2019.02.163] [PMID: 30831167]

[27] Chen, M.; Wu, H.; Li, Z.; Wu, K.; Jiao, Y.; Zhou, C. Preparation of reduced graphene oxide porous beads for lipase immobilization and its application for oil adsorption and glycerolysis reaction *in situ*. *Micropor. Mesopo. Mater,* **2020**, 294.

[28] Agrawal, D.C.; Yadav, A.; Kesarwani, R.; Srivastava, O.N.; Kayastha, A.M. Immobilization of fenugreek β-amylase onto functionalized graphene quantum dots (GQDs) using Box-Behnken design: Its biochemical, thermodynamic and kinetic studies. *Int. J. Biol. Macromol.*, **2020**, *144*, 170-182. [http://dx.doi.org/10.1016/j.ijbiomac.2019.12.033] [PMID: 31843605]

[29] Skoronski, E.; Souza, D.H.; Ely, C.; Broilo, F.; Fernandes, M.; Fúrigo, A.; Ghislandi, M.G. Immobilization of laccase from Aspergillus oryzae on graphene nanosheets. *Int. J. Biol. Macromol.*, **2017**, *99*, 121-127. [http://dx.doi.org/10.1016/j.ijbiomac.2017.02.076] [PMID: 28237573]

[30] Wang, Y.; Wang, Q.; Song, X.; Cai, J. Hydrophilic polyethylenimine modified magnetic graphene oxide composite as an efficient support for dextranase immobilization with improved stability and recyclable performance. *Biochem. Eng. J.*, **2019**, *141*, 163-172. [http://dx.doi.org/10.1016/j.bej.2018.10.015]

[31] Mulko, L.; Pereyra, J.Y.; Rivarola, C.R.; Barbero, C.A.; Acevedo, D.F. Improving the retention and reusability of Alpha-amylase by immobilization in nanoporous polyacrylamide-graphene oxide nanocomposites. *Int. J. Biol. Macromol.*, **2019**, *122*, 1253-1261. [http://dx.doi.org/10.1016/j.ijbiomac.2018.09.078] [PMID: 30219512]

[32] Atiroğlu, V. Lipase immobilization on synthesized hyaluronic acid-coated magnetic nanoparticle-functionalized graphene oxide composites as new biocatalysts: Improved reusability, stability, and activity. *Int. J. Biol. Macromol.*, **2020**, *145*, 456-465. [http://dx.doi.org/10.1016/j.ijbiomac.2019.12.233] [PMID: 31883900]

[33] Dedania, S.R.; Patel, M.J.; Patel, D.M.; Akhani, R.C.; Patel, D.H. Immobilization on graphene oxide improves the thermal stability and bioconversion efficiency of D-psicose 3-epimerase for rare sugar production. *Enzyme Microb. Technol.*, **2017**, *107*, 49-56. [http://dx.doi.org/10.1016/j.enzmictec.2017.08.003] [PMID: 28899486]

[34] Lin, L.L.; Chi, M.C.; Lan, Y.J.; Lin, M.G.; Juang, T.Y.; Wang, T.F. Facile immobilization of Bacillus licheniformis γ-glutamyltranspeptidase onto graphene oxide nanosheets and its application to the

biocatalytic synthesis of γ-l-glutamyl peptides. *Int. J. Biol. Macromol.*, **2018**, *117*, 1326-1333. [http://dx.doi.org/10.1016/j.ijbiomac.2017.11.153] [PMID: 29183740]

[35] Orfanakis, G.; Patila, M.; Catzikonstantinou, A.V.; Lyra, K.M.; Kouloumpis, A.; Spyrou, K.; Katapodis, P.; Paipetis, A.; Rudolf, P.; Gournis, D.; Stamatis, H. Hybrid nanomaterials of magnetic iron nanoparticles and graphene oxide as matrices for the immobilization of β-glucosidase: synthesis, characterization, and biocatalytic properties. *Front. Mater.*, **2018**, *5*, 25. [http://dx.doi.org/10.3389/fmats.2018.00025]

[36] Chaudhary, K.; Kumar, K.; Venkatesu, P.; Masram, D.T. Protein immobilization on graphene oxide or reduced graphene oxide surface and their applications: Influence over activity, structural and thermal stability of protein. *Adv. Colloid Interface Sci.*, **2021**, *289*, 102367. [http://dx.doi.org/10.1016/j.cis.2021.102367] [PMID: 33545443]

[37] Wang, T.; Liu, J.; Ren, J.; Wang, J.; Wang, E. Mimetic biomembrane–AuNPs–graphene hybrid as matrix for enzyme immobilization and bioelectrocatalysis study. *Talanta*, **2015**, *143*, 438-441. [http://dx.doi.org/10.1016/j.talanta.2015.05.022] [PMID: 26078181]

[38] Narwal, V.; Pundir, C.S. Development of glycerol biosensor based on co-immobilization of enzyme nanoparticles onto graphene oxide nanoparticles decorated pencil graphite electrode. *Int. J. Biol. Macromol.*, **2019**, *127*, 57-65. [http://dx.doi.org/10.1016/j.ijbiomac.2018.12.253] [PMID: 30594620]

[39] Wu, X.; Li, M.; Li, Z.; Lv, L.; Zhang, Y.; Li, C. Amyloid-graphene oxide as immobilization platform of Au nanocatalysts and enzymes for improved glucose-sensing activity. *J. Colloid Interface Sci.*, **2017**, *490*, 336-342. [http://dx.doi.org/10.1016/j.jcis.2016.11.058] [PMID: 27914332]

[40] Cai, G.; Yu, Z.; Ren, R.; Tang, D. Exciton–plasmon interaction between AuNPs/graphene nanohybrids and CdS quantum dots/TiO_2 for photoelectrochemical aptasensing of prostate-specific antigen. *ACS Sens.*, **2018**, *3*(3), 632-639. [http://dx.doi.org/10.1021/acssensors.7b00899] [PMID: 29465232]

[41] Liu, J.; Wang, T.; Wang, J.; Wang, E. Mussel-inspired biopolymer modified 3D graphene foam for enzyme immobilization and high performance biosensor. *Electrochim. Acta*, **2015**, *161*, 17-22. [http://dx.doi.org/10.1016/j.electacta.2015.02.034]

[42] Liu, H.; Gao, J.; Duan, C.; Wu, K.; Guo, K. A novel mediator-free biosensor based on hemoglobin immobilized in the Au nanoparticals and TiO_2 nanosheets co-modified graphene nanocomposite. *Mater. Lett.*, **2020**, *275*, 128142. [http://dx.doi.org/10.1016/j.matlet.2020.128142]

[43] Zhu, G.; Agrawal, N.; Singh, R.; Kumar, S.; Zhang, B.; Saha, C.; Kumar, C. A novel periodically tapered structure-based gold nanoparticles and graphene oxide – Immobilized optical fiber sensor to detect ascorbic acid. *Opt. Laser Technol.*, **2020**, *127*, 106156. [http://dx.doi.org/10.1016/j.optlastec.2020.106156]

[44] He, B.; Liu, H. Electrochemical biosensor based on pyruvate oxidase immobilized AuNRs@Cu_2O-NDs as electroactive probes loaded poly (diallyldimethylammonium chloride) functionalized graphene for the detection of phosphate. *Sens. Actuators B Chem.*, **2020**, *304*, 127303. [http://dx.doi.org/10.1016/j.snb.2019.127303]

[45] Centeno, D.A.; Solano, X.H.; Castillo, J.J. A new peroxidase from leaves of guinea grass (Panicum maximum): A potential biocatalyst to build amperometric biosensors. *Bioelectrochemistry*, **2017**, *116*, 33-38. [http://dx.doi.org/10.1016/j.bioelechem.2017.03.005] [PMID: 28384528]

[46] Liu, L.; Zhai, J.; Zhu, C.; Gao, Y.; Wang, Y.; Han, Y.; Dong, S. One-pot synthesis of 3-dimensional reduced graphene oxide-based hydrogel as support for microbe immobilization and BOD biosensor preparation. *Biosens. Bioelectron.*, **2015**, *63*, 483-489. [http://dx.doi.org/10.1016/j.bios.2014.07.074] [PMID: 25129511]

[47] Povedano, E.; Cincotto, F.H.; Parrado, C.; Díez, P.; Sánchez, A.; Canevari, T.C.; Machado, S.A.S.; Pingarrón, J.M.; Villalonga, R. Decoration of reduced graphene oxide with rhodium nanoparticles for the design of a sensitive electrochemical enzyme biosensor for 17β-estradiol. *Biosens. Bioelectron.*, **2017**, *89*(Pt 1), 343-351.
[http://dx.doi.org/10.1016/j.bios.2016.07.018] [PMID: 27450540]

[48] Vlamidis, Y.; Gualandi, I.; Tonelli, D. Amperometric biosensors based on reduced GO and MWCNTs composite for polyphenols detection in fruit juices. *J. Electroanal. Chem. (Lausanne)*, **2017**, *799*, 285-292.
[http://dx.doi.org/10.1016/j.jelechem.2017.06.012]

[49] Hashim, H.S.; Fen, Y.W.; Omar, N.A.S.; Daniyal, W.M.E.M.M.; Saleviter, S.; Abdullah, J. Structural, optical and potential sensing properties of tyrosinase immobilized graphene oxide thin film on gold surface. *Optik (Stuttg.)*, **2020**, *212*, 164786.
[http://dx.doi.org/10.1016/j.ijleo.2020.164786]

[50] Zrinski, I.; Pungjunun, K.; Martinez, S.; Zavašnik, J.; Stanković, D.; Kalcher, K.; Mehmeti, E. Evaluation of phenolic antioxidant capacity in beverages based on laccase immobilized on screen-printed carbon electrode modified with graphene nanoplatelets and gold nanoparticles. *Microchem. J.*, **2020**, *152*, 104282.
[http://dx.doi.org/10.1016/j.microc.2019.104282]

[51] Xie, W.; Huang, M. Immobilization of Candida rugosa lipase onto graphene oxide Fe_3O_4 nanocomposite: Characterization and application for biodiesel production. *Energy Convers. Manage.*, **2018**, *159*, 42-53.
[http://dx.doi.org/10.1016/j.enconman.2018.01.021]

[52] Babadi, A.A.; Wan-Mohtar, W.A.A.Q.I.; Chang, J.S.; Ilham, Z.; Jamaludin, A.A.; Zamiri, G.; Akbarzadeh, O.; Basirun, W.J. High-performance enzymatic biofuel cell based on three-dimensional graphene. *Int. J. Hydrogen Energy*, **2019**, *44*(57), 30367-30374.
[http://dx.doi.org/10.1016/j.ijhydene.2019.09.185]

[53] Yang, Y.; Ye, D.; Zhu, X.; Liao, Q.; Li, J.; Chen, R. Boosting power density of microfluidic biofuel cell with porous three-dimensional graphene@nickel foam as flow-through anode. *Int. J. Hydrogen Energy*, **2018**, *43*(39), 18516-18520.
[http://dx.doi.org/10.1016/j.ijhydene.2018.08.052]

[54] Perveen, R.; Nasar, A.; Inamuddin, ; Asiri, A.M.; Mishra, A.K. Optimization of MnO_2-Graphene/polythioaniline (MnO_2-G/PTA) hybrid nanocomposite for the application of biofuel cell bioanode. *Int. J. Hydrogen Energy*, **2018**, *43*(32), 15144-15154.
[http://dx.doi.org/10.1016/j.ijhydene.2018.06.070]

[55] Yaqoob, A.A.; Ibrahim, M.N.M.; Yaakop, A.S.; Umar, K.; Ahmad, A. Modified graphene oxide anode: A bioinspired waste material for bioremediation of Pb^{2+} with energy generation through microbial fuel cells. *Chem. Eng. J.*, **2021**, *417*, 128052.
[http://dx.doi.org/10.1016/j.cej.2020.128052]

[56] Zhou, W.; Rao, Y.; Zhuang, W.; Ge, L.; Lin, R.; Tang, T.; Wu, J.; Li, M.; Yang, P.; Zhu, C.; Wang, H.; Ying, H. Improved enzymatic activity by oriented immobilization on graphene oxide with tunable surface heterogeneity. *Compos., Part B Eng.*, **2021**, *216*, 108788.
[http://dx.doi.org/10.1016/j.compositesb.2021.108788]

[57] Kamogawa, R.; Issasi, C.S.C.; Sasaki, M.; Matsuda, T. Geotrichumcandidum acetophenone reductase immobilization on reduced graphene oxide: A promising biocatalyst for green asymmetric reduction of ketones. *Biochem. Eng. J.*, **2022**, *177*.
[http://dx.doi.org/10.1016/j.bej.2021.108263]

[58] Noreen, S.; Perveen, S.; Shafiq, N.; Aslam, S.; Iqbal, H.M.N.; Ashraf, S.S.; Bilal, M. Laccase-loaded functionalized graphene oxide assemblies with improved biocatalytic properties and decolorization performance. *Environmental Technology & Innovation*, **2021**, *24*, 101884.

[http://dx.doi.org/10.1016/j.eti.2021.101884]

[59] Rajarathinam, T.; Kwon, M.; Thirumalai, D.; Kim, S.; Lee, S.; Yoon, J.H.; Paik, H.; Kim, S.; Lee, J.; Ha, H.K.; Chang, S.C. Polymer-dispersed reduced graphene oxide nanosheets and Prussian blue modified biosensor for amperometric detection of sarcosine. *Anal. Chim. Acta,* **2021**, *1175*, 338749. [http://dx.doi.org/10.1016/j.aca.2021.338749] [PMID: 34330447]

[60] Xu, J.; Wang, Y.H.; Wei, Z.; Wang, F.T.; Huang, K.J. Significantly improving the performance of self-powered biosensor by effectively combining with high-energy enzyme biofuel cells, N-doped graphene, and ultrathin hollow carbon shell. *Sens. Actuators B Chem.,* **2021**, *327*, 128933. [http://dx.doi.org/10.1016/j.snb.2020.128933]

[61] Moradian, J.M.; Mi, J.L.; Dai, X.; Sun, G.F.; Du, J.; Ye, X.M.; Yong, Y.C. Yeast-induced formation of graphene hydrogels anode for efficient xylose-fueled microbial fuel cells. *Chemosphere,* **2022**, *291*(Pt 2), 132963. [http://dx.doi.org/10.1016/j.chemosphere.2021.132963] [PMID: 34800508]

[62] Yao, L.W.; Ahmed Khan, F.S.; Mubarak, N.M.; Karri, R.R.; Khalid, M.; Walvekar, R.; Abdullah, E.C.; Mazari, S.A.; Ahmad, A.; Dehghani, M.H. Insight into immobilization efficiency of Lipase enzyme as a biocatalyst on the graphene oxide for adsorption of Azo dyes from industrial wastewater effluent. *J. Mol. Liq.,* **2022**, *354*, 118849. [http://dx.doi.org/10.1016/j.molliq.2022.118849]

[63] Hashemi, S.A.; Mousavi, S.M.; Bahrani, S.; Gholami, A.; Chiang, W.H.; Yousefi, K.; Omidifar, N.; Rao, N.V.; Ramakrishna, S.; Babapoor, A.; Lai, C.W. Bio-enhanced polyrhodanine/graphene Oxide/Fe_3O_4 nanocomposite with kombucha solvent supernatant as ultra-sensitive biosensor for detection of doxorubicin hydrochloride in biological fluids. *Mater. Chem. Phys.,* **2022**, *279*, 125743. [http://dx.doi.org/10.1016/j.matchemphys.2022.125743]

SUBJECT INDEX

D

E

F

G

H

I

L

M

N

O

P

R

S

T

U

V

W

X

Made in the USA
Las Vegas, NV
15 November 2026

5bcbc428-da40-45be-ae3c-9a47f4848612R01